Charles Seale-Hayne Library
University of Plymouth
(01752) 588 588

LibraryandITenquiries@plymouth.ac.uk

Ocean Margin Systems

Springer

Berlin
Heidelberg
New York
Hong Kong
London
Milan
Paris
Tokyo

G. Wefer • D. Billett • D. Hebbeln
B.B. Jørgensen • M. Schlüter • T. van Weering

Editors

Ocean Margin Systems

With 213 Figures, 61 in Color, and 31 Tables

 Springer

Editors:

Professor Dr. Gerold Wefer
Universität Bremen, Fachbereich Geowissenschaften,
Klagenfurter Straße 2, 28359 Bremen, Germany

Dr. David Billett
Southampton Oceanography Centre
DEEPSEAS Benthic Biology Group
George Deacon Division for Ocean Processes
European Way, Empress Dock, Southampton, SO14 3ZH, United Kingdom

Dr. Dierk Hebbeln
Universität Bremen, Fachbereich Geowissenschaften
Klagenfurter Straße 2, 28359 Bremen, Germany

Professor Dr. Bo Barker Jørgensen
Max-Planck-Institut für Marine Mikrobiologie
Celsiusstraße 1, 28359 Bremen, Germany

Professor Dr. Michael Schlüter
Alfred-Wegener-Institut für Polar- und Meeresforschung
Am Handelshafen 12, 27570 Bremerhaven, Germany

Dr. Tjeerd C.E. van Weering
Netherlands Institute for Sea Research
P.O. Box 59, 1790 AB Den Burg, Texel, The Netherlands

√ ISBN 3-540-43921-8 Springer-Verlag Berlin Heidelberg New York

Library of Congress Cataloging-in-Publication Data

Hanse Conference on Ocean Margin Systems (2000 : Delmenhorst, Germany)
Ocean Margin Systems / Gerold Wefer ... [et al.]. p.cm. Includes bibliographical references.
ISBN 3540439218 (alk. paper) 1. Continental margins--Congresses. 2. Oceanography--
Congresses. I. Wefer, G. (Gerold) II. Title.
GC84 .H36 2000   551.46--dc21   2002026825

Springer-Verlag Berlin Heidelberg New York
a member of BertelsmannSpringer Science+Business Media GmbH

http://www.springer.de

© Springer-Verlag Berlin Heidelberg 2003
Printed in Germany

The use of general descriptive names, registered names, trademarks, etc. in this publication does not imply, even in the absence of a specific statement, that such names are exempt from relevant protective laws and regulations and therefore free for general use.

Production: PRO EDIT GmbH, Heidelberg, Germany
Cover Design: Erich Kirchner, Heidelberg, Germany
Typesetting: Camera-ready by authors

Printed on acid-free paper   SPIN: 10879859   32/3130/Di   5 4 3 2 1 0

Preface

Ocean margins are the transitional zones between the oceans and continents where most of the sediments derived from the land are deposited. The effective processes here are influenced by a variety of steering mechanisms, from mountain building and climate on the land to tectonics and sea-level fluctuations at the margins of the seas. These areas are also of great importance for the global biogeochemical cycles because, although they only make up about 20% of the ocean's surface, 50% of the global marine production takes place here.

The region of the ocean margins extends from the coastal zone across the shelf and the continental slope to the toe of the continent. More than 60% of all people live in the adjacent coastal land areas, and they have intensively exploited the coastal waters for the extraction of raw materials and nutrients over a long period. In recent times, human activity has spread farther out into the oceans as the margins gain increasing attention as potential centers for hydrocarbon exploration and industrial fisheries. The great commercial potential of these regions, however, is countered by the presence of high potential hazards, for example, in the form of earthquakes and possible tsunamis triggered by slope instability, which can have a direct impact on the densely populated coastal regions.

The ocean margins are a dynamic system in which many processes shape the environment and impact the utilization and hazard potentials for humans. Can the study of ocean margin systems help in devising strategies for coping with this environment? What regulates the long-term development of ocean margins? How do fluids affect the material budgets? Do we have a good idea about the life at and especially below the seafloor? To discuss these kinds of questions we brought together experts from various disciplines of the marine sciences with a strong interest in these systems, to promote discussion between workers in different fields by focussing on a common topic of great interest to society.

The meeting, which took place in Delmenhorst near Bremen in Germany, Nov. 19 to 23, 2000, was arranged in the framework of a "Hanse Conference" within the interdisciplinary program of the Hanse-Wissenschaftskolleg, a foundation set up to promote interdisciplinary studies in collaboration between the universities of Bremen and Oldenburg. The aim of the Hanse Conferences in general is to provide opportunities for experts from different fields of the sciences and humanities to come together and explore the larger framework of topics of common interest. What unites the participants is their desire to look over the fence to neighboring disciplines. Young colleagues who wish to build an interdisciplinary career are particularly welcome.

In conducting the conference, we have attempted to avoid the disadvantages common to many large scientific meetings characterized by information-overload and lack of time for discussion. Instead, we have loosely followed the model of the "Dahlem Konferenzen", introduced by the late Dr. Silke Bernhard. An advisory committee of six scientists from different disciplines met well before the conference to formulate the overall goal and the themes of four discussion groups. This committee was also responsible for producing an initial list of invited participants, a list subsequently expanded through the recommendations of invitees. We aimed for about 40 scientists, complemented by selected graduate students and postdoctoral researchers. The conference was set for four days. Within each of the four theme sections, several participants were asked to provide background papers in their fields, as a basis for discussion. The aim is to have these papers sent as drafts to all participants one month before the conference, to stimulate the formulation of questions and critical comments.

The focus of activity within the Hanse Conferences is discussion, and not presentation of talks. The participants come with the background knowledge acquired through the study of the overview papers prepared for the conference. On the first day of the conference each of the four discussion groups agree on a list of topics derived from the questions and comments that arise from the study of the background papers. The following two days are dedicated to debating these topics, within the four discussion groups. On the fourth day, each group reviews a summary prepared by its rapporteur, who presents the most important results of the discussions. Suggestions for modifications to the summary are incorporated into the final summary, which is presented at the end of the conference to the entire assembly, by each of the rapporteurs. At this meeting, comments are invited by all participants on any of the points raised.

The final proceedings, which are published in this book, begin with a general section containing scientific as well as marine technological overview papers of broad interest. The thematic sections then follow, ordered from physical processes regulating ocean margin development and subsurface material transport to the huge diversity of habitats for benthic life and microbial activity found in these regions.

The final report from each group follows each of the thematic sections, which contain the revised background papers. All papers benefited from peer review. It is hoped that they will be useful in informing the ongoing discussions on preservation, exploration, exploitation, and risk assessment of ocean margins, wherever such debate may take place. We especially hope that high-school and college teachers find much material in these proceedings to enrich their courses in environmental sciences. In the educational realm, a marriage between physical understanding of the Earth's life support systems and an appreciation of history leading to responsibility will be necessary to provide the basis for political action which can deal with the challenge of the sustainable use of delicate marine ecosystems.

The Hanse Conference on Ocean Margin Systems was also planned as part of the development of a Marine Science Plan for Europe, to be drafted by the Marine Board of the European Science Foundation (ESF) in 2001. Focusing on the ocean margins, which form a significant part of the European waters in the North Atlantic and in the Mediterranean, the final group reports formulate recommendations for future ocean margin research within an European Research Area under a joint European Marine Science Plan.

The Convenors
Delmenhorst, November 2000

Contents

Sedimentary Settings on Continental Margins
– an Overview

I.N. McCave

Department of Earth Sciences, University of Cambridge, Downing Street,
Cambridge CB2 3EQ, UK
corresponding author (e-mail): mccave@esc.cam.ac.uk

Abstract: Sedimentation on continental margins bears the strong imprint of the tectonic setting, changes of sea-level and many local processes, including human intervention in sediment supply. Carbonate sediments are mainly created within the depositional basin whereas clastics supplied from upland areas are susceptible to enhancement due to deforestation or cut-off due to damming. Quaternary oscillation of sea-level has occurred with a frequency that has generally prevented supply and dispersal systems coming to equilibrium. Thus shelf shape reflects low-stand of sea-level and sediments dispersed from outer shelves/upper slopes were mainly emplaced at low stand - only in cases of major deltas has delivery overcome post-glacial sea-level rise. Dispersal processes of waves, tidal currents, wind-driven currents, oceanic currents and slope currents, and gravity flows down slopes and in canyons are briefly outlined. Major areas of uncertainty remain in budgets where flood-plain storage is generally unknown, the amount (and grain size distribution) trapped on shelves versus escaping to the ocean is unknown, and the magnitude/frequency structure of modern mass flows in canyons is poorly known.

Introduction

Continental margins are where most sediments on earth are deposited. As much as 90% of the sediment generated by erosion on land is deposited there, particularly in major deltaic cones. The controls of this deposition are many, ranging from the tectonics and climate of the hinterland where sediment is produced, to tectonics and sea-level changes on the margin itself which define the rate at which space is made available for the deposition of sediment. The local wave and tidal regimes cause dispersal of sediment across the shelf, and many factors, including particle size of sediment load and tectonics, control sedimentation systems on continental slopes. From that point mass failure can lead to debris flows, formation of submarine canyons and dispersal via turbidity currents into the deep-sea. Apart from these natural controls many regions of the Earth experience increasing impacts due to human activities, whether this be by deforestation causing increased sediment run-off or dam construction causing its severe reduction. Here I provide a brief introductory overview

to major features of continental margin sedimentary systems, concentrating particularly on those which occur at the present day. Non-geologists are often not conscious of the evolution of these sedimentary systems and continental margins over many millions of years, and that modern processes are to some extent dependent on that history.

Tectonic settings

There is a profound difference in the sedimentation systems occurring on tectonically passive continental margins (such as those found surrounding the Atlantic and western Indian Ocean) and those on active margins surrounding most of the Pacific Ocean where there is active plate subduction (Fig.1). There is another type of margin which, although it is not continental, nevertheless figures importantly in the supply of sediment to the oceans namely the margin of ocean islands. Ocean islands, according to Milliman and Syvitski (1992), have very high sediment yields, both absolute and per unit

From WEFER G, BILLETT D, HEBBELN D, JØRGENSEN BB, SCHLÜTER M, VAN WEERING T (eds), 2002,
Ocean Margin Systems. Springer-Verlag Berlin Heidelberg, pp 1-14

area. Passive continental margins have undergone thermal subsidence following the original separation of continents e.g. Africa from South America, and this subsidence has created space for the accumulation of sediments over more than 100 million years. Seismic data indicate sedimentary piles in excess of 12 km in thickness on some passive continental margins.

Active margins suffer strong tectonic deformation, and sediments which escape the margin to be deposited in the adjacent trench may subsequently be tectonically plastered onto the continental margin. On both active and passive margins carbon-rich fluids may be expelled in connection with either tectonics or compaction.

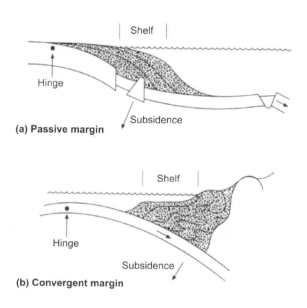

Fig. 1. Principal features and differences between active (Pacific type) and passive (Atlantic type) continental margins (from Johnson and Baldwin, Sedimentary Environments: Processes, Facies & Stratigraphy. Blackwell, 1996).

Clastic versus carbonate sediment supply: Climate and human control

Clastics

Despite considerable recent work on compiling data sources (Milliman and Meade 1983; Milliman and Syvitski 1992; Milliman et al. 1995), the present supply of sediment from the land to the oceans is very poorly known. The reason is that the lowest gauging stations on rivers are above the tidal limit and may be a considerable distance inland from the coast, more than 100 km in significant cases. The lower flood plains, and upper deltaic plains of great rivers are commonly major regions of subsidence, due to both tectonics and compaction, and thus are sites of major sediment deposition. The amount of sediment reaching the sea is therefore considerably less than the quantity estimated via gauging (about 20×10^9 t/a), but by an unknown proportion. Significant amounts of sediment that do reach the sea, especially sand and gravel, do not escape the coastal zone but contribute to the growth of beach ridges and coastal sand dune systems.

A further major unknown is the particle size distribution of the material delivered to the sea. Geological compilations and geochemical estimates of sedimentary materials suggest that clastic sediments (that is sediments produced by breakage of rocks but including clays produced chemically) are dominated by muds (silt plus clay, material less than 63 µm diameter) to the tune of 70-80%. Sand and gravel are minor components. It is not known for any major sediment source what the figure is at the present day. One might expect an overall decrease in average grainsize at present, given the warm climate of the Holocene (last 11,500 years) leading to greater clay production by weathering, compared with the preceding 60,000 years of glacial conditions.

The conclusion to this sorry tale is that we don't know how much material is presently supplied to the sea, but it is probably less than we think (maybe $10-15 \times 10^9$ t/a), we don't know how much of that escapes to the open ocean (maybe 10-20%), and we don't know its size distribution (but it is likely to be finer than the Pleistocene average).

Carbonates

Large areas of continental shelf receive very little terrigenous sediment but are the sites of copious *in situ* production of carbonate material. Obvious examples are the Great Barrier Reef of Australia and many other fringing reef tracts of continental margins and ocean islands. However many other shelves are covered with shell material produced *in situ* in both subtropical areas (e.g. West Florida shelf, Saharan shelf) and temperate to cold water shelves (e.g. South Australian shelf, Hebridean shelves). These "cool-water carbonates" are dominantly of calcite produced by red algae, bryozoans, barnacles and molluscan argonite whereas warm water carbonates are mainly aragonite and high magnesium calcite secreted by microbes, nannoplankton, calcareous algae corals and molluscs (Jones and Desrochers 1992, Fig.2).

The carbonate production rates are such that in warm water settings accumulation can be of the order of 1 m/ka (and for some reefs ten times that rate), whereas for cool water carbonates rates are mainly 1 to 10 cm/ka. This means that some carbonate systems can keep up with the very rapid late Quaternary eustatic changes of sea-level caused by deglaciation, though most do not.

Climate

The amount of sediment produced from a given region on land through weathering in mountainous areas is strongly controlled by geology and climate. Tropical areas where chemical weathering is intense under high temperature and high rainfall produce copious quantities of muddy sediment. This is evident in the deltas of such rivers as the Irrawaddy and the Mekong in south east Asia, and in the great submarine fans of the Amazon and Congo. At the other end of the spectrum Arctic shelves have a much higher proportion of coarse grained material, sand and gravel. Exceptions exist, and rivers draining a hinterland underlain by relatively fine-grained sediment will discharge a dominantly fine-grained material load to the Arctic (as does the MacKenzie in northern Canada.) In 1967 Hayes assessed the distribution of sediments on the inner part of continental shelves, in water depths less than 200 feet (60 m) assuming that this shallower region of the shelf was more likely to reflect modern sediment supply rather than the history of deposition at low sea-level. Hayes produced a simple diagram (Fig.3) showing the major areas of mud dominance on inner shelves to be in regions where the mean annual rainfall is over 1000 mm per year and the mean annual

temperature is over 25°C. Gravel is at a maximum in the cold arid regions of the north, though the gravel maximum, at 15% of the sediment is still less than the dominant sand component. It is not obvious that the gravel recorded by Hayes (1967) is presently being input by rivers; much of it may be reworked gravel deposited by glaciers several tens of thousands of years ago. Between the poles and the tropics, the dominant inner shelf sediment is sand. However on the majority of continental margin areas, namely the slope and rise, mud (silt plus clay) dominates. As we have seen above, climate also controls the distribution of sediments on carbonate shelves through the ecology of sediment-producing organisms. In addition, the higher latitude regions not only lack reefs on the shelves but also have intense wave action which breaks up the biogenic material yielding gravelly sand covering the whole shelf in areas such as the Great Australian Bight and the shelf west of Scotland.

Human control

In many regions human activity has become a geological agent of major influence with an increasing degree of impact on the sediment cycle. Particular aspects of this are:
• Dam construction has resulted in the reduction to almost zero of sediment yield from several major rivers (e.g. Nile, Indus, Colorado and soon perhaps the Yangtze) and many smaller rivers. The Nile, for example has virtually ceased supplying sediment to the coast (Stanley and Warne 1993; Stanley 1996).
• Dredging of small rivers for sand and gravel frequently takes out more sand and gravel than is supplied from upstream resulting in long term depletion of sand in the beds of these rivers which leads to accelerated deposition and an effective cessation of their contribution to the coastal zone. (This has led to severe problems of beach erosion, for example along the Italian Adriatic Coast, (Dal Cin 1983; Sestini 1996)).
• Intensive farming practices and deforestation are leading to increased erosion rates and more frequent flooding with consequent greater sediment transport to estuaries and the sea (Meade 1982).

4

McCave

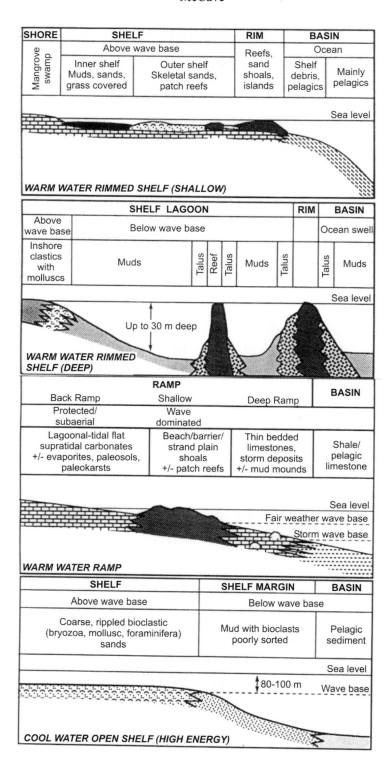

Fig. 2. Carbonate facies on continental margins contrasting warm and cool water and rimmed shelf and ramp settings (From Jones and Desrochers 1992, Shallow Platform Carbonates, in R.G. Walker and N.P. James (eds.) Facies Models: Response to Sea Level Change: Geological Association of Canada. Reproduced with permission. After several authors).

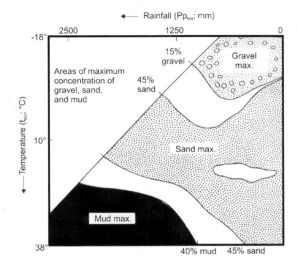

Fig. 3. Maximum contents of gravel, sand and mud in inner shelf sediments as a function of the climatic variables mean annual temperature (°C) and rainfall (mm) (Reprinted from Marine Geology, Vol 5, Hayes, Relationship between coastal climate and bottom sediment type on the inner continental shelf, pp 119, Copyright (1967), with permisson from Elsevier Science.).

Although one might argue that these effects roughly cancel out, this is no consolation to the inhabitants of areas that are either frequently flooded or on the other hand lack the sediment supply necessary to maintain their coasts. Examples of these are Bangladesh on the one hand and the Nile Delta coast on the other. Many aspects of the impact of changing sediment supply and subsidence are presented in Milliman and Haq (1996).

Sea-level change: Disquilibrium sedimentation

Sea-level has been at about its present high elevation for at least the last 5,000 years. The last time that sea-level was as high as this was about 120-130 ka ago. For at least 90% of the last million years sea-level has been lower than it is at present. The maximum lowering at peak glacial times was 120-130 m below the present level. This degree of lowering uncovers effectively all of the continental shelves as the present depth of the shelf-break is at about 130 m to 150 m, corresponding to the depth of significant wave attack on an outer shelf at low sea-level stand. Thus for most of the later Quaternary period shelves were partially uncovered and were the locus of fluvial and aeolian sediment deposition and dispersal. Aeolian dunes were clearly present on the Saharan shelf at low sea-level stand, and many shelves have abundant evidence of river channels, outer shelf deltas and shelf edge beach deposits. Many shelves also contain a record of sea-level rise in abandoned shoreface sand ridges produced as the shoreface retreated to its present position.

At low sea-level sediment was discharged directly into the ocean basins with much reduced trapping in coastal plains. Consequently, for the Quaternary as a whole, lowering of sea-level is by far the most efficient mechanism of sediment transport to the ocean basins. It bypasses the present mainly diffusive tidal and wave driven mechanisms which presently take sediment inefficiently from the coast to the shelf edge and discharge it into the ocean basins beyond. Shelf sedimentation at high sea-level stand is therefore arguably in disequilibrium for most of the time. That is to say the sediment is actively filling-in a very large space made available by the rise of sea-level and is filling it in at a much slower rate than the time available to reach equilibrium before sea-level starts to fall again. Interglacial periods have typically lasted only about ten thousand years. By that standard we would not have much time left before sea-levels should start to fall again, were it not for human interference in the normal course of events! At the present rate of sediment input it would take many of the linear coastlines of the world at least 100,000 years to advance to the outer edge of the present continental shelf. The present shape of the shelves must be seen as the product of low sea-level processes. In the case of some deltas and a few coasts the rate of sediment input has overcome the rise of sea-level to yield a "shelf" that is related to the balance between sediment input from rivers and its removal by marine processes. In consequence the "shelfbreak depth" on the outer part of many deltas lies at a few tens of metres water depth at most. For example, it is at about 11 m on the south side of the Mississippi and 20 m south of the Po at the edge of a prograding coastal prism.

However it is probably unrealistic in any case to speak of "equilibrium sedimentation" as this is an ideal state that is rarely or never attained. Relative sea-level is constantly changing whether due to glacioeustatic or tectonic causes. This unceasing change of relative sea-level determines the patterns of sedimentation found in the sub-surface (Fig. 4). Thus, the present situation may be unusual but it has occurred at regular intervals over the last million years.

Shelf sediment dispersal and depostion

Escape from estuaries and river mouths

To reach the continental shelf, material has to escape from the river/estuary system. There is extensive documentation of the modes of sediment trapping in the two-layer estuarine circulation. This comprises a surface outflow of fresher water and a sub-surface inflow of saline water. Sediment in suspension is concentrated close to the null zone of this circulation - that is the meeting point of the

net downstream and net upstream flow near the bottom. This circulation often leads to net upstream transport of sand in the lower reaches of estuaries, as in the Rhine for example, and to high concentrations and copious deposition of mud a considerable distance upstream from the mouth. Historically this is frequently the place at which port systems were developed just below the lowest bridge-point, for example London Docks on the Thames, the old Rotterdam harbour, the main docks of Bordeaux on the Gironde and Bremen on the Weser. The high dredging costs and modern requirement for deeper draft ships were factors in the decline of these ports in their original locations. In higher flow velocity deltaic systems there is also an estuarine circulation at the delta mouths where the landward bottom flow piles up river transported sand. The resulting bar is cut by the main river channel commonly in more than one place leading to the characteristic bifurcation of channels in deltas. The dynamics of river mouths in general are clearly set out by Wright (1977).

Escape of sediment from river systems usually occurs during times of river flood and comprises very fine sand and mud in suspension as well as

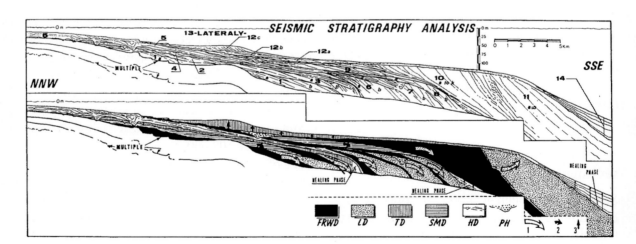

Fig. 4. Seismic stratigraphy of the continental margin of the Gulf of Cadiz after Somoza et al. (1997). This is thought to represent the last 180,000 years. Unit 4 represents the rapid transgression following the penultimate glacial 140 to 130 ka ago and unit 5 the high sea-level stand deposit of the last interglacial (isotope stage 5e). The thin deposits 12 a-c on the shelf and 14 on the slope represent the present interglacial. All the rest represents deposition at lower sea-level than present with 8-11 being the last glacial (isotope stages 4 to 2), demonstrating that most sediment is carried away from the coast to be deposited on the slope at low sea-level. Key: FRWD - forced regressive wedge deposits, LD - lowstand deposits, TD - transgressive deposits, SMD - shelf margin deposits, HD - highstand deposits, PH - incised palaeochannel infill.

coarser material carried much closer to the bed (Wheatcroft 2000). The finer material then enters the shelf circulation and is dispersed via a variety of mechanisms, whereas the material closer to the bed is principally dispersed via waves and tides.

Waves and shelf sediment distribution

The primary effects of waves are to put material from the sea-bed into suspension, or to prevent material that is already in suspension from being deposited. If some time elapses, for example several days, between deposition of fine sediments and a potential resuspension event, the mud may have acquired a cohesive strength in that intervening period which renders it more difficult to erode than fine to medium sand. Frequent wave attack in shallow depths of 5-10 m normally results in that zone being cleaned of fine sediment leaving only sand whose individual particles have a high enough settling velocity to avoid being exported. However, at depths below 10-20 m in regions where there is a significant input of fine sediment from coastal sources the seabed is frequently muddy. The occurrence of fine sediments on the bed of continental shelves is thus the result of a balance between the rate of deposition of fines (directly proportional to their concentration in suspension) and the magnitude and frequency of wave events capable of stirring up the bed sediments. Once mud is put into suspension it will stay there longer than sand as, even when aggregated, it has a lower settling velocity and will be dispersed. Because in almost all coastal situations suspension concentration decreases seawards, there is a diffusive flux of fine sediment in an off-shelf direction. These essential features controlling the distribution of fine sediments result in the typical mud-belt patterns recognised by McCave (1972) (see Fig. 5). In cases with extremely high supply and concentration of mud near-shore, such as are found along the coasts of the countries to the northwest of the mouth of the Amazon, the sea bed is covered with mud all the way up onto the 'beach'. Such muddy coasts in the tropics are commonly colonised by mangroves which tend to trap mud by preventing effective wave action.

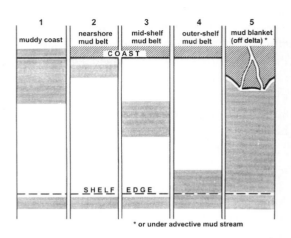

Fig. 5. Schematic representation of five types of spatially distinct shelf mud belts (areas of accumulation) (from McCave 1972).

The distribution of fine suspended sediment on shelves is determined via optical methods. Highly concentrated zones in the water column are found near the surface and bottom, termed the surface nepheloid layer (SNL) and bottom nepheloid layer (BNL). The surface nepheloid layer normally is dominated by organic matter from primary production, whereas the bottom nepheloid layer consists primarily of inorganic material resuspended from the bed. Close to the shore, particularly in river plumes, the surface nepheloid layer tends to be dominated by inorganic material as in the BNL. Along the coast, in water depths of 5-10 m, wave and current action frequently homogenise the water column both in temperature and salinity, and also suspended sediment producing what may be termed the coastal nepheloid layer. The SNL and BNL tend to be of the order of a few tens of metres in thickness. Together they can comprise about half of the shelf total water column. Although bottom nepheloid layers are abundant, relatively thick and concentrated over the outer continental shelf, in most cases that region has a sandy bed. In this region the material within the BNL is in transit to the slope, maintained in suspension by particularly vigorous wave action. Outer shelf areas in the major North Atlantic and North Pacific storm belts, that is north of about 40° N, suffer the impact of long period and high waves with sufficient frequency to clean the bottom of fine sediment down to 150 m. A well known example of wave-generated sand ripples at a water depth of 204 m off the Oregon shelf was documented by Komar et al. (1972). These long period storm waves are capable of resuspending and moving sands and this has an impact on slope deposition as we shall see below.

Tidal sand dispersal

Areas of strong tidal currents tend to have bottom sediments dominated by fine to medium sands. Well-studied areas are the southern North Sea and the English Channel. In these and most of the other classic areas of tide dominance, for example the Bay of Fundy, George's Bank, Nantucket Shoals, the Patagonian Shelf and Bonaparte Gulf off northwest Australia, the sediments which are moved within the modern high energy tidal regime are all relict, that is to say they were deposited at low stand of sea-level and re-worked during the last transgression. At present they are shaped into fields of sand waves and sand banks by the tidal currents. Tidal sand wave fields exist where the peak tidal currents are in excess of 0.4 m/s while areas of linear tidal sand banks occur where peak tidal current speeds are in excess of about 0.9 m/s. There are rather few areas with both substantial sediment input and tidal regimes capable of creating sand waves, but examples on which research has been performed are not common, significant areas being the eastern Yellow Sea off Korea (Park and Lee 1994), mouth of the Amazon (Adams et al. 1986) and the shelf off the Fly River in Papua New Guinea (Harris et al. 1993). Dispersal paths have been mapped for the well-investigated cases of tidal sand transport, in particular the northwest European shelf, where it has been shown that sand transport paths flow perpendicular to the Armorican shelf edge, but are weak or parallel to the shelf edge farther north (see papers in Stride 1982) (Fig. 6). There may well be a causal relationship between the existence of numerous submarine canyons south of the Armorican shelf and high supply of sand to the shelf edge which would yield particularly abrasive turbidity currents.

Wind driven circulation

The most important process for the net dispersal of fine sediment along and across the continental shelf is suspension transport in the general shelf circulation driven by wind and water column density. This may occur during fair weather conditions whenever significant amounts of sediment are put in from rivers, or especially during storms when significant amounts of material are re-suspended from the bed. Regions of significant and well-investigated shelf sediment dispersal under wind-driven currents are for example the plume off the Eel River in California, which yields a mid-shelf mud belt deposited to the north of the river (Wheatcroft 2000). Farther to the north there is another mid-shelf mud belt on the Washington shelf north of the Columbia River. On the western European margin similar mud belts are found to the north of the Douro off Portugal and to the north of the Gironde on the French continental shelf. In all these cases a wind dominated circulation involving stratified shelf waters transports sediment to the north during the season when significant amounts of material are discharged by rivers. Off both the Columbia River and Portugal the circulation is strongly seasonal. In summer when less sediment is discharged weak transport to the south occurs; conversely high winter input and prevailing wind to the north results in deposition of the mid-shelf mud belts (cf. Fig. 5).

Sedimentation under storm surges is of considerable interest but has not been investigated in satisfactory depth. A number of surmises have been made concerning the dispersal of sediment during storm surges which are in conflict with the physical oceanography of continental shelves. In particular, in such a shallow water environment where the frictional bottom boundary layer is thick relative to water column especially during storms, it is most unlikely that a simple geostrophic balance yields good estimates of flow speed and pattern. Nevertheless this is a key feature of a number of geological reconstructions of what is supposed to happen on continental shelves (e.g. Duke 1990). Storm surges are meteorologically forced waves occurring over a period of a few days (Bowden

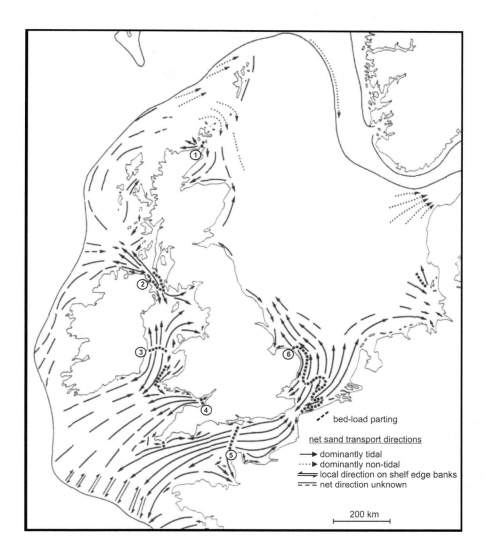

Fig. 6. Net sand transport directions of the relict sands on the N.W. European shelves derived mainly from analysis of sediment bedforms. Numbered zones are bed-load partings (divergences) (from: Offshore Tidal Sands - Processes and Depositis, 1982, p 76, Stride, f.4.10., with kind permission from Kluwer Academic Publishers).

1983). The time history of the dominantly shore-parallel flows associated with storm surges has been well shown in numerical models for the southern North Sea coasts by Davies and Flather (1978) and Flather and Davies (1978). Most numerical models of coasts show an average shore-parallel flow over most of the inner shelf from top to bottom with either a landward surface flow and near-bed offshore flow in the nearshore region or vice versa (e.g. Maldonado et al. 1983). This depends on whether the coast is to the left or right of the flow and which hemisphere (N or S) it is in. A coast-parallel flow with the shore on the right hand in the northern hemisphere will suffer downwelling and offshore transport of sand resuspended on the shoreface. This sand, suspended by intense wave action, may be transported oblique to the shore by wind driven currents of up to 1 m per second or more. Such currents are capable, over a few days, of transporting copious quantities of sand away from the shoreface to the inner shelf (Fig. 7). Such mechanisms do not appear to transport large quantities of sand across to the shelf edge on wide shelves.

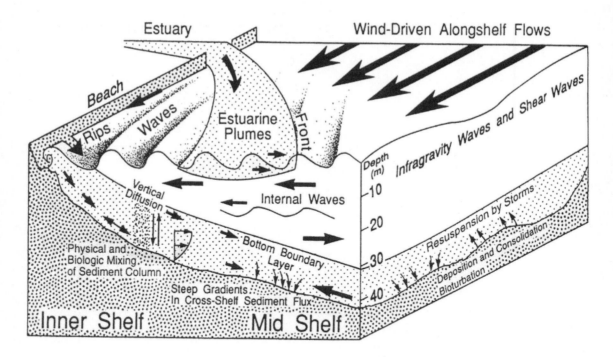

Fig. 7. Summary of mechanisms of shelf sediment transport (from Nittrouer and Wright, Review of Geophysics, Vol. 32, pp 90, 1994, copyright © [1994] by the American Geophysical Union).

Shelves influenced by major oceanic currents

There are a few shelves where sediment dispersal is controlled by intrusion of major oceanic currents. Notable examples are the Natal shelf off southeast Africa where the Agulhas current flows over the outer shelf towards the southeast and creates a large field of sand waves (Flemming 1981). There is a similar occurrence of somewhat smaller extent where the Gulf Stream passes over the continental shelf off Cape Hatteras on the eastern United States. The Kuroshio also affects sedimentation on the outer shelf off northeast Taiwan. These mechanisms are not responsible for moving material from the coastal zone across the shelf but rather for dispersing material that is already on the shelf, moving it along shelf, and perhaps over the shelf edge and onto the slope.

Perhaps surprisingly, considering the long-standing interest in the topic, there is still very little that one can say about modern transport of sand from the shoreface to the shelf edge. In areas of high sediment input, some sort of a shelf mud belt is normally found, with the outer shelf sediments commonly consisting of fine to medium sands with some patches of coarse sand, and the inner shelf sands fining seaward into the mud belt. From this pattern it seems unlikely that there is much diffusive sand transport across the continental shelf from wave resuspension during storms. On some coasts this process is bypassed through major submarine canyons which cut back almost to the coastal zone, for example the Nazare Canyon off Portugal and Scripps Canyon off California, which intercept the longshore transport of sediment and divert it to the deep-sea.

Slope sediment dispersal systems

The shelf-break.

The region comprising the outermost shelf and upper part of the slope is crucial for the dispersal of sediment to the ocean basins because it is within the range of resuspending orbital motions due to large storm waves and also because internal wave and tidal energy is focussed at this region of relatively sharply changing bottom gradient. Long period waves of mid latitude storms are capable of stirring the seabed down to at least 150 m. For example Airy wave theory predicts the bottom orbital velocity associated with a 15 second wave 10 m high at a water depth of 150 m to be 27 cm/s. The longer period but lower amplitude forerunners from storms which can have a period around 20 seconds can affect the seabed to even greater depths (Komar et al. 1972).

Both theory (Huthnance 1981, 1995) and observation (Dickson and McCave 1986) demonstrate that internal wave and tide motions are amplified to the extent that seabed sediment is resuspended where the slope of the parameter given by the ratio $[(\sigma^2 - f^2) / (N^2 - \sigma^2)]^{1/2}$ is close to the slope of the seabed (f is the Coriolis parameter, N is the buoyancy (Brunt-Väisälä) frequency and σ is the frequency of the internal wave or tide motion). Longer period flows on the continental margins may be wind-driven and may involve either upwelling or downwelling according to the wind direction. Several instances of downslope flow in the bottom boundary layer have been recorded (e.g. Pingree and Le Cann 1989). The phenomenon of cascading involves the sliding of dense water layers off the shelf into deeper water, for example water whose density is increased by severe winter cooling over the outer shelf. A front is frequently found over the outer shelf or shelf edge, and intersection of such fronts with the seabed commonly results in resuspension of sediment. Oceanic eddies from major currents offshore impact on the continental slope and contribute to sediment resuspension and seaward dispersal. Finally, in many areas, but particularly well documented along the continental margin of western Europe, a slope current is found (Huthnance 1995). In western Europe this current extends from Iberia to northern Norway. The slope current is associated with sandy bedforms as revealed by side-scan sonar (Kenyon 1986).

It is worth noting that if under present conditions storm waves are capable of occasionally resuspending sediment at water depths down to 150 m, then at glacial maximum lowering of sea-level to minus 120 m the upper slope down to a present water depth in excess of 250 m was subject to winter storm-wave action capable of resuspending fine grained materials. This fact, coupled with the slope current which occupies a zone between about 300 and 600 m, may account for the sandy mud texture of upper slope sediments in many parts of the world and the presence of a transition to mud part way down the slope commonly called "the mud line" (Stanley et al. 1983).

Nepheloid layers

The resuspension of sediment at changes in slope gradient, coupled with eddying motion in the off-slope flow field causes the resuspended material to be dispersed offshore in relatively thin turbid layers termed intermediate nepheloid layers (INL) (Fig. 8). As these layers spread into the ocean basin they rain out their coarser components which are then deposited on the lower slope and continental rise. What is detected optically in the INLs is normally the fine residue of what was originally resuspended, the very fine sand and coarser mud aggregates being lost through fall-out soon after the turbid layer detaches from the seabed. The bottom nepheloid layer (BNL) on the continental slope, as mentioned previously, often has a strong downslope component to its motion and contains large aggregates some of which contain phaeopigments demonstrating rapid transport down to great depths within a few tens of days (Thomsen and Ritzrau 1996).

Along most continental margins the present day processes acting on the outermost shelf and upper slope effect seaward dispersal of fine grained sediment that was originally largely emplaced when sea-level was low. This material then rains out over the continental slope and rise, and some enters the deep-sea circulation system.

Mass wasting and the development of canyons

Fine grained sediment deposited on continental slopes is unstable. Almost all continental slopes display abundant evidence of slope failure and mass movement in debris flows. Pratson and Coakely (1996) demonstrate how such slope failures can

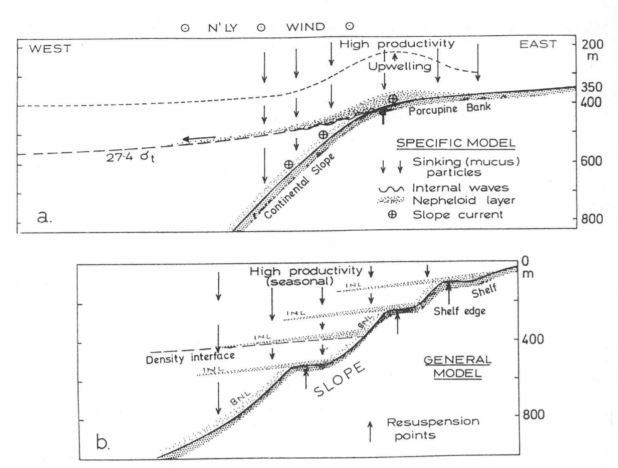

Fig. 8. Sketch showing formation of INL's off Porcupine Bank and a general model for an irregular continental slope where internal wave and tide activity is focussed on breaks in slope. Downslope transport in the BNL continues throughout (Reprinted from Deep-Sea Research, Vol. 33, Dickson and McCave, Nepheloid layers on the continental slope west of Porcupine Bank, pp. 816, 1986, with permission from Elsvier Science).

lead to formation of sediment flows which become turbidity currents and which in turn erode gullies which grow into canyons. Some of these incipient canyons grow headward up the original rill which caused their initial growth and ultimately form canyons which cut into the shelf edge. Other canyons start lower down the slope by a similar mechanism. A further class of canyons is located at the low sea-level stand terminus of major river systems. Some of these have cut back headward to the inner shelf. An extreme example is the Congo Canyon where the water depth at the river mouth is 400 m. Such canyons allow very efficient passage of fluvial material directly from the land to the deep-sea. Material which is deposited in canyons may also

be resuspended by focussed internal waves and dispersed onto the ocean beyond (Gardner 1989).

The slope is thus a region in which many processes operate, from sediment rain-out to mass movement in debris flows to high-speed erosive turbidity currents flowing down submarine canyons. In addition the along-slope geostrophic currents of the deep-sea circulation mould sediments into contour-current deposits, particularly low on the slope and over the continental rise beyond.

Conclusions

Perhaps the most useful conclusion to this brief survey is to emphasise the critical gaps in our knowledge. On the geological time-scales of margin building what is happening at present is of minor importance. On the decadal to millennial scale, however, it may be that the processes of material transport are important in regulating particulate carbon fluxes with implications for global carbon cycling.

Some key areas of our ignorance are:

(i) the unknown storage flux deposited on lower flood plains/subaerial deltas that must be subtracted from gauged fluvial transport to yield supply to the ocean.

(ii) the unknown fraction of fine sediment entering the shelf systems and deposited there versus the fraction which escapes to the deep-sea.

(iii) the total flux of material in suspension leaving the outer shelf and slope, and its components of 'new' material and 'old' material first deposited at lower sea-level. This division into new and old is partly known for a few areas covered by studies of SEEP (Biscaye and Anderson 1994) and OMEX (McCave et al. 2001) but not for margins in general.

(iv) most aspects of turbidity currents in canyons (magnitude/frequency, integrated sediment flux out of canyons via this mechanism) are not known. (This has a bearing on the dispersal of pollutants naturally or anthropogenically deposited in canyons). We probably have a better idea of these quantities for canyons on the geological time-scale than for the present, through analysis of deposits of turbidity currents.

Some progress is being made in understanding fluxes through large scale programmes such as SEEP, ECOMARGE and OMEX (Biscaye et al. 1994; Monaco 1990; van Weering et al. 1998) but they have concentrated on unknown (iii) above, leaving a considerable amount to be done about the other topics.

Acknowledgements

I thank Wolf Berger and Frank Lamy for their work on this paper which has made it much more readable.

References

Adams CE, Wells JT, Coleman JM (1986) Transverse bedforms on the Amazon shelf. Cont Shelf Res 6:175-187

Biscaye PE, Anderson RF (1994) Fluxes of particulate matter on the slope of the southern Middle Atlantic Bight: SEEP-II. Deep-Sea Res II 41:459-511

Biscaye PE, Flagg CN, Falkowski PG (1994) The Shelf Edge Exchange Processes experiment, SEEP-II: An introduction to hypotheses, results and conclusions. Deep-Sea Res II 41:231-252

Bowden KF (1983) Physical Oceanography of Coastal Waters. Ellis Horwood, Chichester

Dal Cin R (1983) I litorali del delta del Po e alle foci dell' Adige e del Brenta caratteri tessiturali e dispersione dei sedimenti, cause dell'arrentramento e previsioni sull'evoluzione futura. Boll Soc Geol Ital 102:9-56

Davies AM, Flather RA (1978) Application of numerical models of the northwest European continental shelf and the North Sea to the computation of the storm surges of November to December 1973. Deut Hydrog Zeitschr Ergänzungheft Ser A, No 14:1-72

Dickson RR and McCave IN (1986) Nepheloid layers on the continental slope west of Porcupine Bank. Deep-Sea Res 33:791-818

Duke WL (1990) Geostrophic circulation or shallow marine turbidity currents? The dilemma of paleoflow patterns in storm-influenced prograding shoreline systems. J Sed Petrol 60:870-883

Flather RA and Davies AM (1978) On the specification of meteorological forcing in numerical models for North Sea storm surge prediction, with application to the surge of 2 to 4 January 1976. Deut Hydrog Zeitschr Ergänzungheft Ser A, No 15:1-51

Flemming BW (1981) Factors controlling shelf sediment dispersal along the southeast African continental margin. Mar Geol 42:259-277

Gardner WD (1989) Periodic resuspension in Baltimore Canyon by focussing of internal waves. J Geophys Res 94:18185-18194

Harris PT, Baker EK, Cole AR, Short SA (1993) A preliminary study of sedimentation in the tidally dominated Fly River Delta, Gulf of Papua. Cont Shelf Res 13:441-472

Hayes MO (1967) Relationship between coastal climate and bottom sediment type on the inner continental shelf. Mar Geol 5:111-132

Huthnance JM (1981) Waves and currents near the continental shelf edge. Progr Oceanogr 10:193-226

Huthnance JM (1995) Circulation exchange and water masses at the ocean margin: The role of physical processes at the shelf edge. Progr Oceanogr 35:353-431

Johnson HD, Baldwin CT (1996) Shallow clastic seas. In: Reading HG (ed) Sedimentary Environments. Blackwell, Oxford, pp 232-280

Jones B, Desrochers A (1992) Shallow platform carbonates. In: Walker RG, James NP (eds) Facies Models. Geol Assoc Canada, pp 277-301

Kenyon NH (1986) Evidence from bedforms for a strong poleward current along the upper continental slope of northwest Europe. Mar Geol 72:187-198

Komar PD, Neudeck RH, Kulm LD (1972) Observations and significance of deep-water oscillatory ripple marks on the Oregon continental shelf. In: Swift DJP et al. (eds) Shelf Sediment Transport. Dowden Hutchinson and Ross, Stroudsburg Pa. pp 601-619

Maldonado A, Swift DJP, Young RA, Han G, Nittrouer CA, DeMaster DJ, Rey J, Palomo C, Acosta J, Balester A, Castellvi J (1983) Sedimentation on the Valencia continental shelf: Preliminary results. Cont Shelf Res 2:195-211

McCave IN (1972) Transport and escape of fine-grained sediment from shelf areas. In: Swift DJP et al. (eds) Shelf Sediment Transport. Dowden Hutchinson and Ross, Stroudsburg Pa. pp 225-248

McCave IN, Hall IR, Antia AN, Chou L, Dehairs F, Lampitt RS, Thomsen L, van Weering TCE, Wollast R (2001) Distribution, composition and flux of particulate material over the European margin at 47°-50° N. Deep-Sea Res II 48:3107-3139

Meade RH (1982) Sources, sinks, and storage of river sediment in the Atlantic drainage of the United-States. J Geol 90:235-252

Milliman JD, Haq BU (eds) (1996) Sea-Level Rise and Coastal Subsidence. Kluwer, Dordrecht

Milliman JD, Meade RH (1983) World-wide delivery of river sediment to the oceans. J Geol 91:1-21

Milliman JD, Rudkowski C, Meybeck M (1995) River discharge to the sea, a global river index (GLORI). LOICZ Core Project Office, Neth Inst Sea Res, Texel

Milliman JD, Syvitski JPM (1992) Geomorphic-tectonic control of sediment discharge to the ocean: The importance of small mountainous rivers. J Geol 100:525-544

Monaco A, Biscaye P, Soyer J, Pocklington R, Heussner S (1990) Particle fluxes and ecosystem response on a continental margin: The 1985-1988 Mediterranean ECOMARGE experiment. Cont Shelf Res 10:809-839

Nittrouer CA, Wright LD (1994) Transport of particles across continental shelves. Rev Geophys 32:85-113

Park SC, Lee SD (1994) Depositional patterns of sand ridges in tide-dominated shallow water environments: Yellow Sea and South Sea of Korea. Mar Geol 120:89-103

Pingree RD, Le Cann B (1989) Celtic and Armorican slope and shelf residual currents. Progr Oceanogr 23:303-339

Pratson LF, Coakley BJ (1996) A model for the headward erosion of submarine canyons induced by downslope-eroding sediment flows. Geol Soc Amer Bull 108:225-234

Somoza, L, Hernandes-Molina FJ, de Andres JR, Rey J (1997) Continental shelf architecture and sea-level cycles: Late Quaternary high-resolution stratigraphy of the Gulf of Cadiz, Spain. Geo-Mar Lett 17:133-139

Sestini G (1996) Land subsidence and sea-level rise: The case of the Po Delta region, Italy. In: Milliman JD, Haq BU (eds) Sea-Level Rise and Coastal Sub-sidence. Kluwer, Dordrecht, pp 235-248

Stanley DJ (1996) Nile delta: Extreme case of sediment entrapment on a delta plain and consequent coastal land loss. Mar Geol 129:189-195

Stanley DJ, Addy SK, Behrens EW (1983) The mudline: Variability of its position relative to shelfbreak. Soc Econ Palaeont Min Spec Publ 33:279-298

Stanley DJ, Warne AG (1993) Nile Delta - recent geological evolution and human impact. Science 260:628-634

Stride AH (ed) (1982) Offshore Tidal Sands. Chapman and Hall, London

Thomsen L, Ritzrau E (1996) Aggregate studies in the benthic boundary layer at a continental margin. J Sea Res 36:143-146

van Weering TCE, McCave IN, Hall IR (1998) Ocean Margin Exchange (OMEX I) benthic processes study. Prog Oceanogr 42:1-257

Wheatcroft R (2000) Oceanic flood sedimentation. Cont Shelf Res 20 (no.16):2059-2294

Wright LD (1977) Sediment transport and deposition at river mouths: A synthesis. Geol Soc Amer Bull 88:857-868

Continental Margins – Review of Geochemical Settings

R. Wollast

Université Libre de Bruxelles, Laboratory of Chemical Oceanography,
Campus de la Plaine, CP 208, Bd. Du Triomphe, 1050 Brussels, Belgium
corresponding author (e-mail): rwollast@ulb.ac.be

Abstract: The ocean margin, including the continental shelf, slope and rise, constitutes an essential boundary between the continents and the ocean basins and represents about 20% of the surface area of the marine system. It is characterized by an enhanced productivity and biological activity due to the input of nutrient from rivers and more importantly, from the transfer of nutrient-rich deep ocean waters. This transfer results from upwelling under favourable local wind conditions or from turbulent mixing at the shelf break due to wind-stress or internal waves mostly of tidal origin. Despite the great complexity of the ocean margin system, compilation of the literature provides coherent data sets indicating that the global primary production on the shelf may reach 6 - 7 GTC yr^{-1} and about 5 GTC yr^{-1} on the adjacent slope, as compared to 28 GTC yr^{-1} for the open ocean. Results on ^{15}N incorporation experiments or nutrient budget calculations indicate that the f-ratio is most probably around 0.35 - 0.4. The export production estimates, mainly obtained from flux measurements in sediment traps deployed on the slope, are often lower than those for the new production, likely due to the rapid degradation of organic matter in the intermediate and deep water column and to the low trapping efficiency. Furthermore, organic matter may be exported from the shelf to the open ocean in surface waters or within the benthic boundary layer by resuspension and off-shore transport. These fluxes may not be recorded by the sediment traps. Exchange of CO_2 with the atmosphere at the margins is complicated by the competition between high inputs of this component from deep ocean water and river water, and its rapid removal by photosynthesis in a very productive area. In addition, a tentative mean global cycle for nitrogen on the shelf is presented and discussed. It confirms on a global scale the major role of deep open ocean water transfer versus continental input of this element. Finally the data collected in the literature demonstrate that the continental margins are privileged areas of production, deposition and burial of $CaCO_3$, a component which has been often poorly considered in existing global carbon cycle assessments in the oceans.

Introduction

Continental margins are important in the context of the global carbon cycle in the oceans and hence in evaluating the ability of the marine system in scavenging anthropogenic CO_2. Most of the models developed so far regarding these questions were restricted to the open ocean, because of the great complexity and variability of the processes occurring at the boundary with the continents. It has, however, been recognized that the role of this boundary in the global carbon cycle can no longer be ignored (Walsh 1988; Berger et al. 1989; Mantoura et al. 1991; and recommendation in SCOR 1992 and 1994).

The coastal zone receives significant amounts of dissolved or particulate organic and inorganic carbon (about 1 GTC yr^{-1}), but little is known about the fate of this material. Compared with the open ocean, the coastal zone is characterized by enhanced biological productivity due to the large input of nutrients, not only from the continents via rivers but also from the transfer of nutrient-rich deep oceanic water across the shelf break (Walsh 1991; Wollast 1991; Galloway et al. 1996). Some of the resulting organic matter can be deposited and buried in the sediments on the shelf and slope or exported to the deep-sea, contributing to the biological pump,

From WEFER G, BILLETT D, HEBBELN D, JØRGENSEN BB, SCHLÜTER M, VAN WEERING T (eds), 2002,
Ocean Margin Systems. Springer-Verlag Berlin Heidelberg, pp 15-31

transferring CO_2 from the atmosphere to the ocean. In addition, the shelf is a very favourable environment for the production of biogenic calcium carbonate that can be accumulated there as coral reef for example, or exported as debris of benthic skeletons and dissolved in deep waters, contributing to the carbonate pump (Milliman 1993). All these processes can modify significantly the partial pressure of dissolved CO_2 in one direction or the other. Continental margin waters thus may become strongly undersaturated or oversaturated with respect to pCO_2 in equilibrium with the atmosphere. Departure from this equilibrium is much more pronounced in the continental margin zone than in the open ocean. Even if the surface area of the continental shelf and slope is limited, the strong gradients occurring in this area can be responsible for fluxes (the chemical pump) much larger than those observed in the open ocean where departure from equilibrium with respect to the atmosphere remains small (Frankignoulle and Borges 2000).

Most of the fluxes of concern are poorly quantified at present, except maybe for some limited areas. Furthermore, the large variability of the ocean margin environments makes it difficult to extrapolate well constrained fluxes to a global basis. In the last decade, there have been, however, several large projects devoted to a better understanding of the carbon cycle at the continental margin in several contrasting areas (Liu et al. 2000b). They provide valuable information concerning the processes and the fluxes occurring within the area and at the boundaries. The aim of this paper is first to review briefly the physical processes occurring at the ocean margins, which are responsible for the transfer of energy, water masses and elements between the shelf, the slope and the open ocean. We will then consider the biogeochemical processes responsible for the organic and inorganic carbon cycle, including the nutrient cycles. Tentative fluxes will be given on a global scale and compared with those accepted for the open ocean. Special attention will be paid to the comparison of fluxes between the shelf, the slope and the open ocean. The exchanges between the continents and the coastal waters will not be discussed here.

Physical and hydrographic properties

Physiography of the margins

The ocean margins considered here are taken in the broad sense and include the area situated between the shore line and the deep ocean. They consist of the continental shelf, the shelf edge, the continental slope and rise. With a total surface area of 74×10^6 km^2, they represent about 20% of the area of the world ocean (Table 1). The shelf break constitutes a pronounced topographic discontinuity separating the continental shelf characterized by a gentle slope (typically 0.1°) from the continental slope with an average slope of about 4°. It occurs typically at a depth of about 130m except in the polar regions where it is often below 350m due to the isostatic pressure of ice on the continent. This bathymetric discontinuity plays a major role in the dynamical processes controlling vertical and horizontal mixing at the shelf edge.

The surface area of the shelf represents 27×10^6 km^2, or about 7.5% of the total surface area of the ocean. It is of comparable size to that of the continental slope (28×10^6 km^2) and rise (19×10^6 km^2). The total length of the shelf break is about 320,000 km. On the average, the continental shelf is about 85 km wide, but it is narrower in the Pacific than in the Atlantic. This is an important characteristic from the point of view of the relative ability of the shelf to retain allochthonous and autochthonous particulate material by deposition and accumulation on the continental plateau. This difference is a result mainly of global tectonics, with wide continental shelves and extensive continental rises in areas of low seismic activity ("passive margins") and narrow shelves and steep slopes along margins that are seismically active. Both passive and active margins are found in all major ocean basins. The widest shelves are present in the Arctic seas.

Another important feature distinguishing the various ocean basins is the land area that they drain, and thus the relative input of fresh water to the margins. Approximately two thirds of the fresh water input from the continents occurs in the coastal zone of the Atlantic ocean. We will see later that

	Pacific		Atlantic		Indian		World ocean	
	area	%	area	%	area	%	area	%
shelf	10.1	5.6	8.2	7.6	3.0	4.1	27.1	7.5
slope	13.5	7.5	12.6	11.7	3.7	5.0	28.1	7.8
rise	4.9	2.7	9.1	8.5	4.2	5.7	19.1	5.3
total margin	28.5	15.8	29.9	27.9	10.9	14.7	74.3	20.6
total basin	180		107		74		361	
land drained	19		69		13		101	

Table 1. Depth zones in the ocean basins in 10^6 km^2 and % of the total surface area of each basin (from hypsometric data of Menard and Smith 1966).

the fresh water input may be an important factor enhancing exchanges between the shelf and the open ocean.

Factors affecting the transport of energy and material

Many dynamical processes interact in the transport of energy and material at the margins, within a wide range of both time and space scales (See Huthnance 1995; Brink and Robinson 1998 for extensive coverage). We will first briefly describe the processes that are responsible for the water flow characteristic of the margins. We will then review the physical factors affecting the vertical mixing of the water masses and the transfer of nutrients from deep ocean water onto the shelf.

The slope acts to impede shelf-ocean exchange; it produces a flow which tends to be steered parallel to the bathymetric contours. This topographically induced flow remains constrained in a narrow band along the shelf edge (Pingree et al. 1999 and references therein). Non-residual tidal velocities up to 20 cm s^{-1} have been recorded along the continental slope of the Celtic sea at depths close to the shelf break. At a depth of 400 m, along slope currents are still at around 10 cm s^{-1}. In contrast, the cross-slope residual currents in the same area are usually much smaller and are about 1-2 cm s^{-1}. Instantaneous tidal currents of the order of 1 m s^{-1}

are, however, often observed perpendicular to the shelf break, but net transport is only possible when the asymmetry in the flow causes local imbalances. Dynamic wind fields can also induce storm surges but on the average the surge currents are smaller than the tidal streams by a factor of about 10 (Huthnance 1995). Fresh water input by river discharge or ice melting in the coastal zone produces vertical and horizontal density gradients. The associated pressure gradients are responsible for the existence of density currents. According to Heaps (cited in Simpson 1998b) the flow, almost parallel to the isopycnals, may reach typically 5 cm s^{-1} in the surface layer, decreasing towards the bottom where a weak onshore flow at the seabed can be observed.

The sea surface water is heated by solar radiation and because of the thermal expansion of sea water, it introduces a stabilizing buoyancy force. In coastal zones, fresh water input from the continent reinforces the buoyancy and induces in addition horizontal density gradients. Surface waves, generated by a variety of mechanisms, forcing currents over the continental shelf and slope, act as a very efficient and active way for vertical mixing of the upper water column. The resulting vertical stratification with a strong density gradient at the pycnocline prevents the transfer of nutrients from the deep water to the euphotic zone. In the open ocean, this stratification can essentially be destroyed by the cooling of the surface water during the winter and by vertical mixing during storm events. On the shelf, deep ocean water can be transferred by upwelling under favourable wind conditions or by vertical mixing mainly due to bottom shear stresses.

Wind forcing can, however, also produce significant offshore transport of surface waters by Ekman motion, inducing upwelling of deeper ocean water onshore to maintain continuity. Such upwelling-favourable winds are equatorward on an eastern ocean boundary and poleward on the western ocean boundary. Wind forcing is also very effective in fjords and bays. Mean upwelling velocities are about 1-2 m per day, but with a high variability on short time scales. Depending on wind direction and intensity, upwelling is typically a seasonal process which can be seriously hindered if strong stratification of the water column is established. This situation occurs in low latitude zones with western boundary currents, where stratification is maximum and only a small fraction of the volume of water upwelled is transferred to the euphotic zone. Upwelling is therefore mainly observed along eastern ocean boundaries. The reverse downwelling process is observed in upwelling areas for winds blowing in the opposite directions. Downwelling can also occur when cold and dense water masses are formed on the shelf. Figure 1 shows the seasonal evolution of upwelling and downwelling observed along the Iberian coast, where upwelling favourable equatorward winds occur mainly in the summer. Advection on the shelf of deep ocean water by upwelling is one of the most efficient ways to supply nutrients to the coastal zone. Figure 2 shows an example of an upwelling event along the Iberian coast indicating the cold water masses in a narrow band along the coast and typical eddies and filaments extending over the slope and the deep ocean due to flow instabilities in the slope current. They may extend 200-250 km offshore and have potential importance in exporting nutrients and organic matter produced on the shelf to the deep ocean.

Vertical mixing, mainly due to tidal effects, is another way to transfer nutrients from the deep ocean waters to the surface waters in stratified shelf seas (Fig. 3). At the margins, the interactions

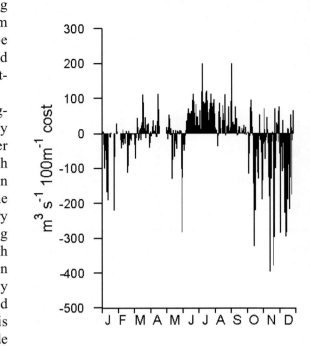

Fig. 1. Daily offshore Ekman transport used as upwelling index along the Iberian coast at 42N-9W during the year 1997. Positive values correspond to upwelling events and negative values to downwelling (by courtesy of S. Groom, Plymouth Marine Laboratory).

of the waves and tides with the slope and continental shelf provide efficient vertical mixing processes of the stratified waters. The tides are indeed a major source of energy for the ocean which may be dissipated in the shelf seas through frictional stresses in the bottom boundary layer (Simpson 1998a). As the flux of energy in the tide crosses the steep topography of the slope and comes onto the shelf, a fraction of this energy is transferred to internal motions and generates internal tides. They propagate slowly onto the shelf transporting a substantial amount of energy, which is dissipated as turbulent kinetic energy and makes an important contribution to vertical mixing in the region of the shelf break. Pingree and New (1995) have shown that the amplitude of the internal waves may reach 60 m in the very energetic tides at the shelf edge of the Celtic Sea (Fig. 4). Their propagation onto the shelf is about 70 km. The resulting distribution of density (Fig. 5) shows a pronounced spreading of the

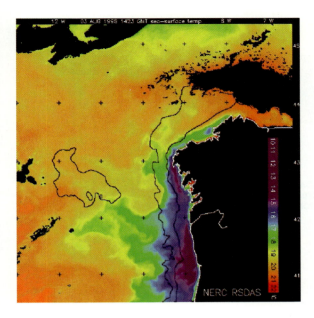

Fig. 2. Remote sensing image of the distribution of sea surface temperature in the upwelling area along the Iberian margin (03 August 1998). The two isobaths correspond to 200 and 2000 m. (AVHRR data were received at NERC Dundee University Station and processed by the Remote Sensing Group headed by Steve Groom at the Plymouth Marine Laboratory).

isopycnals in the shelf break region (Pingree et al. 1982). This vertical mixing is responsible for the transfer of nutrients into the euphotic zone and for the enhanced phytoplanktonic activity which leads to the chlorophyll maximum observed in the shelf break area. The position of the shelf break front can be easily detected by remote sensing, such as in the Gulf of Biscay (Fig. 6) by the presence of a narrow band of cold water along the 200m isobath.

Most of the tidal energy coming onto the shelf is, however, lost by frictional stresses in the bottom boundary layer (Simpson 1998a). These stresses are transformed into turbulent eddies which are responsible for the stirring of the water column. When the water column is stratified, a fraction of this turbulent kinetic energy is used to work against buoyancy forces and produces vertical mixing. The competition between buoyancy input and vertical stirring determines whether or not the water column remains stratified on the shelf. In the absence of horizontal advection and

strong wind, stratification develops only during the heating phase of the seasonal cycle if the ratio of the water depth to the mean tidal velocity is larger than a value depending on the heating rate and the efficiency of tidal mixing. The change from stratified to mixing conditions tends to occur rather abruptly and leads to the tidal mixing front. As shown in figure 6, this front can be easily identified by remote sensing from IR imagery. In the Celtic Sea it is situated at the entry of the Channel (Ushant front) at about 70 km off the shelf break (Fig. 6). The tidal mixing zone is also shown by a pronounced spreading of the isopycnals and a marked maximum in the concentration of chlorophyll due to nutrient input in the euphotic zone. The effect of wind on vertical mixing is much less important than tidal mixing and contrary to internal waves, wind acts mainly on the surface layer of the water column. According to Huthnance (1995) typical energy potentially available for mixing by wind is of the order of $0.01\,W\,m^{-2}$ compared with 0.15-$0.5\,W\,m^{-2}$ for the surface waves and 0.1-$1\,W\,m^{-2}$ for tidal bottom friction.

The large bottom shear stresses induced by tidal forcing are also responsible for the transport of coarse particles along the seabed (bed load). Here again, it is the asymmetry of the flow that results in net transport. The stress asymmetry arises principally from the interaction of the M2 (semi-diurnal lunar) and M4 (quarter-diurnal lunar) components of the tide. The tidally induced stresses are also responsible for the resuspension and transport of fine particles in the water column. These processes may explain why some continental shelves are not necessarily acting as depocenters for river borne or autochthonous particulate material (van Weering et al. 1998).

Primary and new production

The high productivity of the ocean margins compared to the open ocean is a well-established fact (review in Walsh 1988; Berger et al. 1989; Mantoura et al. 1991). In addition, the structure and the length of the food web on the shelf and in the open ocean differ markedly. The food web is very short in the coastal area which implies that the recycling of nutrients is limited and that a large

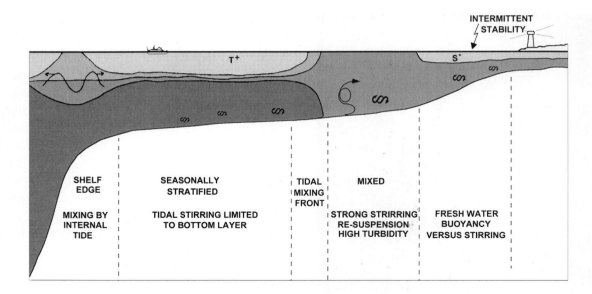

Fig. 3. Schematic representation of the tidal processes at the margins. Intensity of stirring is indicated by the size of the ℅ symbols (from Simpson in: The Sea - Vol 10, Brink, Copyright © (1998). Reprinted with permission of John Wiley & Sons, Inc.

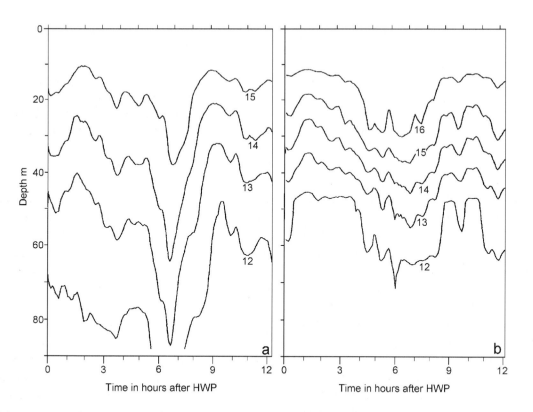

Fig. 4. Evolution of vertical temperature profiles at La Chapelle Bank area, showing the importance of the internal tides (Reprinted from Deep-Sea Research, Vol 42, Pingree and New, Structure, seasonal development and sunglint spatial coherence of the internal tide on Celtic and Armorican shelves and the Bay of Biscay, 245-284, Copyright (1995), with permission from Elsevier Science) **a)** for a station situated at the shelf break and **b)** for a shelf station situated 20 km landwards from the shelf break. HWP denotes high water measured at Plymouth.

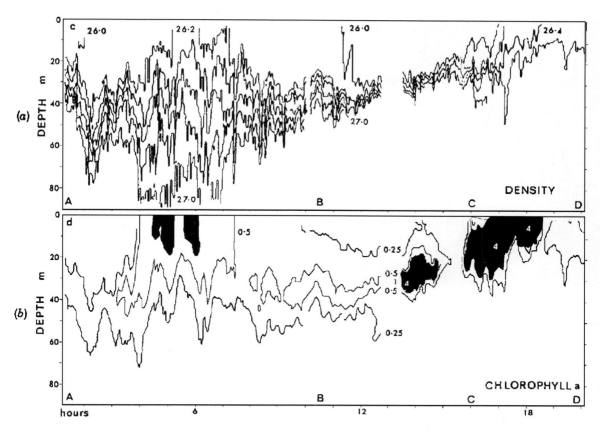

Fig. 5. Density **a)** and chlorophyll concentration **b)** distribution observed during a section from the deep ocean (A) to the shelf of the Celtic Sea (B-D). A first breakdown of the stratification between A and B correspond to the shelf break front. A second breakdown with very high concentration of chlorophyll is observed around C and corresponds to the tidal mixing front (Reprinted from Continental Shelf Research, Vol 1, Pingree et al., Celtic Sea and the Armorican current structure and the vertical distribution of temperature and chlorophyll, 99-116, Copyright (1982), with permission from Elsevier Science).

fraction of primary production might be exported either to the sediments or to the slope. In the open ocean, phytoplankton is dominated by small size organisms and the food web is rather long. Consequently, most of the organic matter is remineralized in the euphotic zone and little is exported to deep waters. Concurrently, only small amounts of nitrate are transferred from the deep water to the surface waters, due to the poor vertical mixing taking place in the open ocean. One can thus distinguish the recycled production which uses ammonia produced by respiration of organic matter as N source, from the new production fuelled by nitrate transferred from the deep water of the ocean to the euphotic zone (Dugdale and Goering 1967). This concept can be quantified by defining the f-ratio which is simply the fraction of the primary

production corresponding to new production. One classical way to evaluate the f-ratio is to measure the uptake of $^{15}NO_3$ and $^{15}NH_4$ during incubation experiments. This method assumes that nitrification in surface waters, N_2 fixation by specialized organisms, and uptake of organic dissolved nitrogen are all negligible. If these assumptions can be reasonably accepted for the open ocean, they are much more critical for the coastal ocean. On the shelf, a large fraction (~30%, Wollast 1998) of the primary production is deposited in the bottom sediments where organic matter is rapidly consumed by benthic organisms. Recycling of organic nitrogen in the sediments is usually efficient and finally most of the deposited nitrogen is restored and transferred to the water column, partly as ammonia and partly as nitrate,

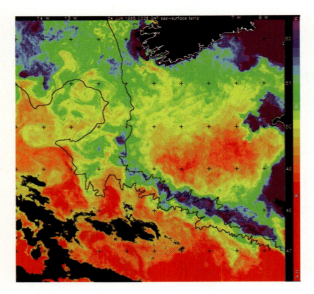

Fig. 6. Sea surface temperature distribution (24 June 1995) along the margin of the Gulf of Biscay (Northeast Atlantic Ocean). Note the close coincidence of the cold water layer with the 200 m isobath and the existence of tidal mixing fronts on the shelf (AVHRR data were received at NERC Dundee University Station and processed by the Remote Sensing Group headed by Steve Groom at the Plymouth Marine Laboratory).

with a small fraction of N lost as N_2 by denitrification. If the water depth is shallower than the mixing depth of the surface water, these nutrients are immediately available for photosynthesis and should be considered as recycled nitrogen. Furthermore, some ammonia may be added to the euphotic zone by rivers and atmospheric input especially in anthropogenically disturbed environments. The use of labelled nitrogen is thus not advisable for the coastal zone (Eppley and Peterson 1979) and some early measurements using ^{15}N are therefore questionable. A better way to evaluate the f-ratio could be to determine the recycling fluxes of the organic matter and thus to estimate (1-f). This is however a very tedious method. It requires not only measurements of the total respiration of planktonic organisms, but also those of benthic organisms in shallow areas, where the recycled nutrients are immediately transferred from the sediments to the surface-mixed layer.

Another approach to evaluate the f-ratio is to consider the export flux of carbon out of the eu-

photic zone. On sufficiently long time scales, the new production must obviously be compensated by an equal export flux to maintain steady state conditions in the system, and thus to avoid depletion or accumulation of nutrients in the coastal zone. There are however some discrepancies in the way how this export flux is estimated. Evaluations are often based on fluxes of organic matter in a sediment trap moored across the slope, but they do not take into account losses by respiration during settling. The same comment is valid for deposition fluxes below the mixing depth. In shallow areas, the export of organic matter to the sediments should take into account the fact that the export flux is not equal to the deposition rate, but to the burial rate. Most of the organic carbon deposited is respired by benthic organisms at the sediment-water interface and is not preserved in the sediments. This subject is extensively discussed by various authors in this volume (Flach, Haese, Lochte and Pfannkuche, Thomsen, Zabel and Hensen). In deeper parts of the margins, the vertical export of organic matter should be estimated from the fluxes across the pycnocline, and the horizontal export to the slope and to the open ocean from the fluxes across the shelf break. Both fluxes are usually inaccessible, except however by modelling or indirectly by mass balance calculations.

Taking into account all the limitations discussed here above and the geographical variability of the ocean margins, it is astonishing to see that the values suggested for global primary production and new production by various authors are not that different. They indicate, in any case, the same trends with similar values for the fluxes. Table 2 shows the mean value and variance of primary and new production calculated by Liu et al. (2000a) from published data in the literature, and the values that they selected, after a critical analysis of the estimations. These authors have also included in their data set, evaluations of the new production and of the export flux. The export flux is systematically lower by a factor of 2-3 than the new production. This is because the export fluxes are mostly measured with sediment traps moored at various depth (often below 500m) or estimated from deposition rates on the slope. These values

Zone	Primary Production		New Production		f-ratio	Reference
	$gCm^{-2}yr^{-1}$	$GTCyr^{-1}$	$gCm^{-2}yr^{-1}$	$GTCyr^{-1}$		
Shelf (mean)	195	7.2	89	2.7	0.35	Liu et al (2000a)
variance	75	2.5	50	1.5	0.23	"
selected value	**215**	**7.8**		see text		"
Open Ocean(mean)	93	31	12.3	4.1	0.13	Liu et al (2000a)
variance	30	10	6	2	0.20	"
selected value	**96**	**32**	**12**	**4**	0.13	"
Shelf	**230**	**6.0**	**83**	**2.4**	0.40	Wollast (1998)
Slope	150	4.8	30	1.0	0.20	"
Open Ocean	94	28	14	3.8	0.15	"

Table 2. Estimates of primary production and new production in the global ocean and margins.

do not take into account the respiration of organic matter during settling through intermediate and deep waters that must be included necessarily in the export term. Berger et al. (1987) have proposed to estimate the f-ratio by using an empirical relation based on observation that there is an obvious relation between the primary production PP (in gC m^{-2} yr^{-1}) and the f-ratio.

$$f = PP/400 - PP^2/340000$$

For the mean primary production of 215 gC m^{-2} yr^{-1} proposed by Liu et al. (2000a) as typical for the shelf, the f-ratio would be equal to 0.40. The values of Liu et al. (2000a) in Table 2 are compared to values estimated during the study by Wollast (1998) (not considered in the review by Liu et al. (2000a)) where a distinction has been made between the shelf and slope, besides the open ocean. The agreement with the selected values of Liu et al. (2000a) are surprisingly good. This gives

us more confidence in the tentative global carbon cycle published earlier (Wollast 1998) which is reproduced in Figure 7. Almost no information is available concerning the fluxes across the boundaries between the shelf, the slope and the open ocean. Therefore, a hypothetical flux of 2.2 GTC yr^{-1} has been estimated between the shelf and the slope area from mass balance requirements. This represents roughly one third of the primary production and appears to be overestimated. Fluxes of organic matter recorded in sediment traps along the slope deployed during various recent ocean margin studies (Liu et al. 2000b) all suggest that only a small fraction of the shelf primary production is exported to the slope. From their review of the literature, Liu et al. (2000a) have estimated that only 0.9 GTC of organic matter are exported from the shelf, which represents only 12% of the coastal primary production. Here again this value may be underestimated, because it does not takes into account the respiration of

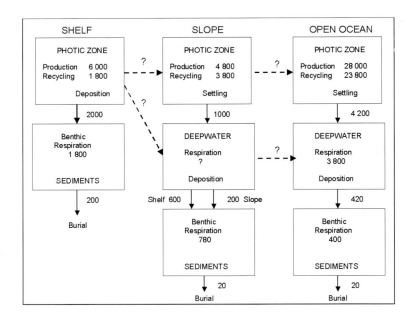

Fig. 7. Tentative global cycle of organic carbon in the oceanic system, expressed in MTC yr⁻¹ (from Wollast 1998).

settling organic particles. These studies also show that there may be a significant lateral transport of organic rich shelf surface water across the slope for example by filaments and eddies in the vicinity of the shelf break. Deposited organic matter may also be resuspended and transported off the shelf within the benthic boundary layer (Thomsen this volume). Neither of these fluxes are recorded in sediment traps and complicate seriously the evaluation of the export fluxes.

In a recent European study of the carbon cycle in the Gulf of Biscay (OMEX), we found a primary production of 200 gC m⁻² yr⁻¹ at the shelf break with a respiration in the water column below the mixing depth of 50 gC m⁻² yr⁻¹. An additional respiration of 20 gC m⁻² yr⁻¹ occurs in the surficial sediments and the export flux to the slope is about 30 gC m⁻² yr⁻¹ (Wollast and Chou 2001). This export represents thus 15% of the primary production at the shelf break but 50% of that primary production leaving the euphotic zone is respired below the mixing depth. In the adjacent area above the slope, 54 gC m⁻² yr⁻¹ representing 34% of the primary production there, is transferred from the euphotic zone to deeper waters. It is believed that some of this organic matter may have been transported from the surface water of the shelf to that of the slope. In the upwelling area of the Iberian margin, covered during the second phase of the European project OMEX, the preliminary results show that the primary production on the shelf is about 360 gC m⁻² y⁻¹. The production is about 270 gC m⁻² y⁻¹ over the slope, of which 200 gC m⁻² y⁻¹ is recycled production fed mainly by organic matter exported from this narrow shelf (Joint et al. 2001).

The export of organic carbon from the shelf to the open ocean remains difficult to quantify from the existing measurements. The analysis of the nitrogen cycle may help us to gain some additional information concerning the exchanges across the shelf break. If a large amount of organic matter with its N and P content is exported to the open ocean, then there must be a reverse flux of nutrients from the deep ocean to the shelf, to compensate the losses from the coastal zone. This will be examined in the next section.

Nutrient fluxes

Besides light, which is the main factor controlling photosynthesis, primary production in oceanic systems is in addition often limited by the availability of nutrients, including iron. The input flux of nutrients in the euphotic zone is thus a basic parameter which can be used to evaluate the potential photosynthetic capacity of the system. It is, however, important to keep in mind that a significant fraction of the primary production is fuelled by nutrients recycled in the euphotic zone by remineralization of organic matter. The new production is related to nutrient input at the boundaries of the system. The nutrient fluxes in the coastal ocean are very complex and include river discharge, submarine groundwater exchange, wet and dry atmospheric deposition, diagenetic processes in the sediments and transfer of deep ocean water across the shelf break by upwelling or vertical mixing. It is the importance of these various fluxes which controls the relative fertility of the shelf compared to the open ocean. The nutrient fluxes at the boundaries of the coastal zone are also responsible for the high values of new production versus recycled production and thus for the ability of the shelf seas to export organic carbon to the open ocean or to the sediments. Our discussion will be limited here to the nitrogen fluxes because this element is probably the limiting nutrient in most parts of the ocean. Nitrogen is also an interesting element because in the open ocean at least, ammonia is the nitrogen species resulting from the respiration of organic matter. In the absence of nitrification in the euphotic zone, it might serve as a tracer to evaluate the recycled production. In contrast, nitrate provided by deep water to the photic zones sustains the new production. As indicated previously, the nitrogen cycle is in fact complicated. N_2 can be fixed by specialized organisms in systems highly depleted in nutrients. Oxidation of ammonia to nitrate can occur in surface waters and compete with phytoplankton uptake. Nitrate can be used by denitrifying bacteria and transformed into N_2 under anoxic conditions, mainly occurring in the sediments. Finally, phytoplankton can use some dissolved organic nitrogen compounds as N nutrient source.

Unfortunately, most of these processes are enhanced in the coastal zone and complicate the interpretation of the data. Thus there are large uncertainties associated with the fluxes related to these processes. We have attempted to quantify the global N cycle for the coastal zone, based on the C cycle and on existing reviews of the processes affecting nitrogen fluxes at the boundaries (Wollast 1991, 1998). The nitrogen fluxes associated with planktonic activity can be easily evaluated from the carbon fluxes by imposing the classical Redfield ratio. We have assumed that the composition of the phytoplankton and that of the detrital organic matter in the euphotic zone is similar and constant (C:N = 6.6). The nitrogen fluxes required to sustain total primary production and new production are then easily calculated, as well as the deposition flux to the sediments and the export of organic particulate N to the open ocean (Table 3 and Fig. 8). The present-day river input has been carefully estimated by Meybeck (1993) and the atmospheric deposition evaluated by GESAMP (1989). Nutrient input by submarine groundwater discharge (Schlüter this volume) is not quantified at present on a global scale, but may be locally comparable to the river flux. The contribution of groundwater flow to the input of nutrient in the coastal zone may exceed river input in highly populated areas where septic tanks are used in very porous soils (Valiela and Costa 1988). In other regions, pumping of ground water in the adjacent coastal plain induces a reverse submarine flow. Given the uncertainties the groundwater flux is neglected in most global budget calculations. Thus, continental inputs, through either river discharge or atmospheric deposition, are totally insufficient to maintain the nitrogen requirements of the planktonic activity. There is, in addition, a consensus concerning the low capacity of nitrogen fixation in the coastal zone where the concentrations of ammonia or nitrate are rarely sufficiently exhausted (Capone 1988; Nixon et al. 1996).

There are, in contrast, large disagreements in the literature concerning the importance of the import flux of N from the open ocean and the denitrifying activity in the coastal sediments. In figure 8 the evaluation of the denitrification flux is based on model calculations which take into account the rate of deposition of organic matter as the main controlling factor (Wollast 1998). The model indicates that about 20% of the N deposited is lost as N_2 by denitrification, most of the depositional flux being recycled as NO_3 and NH_4

	Walsh (1991) Global	Galloway *et al.* (1996) North Atlantic	Wollast (1998) Global
Primary production	40	29	41
New production	22 (f = 0.52)	7 (f = 0.24)	18 (f = 0.44)
River discharge	2.3	1.6	2.3
Atmospheric deposition	neg.	0.32	0.38
N fixation	neg.	0.05	0.58
N burial	neg.	0.29	0.38
Denitrification	1.9	3.4	1.9
Offshore export	22	neg.	15
Deep ocean water input	21	5	14.6

Table 3. Averaged global fluxes of nitrogen (in gN m^{-2} yr^{-1}) in the coastal zone.

and restored to the water column. Only a negligible fraction of N is stored in the sediments by burial. This figure shows the important role of the coupling of pelagic and benthic biological processes on the shelf in the N cycle. In the budget shown in figure 8, the oceanic input flux of N nutrients was calculated to fulfill the mass balance requirements for the N cycle.

Walsh (1991) did a similar exercise but attempted to estimate the N fluxes from the deep ocean to the shelf, by distinguishing and assessing the transfer due to various physical processes. His global budget is thus based on areas dominated by upwelling, western boundary currents and estuarine circulation where fresh water input occurs. These estimates are however affected by a large degree of uncertainty due to the variability of the physical processes involved. Walsh (1991) estimated a global N input flux from the ocean across the shelf break of about 560×10^{12} gN yr^{-1} (or 21 gN m^{-2} yr^{-1}) and assumed that there is no significant accumulation of N in the sediments. To fulfill the mass balance constraints for N, the input is compensated by export of organic N both as particulate and dissolved species. The value of Walsh (1991) for primary production on the shelf is similar to ours and corresponds to an uptake of

40 gN m^{-2} yr^{-1}. Walsh assumes that the f-ratio is 0.52, a value in the upper range of the observations. Even if there are differences in the absolute values, they agree with ours in a sense that the new production and productivity of the coastal zone are mainly supported by the transfer of nutrients from the open ocean. The river input and atmospheric deposition, further enhanced by anthropogenic activities, are secondary sources for the shelf, which are roughly compensated by denitrification and N burial in the sediments.

An attempt to establish a detailed nitrogen cycle in the North Atlantic has been published recently in a special issue of Biogeochemistry (Galloway et al. 1996). The authors first estimate that the rivers carry an equivalent of 3.55 gN m^{-2} yr^{-1} but that a large fraction of the nitrogen is lost by denitrification in the estuaries and thus only 1.6 gN m^{-2} yr^{-1} reaches the coastal zone (Table 3). Atmospheric deposition (0.32 gN m^{-2} yr^{-1}) is very similar to our global estimate, but Galloway et al. (1996) assume that the nitrogen fixation is negligible (0.05 gN m^{-2} yr^{-1}). Burial on the shelf and slope (0.29 gN m^{-2} yr^{-1}) is similar to our global mean but denitrification in the sediment is equivalent to a loss of 3.4 gN m^{-2} yr^{-1}, almost twice our value. The authors have based this value on a

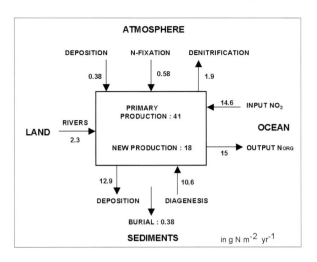

Fig. 8. Tentative global average fluxes of nitrogen in the coastal zone (adapted from Wollast 1998).

model and recognize that the denitrification flux is high compared to the experimental observations of Seitzinger and Giblin (1996). Export of organic nitrogen from the shelf to the open ocean is considered by the authors as negligible (< 2% of the primary production), mainly based on the results of the SEEP experiment (Anderson et al. 1994). An interesting paragraph in the review is devoted to the input of DIN from the deep oceanic waters (Nixon et al.1996). The authors first report several estimates of the nitrate flux from slope waters onto various regions of the North Atlantic continental shelf. They vary by a factor of 20, but Nixon et al. (1996) suggest that most often the fluxes are between 4 and 15 gN m^{-2}yr^{-1} with a few special areas more intensively enriched such as Georges Bank. They have used, on the other hand, a relation between primary production and DIN input for various areas of the world. These authors obtain a total flux of 7 gN m^{-2}yr^{-1}, which gives an input of deep water at the shelf break of 5 gN m^{-2} yr^{-1}, if the river and atmospheric inputs are subtracted. A simple calculation suggests that a total N input equal to 7 gN m^{-2}yr^{-1} corresponds to a f-ratio of 0.24 for a primary production of 165 gC m^{-2} yr^{-1}, assumed by the authors. This gives a very low value for the new production for the shelf area. In this nitrogen budget, the export of nitrogen from the deep ocean is compensated by the rates of nitrogen fixation in the pelagic North Atlantic

ocean and not by the export of particulate organic N from the shelf.

More recently Soetart et al. (2001) developed within the OMEX project a coupled biogeochemical-1D hydrodynamic model describing the N cycle at the margins in the northern part of the Gulf of Biscay. There is no river in the area and the new production of this very broad shelf is sustained only by import of nutrients from the deep waters at the shelf break. These authors obtain an N input from the deep water equal to 14 gN m^{-2} yr^{-1} with a f-ratio of 0.40. Interestingly, their model shows that about 20% of the nitrate consumed by the phytoplankton results from oxidation of ammonia in the euphotic zone. This nitrification leads to an overestimation of the f-ratio based on ^{15}N incubation experiments (results in Table 3).

The discussion presented here on the nitrogen cycle shows that there are large discrepancies concerning the importance of the nutrient fluxes at the margins. Yet the lowest estimates of the N transfer through the shelf break in the North Atlantic ocean exceeds the sum of river discharge and atmospheric deposition, even for an ocean subjected to a large continental supply of fresh water. In other words, the fertility of the coastal zone is largely due to nutrients provided by deep ocean waters and not by continental input.

The CaCO$_3$ Cycle

Less attention has been paid until now to the role of precipitation and dissolution of calcium carbonate in the global carbon cycle. In near shore sediments, the accumulation of inorganic carbon is roughly equivalent to that of organic carbon. In the open ocean, it represents twice the organic carbon preservation (Wollast 1998). The most recent budget for calcium carbonate presented by Milliman (1993), clearly demonstrates the major role of shelf areas in the production of CaCO$_3$ and its preservation in the marine system. Coral reefs are the most productive carbonate habitats with a mean precipitation rate of 180 gC m^{-2} yr^{-1} and a preservation rate of 145 gC m^{-2} yr^{-1}. Banks and embayments in tropical and subtropical areas also have high rates of carbonate production, with an average value of 60 gC m^{-2} yr^{-1}, resulting mainly from benthic algae, molluscs and benthic fora-minifers. The skeletons of these organisms can be easily resuspended and transported across the shelf break to the deep ocean. Another fraction is dissolved during early diagenetic processes related to the respiration of deposited organic carbon. Milliman (1993) estimates that only 50% of these carbonates are on the average preserved and 25% are exported to the slope and rise. For the entire shelf, the mean carbonate production calculated from the data of Milliman (1993) is 10 gC m^{-2} yr^{-1} of which 3 gC m^{-2} yr^{-1} are dissolved and 1.8 gC m^{-2} yr^{-1} are exported to deeper waters. The production above the slope is about 2 gC m^{-2} yr^{-1} and that of the open ocean 1 gC m^{-2} yr^{-1}. The ocean margins are responsible for the production of almost one half (360 MTC yr^{-1}) of the pelagic production of carbonates.

The precipitation of $CaCO_3$ strongly affects the partial pressure of CO_2 in the water column according to the following stoichiometric relations:

$$Ca^{++} + 2\ HCO_3^- = CaCO_3 + H_2CO_3 = CaCO_3 + H_2O + CO_2 \quad (1)$$

According to the first dissociation constant of carbonic acid :

$$K_1 = \frac{a_{HCO_3^-} \cdot a_{H^+}}{a_{H_2CO_3}} \quad (2)$$

a decrease of the activity (a) of HCO_3^- associated with an increase of the activity of H_2CO_3 also induces a pH drop which in turn enhances the partial pressure of CO_2 in surface waters. The dissolution of the skeletons at greater depths has the reverse effect and thus the redissolution of $CaCO_3$ represents a sink for CO_2 in the marine systems. Most of the existing models of the $CaCO_3$ cycle in the ocean are based on thermodynamical considerations and do not describe satisfactorily the behaviour of this mineral in the ocean (Wollast 1998). For example, it is generally assumed that there is no dissolution of $CaCO_3$ in the shelf sediments because of the high degree of oversaturation of surface waters with respect to calcite. In fact, the production of carbonic acid by the respiration of detrital organic carbon results in the dis-

solution of calcium carbonate according to the reverse of reaction (1). Similarly, the measurements of the vertical distribution and sediment trap fluxes carried out during the OMEX project show that carbonate is exported from the shelf to the slope and that it undergoes dissolution during settling, even if the water column is oversaturated with respect to calcite (Wollast and Chou 1998). Here again this phenomenon is probably due to respiration of organic matter in faecal pellets or aggregates.

Conclusions

The ocean margins including the shelf, slope and rise represent only about 20% of the total area of the ocean, but compared with the open ocean, it is the centre of an enhanced activity of physical and biogeochemical processes which cannot be neglected in a global description of the marine system. Compared with the open ocean, various specific physical processes, often related to the presence of the marked bathymetric discontinuity, are responsible for the transfer of deep ocean water to the euphotic zone, in an area extending from the slope to the shore line. This discontinuity produces perturbations of the geostrophic currents and interacts with the propagation of the tides and waves. In addition, adequate wind stresses along the coast induce upwelling, transferring deep slope water onto the shelf. The ocean margins are thus privilege areas where deep nutrient-rich waters can be advected or mixed with surficial nutrient-depleted waters, continuously or seasonally. There is now sufficient evidences showing that these peculiar physical processes are the most important factors controlling the high productivity of the shelf. Even if the coastal zone is directly under the influence of nutrient input of continental origin by rivers and the atmosphere, these sources have only a negligible or a very local influence on the primary production. Submarine transfer of groundwater to the coastal zone which is until now neglected in the calculations, needs to be better quantified on a global scale.

The fate of the organic matter, produced on the shelf, remains however a question of intense debate. It is generally admitted that the food web in the coastal zone is short and thus that a large fraction of the primary production is available for export either by burial in the continental shelf sediments or by transport to the open ocean. For mass balance requirements at long time scales, this export of organic carbon and nutrients must correspond to the new production sustained by nutrients of continental or oceanic origin. The most recent studies (e.g. van Weering et al. 1998; Liu et al. 2000a) all indicate that there is a tight coupling between the benthic and pelagic activity. Most of the organic matter deposited is thus very rapidly respired and only a very small fraction is buried and thus preserved on a long time scale. Deltas and canyons may be exceptions to this rule. Furthermore, the tides generate high bottom stresses which are responsible for resuspension and transport of sediments and possibly for the export of freshly deposited material to the slope and rise (Thomsen this volume). All the recent measurements of the fluxes of organic material collected in sediment traps deployed along the slope show that there is a significant lateral transport from the shelf break to the open ocean (Liu et al. 2000a and b), often probably linked to bottom transport in the benthic boundary layer of resuspended organic matter.

Suspended material or dissolved organic matter can also be exported from the shelf to the open ocean and it seems that this process is especially important in upwelling areas where primary production is very high, the food web very short and the shelves rather narrow. If the organic material is exported to the surface water above the slope, it can be recycled and contribute to an enhanced productivity in that area. Otherwise, this transport across the shelf break constitutes a possible way to export organic matter to deeper waters and act as a sink for carbon at rather long-term scales.

It is difficult at present to quantify the fluxes of CO_2 between the surface waters and the atmosphere at the margins. Deep ocean water and river water are commonly strongly oversaturated with respect to atmospheric CO_2. However, their high nutrient content favours an intensive phytoplankton activity, which decreases the pCO_2 usually well below the equilibrium with the atmosphere. In temperate areas, this occurs only when sufficient light is available. The net budget of the CO_2 exchange with the atmosphere is thus difficult to establish. A recent study of the European shelves (Frankignoulle and Borges 2000) shows however that the coastal zone acts as a sink for atmospheric CO_2 over a one year period. The CO_2 fluxes observed at the margin are about one order of magnitude larger than those reported for the open ocean.

All these processes are poorly quantified at present. Even if the fluxes involved concern only a small fraction of the primary production recorded in the coastal zone, they all together may constitute important components of the organic and inorganic carbon cycle, which cannot be neglected in present and future global budgets.

Acknowledgements

This work is partly based on the results of the OMEX project supported by the European Union MAST program (contract numbers MAS2-CT93-0069, MAS3-CT96-0056 and MAS3-CT97-0076) and the Belgian State - Prime Minister's Services - Science Policy Office in the framework of the Impulse Programme "Global Change" (Contract N° GC/11/009). I thank L. Chou and J.P. Vanderborght for thorough review of the manuscript and D. Bajura and M. Loijens for editorial handling. I thank also the two reviewers, W. Berger and O. Pfannkuche for their critical and very useful comments.

References

Anderson RF, Rowe GT, Kempe PF, Trumbore S, Biscaye PE (1994) Carbon budget for the midslope depocenter of the Middle Atlantic Bight. Deep-Sea Res II 41 (2/3):669-703

Berger WH, Fisher K, Lai C, Wu G (1987) Ocean productivity and organic carbon flux. Part I. Overview and maps of primary production and export production. SIO Reference Series, Scripps Inst of Oceanography, La Jolla, CA 67p

Berger WH, Smetacek VS, Wefer G (1989) Ocean productivity and paleoproductivity: An overview. In: Productivity of the Ocean: Present and Past. Berger WH, Smetacek VS, Wefer G (eds) Dahlem Workshop Reports. J Wiley & Sons, Chichester, pp 1-34

Brink KH, Robinson AR (1998) The Sea - Vol 10: The

Global Coastal Ocean. J Wiley & Sons, Chichester

Capone DG (1988) Benthic nitrogen fixation. In: Blackburn TH, Sorensen J (eds) Nitrogen Cycling in Coastal Marine Environments. J Wiley & Sons, New York, SCOPE 33:85-123

Dugdale RC, Goering JJ (1967) Uptake of new and regenerated forms of nitrogen in primary productivity. Limnol Oceanogr 23: 196-206

Eppley RW, Peterson BJ (1979) Particulate organic matter flux and planktonic new production in the deep ocean. Nature 282:677-680

Flach E (2002) Factors controlling soft bottom macrofauna along and across european continental margins. In: Wefer et al. (eds) Ocean Margin Systems. Springer, Berlin pp 351-363

Frankignoulle M, Borges A (2000)The European continental shelf as a significant sink for atmospheric carbon dioxide. Glob Biogeochem Cycl

Galloway JN, Howarth RW, Michaels AF, Nixon SW, Prospero JM, Dentener FJ (1996) Nitrogen and phosphorus budgets on the North Atlantic Ocean and its watershed. Biogeochem 35:3-25

GESAMP (1989) Atmospheric Input of Trace Species to World Ocean. Reports and studies 38 World Meteorological Organization Geneva 111 p

Haese RR (2002) Macrobenthic activity and its effects on biogeochemical reactions and fluxes. In: Wefer et al. (eds) Ocean Margin Systems. Springer, Berlin pp 219-234

Huthnance JM (1995) Circulation, exchange and water masses at the ocean margin: The role of physical processes at the shelf edge. Prog Oceanogr 35: 353-431

Joint I, Chou L, Figueiras FG, Groom S, Loijens M, Tilstone G, Wollast R (2001) The response of phytoplankton production to periodic upwelling and relaxation events at the Iberian shelf break. J Mar Syst

Liu KK, Iseki K, Chao S-Y (2000a) Continental margin carbon fluxes. In: Hanson RB, Ducklow HW, Field JG (eds) The Changing Carbon Cycle: A Midterm Synthesis of the Joint Global Ocean Flux Study. International Geosphere-Biosphere Programme Book Series, Cambridge, Cambridge University Press pp 187-239

Liu KK, Atkinson L, Chen CTA, Gao S, Hall J, Mcdonald RW ,Talaue McManus L, Quiñones R (2000b) Exploring continental margin carbon fluxes on a global scale. EOS 81(52):641-644

Lochte K, Pfannkuche O (2002) Processes driven by the small sized organisms at the water-sediment interface. In: Wefer et al. (eds) Ocean Margin Systems. Springer, Berlin pp 405-418

Mantoura RFC, Martin JM, Wollast R (eds) (1991) Ocean Margin Processes in Global Change. Dahlem Workshop Reports. J Wiley & Sons, Chichester 469p

Menard HW, Smith SM (1966) Hypsometry of ocean basin provinces. J Geophys Res 71: 4305

Meybeck M (1993) C, N, P and S in rivers: From sources to global inputs. In: Wollast R, Mackenzie FT, Chou L (eds) Interactions of C, N, P and S Biogeochemical Cycles and Global Change. NATO ASI Series 14, Springer, Berlin pp 163-193

Milliman JD (1993) Production and accumulation of calcium carbonate in the ocean: Budget of non-steady state. Glob Biogeochem Cycl 7 : 927-957

Nixon SW, Ammerman JW, Atkinson LP, Berounsky VM, Billen G, Boigourt WC, Boynton WR, Church TM, Ditoro DM, Elmgren R, Garber JH, Giblin AE, Jahnke RA, Owens NJP, Pilson MEQ, Seitzinger SR (1996) The fate of nitrogen and phosphorus at the land-sea margin of the North Atlantic Ocean. In: Howarth R (ed) Nitrogen Cycling in the North Atlantic Ocean and its Watersheds. Biogeochem 35:141-180

Pingree RD, Mardell GT, Holligan PM, Griffiths DK, Smithers J (1982) Celtic Sea and the Armorican current structure and the vertical distributions of temperature and chlorophyll. Cont Shelf Res 1 (1):99-116

Pingree RD, New AL (1995) Structure, seasonal development and sunglint spatial coherence of the internal tide on the Celtic and Armorican shelves and the Bay of Biscay. Deep-Sea Res 42 (2):245-284

Pingree R.D, Sinha B, Griffiths CR (1999) Seasonality of the European slope current (Goban Spur) and ocean margin exchange. Cont Shelf Res 19: 929-975

SCOR (1992) Joint Global Ocean Flux Study Implementation Plan. JGOFS Report No. 9. SCOR, Baltimore

SCOR (1994) Report of the JGOFS/LOICZ Task Team on Continental Margin Studies. JGOFS Report No. 16. SCOR, Baltimore

Schlüter M (2002) Fluid flow in continental margin sediments. In: Wefer et al. (eds) Ocean Margin Systems. Springer, Berlin pp 205-217

Seitzinger SP, Giblin AE (1996) Estimating denitrification in North Atlantic continental shelf sediments. Biogeochem 35:235-259

Simpson JH (1998a) Tidal processes in the shelf seas. In: Brink KH, Robinson AR (eds) The Sea - Vol 10: The global coastal ocean: Processes and methods. J Wiley & Sons, Chichester pp 113-150

Simpson JH (1998b) The Celtic seas coastal segment. In: Brink KH, Robinson AR (eds) The Sea - Vol 10: The global coastal ocean: Processes and methods. J Wiley & Sons, Chichester pp 569-698

Soetart K, Herman PJM, Middelburg JJ, Heip C, SmithCL,

Tett P, Wild-Allen K (2001) Numerical modelling the shelf break ecosystem: Integrating field measurements of chlorophyll, primary production, zooplankton, nitrate, ammonium, oxygen and sediment properties. In: Wollast R, Chou L, Avril B (eds) Ocean Margin Exchange in the Northern Gulf of Biscay: OMEX I. Deep-Sea Res II, 48:3141-3177

Thomsen L (2002) The benthic boundary layer. In: Wefer et al. (eds) Ocean Margin Systems. Springer, Berlin pp 143-156

Valiela I, Costa JE (1988) Eutrophication of Buttermilk Bay, a Cape Cod coastal embayment: Concentrations of nutrients and watershed nutrient budgets. Environ Manag 12:539-551

Van Weering TjC, Hall IR, de Stigter HC, McCave IN, Thomsen L (1998) Recent sediments, sediment accumulation and carbon burial at Goban Spur, NW European Continental Margin (47-50°N). Progr Oceanogr 42:5-35

Walsh JJ (1988) On the Nature of Continental Shelves. Academic Press, London 520p

Walsh JJ (1991) Importance of continental margins in the marine biogeochemical cycling of carbon and nitrogen. Nature 350:53-55

Wollast R (1991) The coastal organic carbon cycle: Fluxes, sources and sinks. In: Mantoura RFC, Martin JM, Wollast R (eds) Ocean Margin Processes in Global Change. Dahlem Workshop Report, J Wiley & Sons, Chichester pp 365-381

Wollast R (1998) Evaluation and comparison of the global carbon cycle in the coastal zone and in the open ocean. In: Brink KH, Robinson AR (eds) The Sea - Vol 10: The global coastal ocean: Processes and methods. J Wiley & Sons, Chichester pp 10:213-252

Wollast R, Chou L (1998) Distribution and fluxes of calcium carbonate along the continental margin in the Gulf of Biscay. Aquatic Geochem 4: 369-393

Wollast R, Chou L (2001) The carbon cycle at the ocean margin in the Northern Gulf of Biscay. In: Wollast R, Chou L, Avril B (eds) Ocean Margin Exchange in the Northern Gulf of Biscay: OMEX I. Deep-Sea Res II, 48:3265-3293

Zabel M, Hensen C (2002) The importance of mineralization processes in surface sediments at continental margins. In: Wefer et al. (eds) Ocean Margin Systems. Springer, Berlin pp 253-267

Imaging the Subsurface with 2-D and 3-D Seismic Data

D. Ristow[1*], K. Hinz[2], J. Hauschild[3], T. Gindler[1],
A. Berhorst[1], C. Bönnemann[2]

[1]*GEOMAR, Forschungszentrum für Marine Geowissenschaften,
Wischhofstr. 1-3, 24148 Kiel, Germany*
[2]*BGR, Bundesanstalt für Geowissenschaften und Rohstoffe,
Stilleweg 2, 30655 Hannover, Germany*
[3]*Universität Bremen, Fachbereich Geowissenschaften, 28359 Bremen, Germany*
* corresponding author (e-mail): sschenck@geomar.de*

Abstract: In regions of complex geology, the corresponding seismic time sections show diffraction events, distorted positions of reflectors, smeared lateral discontinuities and uncertain amplitudes of seismic reflections with the consequence that the interpreter cannot automatically reconstruct and identify the cross-section through the earth from the seismic time section. The purpose of migration (imaging of seismic data) is to reconstruct structural depth sections from seismic time sections and to produce a geological image of the subsurface. 'Poststack migration' is an imaging process that uses stacked averaged seismic data. Stacking reduces the amount of data drastically, and thus subsequent migration of stacked data is less expensive than prestack migration. On the other hand, severe problems occur where the standard stacking procedure can no longer be applied, e.g. when the reflection surfaces are strongly curved or there are strong variations of velocities. The 'prestack seismic migration process' is a reconstruction of subsurface structures directly from measurements at the surface of the earth. The advantages of this process particularly appear in faulted steep dip regions with extreme flexures. Furthermore, this process allows an iterative estimation of the migration velocity field (macro velocity field), which is needed as an important parameter field for the migration process itself. 'Depth migration' is a transformation, mapping, and imaging of measured seismic time data into a migrated depth section. It yields a reliable image of the subsurface, provided that the migration velocity field (macro velocity field) is correct. 'Time migration', in principle, represents a special type of depth migration, where the depth scale z is scaled by a pseudo-time scale τ using a replacement velocity. Most time migration schemes do not take into account the refraction of waves at the boundaries, hence, these schemes fail to image data from laterally inhomogeneous media. On the other hand, time migration schemes show low sensitivity to migration velocity errors and can be used as an intermediate step in seismic data processing for imaging seismic data. The conventional seismic 2-D technique provides an erroneous image of the subsurface, if the subsurface varies perpendicularly to the 2-D seismic recording direction. In this case, a 3-D technique has to be applied. The 3-D migration process plays a fundamental role in the 3-D processing sequence. It provides a reliable image of the earth and avoids possible misinterpretation which may occur when 3-D structures are explored with 2-D methods. 3-D poststack time migration is often used in practical work because this first result is needed for first interpretation and to check the effect of 3-D processing and 3-D imaging. Nevertheless, this type of imaging fails to give a reliable image of the subsurface in the case of complex geological areas, e.g. complex overburden structures. 3-D poststack depth migration needs the same computing time as 3-D prestack time migration, but yields a reliable depth image of the subsurface provided that the conditions for 3-D stacking are given and a reliable 3-D macro velocity field is available. 3-D prestack depth migration has long been recognized as the appropriate imaging procedure for complex structures and complex velocities. It requires that the velocity field is estimated using the migration procedure itself in an iterative manner, however, therefore this migration process consumes an enormous amount of computing time. In this paper, we focus on the description of a 3-D imaging process which uses a special type of 3-D prestack depth migration only to estimate the 3-D macro velocity field and then applies a 3–D poststack depth migration scheme. This scheme is applied to a 3-D data set acquired at the Costa Rica convergent margin.

From WEFER G, BILLETT D, HEBBELN D, JØRGENSEN BB, SCHLÜTER M, VAN WEERING T (eds), 2002,
Ocean Margin Systems. Springer-Verlag Berlin Heidelberg, pp 33-55

Introduction

Role of seismology and seismics

Seismology and seismics are based on data called seismograms, which are records of mechanical vibrations of the earth. Seismology uses natural seismicity to image the earth's subsurface, whereas seismics use sources like airguns, vibrators or explosives to generate seismic waves for imaging.

Seismology is at an extreme of the whole spectrum of earth measurements. It considers mechanical properties of the earth and offers a means by which investigation of the earth's interior can be carried out to the greatest depths. Its resolution and accuracy are higher than are attainable in any other branch of geophysics such as gravity and magnetics. However, the earth damps seismic waves of higher frequencies with increasing traveltimes, so that resolution is reduced with increasing depth.

A relatively high resolution and accuracy can be obtained from seismics because seismic waves have the shortest wave length of any kind of waves that can be observed passing through internal structures of the earth. Seismic waves undergo the least distortion in waveform as compared with other geophysical observables, such as gravity or electromagnetic phenomena.

Seismology contributes to our knowledge of the present state of the earth's interior. Because of its emphasis on current tectonic activity, seismology is used by academic and research institutes.

Within the last 40 years, seismics have found wide and intense applications in exploration for oil and gas and mineral prospecting. This led to the development of powerful and modern exploration seismic methods which was also introduced by academic and research institutes and which is discussed in this paper with respect to imaging.

Historical development of seismic exploration technology

It is impossible to discuss exploration technology without discussing seismics. It is widely agreed that the seismic industry was one of the first to enter the digital revolution in the 1960s.

With seismic acquisition, the systems became then able to record more data by increasing the number of channels and fold of coverage. The field operations were made easier, and at much less cost than before.

With digital data-processing, the digital revolution needed powerful digital computers to analyse the recorded seismic data. Many of the data-processing algorithms including deconvolution, velocity analysis, multichannel filtering and automatic filtering to improve signal/ noise ratio and even the first types of migration were designed in the 1960s and 1970s. At that time, the seismic industry was already pushing the computer industry to build faster computers, e.g. for special arithmetic operations (inner-product). This demand led to the design of array-processors and later to vector-computers.

In the 1980s, the seismic industry took another big step forward; it was now beginning to provide companies with 3-D images of the subsurface. It is widely agreed that 3-D seismics has been the most important factor in improving success rates and lowering risk in exploration for oil and gas. It took years of experimental research to test numerous methods to accurately estimate an earth model in depth and to use it to create a 3-D image of the earth by 3-D depth migration. In the 1990s, 3-D prestack depth migration was successfully introduced into 3-D seismic data processing and, once again, the computer industry was challenged to provide cost-effective solutions for numerically intensive applications with large input-output operations (Yilmaz 1987).

Migration

Migration and imaging

Imaging and mapping by acoustic waves is called acoustic imaging and is being used e.g. in acoustic holography, non-destructive testing of materials, acoustic microscopy, tomography, medical ultrasonic imaging, underwater acoustic imaging and seismic migration. In all these procedures, emitters and receivers are arranged outside the object to be explored, and reflection data or transmission data as measured at the surface are recorded.

The aim of the reconstruction process is to image the spatially varying properties of the object. The terms reconstruction, imaging, mapping and migration are synonyms for the same process. The term "migration" in seismic data processing is historically founded and considers only one aspect, namely the correction of distorted reflector positions. Another aspect of migration is the removal of diffractions by focussing the diffracted energy into the image of the diffractors in order to improve lateral resolution. The term "migration" is not appropriately chosen and can be misleading (it was already claimed earlier by the petroleum engineers to describe the movement of hydrocarbons from their place of origin to the reservoir).

The purpose of migration

Seismic time sections represent cross-sections through the earth's subsurface, if the subsurface is given by parallel homogeneous layers. On the other hand, in regions of complex geology we must accept the seismic section as a wavefield recorded at the surface of the earth. In regions of complex geology, the corresponding seismic sections show diffraction events, distorted positions of reflectors, smeared lateral discontinuities and uncertain amplitudes of seismic reflections with the consequence that the interpreter cannot automatically reconstruct and identify the cross-section through the earth from the seismic time section.

The purpose of migration is to reconstruct structural depth sections from seismic time sections. It should be noted that a "depth section" is a section through the three-dimensional structure, whereas a "time section" is the set of all time functions measured on a line.

The migration process removes diffractions and concentrates the diffracted energy at the position of the diffractor image. The migration process shifts the distorted position of reflectors to the geometrically correct position (see Fig. 1) (see also Berkhout 1984, Yilmaz 1987).

The migration process is primarily directed at the structural reconstruction of the subsurface. The extraction of lithologic information with help of the migration process (true amplitude migration) is a topical research problem for the present as well as the future. It is briefly discussed below in this paper, 'The outlook for seismic technology beyond 2000'.

Some terms in seismic data processing

CMP and offset vector - A seismic time series - initiated by a source and recorded by a receiver - is called a "seismic trace". The vector from the source to the receiver is the offset vector. The midway point between shot and receiver is the "common mid point" (CMP point). Every seismic trace is associated with an offset vector and a CMP point. The set of all traces which have the same CMP point is called CMP gather. The absolute value of the offset vector is the source-receiver distance or simply the offset. In general, for every CMP point there are many traces with distinct offsets. The number of traces is called 'coverage'.

Stacking velocity, NMO correction, CMP stacking - The traveltime curve for a horizontal reflector in a medium with a constant acoustic velocity is described by a hyperbolic curve $t^2 = t_0^2 + h^2/v^2$, where t is traveltime, h is offset, v is medium velocity and t_0 is zero offset traveltime for $h = 0$.

In the case of plane dipping layers, the CMP traveltime curve for a reflection point can be well approximated by a hyperbolic curve, but now the velocity is defined as "normal move-out velocity" (NMO velocity) or "stacking velocity".

A transformation, using stacking velocity, of a non-zero offset trace into its corresponding zero-offset trace is called "normal move-out correction". A summation of all NMO-corrected traces, associated to a common CMP point, is called "CMP stacking". The stacked trace is approximately a zero-offset trace. The stacking velocity - to be used for all CMP traces with the NMO correction process - is determined in such a way that the result of stacking has a maximum signal-to-noise ratio. An example for NMO correction can be found below (see Fig. 6 and Fig. 7).

Poststack migration, prestack migration

Poststack Migration - 'Poststack Migration' is an imaging process using CMP stacked seismic data.

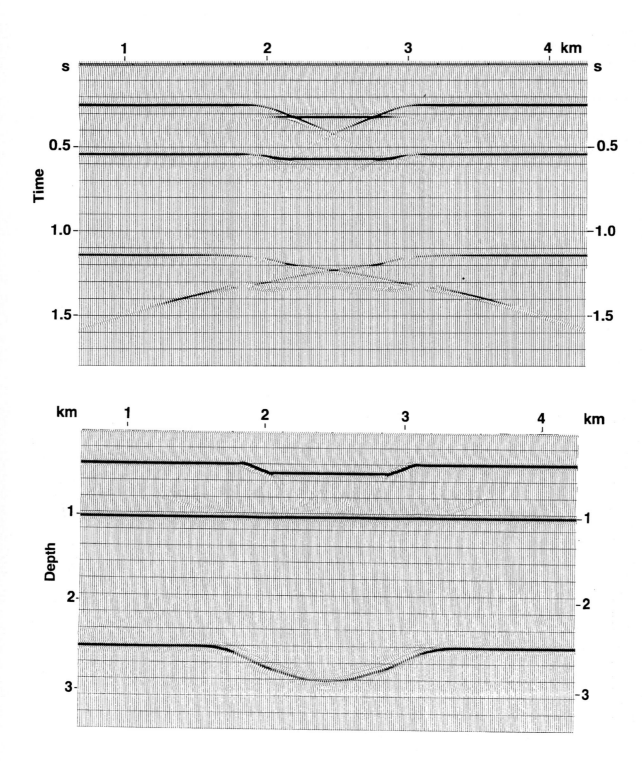

Fig. 1. 2-D synthetic data set. Top: zero-offset data set, vertical axis = recording time. Bottom: Data set after 2-D depth migration, vertical axis = depth.

Usually, the full process consists of many separate steps:

• Velocity analysis of CMP data to estimate stacking velocities.

• 'DMO' process. The DMO (Dip-Move-Out) algorithm modifies the single traces so that the stacking process generates zero-offset traces in an optimized way.

• Normal move-out correction and stack to simulate zero-offset data.

• Migration of the stacked data (poststack migration).

The stacking process is founded on the assumption that the earth's subsurface structure is simple. The assumption of plane dipping layers leads to stacking procedures that require only simple hyperbolic moveout corrections

Stacking reduces the amount of data drastically, and thus subsequent migration of stacked data is less expensive than prestack migration. On the other hand, severe problems occur where the standard stacking procedure can no longer be applied, e.g. when the reflection surfaces are strongly curved or there are strong variations of velocities. For example, moveout across a CMP gather may not be even approximately hyperbolic. For data beneath zones of complex structure, the stacking process should be avoided. In this case, the process 'prestack migration' should be applied.

Prestack Migration - The general seismic migration process is a reconstruction of subsurface structures directly from measurements at the surface of the earth, where the measurements are characterized by a non-zero distance between source and receivers. This process is called 'prestack migration'.

The advantages of this process particularly appear in faulted, steep dip regions with extreme flexures. Furthermore, this process allows an iterative estimation of the migration velocity field (macro velocity field) which is needed as an important parameter field for the migration process itself.

On the other hand, though, prestack migration schemes are very time-consuming on the computer and increase the computing time by a significant factor (say 10 to 1000) in comparison to poststack migration.

Time migration, depth migration

'Time migration' was the only standard migration type in the 1960s and 1970s and was then improved by the more accurate 'depth migration', which is today's standard level in imaging seismic data.

Depth migration - The transformation, mapping, and imaging of measured seismic time data into a migrated depth section is called 'depth migration'. Here, travel time t in the data set is removed, and depth z is introduced. The importance of depth migration is that it provides the correct geometry of subsurface structures which is necessary for reliable geological interpretation.

Depth migration should be the final process of seismic data processing. It yields a reliable image of the subsurface, provided that the migration velocity field (macro velocity field) is correct. The well-known extreme sensitivity of a depth-migrated image to velocity errors has led to the development of prestack migration schemes that involve a sufficiently accurate estimation of macro velocity fields.

Time migration - In principle, 'time migration' represents a special type of depth migration, where the depth scale z is replaced by a pseudo-time scale τ. A "replacement velocity" $c = c(z)$ is usually introduced in time migration. Its role is to transform depth to a vertically stretched time scale τ.

Most time migration schemes do not take into account the refraction of waves at the boundaries. Hence, these schemes fail to image data from laterally inhomogeneous media. Time migration is the wrong process to use in such complicated areas. On the other hand, time migration schemes show low sensitivity to migration velocity errors and can be used as an intermediate step in seismic data processing for imaging seismic data.

3-D seismics

2-D and 3-D data acquisition - If the CDP points of all seismic traces form essentially a set of one-dimensional manifold, then 2-D data recording is performed. If the CDP points of all seismic traces are uniformly distributed over an area, the 3-D data acquisition is realized.

2-D seismics versus 3-D seismics - The conventional seismic 2-D technique provides an erroneous image of the subsurface, if the subsurface varies perpendicularly to the 2-D seismic recording direction. Many events in a 2-D stacked seismic section or a 2-D migrated section are caused by diffractors and reflectors outside the vertical plane through the seismic line. 2-D seismic processing is justified, if the 2-D direction is perpendicular to the strike direction at every point of the vertical plane through the 2-D line. This ideal condition is never met in practical field-work but can sometimes be approximated by a well-designed field technique.

New developments in field technology, increased storage capacity and speed of digital computers, and the design of fast and new algorithms for 3-D data processing initiated a rapid increase of 3-D seismic work within the last twenty years. 3-D seismic exploration is a combination of 3-D data acquisition, 3-D data processing and 3-D data interpretation in order to obtain reliable vertical images through the three-dimensional subsurface (see also Yilmaz 1987). The 3-D migration process plays a fundamental role in the 3-D processing sequence, it provides a reliable image of the earth and avoids interpretation errors that occur when 3-D structures are explored with 2-D methods.

The benefits of 3-D migration are:
• Accurate positioning of reflectors. The 3-D migration process provides the most suitable approximation to the true 3-D subsurface structure by accurately positioning the reflectors.
• Contraction of diffraction events. The diffraction events are collapsed to the right image of the diffractors.
• Reliable and detailed structural interpretation. The interpretation is based on the sequential interpolation of 2-D sections through the 3-D data volume, yielding accurate mapping of events.
• Improvement of fault recognition and identification. Fault recognition is difficult in 2-D exploration and 2-D processing if the line crosses a diffracting edge obliquely.

How does the migration algorithm work?

In principle, there are two different interpretative versions to describe the migration process. They are equivalent, both are based on different mathematical formulations of the acoustic wave equation.

Downward continuation and imaging (Claerbout 1976) - The acoustic wave equation is used to downward continue sources and receivers by a small depth increment simulating the case that the shooting and recording line lie below the surface. At the beginning, all receivers are downward continued to a new depth level, whereby each shot-gather (set of all seismic traces originating from the same shot) is handled separately. Then, following the principle of reciprocity (Claerbout 1976), shots and receivers are exchanged, new shot-gathers are built up and downward continued to the same depth level as in the step before. This concludes the downward continuation of both shots and receivers to a new depth level. This procedure is repeated for as many depth-steps as necessary.

A reflector present at a certain depth in a downward continued wave field would have all its reflected energy restricted to zero traveltime and zero source-receiver offset. An application of these two imaging conditions - zero time and zero offset - during progressive downward continuation forms a fully migrated depth section (Claerbout 1984).

Summation procedure (Schneider 1978) - A point diffractor in the earth will scatter seismic energy, causing a detectable signal in our recorded data. The timing of the recorded event will depend on the subsurface location of the diffractor.

Consider an output location at some point in image space (x, z). For any given input trace, we ask the question: If there was a point scatterer at this output location, at what time would it be recorded at the receiver location? This traveltime is the sum of the traveltimes from the source to the scatterer and back to the receiver.

To obtain the migrated signature of the scatterer, we compute this traveltime and use it to sum the input-trace amplitude into the output depth-point location. This process is repeated for all input traces. If an event is real at an output location, then the amplitudes will sum constructively to create a migrated image. If an event is not at an output location, the amplitudes will sum deconstructively to produce a small output to be interpreted as noise.

Up to now, we have considered one point diffractor only. Since a reflector in the earth can be considered a collection of point diffractors, by looping over a grid of subsurface points and checking each we can build up an image of the subsurface.

Migration algorithms

There are three desirable features a migration program should have: It should handle
• steep dips of dipping reflectors and of diffraction events,
• velocity variations, both vertical and lateral.
• It should be cost-effective to handle large amounts of data.

One reason for the wide range of migration algorithms used in seismic data progressing today is than none of the algorithms fully meets the important criteria stated above. We briefly discuss the two most important methods.

Finite-difference algorithms (FD methods) (Claerbout 1984) - FD methods are based on the downward continuation process of wave fields. Finite-difference algorithms can handle all types of velocity variations, but they have different levels of dip approximations. The higher the dip performance, the longer the computing time, hence a compromise is necessary. FD schemes yield results with the highest possible resolution, but the schemes can only be applied when the data are regularly sampled in the acquisition plane.

Kirchhoff algorithm (Schneider 1978) - In principle, the Kirchhoff algorithm can be described by the summation procedure. Migration algorithms based on the integral solution to the scalar wave equation, commonly known as Kirchhoff migration, can handle all dips up to 90 degrees. Kirchhoff migration handles seismic data, which are not regularly sampled in the acquisition plane. The

disadvantage is that many Kirchhoff migration schemes do not show the same resolution as FD schemes.

The macro velocity field

The acoustic wave equation used for migration includes an important parameter: the 3-D velocity function $v = v(x,y,z)$ as a function of surface coordinates x and y and of the depth coordinate z. The 2-D function reads $v = v(x,z)$.

The most severe problem of seismic depth migration is the estimation of these migration velocities: Whereas in other imaging systems (e.g. medical imaging) the velocity is nearly constant, the seismic velocity in complicated subsurface geological regions changes in vertical and lateral direction. The seismic migration velocities (macro velocity field) can be estimated from seismic measurements with some effort, but this parameter-estimation technique makes seismic data processing a non-straightforward process.

The term 'macro velocity field' has been chosen indicating the contrast to 'micro velocity fields' which are measured, for example, in boreholes. The purpose of the macro velocity field is to describe velocity contrasts at the reflectors and velocity gradients in vertical and lateral direction within the layers (Berkhout 1984).

A successful application of depth migration strongly depends on the accuracy of the migration -velocity field. Even minor velocity errors can severely degrade the migrated depth section.

Migration strategies

2-D data (Table 1)
• 2-D poststack time migration - a very fast process on the computer - is the least sensitive to velocity errors, and it often yields results acceptable for a first interpretation. In general, the migration velocity field is simply derived from the stacking velocity field.
• 2-D poststack depth migration is needed when there are strong lateral velocity variations, which are generally associated with complex overburden structures. 2-D poststack depth migration needs an

	2-D Poststack Time Migration	2-D Poststack Depth Migration	2-D Prestack Depth Migration
Preconditions	stacked data	stacked data - macro velocity field v (x, z)	
Migration velocity field	stacking velocities and add. information	macro velocity field v (x, z)	macro velocity field v (x, z)
Use, benefits	first result for interpretation, limited accuracy	optimized image from stacked data	2-D macro velocity field estimation - optimized image from complex data
Computing time	low	low	medium
Role in processing	intermediate result	Final standard processing	final standard processing
Algorithm (computer)	FD or Kirchhoff	FD or Kirchhoff	FD or Kirchhoff

Table 1. 2-D migration.

accurate 2-D macro velocity field, which is derived from the 2-D prestack depth migration scheme. Hence, 2-D prestack depth migration can be applied directly and 2-D poststack depth migration can be omitted.

• Today, 2-D prestack depth migration is 'state of the art' and can be realized by modern seismic processing systems. The computing time for iterative processing lies within the range of modern workstations, which are available to research institutes and university departments. 2-D prestack depth migration is applied in an iterative manner to derive the 2–D macro velocity field and to finally produce a 2-D image of the subsurface.

3-D data (Table 2) - The recommendations for migrating 2-D data cannot simply be copied for the 3-D case, because the huge amount of 3-D seismic data requires much more computing time.

3-D poststack time migration - This type of 3-D imaging is often used in practical work because this first result is needed for first interpretation and to check the effect of 3-D processing and 3-D imaging. In practical work, a lot of effort is spent on estimating a more reliable migration velocity field for time migration from the 3-D stacking velocity, using, for example, additional information from bore hole data, 2-D seismic lines and geological knowledge about the subsurface. Never-

theless, this type of imaging fails to give a reliable image of the subsurface in the case of complex geological areas, e.g. complex overburden structures.

3-D poststack depth migration - This type of migration needs the same computing time as the 3-D prestack time migration, but yields a reliable depth image of the subsurface, provided that the conditions for 3-D stacking are met (e.g. hyperbolic NMO curves) and a reliable 3-D macro velocity field for 3-D depth migration is available.

3-D prestack depth migration - 3-D prestack depth migration has long been recognized as the appropriate imaging procedure for complex structures and complex velocity distributions. A key concern in 3-D prestack depth migration is the accuracy of the velocity field. If the velocity field is simple, 3-D prestack depth migration is not necessary. If, however, severe lateral velocity variations exist, it is difficult to estimate the migration velocity field. The velocity field must be estimated using the migration procedure itself in an iterative manner.

The migration process consumes an enormous amount of computing time. In recent years, however, advances in supercomputer capability have made the process economical and practical. Today, exploration companies and oil industry use powerful vector computers and parallel super

	3-D Poststack Time Migration	3-D Poststack Depth Migration	3-D Prestack Depth Migration
Preconditions	stacked data	stacked data - macro velocity field v = v (x, y, z)	
Migration velocity field	stacking velocities, additional information, ray tracing programs for corrections	macro velocity field v (x, y, z)	macro velocity field v (x, y, z)
Use, benefits	first result for interpretation, ray tracing programs for corrections	optimized image from stacked data	3-D macro velocity field estimation - optimized 3-D image from complex data
Computing time	high	high	extremely high
Role in processing	intermediate result, sometimes final result	final standard processing	standard processing for 3-D macro field estimation, target-oriented processing
Algorithm (computer)	FD or Kirchhoff	FD or Kirchhoff	mostly Kirchhoff

Table 2. 3-D migration.

computers. Geoscience institutes and academic research institutes, however, do not have the required computer hardware and are therefore forced to approximate 3-D prestack depth migration.

In this paper, a 3-D imaging process is described where a special 3-D prestack depth migration (two-pass procedure) was applied only to estimate the 3-D macro velocity field which was then fed into a 3-D poststack depth migration scheme to image the 3-D data set.

Estimation of the 2-D macro velocity field

Migration velocity analysis takes advantage of the sensitivity of prestack depth migration to the velocity model, and uses the migration error to update the velocity model. Many approaches to prestack migration velocity analysis have been developed, such as depth - focusing analysis (DFA) (Yilmaz and Chambers 1984, McKay and Abma 1992). Although the focusing equations were established for a constant velocity medium and a flat, horizontal reflector, focusing analysis has been used extensively for velocity model-building, even in the presence of severe lateral velocity variations.

Jeannot, Faye and Denelle (1986) used the method of prestack depth migration for depth-focusing analysis, where velocity estimation is performed by repeated prestack depth migration to converge on a final velocity field. The wave equation is used to propagate the surface-recorded wavefield to a slightly greater depth. By this downward continuation process, a situation is simulated where the shots and the receivers lie on a deeper acquisition line or plane. A reflector present at the current depth would have all of its reflected energy restricted to zero traveltime ($t = 0$) and zero source-receiver offset ($h = 0$). An application of these two imaging conditions - zero time and zero offset - during progressive downward continuation of shots and receivers forms the migrated depth section (MacKay and Abma 1992).

Fig. 2a illustrates the migration of data from a horizontal reflector within a CMP-gather. The wave equation is used to propagate the surface-recorded wavefield, $P(x, h, t, z=0)$, shown in panel A down into the subsurface to depth z_B (panel B).

The downward-continued data, $P(x, h, t, z=z_B)$, appear as if they had been recorded at depth z_B, and the new $t = 0$ (dashed line) is defined by intersection of this panel with the imaging line. Two important features of the wavefield in panel B should be noted. First, the data have partially collapsed to zero offset. Second, the traveltime to the reflector (from the new $t = 0$) is reduced. Both of these observations are consistent with the data having been recorded from a surface closer to the reflector. Repeated downward continuation to the depth of the reflector (panel D) results in all the reflected energy $P(x, h, t, z=z_D)$, focusing to zero offset. Further, the reflected energy appears exactly at the local $t = 0$. Invoking the zero-time and zero-offset imaging conditions thus yields the correctly migrated image, $P(x, h = 0, t = 0, z = z_D)$. (MacKay and Abma 1992).

Fig. 2b illustrates migration with a too fast velocity. Note that the imaging line is less steep than in Fig. 2a , since a higher velocity implies less vertical traveltime for the same migration depth. Fig. 2b also shows that during downward continuation the reflected energy focuses best to zero offset (panel C) at a focusing depth that is shallower than the reflector's real depth (panel D). The focused energy in panel C is not, however, at zero time for this depth and will not be extracted into the migrated image. The focused energy is at time $t = 0$ when the migration depth is reached (panel E), but in this case, the data are clearly overmigrated.

This example illustrates that when migration velocities are inaccurate, the zero-time, zero-offset imaging conditions do not extract the best-focused data. Nevertheless, this information can be used to correct the migration velocities.

Both the migration depth, as dictated by the migration velocity field, and the depth at which the data best focus may be viewed. As shown by Jeannot et al. (1986) for depth migration, differences between the migration and focusing depths can be used to update the velocity field for another iteration of prestack migration (MacKay and Abma 1992).

a)   b)

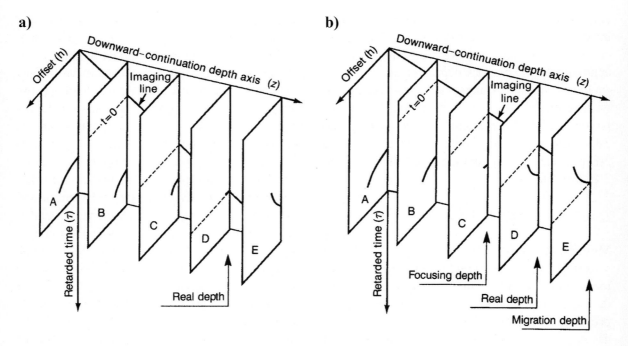

Fig. 2. a) CMP gather during the downward continuation process (after Jeannot et al. 1986): Using the real velocity. **b)** - using a too fast velocity.

Top-down approach - As errors at any particular depth affect the imaging of all deeper data, the velocity model is generally built from the top down, progressing to deeper layers as we reach satisfactory velocities. The weakness of the top-down approach is that accumulated errors in the estimation of velocities for shallower layers can adversely affect the accuracy of the estimation for underlying layers.

3-D prestack depth migration (two-pass procedure) for the estimation of the 3-D macro velocity field

In principle, full 3-D prestack depth migration (one-pass migration) is an effective way to image complex 3-D structures with significant lateral velocity variations and to estimate the 3-D macro velocity field. A typical 3-D survey, however, involves millions of traces, and 3-D prestack depth migration is a massive computation and data manipulation task. Yet the process of a full 3-D prestack depth migration needs very high computing time and can only be realized with the latest computer technology, which is beyond the capacity of academic research institutes.

Computing time can be saved by approximating the 3-D prestack depth migration by a "two-pass procedure". There are at least two different algorithms to realize the "two-pass procedure" (Gindler 2000). One of these algorithms - "Crossline Migration to Single Offset" (CMSO) - will be presented here in some detail.

Using 3-D prestack migration as a "two-pass procedure", it is possible to reduce the computing time to a reasonable level. The first pass - crossline time migration - is considered as a preconditioning process putting energy scattered in all directions into the correct 2-D plane and thus transforming the 3-D data volume into a set of parallel true 2-D lines. Crossline migration eliminates sideswipe and creates an ideal 2-D line that can then be migrated in the inline direction.

In the second pass, the 2-D lines can be depth migrated in inline direction including an iterative 2-D velocity analysis procedure. The final 2-D macro velocity field can be interpreted as a 2-D section through the full 3-D macro velocity field. Hence, an estimate for the 3-D macro velocity field is obtained by combination and interpolation of all 2-D macro velocity fields derived from a certain set of 2-D lines.

Crossline migration to Single Offset (CMSO) (first pass) - This kind of crossline migration technique is a velocity-dependent procedure and is based on an idea of Devaux et al. (1996). For 3-D prestack data in the case of constant velocity, the 3-D impulse response is an ellipsoid with foci at shot and receiver (Fig. 3). The distance from S to R across any reflection point of the ellipsoid is constant $= c * t$, where c is a constant velocity and t the traveltime. The contribution of this ellipsoid to the image in the inline plane is given by amplitude distribution along the elliptical curve which will be formed by the intersection of the ellipsoid with the chosen inline plane (Fig. 3) (Gindler 2000).

This elliptical curve can be interpreted as a 2-D impulse response where the ellipse is defined by a new shot and receiver pair at the foci of the ellipse with the new midpoint M', offset h' and two-way traveltime t' (Fig. 3). The distance S'R', the new CMP point M' and the new constant $c * t'$ can be computed by geometrical relations.

In other words:
Given are: M, S, R, t, $2 * h$, φ (3-D time series)
To be computed: M', S', R', t', $2 * h'$ (2-D time series) (Explanations of abbreviations see Fig. 3).

This way, a 3-D amplitude for time t and its 3-D parameters is mapped to a 2-D amplitude for time t' and its corresponding 2-D parameters. This process generates from each input sample one output sample, therefore we call this technique "Crossline Migration to Single Offset" = CMSO.
Example from synthetic data:

The synthetic 3-D model (Fig. 4) consists of a subsurface with constant velocity and 5 line diffractors with different dippings. By seismic modelling methods we have computed about 200 000 synthetic seismic traces simulating 3-D seismic data acquisition for that 3-D model. We choose the x - axis (inline axis) of the model to create crossline migrated (first pass) 2-D pre-stack data.

Fig. 5 shows the original CMP input gather 41 (CMPx position = 1000 m, CMPy position = 0 m).

CMSO-method

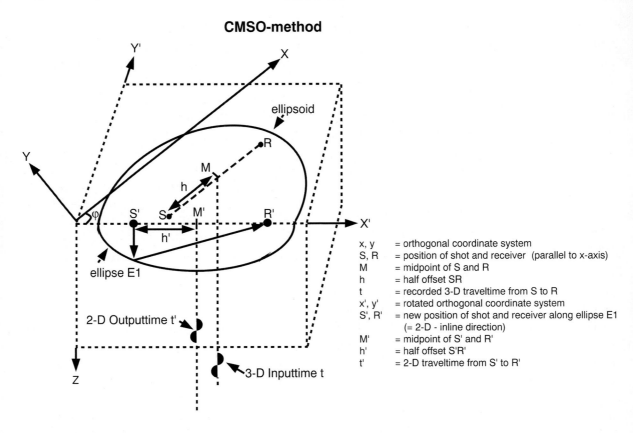

x, y = orthogonal coordinate system
S, R = position of shot and receiver (parallel to x-axis)
M = midpoint of S and R
h = half offset SR
t = recorded 3-D traveltime from S to R
x', y' = rotated orthogonal coordinate system
S', R' = new position of shot and receiver along ellipse E1
        (= 2-D - inline direction)
M' = midpoint of S' and R'
h' = half offset S'R'
t' = 2-D traveltime from S' to R'

Fig. 3. Principle of the CMSO method (Crossline Migration to Single Output = CMSO).

Dip α [degr.]	t_0[s] at origin	t_0[s] at CMP 41	y-position [m]	Line diffractor Nr.
45	0.57	1.13	0	1
0	1.60	1.60	0	2
20	1.88	2.15	0	3
15	1.23	1.42	500	4
0	2.43	2.43	500	5

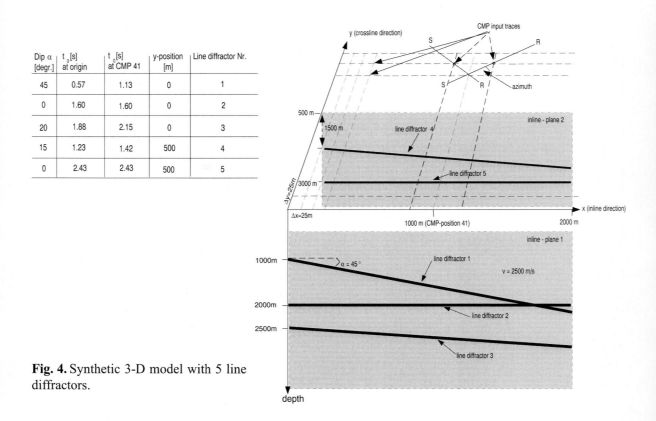

Fig. 4. Synthetic 3-D model with 5 line diffractors.

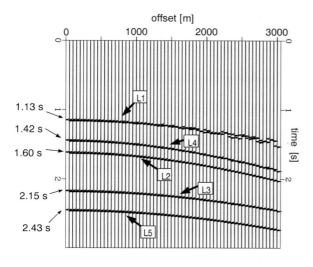

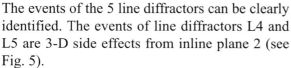

Fig. 5. CMP gather 41 before the CMSO process. Events L4 and L5 are side effects. Events L1, L2, L3 correspond directy to line diffractors below the CMP line y = 0. The position of CMP number 41 is indicated in Fig. 4 for y = 0.

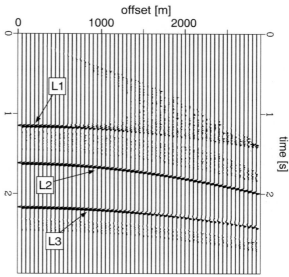

Fig. 6. CMP gather 41 after the CMSO process. The side effects L4 and L5 have been removed.

The events of the 5 line diffractors can be clearly identified. The events of line diffractors L4 and L5 are 3-D side effects from inline plane 2 (see Fig. 5).

Fig. 6 shows the CMP gather 41 after the CMSO process. The 3-D side effects have almost been removed. The data are directly transformed into 2-D prestack data. In the second pass, the data can now be processed in inline direction by a 2-D prestack depth migration analysis to get a 2-D macro velocity field in a 2-D depth section. Both sections were vertical cuts though the 3-D sub-surface model and the 3-D macro velocity field. For illustration only, we have normal move out corrected the single events in Fig. 6, using dip-dependent stacking velocities, to obtain a NMO-corrected data set in Fig. 7.

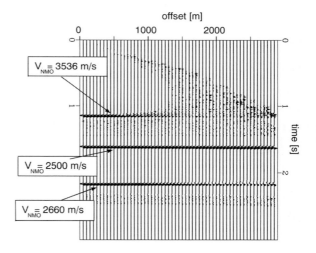

Fig. 7. CMP gather 41 after Normal Move Out (NMO) correction.

3-D marine seismic example

3-D seismic survey

Offshore Costa Rica, three oceanic crustal segments generated at the Cocos-Nazca spreading center enter the subduction zone (von Huene et al. 1999) (Fig. 8). These segments are from the south-east to the north-west: The thick-crusted Cocos Ridge segment, a seamount-covered crustal segment 19.5 - 14.7 Ma old, and the crustal segment off Nicoya Peninsula showing relatively smooth crustal morphology like the contiguous East Pacific Rise (ERP) - generated crust.

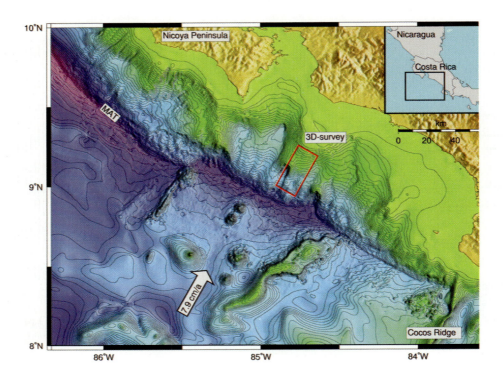

Fig. 8. Morphological overview on the Costa Rica continental margin. The 3-D survey is marked by a red line. The contour interval is 100 m. MAT = Middle American Trench (modified from v. Huene et al. 1994).

The basic structure of the Pacific continental margin of Costa Rica is a wedge-shaped margin of high-velocity rock characterized by high amplitude reflections at the top and at the base, covered by a sedimentary apron, fronted by a small accretionary prism and underlain by a distinct sequence of high-amplitude - low-frequency reflections. The lateral modification of the basic structure has been explained by subducting seamounts. Since also alternate processes such as intensive underplating of basal removed material from the margin wedge and/or tectonism associated with transform faults have been inferred for the variability of the margin wedge and the slope morphology, the Federal Institute for Geosciences and Natural Resources has complemented the seismic coverage by a 3-D seismic survey.

The selected 3-D area is located near to the seamount-covered crustal segment approximately between 08°58'S and 09°14'S/ 84°40'W and 84°50'W, and has a 30°E orientation (Fig. 8). The water depth within the 3-D area varies between 400 m and 2500 m. The southern, i.e. seaward limit of the 3-D area runs approximately 10 km landward off the 3500-m-deep Middle American Trench. The westernmost line of the 3-D area traverses along the middle axis of a ca. 5 - 6 km wide and up to 450 m deep furrow which is interpreted as a path of disruption in the upper plate from a seamount on the subducting lower plate (von Huene et al. 1995). The easternmost line traverses an area with smooth slope morphology.

Data acquisition

The 3-D data were acquired by the survey vessel *M.V. GECO LONGVA* in an area of 30 km length (inline direction) and 15 km width. A 130 l airgun array was shot at intervals of 50 m. The data were recorded with two streamers of 3000 m length each and 200 m distance from each other. Thus every sail line resulted in two CMP lines with 100 m nominal distance. The receiver group spacing was

25 m, there were 120 channels per streamer. The record has a length of 13 s with a sampling rate of 4 ms.

Processing

Time domain processing - Soon after the survey, the data were processed to a 3-D poststack time migration. The processing sequence consisted of the following steps:
• 3-D binning with a bin length of 25 m in inline direction and 100 m in crossline direction
• spherical divergence correction
• deconvolution
• multiple attenuation in the *f-k* domain
• stacking velocity analysis (500 m * 500 m grid)
• 3-D stacking
• poststack interpolation to 50 m inline spacing, resulting in a revised bin size of 25 m * 50 m.

Each processing step involved extensive parameter tests. The result improved the signal-to-noise ratio and facilitated the identification of structural boundaries. The 3-D time migration, however, includes some principal deficiencies. In the case of strong lateral velocity variations, which is the case in the area under investigation, insufficient inclusion of ray effects may result in incorrect images. Therefore the project involved a triple approach for a better imaging and understanding of the structure:
• Building up a 3-D macro velocity model for a 3-D poststack depth migration. In addition, the derived velocities are useful for lithologic interpretation.
• Accomplishing a 3-D poststack depth migration (Ristow and Rühl 1997).
• Use of the seismic data, the depth information and the velocities for geological interpretation (Berhorst 1999).

All these steps are intertwined during the whole work process, i.e. from the beginning a preliminary interpretation is needed to determine the boundaries of the macro velocity model.

3-D depth imaging

One of the main aims of the project was the generation of a 3-D depth domain seismic image. Due to the high requirements for computing resources, a full 3-D prestack depth migration was beyond the possibilities of the project. Therefore, a 3-D poststack depth migration was the method of choice for imaging. To get a first depth migrated data cube, a preliminary velocity model ("Model A") was created (Fig. 10). It is based on the stacking velocities, from which interval velocities were computed by applying the Dix equation according to Yilmaz (1987) followed by a calibration of the velocities. Significant improvements were obtained by the derivation of two 3-D macro velocity fields based on 3-D prestack depth migration (two-pass method) combined with a 2-D migration in inline direction with a focusing analysis. 3-D macro velocity fields B and C were obtained by producing 12 and 23 seismic 2-D lines respectivelyby 3–D prestack depth migration (two-pass method) (see Fig. 11 and Fig. 12).

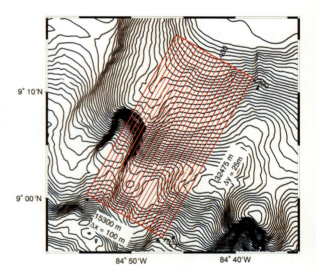

Fig. 9. Map of the 3-D survey and the 12 (distance = 1400 m) resp. 23 (distance = 700 m) crossline-migrated inlines. CMP numbers run in inline direction from 1021 to 2320, increment: 100 m.

Figs. 13a, 13b and 13c show two-dimensional sections through the macro velocity fields. Field A (Fig. 13a) is dominated by the behaviour of the stacking velocity field. Only the sedimentary apron is visible, the margin wedge is hardly present in velocity field A. In contrast, in field B

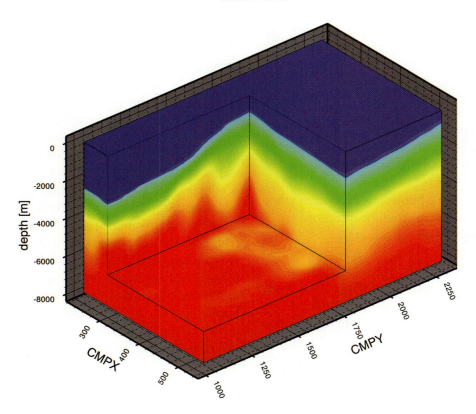

Fig. 10. 3-D volume of the macro velocity field A, derived from stacking velocities. Increment CMPY = 25 m, increment CMPX = 50 m.

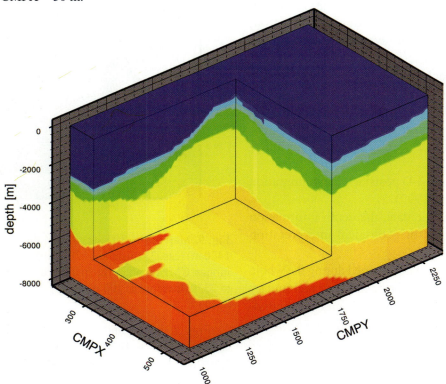

Fig. 11. 3-D volume of the macro velocity field B, derived from 12 crossline-migrated inlines. Increment CMPY = 25 m, increment CMPX = 50m.

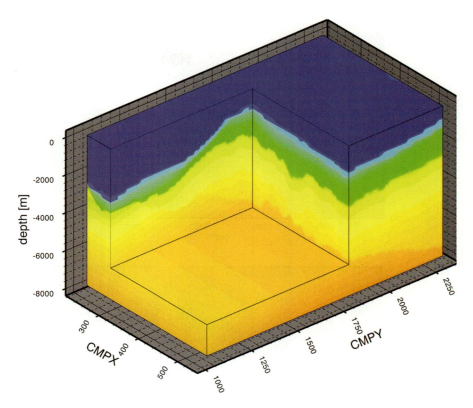

Fig. 12. 3-D volume of the macro velocity field C, derived from 23 crossline-migrated inlines. Increment CMPY = 25m, increment CMPX = 50m.

(Fig. 13b) the velocity layers follow geological units. The margin wedge is characterized by velocities of more than 4000 m/s. The velocities of the units below the wedge should not be used for interpretation (Hauschild et al. 1999). Taking field B as an initial model for focusing analysis, field C (Fig. 13 c) was obtained. Macro velocity field C (Fig. 13 c) shows a higher lateral resolution in comparison to field B (Fig. 13b) and yields more reliable velocities for greater depths, because field C was derived down to 12 km (in Fig. 13c, field C is only shown down to 8 km).

Model A versus model B - Figs. 14a and 14 b show a comparison of 3-D poststack depth migration with field A and field B, respectively. There are 250 3-D migrated inlines after 3-D poststack depth migration. One single section is shown. The maximum depth is 8 km. Both versions already show the major geological units, namely from top to bottom: The 1000 m to 2000 m thick sedimentary apron, the margin wedge and the

plate boundary beneath, represented by a band of high amplitude - low frequency reflections. Comparing the results of both migrations, some improvements are visible in the 3-D poststack depth migration using macro velocity model B:

• The general quality of the reflections is improved.
• The imaging of several features is improved, especially the top of the margin wedge, the landward dipping series of reflectors within the sedimentary apron, and the reflective plate boundary sequence.
• The effect of velocity pull-ups due to insufficient velocities is reduced. This is visible e.g. across the plate boundary reflective sequence.

Model C versus model B - By increasing the number of crossline migrated 2-D lines from 12 (model B) to 23 (model C) and reducing the crossline distance of the 2-D lines (see Fig. 9), a more reliable and accurate velocity model C (see Fig. 12) could be obtained. The reasons for increasing accuracy are:

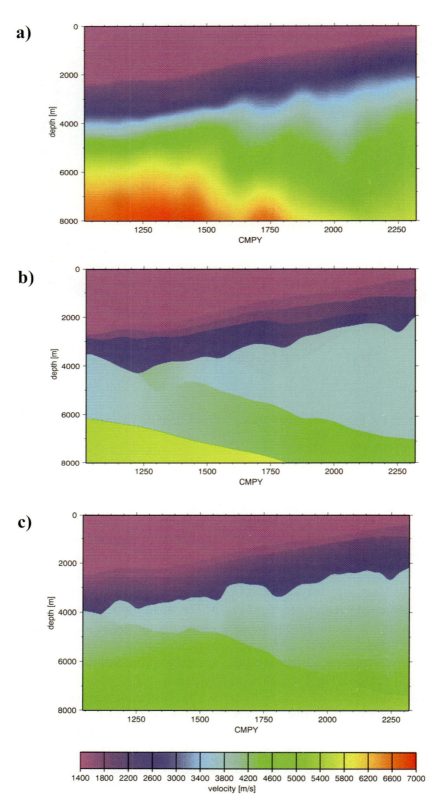

Fig. 13. a) Inline section 400 of the 3-D macro velocity field A (CMPX = 400, CMPY variable, increment CMPY = 25 m). **b)** Inline section 400 or the 3-D macro velocity field B (CMPX = 400, CMPY variable, increment CMPY = 25 m). **c)** Inline section 400 of the 3-D macro velocity field C (CMPX = 400, CMPX variable, increment CMPY = 25 m).

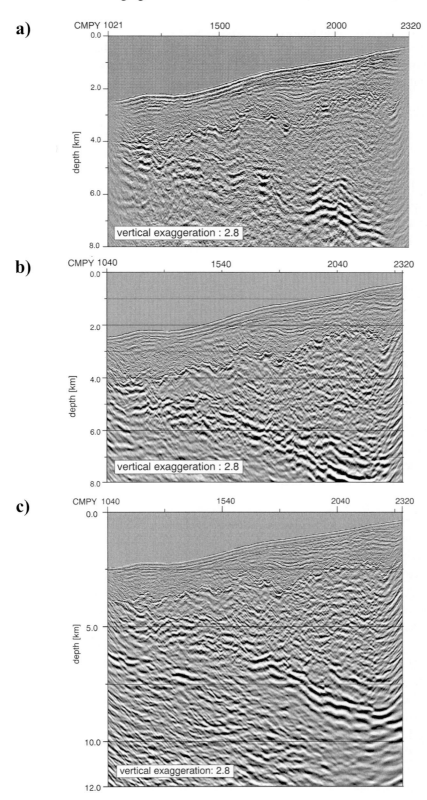

Fig. 14. a) Inline section 400 of the 3-D poststack depth migrated data volume using macro field A. Increment CMPY = 25 m. **b)** Inline section 400 of the 3-D poststack depth migrated data volume using macro field B. Increment CMPY = 25 m. **c)** Inline section 400 of the 3-D poststack depth migrated data volume using macro field C. Increment CMPY = 25 m.

• The result of model B was taken as an initial velocity field for iterative processing.
• 2-D focusing analysis includes geological interpretation and is facilitated by nearby parallel 2-D lines.
• The target 3-D macro velocity field can be built more reliably from more 2-D macro velocity fields by interpolation.

In summary, the accuracy of field C was more complete and more reliable than that of field B because more information was available (Gindler 2000). Even reflections from deeper zones could be imaged, yielding a macro velocity C for up to 12 km. The result of 3-D poststack depth migration with field C shows further improvements in comparison to field B (Fig. 14c).

Interpretation

The full 3-D depth migrated data volume was fed into a 3-D interpretation system and interpreted by BGR Hannover (Hinz et al. 1999) The greatly improved images of the structural architecture of a small part of the active convergent margin landward of the Middle American Trench off Costa Rica, produced by the 3-D acquisition and processing techniques discussed before, clearly demonstrate a subdivision of the margin wedge into a seaward (lower) and a landward (upper) wedge. The upper margin wedge very probably represents material continuing with the adjacent onshore Nicoya Complex consisting of mid-ocean ridge basalt (Jurassic), deep-sea sedimentary rocks and of a late Cretaceous basaltic suite. The thinned lower margin wedge is characterized by a more irregular relief and by a different internal seismic image suggesting a different nature of the lower margin wedge. We interpret the lower margin wedge as consisting of an oceanic crustal section formed from the Galápagos plume tail which was subsequently detached during collision of the Eocene-aged crust and attached against and to some extent beneath the Carribean Nicoya Complex, i.e. the upper margin wedge.

The reflective plate boundary sequence is thought to represent subducted trench sediments and crushed basement from the upper and lower plate (e.g. von Huene et al. 1999). Where the lower plate elevations or seamounts tunnel into the plate interface as in the western part of the 3–D area, they do not breach and do not dismember the lower margin wedge. The 3-D seismic survey has greatly improved the correct images of the complex structures of the active convergent continental margin off Cost Rica and has offered some new explanations regarding their nature and development.

The outlook for seismic technology beyond 2000

4C acquisition and processing

In the recent years, the development of marine four-component (4C) acquisition and processing technology has made it possible to use the technique in most geographical areas and in a wide range of water depths. Applications have been developed like multi-component seismics, which take into consideration shear waves as well as compressional waves. In principle, there are two different measurements for the same rock type to get additional information. Shear waves also have an additional property - they can be polarized, and the shear waves can be related to the pore structure and fractures. P-waves can give reservoir engineers a good image of the geometric shape of the reservoir, but shear waves and polarization provide a new parameter: the exact movement of the fluids.

Using the 4C seismic method, subsalt and sub-basalt targets are sometimes better imaged. The oil-water contact and the top and base of a reservoir unit - that could sometimes not be delineated using only P-waves - can be detected. By this method it is tried to identify fluid types in reservoir rocks, discriminate sand from shale and map hydrocarbon saturation.

A new field of imaging seismic data is under current research. P-waves and S-waves are being used for imaging: full-vector wavefield imaging. Full-vector wavefield imaging is expected to enable geoscientists to fine-tune the resolution of a reservoir from its current limits to a much finer resolution. Vector wavefield imaging will increase the reservoir resolution so that we can see smaller and smaller volumes in the rock.

Amplitude - Variation - with - Offset (AVO)

Since 1980, geoscientists have been using seismic amplitude analysis, which goes by the name amplitude-variation-with-offset analysis (or, commonly, AVO) to detect gas sands and oil sands, to identify certain types of lithology, and even to identify porosity and other finescale properties of the rocks that have caused the seismic response.

It is widely accepted that under proper circumstances, AVO has proved to be a useful prospecting tool. But in other situations, the technique may simply fail. One reason for the failure of AVO analysis arises when we apply it on unmigrated data because the anomalies might be corrupted with diffraction energy and the variation of amplitudes is caused by wave propagation effects instead of lithologic parameters.

True-amplitude migration

Most migration algorithms described in this report will produce an image of the subsurface, the amplitudes in the resulting image will be somewhat arbitrary. In practice, some sort of automatic scaling or gain control (AGC) is applied to the result to normalize the image.

The only goal of the migration schemes discussed in this paper is to produce a structural image of the subsurface. According to current research projects, it is possible to perform a more sophisticated migration, one that should produce meaningful amplitudes. This type of migration can then produce a "true amplitude" image (for example, one in which the amplitudes are proportional to the subsurface reflectivity). A true-amplitude image can be used for more detailed analysis than are possible for a mere structurally correct image: comparing relative amplitudes of reflectors at different depths and lateral positions, studying how amplitude varies with offset.

Imaging methods independent of the macro velocity model

Parameter updating (Hubral et al. 1996) - The basic idea of a macro-model independent concept is an extension of the familiar NMO/ DMO/ CMP stacking to more general procedures designed to

stack data from arbitrary source-receiver locations. The traveltime moveout of paraxial rays is formulated in terms of parameters that refer to the central ray and form the theoretical basis of new stacking algorithms. The parameters necessary to perform the stack are derived from a coherency analysis applied directly to the multi-coverage seismic data without any a priori assumptions. Finally, the estimated parameters can be used to construct a reliable macro-velocity model for subsequent imaging purposes. Furthermore, a true-amplitude migrated section can be constructed.

Operator updating (Berkhout 1997) - The basic idea is to replace velocity updating by operator updating. The concept of operator updating is based on the principle of equal traveltime. It requires that the time-reversed focusing operator (based on the underlying velocity model) and the focus point response (based on the underlying seismic meas-urements) must have equal traveltime (Berkhout 1997).

A replacement of velocity updating by operator updating also shows that seismic imaging can be carried out without any velocity information, it does not depend on the way the subsurface has been parameterized. From the entire distribution of verified focusing operators, a velocity model is derived with the aid of traveltime tomography. The inversion of updated focusing operators leads to accurate and highly resolved velocity-depth models.

Seismic imaging can be taken beyond depth migration by application of two separate dynamic focusing steps, allowing the construction of an accurate (angle-dependent) image of the subsurface (Berkhout 1997).

Time-lapse seismic, 4-D

Reservoirs are monitored by 3-D seismic data, the fourth component means "monitoring time". A set of time-lapse 3-D seismic survey data constitute the basis for the 4-D seismic method, which can track flow paths and fluid distribution in our reservoirs throughout their lifetime. Time-lapse or 4-D seismic is realized at large oilfields which have permanently installed instrumentation at the surface and within the reservoir to take measurements, including seismic, periodically for monitoring purposes. As time-lapse surveys accumulate over a field, reprocessing of all acquisition snapshots is required, ensuring that all differences of monitored datasets are processed.

Anisotropic imaging

It is recognized that the earth is inhomogeneous and anisotropic. In the last 30 years, research has concentrated on imaging inhomogeneous media, neglecting anisotropy. It has been found, however, that isotropic migration algorithms applied to seismic data from anisotropic media under certain conditions produce large position errors for steep reflectors.

Anisotropic depth migration may result in significant differences in structural positioning in comparison to conventional isotropic depth migration. By accounting for anisotropy, we can map fractures, and increase the accuracy of our velocity estimation and imaging techniques.

Outlook on desirable future work

Reprocessing of old 2-D seismic data, including 2-D imaging

Within the last 20 years, university and research institutes have acquired many 2-D seismic data. These data were processed by data processing systems which were 'state-of-the-art' at that time. Computer technology, in terms of hardware and software, is the driving force for processing and interpretation. New and improved software has been developed, taking into account the increased power of hardware. Advances in data processing technology combined with new visualization and interpretation techniques facilitate the extraction of more useful and more reliable information from seismic data.

Some examples of advanced data processing:
• Improved filtering for enhanced resolution.
• Removal of undesired noise, including multiples.
• Extraction of lithologic parameters.
• Improved imaging of the subsurface.

3-D seismics for research institutes, including 3-D imaging

In contrast to the achievements of exploration technology, there are only a few 3-D seismic projects which have been carried out by academic institutes within the last years. Up to now, the high cost may have prevented the institutes from applying such powerful technology.

Increased computer power and improved data processing systems have made 3-D processing, 3-D imaging and 3-D interpretation available to geoscience institutes. Though technological advancements in 3-D data acquisition have dramatically increased not only the data quality but also the volume of data, the cost for 3-D data acquisition is still very high.

Some solutions could be:
• Groups from geoscience institutes should share the cost of acquiring new data.
• Geoscience institutes and industry (oil or exploration) should cooperate for new 3-D projects.
• Within the last twenty years, oil industry and exploration companies have collected a large amount of 3-D seismic data volumes. After defining new goals, a reprocessing of these data could be initiated (example: 3-D mapping of gas hydrates).

Acknowledgements

The investigation reported and especially the processing for correct imaging of the structures within the 3-D area on the convergent continental margina of Costa Rica were supported by grants of the Deutsche Forschungsgemeinschaft (DFG). We are indebted to C. Ranero, R. von Huene, H. Meyer and Chr. Reichert for constructive discussions.

References

Berhorst A (1999) Die Auswirkungen verschiedener Geschwindigkeitsmodelle und Migrationsmethoden auf das seismische Imaging von 3D - Daten, Diploma Thesis, Christian-Albrechts-Universität zu Kiel

Berkhout AJ (1984) Seismic Migration, Part B, Practical Aspects. Elsevier, Amsterdam 274 p

Berkhout AJ (1997) Pushing the limits of seismic imaging. Geophys 62: 937-969

Claerbout JF (1984) Imaging the Earth's Interior. Blackwell Scientific Publication, Palo Alto, California 348 p

Devaux V, Gardner GHF, Rampersad T (1996) 3-D prestack depth migration by Kirchhoff Operator Splitting. 66th Ann Int Mtg, SEG Exp Abstr: 455-458

Gindler T (2000) Untersuchung und Vergleich von zwei verschiedenen 3D - Prestack - Migrationsverfahren als Two-Pass-Verfahren zur Ableitung eines 3D - Makrogeschwindigkeitsfeldes. Ph D Thesis, Christian -Albrechts-Universität zu Kiel

Hauschild J, Gindler T, Ristow D, Berhorst A, Bönnemann C, Hinz K (1999) 3D - Makro -Geschwindigkeitsbestimmungen und 3D - Tiefenmigration des seismischen 3D - Costa Rica - Datensatzes. GEOMAR REPORT No. 91, Kiel

Hinz K, Bönnemann C, Ranero C, Ristow D, Hauschild J, Gindler T (1999) Three-dimensional detailed imaging of prominent structures in reflection seismics at the active continental margin off Costa Rica. BGR Report 0119373, BGR Hannover

Hubral P, Schleicher J, Tygel M (1996) A unified approach to 3-D seismic reflection imaging - Part II: Theory. Geophys 61: 759 - 775

Jeannot JP, Faye JP, Denelle E (1986) Prestack migration velocities from depth focusing analysis. 56th Ann Int Mtg, SEG Exp Abstr: 438 - 440

Jeannot JP, Berranger I (1994) Ray-mapped focusing: A migration velocity analysis for Kirchhoff prestack depth imaging. 64th Ann Intern Mtg, SEG Exp Abstr: 1326 - 1329

MacKay S, Abma R (1992) Imaging and velocity estimation with depth-focusing analysis. Geophys 57:1608-1622

Ristow D, Rühl T (1997) 3-D implicit finite-difference migration. Geophys 62: 554 - 567

Schneider WA (1978) Integral formulation for migration in two and three dimensions. Geophys 43:49 -76

von Huene R, Flüh ER (1994) A review of marine geophysical studies along the Middle American Trench off Costa Rica and the problematic seaward terminus of continental crust. In: Seyfried H, Hellmann W (eds) Profil 7. Univ of Stuttgart, Stuttgart Germany, pp 143 - 159

von Huene R, Ranero C, Weinrebe W, Hinz K (1999) Quaternary convergent margin tectonics of Costa Rica, segmentation of the Cocos plate, and Central American volcanism. Spec Pap Geol Soc Am 295: 291 - 308

Yilmaz O, Chambers R (1984) Migration velocity analysis by wave field extrapolation. Geophys 49: 1664 - 1674

Yilmaz O (1987) Seismic Data Processing. Soc of Exploration Geophysicists, Tulsa, OK

State of the Art and Future Prospects of Scientific Coring and Drilling of Marine Sediments

D. Hebbeln

Universität Bremen, Fachbereich Geowissenschaften, Klagenfurter Straße, 28359 Bremen, Germany
corresponding author (e-mail): dhebbeln@uni-bremen.de

Abstract: The sampling of sediments from the sea floor is an integral part of marine geological research. However, as the sea floor is not directly accessible, marine geologists have to rely on instruments which they have to lower to the sea floor from seagoing vehicles, including e.g. vessels and drilling platforms. This paper aims to summarise the present state of the art of these sampling methods and tries to give an outlook on future prospects of scientific coring and drilling of marine sediments.

Introduction

Sampling of the sea floor has been an integral part of marine geological research since its beginnings. The sediments of the sea floor contain the most complete and undisturbed record of the environmental conditions that prevailed on Earth. Analysing long sedimentary sequences retrieved from the bottom of the ocean enables the reconstruction of the history and evolution of the Earth's environment, which is essential to understand the functioning of this natural system. In addition, most of the active processes that create and shape the crust of the Earth are confined to the marine realm, where the production of oceanic crust along the mid-ocean ridges and its subduction in the deep ocean trenches result in plate tectonics and continental drift. Thus, the investigation of the dynamics of the Earth's environment and of the Earth's interior requires the sampling of material from the sea floor and its underlying layers.

Scientific sampling of the sea floor began in the 19th century as a by-product of water depth measurements, when a small tube was put beneath the weight of the logging wire. The same basic principle of pushing a tube into the sediment by putting a heavy weight on top of it is still the standard method for sampling the sea floor. However, improvements to this technique, including addi-

tional mechanical forces to push it into the sediment, finally resulted in the present ability to retrieve sediment cores up to a length of 60 m. Since the late 1960s scientific drilling into the sea floor has become another standard technique, which provides the possibility to sample the sea floor to a sub-bottom depth of over 2000 m. These techniques as well as new developments which will be available in the near future, as e.g. riser drilling, are reviewed here.

Coring and drilling technologies

Surface sediment sampling by box corer and multicorer

While studying on the one hand active processes at the seafloor and on the other hand the paleoenvironment, careful sampling of the uppermost sediment layer serves two aims: (1) to allow the establishment of a relationship between the prevailing environmental conditions and the sediment composition in order to investigate the response of the sea floor to actual forcing processes and to provide a ground-truthing for paleo-proxies and (2) to complete the recovered sediment sequence as sampling by long coring or drilling in most cases is associated with

From WEFER G, BILLETT D, HEBBELN D, JØRGENSEN BB, SCHLÜTER M, VAN WEERING T (eds), 2002, *Ocean Margin Systems.* Springer-Verlag Berlin Heidelberg, pp 57-66

the loss of the uppermost part of the sediment column.

The most widely used tools to recover an undisturbed sediment surface are box corer and multicorer, which are in contrast to various grab samplers, which are fast and easily operated tools to collect surface sediments, however, thereby always disturbing the sediment surface. The box corer basically consists of a box (from 20 cm to 20 cm up to 50 cm to 50 cm in size; sometimes also cylinders are used), a frame and a shovel (Reineck 1963) (Fig. 1). When the frame reaches the sea floor the box is pushed into the sediment by some weights on top of it. Upon pulling the wire from the vessel the shovel is released and seals the bottom of the box, keeping the sediments within the box. This instrument provides an almost undisturbed sediment surface by a total recovery of up to 50 cm. However, due to the relatively large surface of the box, the water overlying the sediment can slightly sweep back and forth during recovery and thus, affect the sediment surface to some degree (e.g. concentrating the fluff layer or the finest sediment fraction at only one side of the box).

The multicorer follows the same basic principle, with the major difference that the one big box of the box corer is replaced by a number (6 to 12) of small-sized plastic tubes with diameters reaching from 6 cm to 9 cm (Fig. 2). Each tube has its own

plate (~ shovel) sealing it at the bottom during recovery. The major advantages of the multicorer are the immediate availability of several subsamples from one site and especially the extremely good preservation of the sediment surface (Fig. 3). This is mainly due to the small surface area of the individual tubes, which prevents any sweeping of the water overlying the sediment, which is the actual bottom water and which can thus also be sampled e.g. for its chemical composition.

Fig. 2. Multicorer during deployment onboard RV *Sonne*. Photo courtesy by P. Wintersteller.

Fig. 1. Box corer during deployment onboard RV *Poseidon*. Photo courtesy by A. Freiwald.

Fig. 3. Example of a pristine sediment surface retrieved by a multicorer. The width of the image corresponds to 7 cm. Photo courtesy by P. Wintersteller.

Gravity corer

The gravity corer is the most simple device to obtain sediment cores from the sea floor. It consists of a variable weight (up to 3.5 to), a steel tube (up to ~20 m) and a core catcher (Fig. 4). Normally a plastic tube or "liner" is put into the steel tube, which finally will contain the sampled sediment. The liner can easily be extracted from the steel tube and stored until detailed subsampling of the core is possible. The gravity corer is lowered from the ship by a steel wire to the sea floor and penetrates the sediment forced only by its own weight. The core catcher at the bottom of the steel tube is a simple device to cut the sampled sediments from the underlying sediments to prevent that the

sediments slide out of the tube when the corer is recovered. The gravity corer is well suited for sampling soft sediments. Thin layers of coarser materials, such as ash layers or turbidites can also been recovered, while predominantly coarse sediments often will not be penetrated.

Kasten corer

To extend the amount of available sample material the kasten corer, a modified gravity corer, has been developed (Kögler 1963). The main difference to the normal gravity corer described above is the replacement of the steel tube with a diameter of ~12 cm by an up to 15 m long box with a sampling area of 15 x 15 cm or 30 x 30 cm (Fig. 4). However,

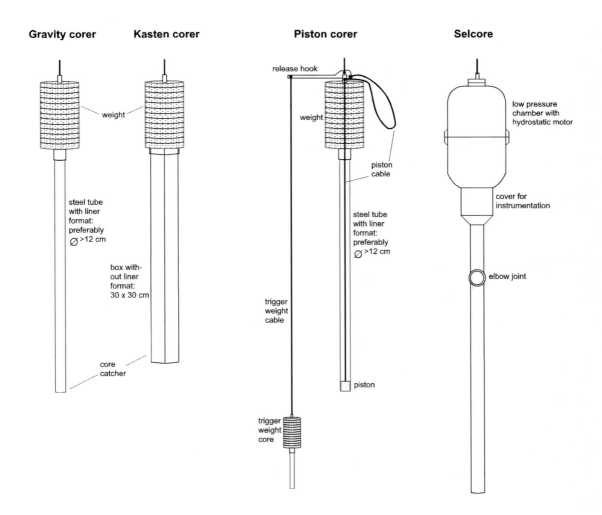

Fig. 4. Basic description of gravity, kasten, piston corer and Selcore.

the kasten corer does not contain a liner, so that the cores must be subsampled immediately after recovery. Due to its design the kasten corer provides sediment samples of extremely good quality. Like the standard gravity corer, the kasten corer is best suited for sampling soft sediments.

Piston corer

Another frequently used corer is the Kullenberg piston corer. Nowadays many different designs of this corer exist (one example is shown in Fig. 4), however, all following the original principle developed by Kullenberg (1947). In general, the piston corer looks like a normal gravity corer with an additional piston placed inside the core. When the corer penetrates the sediment the piston remains at the sediment surface while the steel tube is pushed downward by the weight. The resulting low pressure offsets the friction between sediment and steel tube allowing a much deeper penetration of the piston corer compared to the gravity corer. However, the need for a trigger core to release the piston corer (Fig. 4) makes its handling quite complicate and time-consuming. The trigger core reaches the sediment surface first and is used to retrieve a short core (Fig. 5) usually providing a less disturbed surface as can be got with the piston corer. When reaching the sea floor the trigger core releases the piston corer, which then falls free to the seafloor to penetrate it (Fig. 5). A detailed discussion of the advantages and the problems related to the use of a piston corer and recommendations for its operation are given by Mienert and Wefer (1995). Like the standard gravity corer, the piston corer is also mostly used to sample soft sediments. To allow the sampling of pressure dependent materials/organisms (e.g. gas hydrates or microbes) a pressure chamber piston corer has been developed, which keeps the sample in the liner under *in situ* pressure and temperature conditions (Tab. 1).

Giant piston corer

The longest sediment cores obtained by these relatively simple coring techniques have been retrieved by piston corers, namely with the French giant piston corer (CALYPSO corer) onboard the

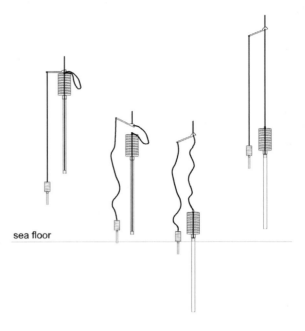

Fig. 5. Operational principle of the Kullenberg-type piston corer.

RV *Marion Dufresne* within the International Marine Past Global Changes Study (IMAGES, Tab. 1). The maximum core length taken with this corer is 59 m and cores of ~35 m have been taken on a regular basis.

Comparison of the efficiencies of gravity corer and piston corer

As soft sediments at the sea floor are easily disturbed, also these coring techniques often result in disturbed sediment sequences. While all these core types are affected by sediment compaction, the piston core can also exhibit extension structures due to the low pressure sucking (Bassinot et al 2001; Skinner and McCave 2001). Besides the general principle (piston vs. gravity corer) also the diameter of the core itself can seriously affect penetration and recovery. A comparison of three different sediment cores taken by different tools at the same position in the Nordic Seas reveals such effects (Fig. 6). The two piston cores (PS 1320-2 with a diameter of 8.4 cm and PS1320-1 with a diameter of 12 cm) yielded a higher recovery (8.7 m and 6.7 m, respectively) compared to the gravity core (PS1296-4 with a diameter of 12 cm), which only got 5.3 m. However, the stratigraphic

Keywords	Internet Links
Pressure Chamber Piston Corer	http://www.bgr.de/b323/autoklav.htm
International Marine Past Global Changes Study (IMAGES)	http://www.images.cnrs-gif.fr
Selcore	http://www.selantic.com
Portable Remotely Operated Drill (PROD)	http://www.bgt.com.au
Ocean Drilling Project (ODP)	http://www-odp.tamu.edu
Hydrate Autoclave Coring Equipment (HYACE)	http://www.tu-berlin.de/fb10/MAT/hyace/welcome/hyace.htm
ODP Long Range Plan	http://www.oceandrilling.org/Documents/LRP/LRP.html
Integrated Ocean Drilling Program (IODP)	http://www.iodp.org
Conference Report on Cooperative Riser Drilling (CONCORD)	http://www.jamstec.go.jp/jamstec-e/reports
OD21 Program	http://www.jamstec.go.jp/jamstec-e/odinfo
Report on the Necessary Capabilities of a New Non-Riser Drillship (COMPLEX)	http://www.oceandrilling.org/complex
Nansen Arctic Drilling (NAD)	http://www.joi-odp.org/NAD

Table 1. Internet links to various items discussed in the text.

range sampled with these three corers did not show such strong differences. A comparison of the water content, which can easily be correlated between these three cores (Fig. 6), revealed that with increasing depth individual core sections in the small piston corer (PS1320-2) are longer than in the big piston corer (PS1320-1). The same sections in the gravity corer are even slightly shorter resulting in huge problems in calculating e.g. sedimentation rates at this site. The correlation problems between the gravity corer and the piston corers between 400 to 500 cm core depth most likely reflect strong compaction in the gravity core. Thus, the original sediment sequence will mostly be disturbed by the coring process to various degrees, which to a large extent depend on the design of the coring instrument. When interpreting the collected sediments this has to be kept in mind, especially if the results of a number of different cores are synthesised. For example, a regional assessment of sedimentation rates obtained from a number of sediment cores taken by different techniques or core designs can result in serious misinterpretations.

Nevertheless, due to their relatively low prices and to their easy handling onboard almost all kinds of vessels, gravity and piston corers are still the workhorses in scientific coring of the deep-sea floor. A way to overcome the above mentioned problems, while still using such simple instruments,

might be the future development of a new generation of core logging instruments. Attached to the outer side of the core barrel, such logging devices could allow a detailed comparison of the almost undisturbed sediment sequence outside the core barrel to the compressed and/or stretched sediment core inside. Based on such a comparison the recovered sediment sequence could be corrected for any stretching and compression, resulting in the possibility to calculate true sedimentation and accumulation rates.

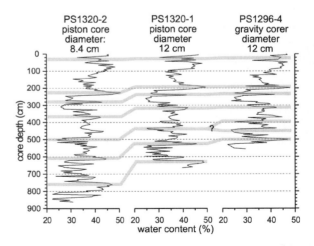

Fig. 6. Comparison of the water contents of three sediment cores retrieved by different coring tools from 77°59'N and 0°34'E in the Fram Strait, northern Nordic Seas. The grey lines link identical core sections.

Vibro corer

As pointed out above, the tools mentioned so far are mainly used for sampling soft sediments. Coarse and/or partly consolidated sediments are difficult to sample with the described "one shot" approach, where all available energy is spent when the corer hits the sea floor. To allow the penetration of coarse and/or consolidated sediments multiple push corers are used. The most common is the vibro corer, which also exists in numerous designs and which is mainly used in relatively shallow waters (<500 m). In very shallow waters (and lakes) vibro corers are often used from floating platforms, while in "deeper" waters and under rougher sea going conditions deployable vibro corers are preferred. These basically consist of a frame containing a core tube or box and a vibro-hammer pushing the core tube into the sediment, while the whole coring process can be remotely controlled from the vessel. Core diameters are variable in the cm range and core lengths are mostly restricted to a few meters.

Selcore

A multiple push device for sampling also the deep-sea floor by this technique is the Norwegian Selcore (Fig. 4, Tab. 1). The Selcore uses the large energy potential offered by the difference between ambient hydrostatic pressure at depth and atmospheric pressure kept in a sealed reservoir in the core barrel. Selcore operates in depths between 100 m and 6000 m and it is lowered to the sea floor like a conventional gravity corer. When it hits the sea floor a hydrostatic motor is automatically activated. The mass of the steel low pressure tank serves as the hammer head at 30-50 strokes per minute. As mentioned above, the energy is derived from the differential pressure between the low pressure tank containing air with atmospheric pressure and the ambient hydrostatic pressure at the sea floor. In depths >1000 m more than 400 strokes represent a 100-fold increase in energy available for penetration compared to a gravity corer of the same dimensions. Core lengths of up to 30 m can be taken with the Selcore in soft sediments as well as in high shear strength sediments.

BGS Rotary Rockdrill

To sample harder rock formations outcropping at the sea floor the British Geological Survey (BGS) designed and built a 1 m- and a subsequent 5 m-penetration subsea rotary rock drill (Skinner 1998). This microprocessor-controlled instrument, which also incorporates a vibrocoring option, is a well proven tool which is used in up to 2000 m water depth in a variety of geological settings. Core penetration logging and full core barrel retraction allows careful and close monitoring of sample progress and quality – when penetration stops so does rotation (or vibration) thus preserving the collected sample. Pull-out using a controlled seabed winch instead of a ship winch and ship heave allows for better core retention and again core quality.

Portable Remotely Operated Drill (PROD)

With the coring techniques described above the maximum possible core length is in the order of 60 m under perfect conditions, on a regular base the limit will be approximately 40 m using a giant piston corer. Significantly longer cores, up to several 1000 m can be obtained by scientific and commercial drilling vessels (see below), however, at enormous expense. To bridge the gap between the conventional coring techniques and deep drilling the Portable Remotely Operated Drill (PROD) has been developed (Davies et al. 2000) (Fig. 7, Tab. 1).

PROD is a sea floor lander system, tethered to the ship, which can core or drill up to 100 m into the sea floor. It is essentially a land drilling rig and a remotely controlled vehicle (ROV) combined, which can be deployed from sea-going ships of opportunity on an umbilical that integrates strength, control and communications. The electro-hydraulic system is locally controlled by a multi-processor machine control, which receives commands from the shipboard control. The system utilizes hydraulic piston coring (HPC) and diamond-bit drilling to allow soft sediment as well as hard rock sampling. PROD stores drilling rod, casing, and piston core barrels (all in 2.75 m sections) on two carousels that may be loaded for

Fig. 7. PROD during test deployment in the port of Pusan, Korea. With a total height of 5.8 m, a footprint on deck of 2.4 m by 2.4 m and a weight of 10 tons it can be used from many ships of opportunity as the barge seen here. Photo courtesy by T. Freundenthal.

each job to any proportions of these tools needed, which can be selected on the fly.

PROD potentially solves several problems inherent in both traditional coring methods and drilling from a surface ship. Gravity and piston corer can only be used in relatively soft sediments and are limited by the length of a single core barrel. PROD eliminates this problem by applying conventional rod drilling techniques used on land to the marine environment. With this technology, drill pipe is added sequentially in 2.75 m lengths, and coring proceeds in similar increments. The primary innovation with PROD is the ability to rack drill pipe and rods, and recovered cores, in the carousels on the sea floor lander. In addition, PROD solves the problem of ship heave, which still affects ODP style drilling operations from a surface ship, which often suffer from ship heave induced problems of core recovery and core disturbances. As PROD

lands on the sea floor, decoupled from the ship, it is not influenced by heave.

Ocean Drilling Project (ODP)

The base for scientific drilling in the deep ocean was laid with the initiation of the Deep-Sea Drilling Project (DSDP) in 1964. Between 1968 and 1983 the famous drilling vessel *Glomar Challenger* provided among other things important evidence to support the theory of plate tectonics. In 1983 the DSDP was succeeded by the Ocean Drilling Program (ODP, Tab. 1), which since 1985 uses the larger and more advanced drilling vessel *JOIDES Resolution*. Until 1999 the *JOIDES Resolution* had recovered over 170 km of cores with deep-sea sediments and rocks from water depths down to 5,980 m and sub-bottom depths down to >2,000 m. It is still the most important platform for deep scientific drilling in the oceans.

Equipped with a dynamic positioning system it can keep exactly its position above the drilled hole despite wind and waves, permitting drilling in water as deep as 8,235 meters. A heave compensator in the derrick acts as a giant shock absorber, so that the up and down movements of the ship are not transferred to the drill pipe. As the various drilling tools cut through layers of sediment and rock, cores of sub-seafloor material in segments as long as 9.5 meters are collected in tubes. The tubes are attached to a wire cable allowing the crew to pull the core up through the drill pipe.

The main coring/drilling tools used onboard the *JOIDES Resolution* are (a) the Advanced Piston Corer (APC), (b) the Extended Core Barrel (XCB) and (c) the Rotary Core Barrel (RCB) (Fig. 8). The APC is a hydraulically activated coring tool following the standard piston core principle that is designed to recover undisturbed core samples from soft sediments down to ~250 m beneath the sea floor with an average recovery of 98%. In water depths of ~2,500 m one APC core (standard length: 9.5 m) can be recovered approximately every 45 minutes. The XCB is used to sample semi-indurated sediments too stiff to be recovered by APC. It has an average recovery of >60%. The XCB is a modification of the rotary coring system

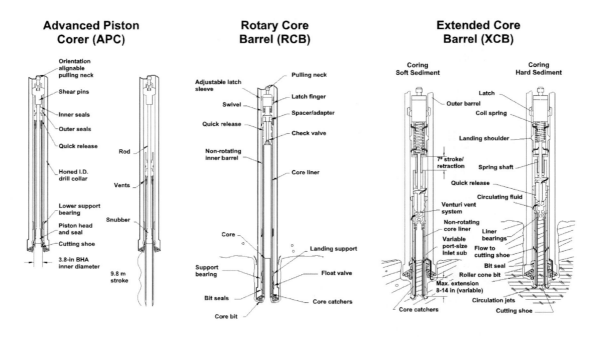

Fig. 8. Overview about the main drilling tools used onboard RV *JOIDES Resolution (*modified from *http://www-odp.tamu.edu/dsd/TOOLS/TOOLS.htm)*.

used in the oil industry. It relies on rotation of the drill string to advance the hole and cut the core sample. To drill deeper down into lithified sediments and basement rocks the RCB is used. It follows the same technological principle as the XCB, however, it has a more rugged design to be able to drill into very hard sample material.

Additional available drilling tools onboard the *JOIDES Resolution* include the Motor Driven Core barrel (MDCB) and the Pressure Core Sampler (PCS). The MDCB allows to extend APC/XCB holes to greater depths and into more indurated sediments and it also can improve recovery in formations which are difficult to core, such as hard, fractured crystalline rock, interbedded formations, and shallow water reef complexes. The PCS is capable of retrieving core samples from the sea floor while maintaining near *in situ* pressures up to ~690 atmospheres.

Hydrate Autoclave Coring Equipment (HYACE)

A recent development to improve the retrieving of pressurized samples is the EU-sponsored HYACE

instrument, standing for HYdrate Autoclave Coring Equipment (Tab. 1). It is designed to sample marine sediment at downhole conditions and bring it on board while maintaining as many downhole parameters as possible. Particular targets are gas hydrate bearing sediments of the continental slopes. Sampling tools will include push, percussion and rotary corers. The rotary corer will be driven by a downhole motor instead of the top drive onboard the research vessel/drillship, thus, allowing the use of the instrument from a number of platforms.

In April/May 2000 a successful onshore test with HYACE was carried out followed by the first offshore test onboard *JOIDES Resolution* during ODP Leg 194 in January 2001 in ocean sediments on the Marion Plateau, NE Australia. The percussion corer produced good core recoveries and also showed the pressure maintenance capacity of the novel autoclave system. Also the rotary corer was deployed in 4 test runs which resulted in partly successful coring and autoclave functions. Significant progress was made in developing efficient handling procedures and items for further operational improvements were clearly identified, giving rise to the expectation that HYACE will be available in the (near) future for standard sampling, especially of gas hydrate bearing sediments.

Integrated Ocean Drilling Program (IODP)

In its present form ODP will continue to drill the deep ocean until 2003. For the time after 2003 the ODP Long Range Plan (Tab. 1) pointed out the need to have access to at least two deep-sea drilling platforms - one with capabilities of the present *JOIDES Resolution* and another with deep water well control (riser) capabilities. In addition, Japan proposed in 1994 an Ocean Drilling Program for the Twenty-First Century (OD21, Tab. 1) which includes plans for construction of a drilling vessel with such riser capabilities. In the meantime both plans merged into a vision for an Integrated Ocean Drilling Program (IODP, Tab. 1) based on two or more deep-sea drilling platforms. In July 1997 the Conference on Cooperative Riser Drilling (CONCORD, Tab. 1) took place in Tokyo during which fundamentally important scientific problems have been identified, which can only be addressed by drilling deeper into the earth.

To drill deeper into the Earth and to drill critical sites in unstable formations (as e.g. seismic active zones, hydrocarbon-bearing rocks, gashydrates) a riser or riser-type technology is essential. It involves circulation of dense, viscous drilling fluids, which are important for well stability and sub-surface fluid control. The new drilling vessel will initially be equipped with a riser, that will enable controlled drilling and rock recovery in 2,500 m to 3,000 m water depth and up to 7,000 m beneath the seabed. In a later phase these capabilities are recommended to be extended to work in depths down to 4,000 m.

While the current program ODP operates just one vessel, IODP will be a multi-platform program based primarily on two ships. Besides the new riser vessel being built by Japan as part of the OD21 Program a new non-riser ship sponsored by the United States is being planned. The discussion about the necessary capabilities of such a vessel are summarized in the COMPLEX report (Tab. 1). Additional platforms (e.g. for shallow water drilling, for drilling in ice covered regions as projected

in the Nansen Arctic Drilling Program (NAD), see Tab. 1) have also been recommended in the scientific planning. With the multi-platform approach, scientists will be able to research more sites globally with advanced capabilities and therefore start to address questions that require specialized techniques and/or broad, global knowledge of the Earth to be answered.

Outlook

Also in the future the present-day workhorses for the scientific exploration of the ocean floor like the conventional coring techniques and the ODP-style drilling will certainly provide important new insights into the dynamics of the Earth's environment and the Earth's interior. However, a major step forward can be expected from the new technological developments like the PROD and the riser vessel. PROD will bridge the gap between conventional piston coring and deep drilling and, thus, signifi-cantly increase the number of long (~100 m) sedi-ment cores, which will help to decipher the history of the Earth's environment through the youngest part of Earth's history. With the new riser vessel this record can be extended far back in time well into the Mesozoic era. In addition, the capability to drill into critical settings like e.g. seismic active zones will contribute in an unprecedented way to our understanding of the internal dynamics of our planet.

References

Bassinot F, Széréméta N, Kissel C, Balut Y, Labeyrie L, Pagel M, Guibert K (2001) New insights into the stretching of sedimentary series collected through piston coring: Evidences, effects on flux estimates and possible corrections. EUG XI Conference, Strasbourg, April 8-12 2001, Abstract Volume, p 145

Davies PJ, Williamson M, Frazer H, Carter J (2000) The portable remotely operated drill. APPEA Journal 40

Kögler F (1963) Das Kastenlot. Meyniana 13: 1-7

Kullenberg B (1947) The piston core sampler. Svenska hydr Biol Komm Skr 3 ser: Hydrografi Bd 1 H2

Mienert J, Wefer G (eds) Proceedings to the International Congress on Coring for Global Change (ICGC '95) held in Kiel 28-30 June, 1995. GEOMAR Report 45, 83 p

Reineck HE (1963) Der Kastengreifer. Natur und Museum 83: 102-108

Skinner AC (1998) Acquisition of offshore geophysical and geological data for regional and environmental surveys. Marine Georessources and Geotechnology 16: 53-74

Skinner L, McCave IN (2001) Extensional distortion in a Calypso piston core from the Iberian margin. EUG XI Conference, Strasbourg, April 8-12 2001, Abstract Volume, p 147

New Technologies for Ocean Margin Studies
– Autonomous Instrument Carrier Systems

C. Waldmann[1*] and R. Lampitt[2]

[1]*Universität Bremen, Fachbereich Geowissenschaften, 28359 Bremen, Germany*
[2]*Southampton Oceanography Center, George Deacon Division, Midwater Biology Team,
European Way, Southampton SO14 3ZH, UK*
* corresponding author (e-mail): waldmann@marum.de

Abstract: Processes that determine the environmental structure of ocean margins are calling for multidisciplinary observations. Current knowledge has been gained mainly through ship board observational programs. But this approach is not sufficient to account for the variability of the processes involved in the ocean margin environment. For instance the importance of intense events of short duration compared to slowly but permanently ongoing processes is not well understood up to now. Therefore the main focus of future observational program will be on improving the spatial and temporal coverage of the measured parameters and enhancing the long-term capability of the system.This paper describes already existing or upcoming technologies that will allow for more intensive investigations of water column properties. Moored profiling platforms that are able to host a complete suit of physical, biological and chemical sensors are introduced as efficient observational tools to fulfil the envisaged goals. Common to all described systems are the need for adequate energy supply, autonomous operation and bi-directional data exchange to enable event triggered sampling. The tradeoffs between different technological approaches addressing the mentioned scientific needs and the consequences for their applicability are described.

Scientific needs

Ocean margins constitute the transition regions between the deep-sea and the continental shelves and as such they represent a conceptual boundary between these two very different environments. They are however far more than a boundary and demonstrate their own physical and biogeochemical characteristics which transform material advected from either side of the region. They have been the focus of a variety of studies over the past two decades (e.g. SEEP 1 & 2, ECOMARGE, KEEP, OMEX 1 & 2, see Biscaye et al. 1988; Biscaye et al. 1994; Biscaye and Anderson 1990; Monaco et al. 1990; Wang and Wong 1992; Van Weering et al. 1998) but in spite of this very large degree of effort the uncertainties in our estimates of the role of these regions are still unacceptably large. Other contributions to this volume describe some of the outstanding questions pertaining to ocean margins

but one of the resounding themes that will undoubtedly emerge is that the spatial and temporal variability in these regions is usually too high to facilitate interpretation of ship based observations of the water column. The situation for the benthos is of course somewhat different and in this environment there is a significant degree of spatial and temporal integration resulting in much less variability. In the water column above, the water flows are complex and often fast with along slope residual currents superimposed on cross slope tidal oscillations. Upwelling of nutrient rich water often enhances productivity above the slopes and near bottom downslope currents can transport large quantities of solid material away from the shelves and into deep water.

One of the overriding themes from all the observations available to date is that the temporal

From WEFER G, BILLETT D, HEBBELN D, JØRGENSEN BB, SCHLÜTER M, VAN WEERING T (eds), 2002,
Ocean Margin Systems. Springer-Verlag Berlin Heidelberg, pp 67-77

and spatial resolution and coverage of data gathering has been inadequate. To overcome this problem, technological developments are required to provide reliable autonomous instrument carriers providing near real time data on a wide range of physical and biogeochemical properties of the water column. This field is still in its infancy but is developing fast in readiness for the next major offensive into this important oceanographic regime.

Comparison of existing and future autonomous measuring systems

From the scientific requirements outlined above and elsewhere in this volume it is clear that insights into ocean margins demand long term observations with frequent repetitive sampling throughout the water column over an array of sites. There are two broad classes of such systems, the first being "Lagrangian" in function with instrument carriers drifting passively with the current, possibly profiling to the surface to make measurements throughout the water column and dumping the data via satellite link whilst at the surface. Such systems are already in action and the ARGO program of floats provide a good example of what can be achieved albeit with a very limited array of sensors attached to the floats. The major disadvantage with such systems is that the sensors can soon drift out of the region of interest and as far as studies on ocean margins are concerned this problem seems to preclude their use. The second category are systems which are fixed at one location and so may be considered as "Eulerian". In this case, sensors with their own data loggers incorporated may be attached to a mooring which is recovered after a period of weeks or months for data extraction and redeployment. Such systems have been used for many years to record current speed and direction and are being enhanced to cover a variety of other variables. The success of the TOGA/TAO array to predict the occurrence and the intensity of El Niño events shows the importance and practicability of the mooring concept (Fig. 1).

The measurements that have to be addressed in the study of ocean margins go beyond water currents and for the future multidisciplinary sets of instruments have to be deployed to investigate the variety of processes involved. That means additional measurements will be needed for CTD, fluorescence, irradiance, nutrient concentration, dissolved oxygen, zooplankton, and particle concentration and characteristics. To achieve a sufficient high vertical resolution using sensors fixed at one depth, a large number of sensors are required with the associated high costs. Furthermore variations in drift between different sensors reduce the reliability of not only the individual records but also the shape of the entire vertical profile. Profiling instrument carriers are considered to be a cost effective alternative. These have the advantage of giving a continuous high-resolution profile from a single set of sensors and furthermore remove the problems of inter-calibration between the various sensors on a particular mooring.

There is increasing evidence that episodic events have a major impact on physical and biogeochemical processes of ocean margin systems. These events demand more intensive sampling during their occurrence and there is therefore a strong need to be able to communicate with the deployed observatories in real time or in quasi real time. This allows the retrieval and first evaluation of the collected data and adaptation of the sampling schedule to suit the ambient conditions.

From the operational perspective, a serious problem which affects long-term observations is bio-fouling during which organisms of various types settle onto structures and impair their function. At present there is no satisfactory solution for preventing this attack on sensors, an effect which is most pronounced in the euphotic zone. Profilers can however reduce this problem by remaining for most of the time below the euphotic zone only accessing this region for the limited time required for taking measurements.

Within the context of future research into ocean margins, we believe that the principles of Eulerian profilers are those which should be adopted and we now discuss some aspects of this particular approach.

Basic principles

Moored profilers are platforms that can carry different sensor systems. These platforms contain the

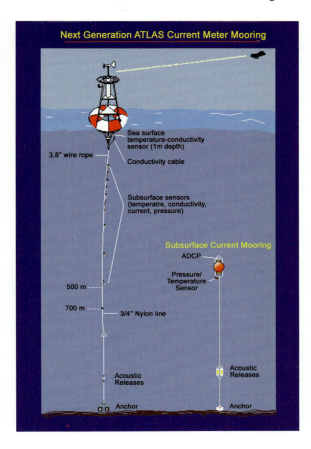

Fig. 1. The current meter ATLAS mooring - the basic building block of the TOGA/TAO array.

propulsion system that enables the movement along the mooring line, some basic housekeeping sensors like pressure and attitude sensors to keep track of the motion and a central controller system. Control software is required which can handle all possible operational states including failure. All sensor systems that are adapted are interrogated by the central controller system and the data are stored inside. Ongoing developments will allow quasi real time access to the collected data in the near future.

Moored profiling systems have been well known as a method to obtain upper ocean time series data since the 1970s (Van Leer et al. 1974). Although the basic ideas are simple and have been realised in different ways, only sparse experience is available up to now. The reason why the approach has not been continued is probably that the benefits in terms of data were not perceived to justify the considerable development costs. Instead the dis-

crete instrument approach has been adopted most frequently. Due to the rapid changes in the fields of electronics and microprocessors, the appearance of new sensor systems and the recognised scientific requirement for the data the opportunities for success are much greater now.

Currently different developments have been started that not only address the needs of physical oceanography but also include a variety of multi-disciplinary sensors (Provost and du Chaffaut 1996; Doherty et al. 1999; Waldmann 1999). These efforts also aim at extending the depth range of these profiling systems to full ocean depths.

Basic problems of all profiling systems, reliability and the efficient use of the available energy are similar to autonomous underwater vehicles (AUVs). Moored profilers in comparison to floats have to overcome friction on the mooring line. This impairs reliability and energy efficiency significantly. In strong currents the friction calls for additional energy for moving the system up and down while at the same time bio-fouling on the mooring line can completely prevent the system from moving. These issues also call for some intelligent control software that is able to detect the sources of errors and decide on a solution. Figure 2 shows the basic configuration of a profiler mooring. The length of the continous mooring line defines the range covered. The bottom stop is chosen as the resting point between profiles.

Mainly due to the problem of the limited amount of energy that can be made available to the system, four different propulsion principles using a fixed mooring line have been considered to date.

Use of currents

Similar to kites in air, systems have been designed to employ water currents for their profiling motion. By adjusting the profile of the wings an upward and downward motion can be achieved (Eckert et al. 1989).

Although this principle seems to be quite attractive because it uses the available ambient energy it is on the other hand difficult to make this principle work reliably. Not only do the current directions change along the water column but the magnitude can vary substantially. In contrast to air

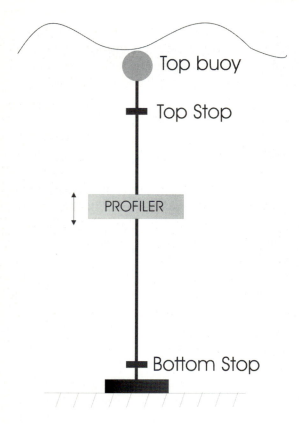

Fig. 2. Basic profiler mooring.

kites with underwater systems the drag is significantly higher so that the system can not adjust quickly enough to changing current directions. Although using this principle in deep-sea deployments seems prohibitive, in shallow water regions with well defined currents or in lakes it could be attractive.

Energy from waves

Wave action on a surface buoy can be transferred down the mooring line causing an up and down motion of the line. The inertia of the profiler selectively uses one direction of this motion by employing a ratchet mechanism. The system will progressively move down or up the mooring line depending on the provisions made at the time of deployment (Fig. 3). This mechanism is just used for one direction of motion. For the other direction the profiler can make use of its buoyancy (positive or negative) (Fowler et al. 1997).

This concept is very attractive because it requires no on-board energy source. Additionally it

already includes some braking mechanism that could be useful for doing chemical measurements. These measurements are often slow and cannot be done in a continuous-profiling mode. On the other hand it is difficult to determine whether this principle could be applied to instruments extending below about 200 m as the wire motion will probably be damped out at longer distances. This will depend on the stiffness of the mooring line and on the hydrodynamic drag on the profiler.

Buoyancy Change

By changing the weight or the volume of a profiling system a buoyancy change is induced and accordingly the system can go up or down the mooring line (Eriksen et al. 1982) (Fig. 4). Basically it is the same principle as it is used for floats such as ALACE or MARVOR.

In summary one can describe this principle as self-contained. It is not explicitly dependant on the condition of the mooring line and uses the mooring line only as guidance. Additionally the smooth and monotonic motion guarantees a continuous flushing of the sensors. That means that high quality data can be recorded which is especially true for current measurements. On the other hand this principle may be particularly sensitive to bio-fouling. The available force (~10 N) to push the obstacle away is too low. Furthermore the energy efficiency is less compared to other principles. To keep the platform moving to the end of its defined range additional energy is required that is lost at the end of the profile.

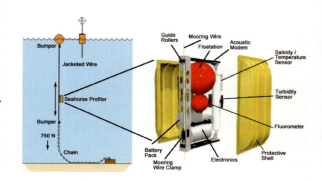

Fig. 3. Profiler driven by wave power.

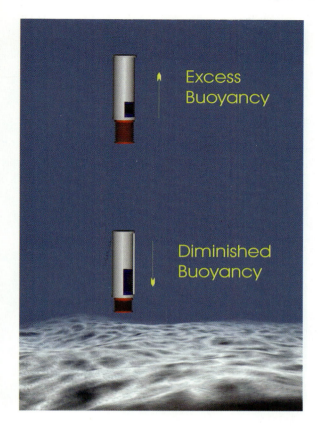

Fig. 4. Buoyancy change by pumping oil between an inner and an outer bladder.

Nevertheless this principle is one of the most promising candidates to sample the whole water column. It is expected that one way to reduce bio-fouling of the instrument and its carrier is to reduce the time spent in the euphotic zone. This approach has already been proved with a system that has been tested in the Florida Strait (Van Leer et al. 1974). Again the lack of experience from long-term deployments makes a final conclusion difficult.

Traction principle

A wheel driven propulsion is made possible by using the friction between the mooring line and traction wheels (Doherty 1999).

This crawler type method has (Fig. 5) proved to be quite succesful. Initial deep-sea deployments have shown the basic functionality and delivered the first scientific results. This method is efficient in terms of energy expenditure as long as there is limited bio-fouling of the mooring line. In this case

the energetic costs come close to the theoretical predicted values dominated by the drag of the profiler body and the buoyancy change due to the different compressibility of the system compared to the surrounding seawater. The main drawback is the close interconnection between the platform and a mooring line which will move vertically if the topmost buoyancy carrying the mooring is within the region of wave action. Internal waves many tens of meters below the surface may also have such an undesirable effect. That results in a substantial back and forth movement that impacts on the quality of the data especially for sensors like current meters and the optical plankton counter.

Common to all principles is the possibility to extend the applicability by adapting additional sensors or setting up a quasi real time data communication via a surface buoy. The most interesting parts of the water column for many scientific questions are the upper 50m and the lower 50 m, regions which are difficult to reach with the current designs. For this purpose special buoy or anchor designs have to be considered.

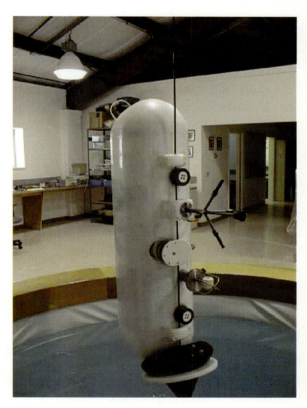

Fig. 5. The CRAWLER of McLane Research Laboratories Inc.

Underwater winches

There are other alternatives, which should be mentioned here. The idea of an underwater winch that reels and unreels a fixed length of wire to bring an instrument package to the surface (Fig. 6) has been contemplated although so far it has not lead to an operational system. The main problem lies in the precise reeling of the different layers of wire and the extended need for energy. One also has to take into account that additionally to moving the profiler up and down a considerable amount of energy is needed for reeling and unreeling the cable against the drag. Nevertheless there are application areas where underwater winches seem to be the best solution. For example in areas with intense ship traffic fixed surface deployments may be damaged. This is the case in a significant number of ocean margins. Furthermore in the Arctic regions where the characteristics of the surface water are of paramount importance, icebergs may cause similar damage.

Therefore new approaches to overcome the technical problems of this principle should be encouraged.

Lagrangian profiling floats

Free profiling systems such as the French MARVOR, the American ALACE or the German APG have been successfully deployed and used for many years and this has made these devices a standard tool in oceanography. They are very attractive tools to study processes from a Lagrangian perspective and have the advantage of reaching the surface undisturbed by the mooring wire. The main drawbacks are that they cannot be kept in the region of interest especially in regions of strong currents and their design allows only a limited amount of payload.

Gliders

Gliders are free floating systems that receive their energy for locomotion from the buoyancy change that they can produce (Fig. 7). They are derivatives from the Lagrangian float development and can be thought of as a slimmed AUV.

Fig. 6. Profiling instrument carrier using an underwater winch.

Fig. 7. The SLOCUM glider of Webb Reseach Corporation.

These platforms could act as a virtual mooring doing profiles by spiralling up and down the water column. At the surface the system sends its data via satellite link and updates its position. While resting on the ocean floor no energy is needed for keeping the system in position.

These future systems are very promising and there are already a few groups in the US that pursue this development concept. The success will strongly depend on the performance of the navigation and control system and the capability of adapting additional measuring units. The payload of these systems will be very limited.

Current achievements

Up to now autonomous profiling instrument carriers are in limited use or are still under devel-

opment. But the results that have been obtained look very promising. For example, the smooth up and downward motion of the buoyancy driven system has been demonstrated on a deployment from RV *Meteor* with the so-called TRAMP systems (see Figure 8, Skoglund 1998). The well mixed surface layer results from strong wave action during the time of measurement. It is also displayed in the enhanced movement of the profiler when it reaches this region.

This feature of the buoyancy driven profiler makes it an ideal platform for current measurements. On the other hand the time period to sample a depth interval of 1000 m lies in the range of 1 hour. This means that there will be some aliasing in resolving the effect of tidal currents .

A monotonic profiling behaviour is also advantageous for measurements which require a con-

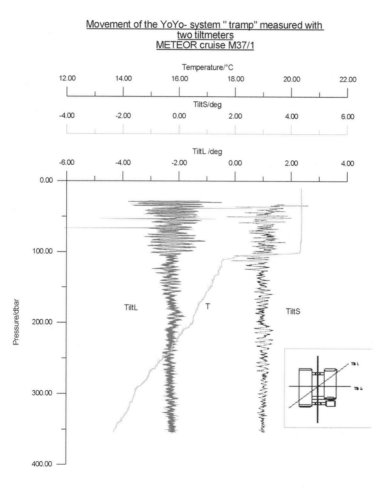

Fig. 8. Temperature profile taken with the TRAMP system (Skoglund 1998) showing the smooth motion of the system. The inserted picture describes the axes of tilt.

tinuous unidirectional water flow such as the optical plankton counter. Any back and forth movement would seriously compromise the measurements. A similar effect is impairing conductivity measurements.

Another example of the performance of a buoyancy driven system in deep water is shown in figures 9 and 10. Two successive profiles were collected where again the up and down motion appeared to be very smooth and at selected depths the data agreed within 0.01 K. An important feature is the constancy of the measuring results in selected depth horizons giving an indication of possible drift of the sensors.

Turning now to the power requirements of a buoyancy driven system the typical amount of energy needed for a single profile is about 50 Wh for the system developed at the University of Bremen (Waldmann 1999). With a typical Lithium battery pack of 3 KWh 60 profiles could be achieved. This compares to about 10 Wh for the Crawler system. The performance of the buoyancy driven system can be improved substantially but the CRAWLER will probably always be better from the standpoint of energy usage by a factor of about 2-3. This is related to the already mentioned additional power needed to keep the system moving till it reaches the final stop and the additional volume that has to be held to accommodate the oil that induces the volume change. This additional volume results in an additional buoyancy change and drag.

The CRAWLER has been the first system that delivered a dataset over an extended time interval. The results are shown in figure 11.

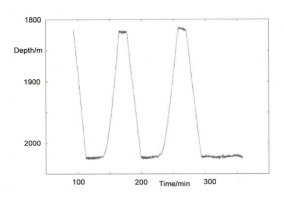

Fig. 9. Depth-time plot showing the profiling behaviour of the Bremen system (Waldmann 1999).

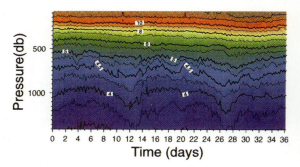

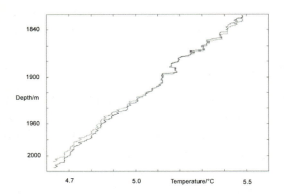

Fig. 10. Two successive profiles taken with the Bremen profiler (Waldmann 1999).

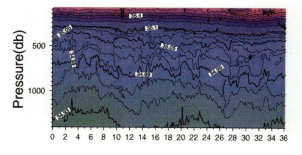

Fig. 11. Time series contoured data recorded with the McLane Crawler on the New England Slope (Doherty et al. 1999, reprinted with permission of the American Meterological Society).

Future technological objectives

What are the future design goals and what technology is needed? From a scientific standpoint the ideal system should be able to accomplish two complete profiles per day over at least six months. It should act as a platform for sensors like CTD, current meter, fluorometer and special optical and acoustical sensors. The most interesting parts of the water column, the surface and the bottom should be covered as well. A quasi real time bi-directional data link should be established to be able to combine the available satellite images with subsurface measurements and, on the occurence of extraordinary events to adapt the sampling process accordingly. The profiler itself should have the capability to cover the full depth range down to 6000 m. The size of the platform should be kept small but there is a limit given by the size of the sensors that have to be carried.

At present it seems not to be possible to achieve all goals at the same time. But by splitting off tasks for instance by using two profiling systems the aims become more realistic. Only the buoyancy principle and the crawler can cover the full ocean depth range. Both realisations are self-contained.

The addition of additional sensor systems leads to some technological problems. To make the platform as universal as possible all external systems should be neutrally buoyant and energetic autonomous. Otherwise too much volume as housing for extra batteries has to be provided and often will be not used. On the other hand it seems to be unrealistic that all additional sensor systems can be made zero buoyant for the whole depth range, i.e. have the same compressibility as the surrounding water. Other measures like using materials that have a higher compressibility than seawater to compensate for the behaviour of other materials also contributes significantly to the overall volume. This again sets some limitations on the miniaturisation of the platform.

Description of a possible approach

To preserve the concept of a universal instrument carrier one has to reconsider the idea of the pro-filing system. It seems to be unavoidable to design a docking station for the profiler independent of the principle used. This docking station will serve as a communication link and at the same time would allow recharging of batteries. The amount of energy that has to be carried with the system can thus be reduced to that required for just one profile leading to a substantial reduction in weight and volume. According to the above-mentioned specifications there will be a time period of several hours available for the recharging process so that new battery sources with very high energy densities, such as seawater batteries, could be used. An inductive coupling between the profiler and the docking station could fulfil two tasks: First, one could read out the collected data and second, the energy could be transferred. This technical solution provides the maximum flexibility in the choice of components for the mooring.

Current developments are addressing the issue of online data access. Satellite communication links are sufficient to handle the anticipated amount of data (Meinecke and Ratmeyer 1999) although the availability of the service is sometimes unsatisfactory. Delays between sending a command and receiving the response can under unfavourable conditions last several hours.

In coastal areas one could also consider the use of fibre optic communication cables and Cellnet phone channels. The main obstacle with fibre optic cables are the costs due to both the initial seabed survey and the deployment itself. On the other hand the energy efficiency makes this kind of transmission very attractive. A design study of the described system is shown in figure 12.

Conclusions

Moored profilers appear to be an attractive tool for future multidisciplinary observational programmes. They have the potential to become a standard component for buoy systems such as the ATLAS moorings for the TOGA or PIRATA arrays.

As the basic building blocks are already available and simply have to be adapted for this goal, operational systems can be anticipated in the near future. The amount of energy needed for the system

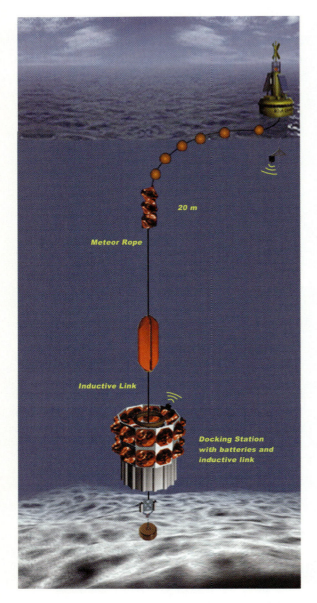

Fig. 12. Design study for the combination of a profiler with a docking station.

itself and the additional sensors right now limit the versatility of this tool. Furthermore the size of the different sensors has to come down substantially as the payload volume and weight set the limits as well.

There are good reasons to pursue the development of profiling instrument carriers be it moored or free floating:

• Resolution - Profiling systems enable measurements that have high spatial and temporal resolution.

• Bio-fouling - Letting the platform rest well below the euphotic zone will reduce substantially sensor bio-fouling.

• Sensor drift - With one set of sensors the complete depth range of interest can be covered. Divergent drift effects that are difficult to account for using individual sensor arrays are not present. Additionally with profiling systems depth regions may be reached where the calibration of the sensors can be checked.

• Water flow - Some sensors such as the optical plankton counter require continuous flow of water. For such sensors it is essential to have a monotonically moving platform.

References

Biscaye PE, Anderson RF, Deck BL (1988) Fluxes of particulates and constituents to the eastern United States continental slope and rise: SEEP-I. Cont Shelf Res 8: 855-904

Biscaye PE, Anderson RF (1990) Transport and transfer rates in the waters of the continental shelf and slope. SEEP Report US Department of Energy 78 p

Biscaye PE, Flagg CN, Falkowski PG (1994) The shelf edge exchange processes experiment, SEEP-II: An introduction to hypotheses, results and conclusions. Deep-Sea Res II 41:231-252

Doherty KW, Frye DE, Liberatore SP, Toole JM (1999) A Moored Profiling Instrument, J Atmos Oceanic Techn 16:1816-1829

Eckert AC, Morrison JH, White GB, Geller EW (1989) The autonomous ocean profiler: A current driven oceanographic sensor platform. IEEE J Ocean Engin 14:195-202

Eriksen CC, Dahlen JM, Shillingford Jr JT (1982) An upper ocean moored current and density profiler applied to winter conditions near Bermuda. J Geophys Res 87(C10):7879-7902

Fowler GA, Hamilton JM, Beanlands BD, Belliveau DJ, Furlong AR (1997) A wave-powered profiler for long-term monitoring. Proceedings Oceans 97, MTS IEEE Conference, 6-9 October 1997, Halifax, NS, Canada pp 225-229

Meinecke G, Ratmeyer V(1999) A bi-directional link into the deep-sea - The DOMEST project, Second International Conference on EuroGOOS, 10-13 March 1999, Rome, Italy

Monaco A, Biscaye P, Soyer J, Pocklington R, Heusner S (1990) Particle fluxes and ecosystem response on a continental margin: The 1985-1988 Mediteranean

ECOMARGE experiment. Cont Shelf Res 10:809-839

Provost C, du Chaffaut M (1996) YOYO profiler: An autonomous multi-sensor. Sea Techn 37:39-45

Skoglund S (1998) The shallow water profiler TRAMP. Internal Report, University of Copenhagen, Denmark

Van Leer J, Düing JW, Erath R, Kennelly E, Speidel A (1974) The Cyclesonde: An unattended vertical profiler for scalar and vector quantities in the upper ocean. Deep-Sea Res 21:385-400

Van Weering TCE, McCave IN, Hall IR (eds) (1998) Ocean margin exchange (OMEX 1) benthic processes study. Progr Oceanogr 42 (1-4)

Waldmann C (1999) A newly designed deep-sea YOYO-profiler for long term moored deployment. Second International Conference on EUROGOOS, 10-13 March 1999, Rome, Italy

Wang DP, Wong GTF (1992) KEEP (Kuroshio Edge Exchange Processes). Terr Atmos Ocean Sci 3:225-447

Autonomous Underwater Vehicles:
Are they the Ideal Sensor Platforms for Ocean Margin Science?

S. McPhail

Ocean Engineering Division, Southampton Oceanography Centre,
Empress Dock, Southampton, SO14 3ZH, UK
corresponding author (e-mail): Stephen.D.Mcphail@soc.soton.ac.uk

Abstract: Autonomous Underwater Vehicles (AUVs) are fast becoming accepted as very useful data gathering platforms within the marine science community throughout the world, as the range and depth envelopes are being pushed, by developments in battery technology, propulsive efficiency, and pressure vessels technologies. It is already accepted that AUVs can bring great benefits in data quality and cost, in for example geophysical surveys for oil and gas exploration. But within the science community there is the perception that AUVs are expensive, complex and risky to use. Is this a fair representation, or is it based on outdated prejudices? This paper examines the advantages and disadvantages of the use of AUVs as platforms for Ocean Margin surveys, compared to conventional towed instruments, drawing on examples of AUVs currently being used throughout the world. It illustrates the development and use of a scientific AUV, Autosub, during the past four years. How has it developed to overcome technological problems, such as launch and recovery, and achieving greater depth and range, and how have the engineers coped with the integration of many different types of sensor? It discusses some possible reasons why AUVs are not more generally used for ocean surveys.

Introduction

The oil and gas exploration industry is becoming very interested in using Autonomous Underwater Vehicles (AUVs) for deepwater geophysical surveys, as the number of deepwater field developments are increasing, and site surveys are needed for early investigations. Within the industry it is widely accepted that the conventional technology of cumbersome towed and Remotely Operated Vehicles (ROVs) cannot deliver the quality of results desired. AUVs are seen as the economical solution (McNeill 2000). The military is also interested in using AUVs for many applications particularly Mine Countermeasures, and intelligence gathering (Dunn 2000). AUVs have also demonstrated their ability to carry out useful scientific missions for physical oceanography (Smith et al. 1996), geological survey (Bradley et al. 2000), and fisheries research (Fernandes et al. 2000).

Recent developments in range and depth capability of AUVs, using high strength composite materials for the pressure vessels, and light weight, high energy density, rechargeable batteries make them look like the ideal choice for many Ocean Margin survey missions, as confirmed by the proceedings of this Workshop into Ocean Margin Systems. In the group 2 report (Fluid flow, Kukowski et al. this volume) it was identified that AUVs and other mobile platforms could be used for both imaging of the seafloor (e.g. to identify mud-volcanoes), and with CTD and chemical sensors, to detect and quantify seafloor fluid emissions. AUVs are seen as a valuable survey tool at a variety of scales ranging from determination of biogeochemical fluxes in the water column; detailed optical and sonar analyses to classify seabed types; to medium scale bathymetric mapping (group 3 report: Margin building - regulating processes, Thomsen

From WEFER G, BILLETT D, HEBBELN D, JÖRGENSEN BB, SCHLÜTER M, VAN WEERING T (eds), 2002,
Ocean Margin Systems. Springer-Verlag Berlin Heidelberg, pp 79-97

et al. this volume). In direct optical survey mode, AUVs can provide the detailed images over a sufficiently large area to provide a synoptic view of variability in ocean margin ecosystems as identified by group 4 (Life at the edge, Rogers et al. this volume).

So what is preventing the wide-scale adoption of AUVs as platforms for Ocean Margin science? What are the advantages of using AUVs for Ocean Margin science compared to conventional towed instruments; what are the issues; and what are the remaining technological challenges?

Advantages of AUVs compared to tethered vehicles:

There are several advantages of using AUVs compared to more conventional towed, tethered, remotely operated, or ship based technologies.

Data productivity

Autonomous underwater vehicles, when operated in completely unattended mode, can lead to increases in the productivity of a science campaign because the mother ship is free to do other tasks while the AUV is deployed. An example of this mode of operation was during the Autosub field campaign in the North Sea when the AUV carried out fish surveys, completely unattended by the support ship, Scotia, which simultaneous carried out its own survey, at times over 100 km away (Fernandes et al. 2000).

AUVs can operate from a smaller support ship (depending upon the size of the AUV). For geophysical surveys there is a requirement for a platform to fly at some distance off the seabed, with geophysical instruments such as sidescan, bathymetric, or subbottom-profiling sonars. Conventionally this involves the use of a large support ship with the facilities to handle the tow cable, which in addition to towing the platform must also provide electrical energy and data communications. With deep water Remotely Operated Vehicles (ROVs), the deep-water umbilical winch and handling system can weigh several times as much as the ROV. Another important factor is that for most ROV operations, the support vessel needs a dynamic positioning system so that it can either keep on station very accurately, or follow (slowly) a pre-determined course. This is not needed for AUV operations, with consequently considerable savings.

It is even possible to consider using an AUV without any support ship. Options for ship independent deployment, include:
• Deployment from fixed platforms (either purpose built, or existing e.g. an oil production platform).
• Deployment from shore. This has already been demonstrated with the Autosub AUV, which operated out of a small harbour in Scotland (Millard et al. 1998) and the Martin AUV, which is routinely launched from a pier for field tests in a shallow creek (Bjerrum and Ishoy 1995).
• Deployment from the air. This is perhaps a longer term goal, but one which is worth pursuing as it offers much quicker deployment to remote areas. At the present helicopter launch and recovery seems to be the most practical option.

For deep operations, one of the most compelling arguments for using an AUV is that the survey speed can be very much greater than with a towed instrument. There are two reasons for this. The tow speed for a deep tow is typically limited to between 0.5 and 1 m/s by the considerable hydrodynamic drag and lift on the tow cable. With a deep tow, the cable length generally exceeds the water depth by 50% or more, so that the towed platform is often several km behind the ship. The consequence of this is that turns at the end of survey lines can be very slow, taking several hours. AUVs typically run at between 1.5 and 2.5 m/s, and have turning circle diameters of 20 to 30 m. The outcome of this is that surveys carried out with AUVs can have coverage rate of 3 to 4 times that of towed vehicles.

Another major advantage of AUV operations is that once an AUV is launched, its operations can be weather independent. Severe weather can stop towed or ROV operations, as either the data quality may become too poor (due to coupling of the ships motion to the towed platform), the ships safety may be compromised by trying to continue operations, or due to wind forces the ship may be unable to maintain the intended course at the necessarily slow survey speeds.

Data access

AUVs can provide data from areas which are inaccessible to other platforms, including ships, towed instruments and ROVs. The classic example is under sea ice, which has until recently been almost completely inaccessible to researchers, except for some military submarine missions under the Arctic sea in cover. The exception is the Theseus AUV, which was developed by International Submarine Engineering (ISE) in Vancouver, Canada. It has lead the way in missions under the Arctic ice north of Canada (McFarlane 1997). In April 1996 the 8.5 tonne AUV completed a 350 km round trip, and successfully laid a 150 km fibre optic cable under sea ice for the US government. This was an outstanding achievement for the time, although it should be appreciated that Theseus was developed for this very mission specific task (although it could be easily modified to carry out science missions), and to date there have been no science missions of any significant distance under either sea ice or under ice shelves. This is set to change. There are several proposals to operate AUVs under ice, including the Autosub Under Ice programme (described later), and the Italian SARA project to develop AUVs for operation in Antarctica (SARA 1999). Clearly, such missions are not possible with a towed vehicle.

Another area which is difficult to survey with a towed vehicle is the near surface layer. Several AUVs have demonstrated surveys within a few metres of the sea surface. The Florida Atlantic University Ocean Explorer AUV, fitted with turbulence probe, CTD, and 1200 kHz ADCPs has made extensive physical oceanography measurements within 3 m of the surface (Dhanak et al. 2000). Autosub, (as described later on in this paper), has carried out long (33 hour, 160 km), missions at depths of 2 m to 6 m, off the Scottish west coast, with turbulence probe, bubble detecting sidescan sonar, and acoustic resonator for measuring bubble size. These measurements would be difficult if not impossible for ship based platforms.

Data quality

AUVs provide a more stable platform in pitch roll and yaw than towed or ship based instrument platforms (Fig. 1). This is important for surveys which use imaging type sensors, either sonars such as side-scan sonar and interferometric or multibeam bathymetric system, or optical imaging systems such as photography or laser line scan systems. Offsetting some of the high cost of the AUV navigation, the high quality pitch, roll, and yaw data output by such navigation devices is useful if not essential for correct operation of a bathymetric sensor systems.

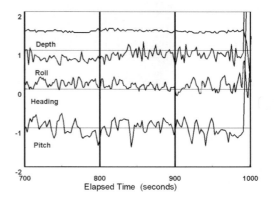

Fig. 1. Autosub depth and attitude stability. Five minute plot of depth (m), pitch (degrees), roll (degrees) and heading (degrees) for Autosub during a 12 m constant depth run off Oban, Scotland, in April 1999 (depth is offset for plotting). Standard deviations are: Depth 0.03 m; Pitch 0.19 degrees; Heading 0.1 degrees; Roll 0.13 degrees. The plot illustrates the good attitude and depth stability possible with an AUV.

The HUGIN vehicle is an Unmanned Underwater Vehicle UUV which has been used for high quality bathymetric surveys using a Simrad EM 3000 multibeam echo sounder (Strokersen and Indreeide 1997) (Fig. 2). The distinction between a UUV and an AUV, is that it is implicit that UUV is used in conjunction with a support ship, from which it can be tracked and controlled via an acoustic command link.

The Maridan MARTIN AUV has demonstrated the efficiency and quality with which tight "lawnmower" pattern surveys can be carried out and navigated using an AUV (Pind 1999) (Fig. 3). In 1999 MARTIN carried out a survey for De Beers Marine for exploration of diamond fields 30 km off the coast of Namibia in 70 to 80 m water depth.

Fig. 2. The HUGIN II UUV being recovered into its launch and recovery ramp, which is built into the a standard 20 foot shipping container. HUGIN II is used for seabed imagery and bathymetry surveys using Simrad 2000 bathymetric sidescan sonar (200 kHz), and a 2 to 16 kHz sub-bottom profiler.

Fig. 3. The Maridan Martin M150 AUV on board the De beers Marine mothership "*Zealous*" during preparation for launch off the coast of Namibia in March 1999. The survey demonstrated the effectiveness of an AUV for diamond exploration (Photo by courtesy of De Beers Marine Pty).

The survey demonstrated the ability of an AUV to carry out a high definition sidescan sonar survey in a "lawn-mower" pattern, with line spacing of only 25 m. The AUV, navigated by an Inertial Navigation System (INS), with velocity input from a 1200 kHz Doppler sonar, demonstrated a navigation accuracy of better than 0.1% of distance travelled (less than 1 m error for every 1 km travelled).

AUVs are also able to detect using sonar and automatically respond to changes in the water depth, and so it is possible to carry out near to seabed surveys in rough terrain. Generally, due to the tow cable, towed systems have a very large inertia, and so cannot respond quickly to seabed obstacles. Consequently they cannot be used for close to seabed surveys, unless the platform is engineered to survive repeated collision with the seabed.

Issues with use of AUVS

Range and depth envelope limitation

There are a number of issues, or at least perceived issues for the use of AUVs, including cost, navigation, possibility of loss, safety, and risks associated with launch and recovery. But by far the most problematic set of issues are those related to: energy, power, range, and depth capability. They are strongly related. Their relationship can only sensibly be expressed algebraically.

The total power used by the vehicle is the sum of the propulsion power and the hotel power (used to power systems and sensors) (Hoerner 1962):

$$P_T = \frac{V^{2/3} \times C_d \times U^3 \times \rho_s}{2 * \xi_m * \xi_p} + P_H \qquad (1)$$

Where:
P$_T$ is the total power needed from the vehicle battery
P$_H$ is the hotel power
V   is the volume of the vehicle
C$_d$ is the hydrodynamic drag coefficient of the vehicle (based on V$^{2/3}$)
ξ_m is the efficiency of the motor

ξ_p is the efficiency of the propeller
U   is the speed of the vehicle through the water
ρ_s is the density of seawater
Note that propulsion power is proportional to speed cubed.

Further, the energy stored in the batteries:

$$E = V \times \rho_e \times K_e \times \rho_s \qquad (2)$$

Where:
E   is the total amount of energy stored on the vehicle
ρ_e is the energy density of the batteries (Joules / kg).
K$_e$ is the fraction of the vehicle displacement which provides buoyancy for battery payload. It is a basic quality measure, in terms of how efficient the mechanical implementation is at providing buoyancy. Pressure vessels made with higher strength materials, or lower density syntactic foams will give higher K$_e$. K$_e$ will generally decrease with increasing depth specification.

The range is easily calculated from (1) and (2) as:
Range = (Energy/Power) x Speed.

Hence:

$$Range = \frac{V \times \rho_e \times K_e \times \rho_s \times U}{\frac{V^{2/3} \times C_d \times U^3 \times \rho_s}{2 * \xi_m * \xi_p} + P_H} \qquad (3)$$

Under conditions where the hotel power is small compared with the propulsion power, equation (3) can be simplified:

$$Range = \frac{V^{1/3} \times \rho_e \times K_e \times \xi_m \times \xi_p \times 2}{Cd \times U^2} \qquad (4)$$

So to increase the range we can:
• Increase the vehicle length (proportional to V$^{1/3}$ for fixed shape vehicle - but note that the doubling of length will increase the vehicle volume and mass by 8 times).

84 McPhail

- Increase the energy density, ρ_e, of the energy source (but higher energy density batteries generally cost more, see section on batteries below).
- Increase K_e, the fraction of the vehicle displacement available for the energy source. Unfortunately, most energy sources (with notable exceptions of pressure balanced lead-acid batteries, or aluminium/seawater semi fuel cells), need to operate in a dry, 1 atmosphere pressure vessel. For deeper operation, this pressure vessel must necessarily have thicker walls hence less battery mass can be floated. For this reason existing or proposed deep diving AUVs generally use high technology, high strength materials. Autosub uses carbon fibre composite pressure vessels with a safe operating depth rating of 1600 m. As well as, and alternatively, the parts of the vehicle which are not associated with energy storage can be made lighter, or more buoyant. One possible route is to use lightweight, so called syntactic foams (containing small hollow glass spheres in a plastic matrix), to form the structural elements of the vehicle. This option is particularly attractive for deep diving operations because at depths over 4000 m, even high technology pressure vessels offer little buoyancy advantage over syntactic foams.
- Increase the propeller efficiency, and increase the motor efficiency. Again there is room for improvement. In practice the total combined efficiency rarely exceeds 50%.
- Decrease the drag coefficient. There is surprisingly large scope for this, as calculations suggest that most operating AUVs have drag coefficients considerably larger than would be suggested by published data on the hull forms. For example, the drag of Autosub appears to be at least 50% greater than tow tank tests of a smooth scale model would suggest. The (difficult) challenge is to mechanically integrate sensor packages so that a they do not significantly increase the overall drag of the vehicle.
- Decrease operating speed. In practice, battery powered AUVs rarely go faster than 2 m/s. But by reducing the speed below 1.5 m/s some of the survey speed advantages of using AUVs are compromised, and in addition at lower speeds currents can cause control problems. There is a speed, the

so-called optimum speed, at which the AUV achieves maximum range. Figure 4 is an evaluation of equation (3) for Autosub speeds of 0.4 to 2.0 m/s. It shows an optimum speed at 0.65 m/s.

So there is an opportunity to improve the range performance (and operating cost), by efficiency improvements in the system. For the Autosub AUV the scope for improvement is at least 100%, made up of a 50% increase in propulsive efficiency, and a 50% drag reduction. Optimised vehicle designs could give better figures for specific energy density, or alternatively allow an increase in depth capability, while maintaining range.

Sensor power limitation

The use of batteries for energy limits the amount of power available for sensor systems. Most notably this can be a limiting factor for lighting for video and camera systems (for most active sonar sensors, the average power used is quite low, usually only a few Watts). However, although video is an essential tool for ROV operations, and needs power intensive lighting it is not an appropriate sensor for AUVs. For optical survey, there is

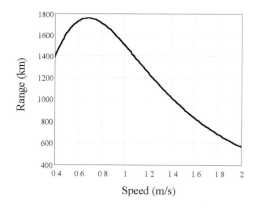

Fig. 4. Range of Autosub AUV as a function of speed. The plot is evaluated from equation (3) in the text with parameters known or measured for the Autosub AUV. Note that there is an optimum speed in terms of range, of 0.65 m/s. At much higher than the optimum speed, range is proportional to $1/\text{speed}^2$. At lower than the optimum speed, range is proportional to speed/(hotel power). The parameters used were : $P_H = 80$ W; V = 3.5 m³; $C_{dv} = 0.056$; $\xi_m = 0.70$; $\xi_p = 0.75$; $\rho_e = 430$ kJ/kg; $K_e = 0.2$ (see text for explanation of symbols).

no advantage in scanning the same part of the seabed at 50 frames a second (as video does); rather much slower scan rates are needed, either using laser scan systems, or flash guns. Both use considerably less power than video lights. For medium size AUVs such as Autosub MARTIN, or HUGIN which have propulsive power requirements of 600 W and 400 W, respectively, provision of up to 200 Watts for scientific sensors does not severely constrain the range achievable.

Energy supply for AUVs

The provision of an affordable, reliable, energy source with high energy density has been, and remains the Holy Grail for AUV operators (Table 1). Whereas some AUV operators have experimented with fuel cells such as the aluminium/oxygen fuel cell which has been used by the HUGIN AUV (Storkersen and Indreeide 1997) in practice, most operators are using or plan to use either primary batteries, or rechargeable batteries. For short missions (up to 10 hours, 50 km), off-the shelf sealed lead-acid or nickel cadmium rechargeable cells give adequate performance (with energy densities of about 110 and 140 kJ/kg, respectively), at low initial cost, and very low per mission cost. The technology is well proven and understood. Sealed lead acid rechargeable batteries were used in the early Autosub trails (Millard et al. 1998). Nickel cadmium cells have been used by several AUV operators, including the Martin AUV (Bjerrum and Ishoy 1995), and the Florida Atlantic Ocean Explorer and Ocean Voyager AUVs (Smith et al. 1996). However, to fully exploit the potential of AUVs, with endurance of at least 24 hours, and

ranges in the hundreds of km, higher energy density batteries are needed.

Until recently silver zinc rechargeable cells were the only practical rechargeable battery option with high energy densities. Although they can provide respectable energy densities of up to 720 kJ/kg, deep discharge of the cells severely limits the number of cycles. Silver zinc cells also suffer from a shelf life limitation (1 or 2 years), and so economical use requires that such cells must be used intensively. For these reasons, together with the care which must be taken with charging and discharging, these cells are becoming less popular for AUV operations. Lithium ion cells (as used in some lap top computers, and camcorders), are becoming more popular for use on AUVs. They are used or are planned to be used by most AUV operators. For example, lithium ion batteries are currently used on both of the Woods Hole Oceanographic Institution (WHOI) AUVs: ABE (Bradley et al. 2000), and REMUS (Von Alt 2000).

Navigation of AUVs

AUV navigation requirements will obviously depend upon the application. Whereas for deep oceanographic surveys, dead reckoning errors of 1 km are often acceptable (the errors can usually be substantially improved upon by post-processing of the navigation data), for geophysical site surveys the positional accuracy of 10 m or better may be required.

The former requirement can be satisfied by the AUV surfacing and acquiring GPS fixes at suitable intervals (e.g. every 6 hours). Indeed, for

Battery Type	Energy Density (kJ/kg)	Range (km)	Maximum of Charge / Recharge Cycles	Cost per km of mission ($ US)
Lithium Ion	600	750	800	1.1
Silver Zinc	575	718	80	2.8
Manganese Alkaline	490	611	1	11.4
Nickel Metal Hydride	326	407	1000	0.7

Table 1. Energy density and costs of some candidate AUV battery systems. Energy densities and cost associated with using rechargeable batteries, and manganese alkaline primary batteries. The cost per km is based upon an AUV with a speed of 2 m/s, a total power use of 800 Watts, and a maximum practical number of charge / recharge cycles of 200. Range is for 500 kg of batteries. From Griffiths et al. (1999) and Vincent (1999).

physical and biogeochemical oceanography in the upper part of deep oceans, this is really the only viable option, as the Doppler logs needed to give more accurate ground referenced dead-reckoning have insufficient range to reach the seabed. Hence dead reckoning relative to the water, with occasional GPS fixes is the only option. In this mode of operation, the big issue is whether the vehicle can successively acquire a GPS fix, in the majority of sea and wind states. Autosub (Meldrum and Haddrell 1994) has developed in a joint programme with Navstar (UK) Ltd. a novel GPS receiver able to acquire GPS fixes despite periodic wash-over of the antenna.

For greater navigation accuracy, there are two approaches commonly used. The WHOI Autonomous Benthic Explorer (ABE), which has carried out geophysical studies of ocean ridge systems (e.g. the Juan de Fuca Ridge off the NW US), is capable of operating in depths of up to 5000 m, and uses an array of acoustic transponders deployed around the working area (Bradley et al. 2000). This approach can give good accuracy of 2 m standard deviation, minimises the complexity and cost needed for the on-board systems, and is appropriate where a detailed accurate survey of a small area (a few tens of km^2) is needed. However, for larger area surveys, the cost of laying and recovering the transponder net can become prohibitive. ABE is generally used in complimentary missions with the manned submersible ALVIN, (ABE operates at night, ALVIN can only operate during the day), and so these costs are shared between projects. One way to reduce the deployment and surveying costs of such long base acoustic nets is to use a system such as the ORCA GPS Intelligent Buoy. An array of 3 more free floating, or anchored acoustic receiver buoys are deployed, each of which carries a differential GPS receiver for location, and a hydrophone. The AUV carries an accurately timed 30 kHz acoustic pinger. From the arrival time of the pings at the buoys, and knowing the buoys GPS positions, the position of the AUV can be calculated. In practice the buoys communicate with a support vessel by VHF radio (Thomas and Petit 1997).

The other approach, which is more in keeping with the ideal of autonomous operations is for the AUV to dead reckon more accurately between the position fixes using a Doppler Velocity Log (DVL) (principle is that the Doppler shift of sonar echo from seabed is proportional to the vehicle speed). The DVL gives velocities in the vehicle's frame of reference (forward and sideways speeds). To navigate, this velocity must be rotated into the geographical frame of reference (north and east speeds) using the output of a heading sensor. Cheaper systems (such as Autosub's current system) use a magnetic compass for the heading data, but the limited accuracy of a compass leads to navigation drift rates of at best to 1% of distance travelled. Use of either Fibre Optic Gyro (FOG), or Ring Laser Gyro (RLG) based Inertial Navigation Systems (INS), can improve the accuracy to 0.1 % of radial distance travelled from the start position, although it should be stressed that this accuracy is only achieved when the Doppler velocity data is available (drift rates of medium accuracy INS systems when unaided by Doppler velocities are of the order of 20 cm/sec). INS systems with sufficient accuracy are becoming more affordable, at $70,000 or less, and so in the future all but the smallest class of AUVs would be expected to use INS plus DVLs for navigation.

As with any sonar instrument, long ranges imply the use of low frequencies (and hence longer wavelengths) as the attenuation of sound in seawater is strongly dependent upon frequency. To achieve the necessary narrow beam-widths for accurate velocity measurements, the aperture of the sonar transducers must be many times the wavelength of sound. The outcome of this is that to achieve long range for our DVLs, we must use large transducer heads. For Autosub (a 3.7 m^3 AUV), a 300 kHz DVL, with 200 m range is the largest that can be accommodated. Smaller AUVs must use smaller, higher frequency DVLs. For example the MARTIN AUV (1.8 m^3), uses a 1200 kHz DVL, with a range of only 30 m. The maximum range is important as it affects the type of survey that can be carried out. There are two ways to improve upon this. One is to use more advanced technology to form the four beams needed by the DVL. Use of a flat phased array, where the entire available aperture is used in the synthesis of each of the four beams means that a lower operating frequency, and hence a longer range can be achieved with a given transducer size. Alternatively, there is promise in the use of a different principle. Correlation Velocity Logs (CVLs) could in principle provide ranges of over 3000 m, but at the present time they are unproven for use in AUVs.

Even using DVL / INS based navigation capable of 0.1% of distance travelled accuracy there still remains the problem of how to maintain the navigation accuracy when the vehicle must descend through 3000 m without any DVL information. Simple calculations show, that if the vehicle can descend at 1 m/s, and the sea currents are 5 cm/s, then the navigation uncertainty in position at the end of the dive is: 3000 x 0.05 = 150 m. This would exceed the allowed errors for many types of survey.

In some regions, e.g. working on the continental slope or continental rise, the problem could be overcome by starting the mission near the shelf break, with the DVL within bottom track range, and then descending down the slope to the survey area, maintaining DVL bottom lock all the way. But where this is not possible, it will be necessary to locate accurately the vehicle at the start of the mission using an acoustic system. Ultra Short Base Line (USBL) systems based on the support ship, used with care to eliminate systematic bias, can locate the vehicle to an accuracy of 0.2 % of slant range to the vehicle. This would give the required accuracy of better than 10 m at the start of the mission. If the vehicle needs good real-time navigation, then this position fix would be acoustically telemetered to the vehicle. By processing the navigation data after the mission (post-processing) it will be possible to measure and correct some of the sources of navigation errors (such as misalignment of the DVL), and so it will be possible to produce higher accuracy navigation data. It should be possible to maintain post processed position accuracy within 10 m, if the AUVs position is determined using the USBL at least every 24 hours.

Summary of AUV attributes

Table 2 shows a summary of attributes for four AUVs which are currently carrying out commercial or scientific surveys.

Bottom crawlers

For some areas of study, particularly when a small area of the seabed must be surveyed over a long period, then bottom crawler AUVs may be appropriate. The same technology could also in principle be used to move moorings from one site to another. To date there have been few examples of use of such technology. One is the Scripps Rover, a 6000 m rated crawler which is used for carrying out benthic measurements as e.g. *in situ* oxygen, water sampling, and photography at up to 30 sites, separated by typically 100 m (Smith et al. 1997) (Fig. 5). Energy is supplied by 360 D type alkaline manganese primary batteries, and it has an endurance of 6 months.

In the context of comparing ROVs with AUVs it should also be noted that bottom crawler Remotely Operated Vehicles (ROV) are used extensively in the oil and gas industry for inspection work and trenching work (for burial of pipe lines and cables). Examples are the Slingsby Olympian Trencher and inspection ROV, or the Sonsub Flexjet II (Trencher).

Autosub - a case study

Autosub, an AUV developed by the Sothampton Oceanography Centre, has a substantial track record. To date it has carried out 240 missions, with a total track length of 2800 km, including >600 km run completely unattended by the support ship. It has dived to 1000 m, and has run in constant altitude mode to within 2 m of the seabed. In the last two years it has executed a busy science programme, with 5 different science campaigns, hosting 20 different scientific instruments and it has been used by marine scientists from all over the world.

The Autosub AUV was designed, built and is operated from the Southampton Oceanography Centre in the UK. From the earliest beginnings of the project back in 1988, Autosub was conceived of as a general purpose autonomous platform, able

	Manufacturer	Range (km)	Cruise Speed (ms^{-1})	Depth (m)	Volume (m^3)	Length, Width (m,m)	Energy Source	Missions and Sensors
Autosub	Southampton Oceanography Centre UK.	1000	1.4	1600	3.5	6.7, 0.9	Manganese Alkaline. 5040 D type cells	General purpose science missions, profiling or bottom following. Physical, biogeochemical, sonars, water samplers, ADCPs.
Hugin II	Kongsberg Simrad AS., Norway.	260	2.0	6000	1.2	4.8, 0.8	Sea Water Battery	Commercial explorations and survey. Sidescan, bathymetry, sub-bottom profiling, fish survey. Usually operates in UUV mode remotely controlled by acoustic link.
Martin M150	Maridan AS., Denmark.	121	1.4	150	1.5	4.5, 1.5	Lead Acid Rechargeable	Commercial survey and exploration. Sidescan, bathymetry, sub-bottom profiling.
ABE	Woods Hole Oceanographic Institution, USA.	59	0.6	5000	0.7	3, 1.5 (Three hulls)	Lithium Ion Rechargeable	Optical imaging, video, chemical sensors. Mainly mid ocean ridge environment. Can hover, land, and in sleep mode can remain for 6 months on the seabed.

Table 2. Performance attributes for 4 medium sized survey AUVs.

Fig. 5. The Scripps Rover. An example of a crawler type of AUV. Rover is designed to make measurements of the sediment community oxygen consumption over a small area. It carries *in situ* pore water oxygen probes, takes photographs of the sediment surface, and measures near bottom currents. At the end of the deployment it takes a water sample for subsequent oxygen sensor calibration. Propelled by two 100 Watt caterpillar drives, it can be deployed for up to 6 months, moving to up to 30 different sites, each separated by typically 100 m.

to carry a wide range of possible science sensor sub-systems, and to operate in a variety of mission scenarios, from e.g. upper ocean hydrographic profiling, to deep-sea benthic surveying (Mc Cartney and Collar 1990). Unlike many AUVs used for research purposes, it was always the intention that Autosub would ultimately carry out long missions of thousands of kilometres, completely unattended by any support ship. To achieve the necessary range, Autosub is a relatively large AUV, 6.8 m long and 0.9 m in diameter, and with a total displacement of 3.7 m³ (Table 3). Considerable effort (through design, methods and testing) has been spent in achieving the necessary system reliability (McPhail 1993) and new technologies have been developed where necessary, e.g. to obtain GPS fixes in conditions where seawater periodically washes over the antenna (Meldrum and Haddrell 1994). The sensor suit is not fixed, as the vehicle has generous volumes in the nose and rear sections for the mounting of sensors, and the design emphasis has been to simplify as much as possible both the mechanical and the systems integration of sensor packages.

Autosub system description

As shown in Figure 6 Autosub has a traditional "fat" torpedo shape, formed from glass fibre reinforced plastic panels attached over an aluminium space frame. This arrangement gives considerable flexibility for the mounting of sensors (it is not expensive to provide replacement panels, and then have holes cut for sensors). For deployment Autosub is generally ballasted to be 10 to 15 kg buoyant. This small amount of buoyancy just allows the vehicle to float when stationary at the surface and ensures that the vehicle will float to the surface in the event of power loss when the vehicle is submerged. For extra reserve there is a 30 kg emergency abort weight, which may be released either by the on-board emergency abort system (if e.g. a leak was detected, or the vehicle has dived too deep), or by an acoustic command from the support ship). The abort system is fail safe; the weight is held on by an electromagnet so that in the event of a power failure the weight is dropped.

As the Autosub was designed for unattended and long missions, multiple redundant systems are used for recovery (e.g. two ARGOS satellite beacons, mounted at opposite ends of the vehicle, two flashing lights, and two 12 kHz acoustic transponders).

The cylindrical central section of the vehicle provides the majority of the dry space and the buoyancy of the vehicle by an arrangement of seven carbon fibre composite tubes, which are held within a matrix of syntactic foam. Four of the tubes contain the energy supply, which are up to 5040 D size Manganese alkaline primary cells. The other tubes are either empty, or contain system electronics. This arrangement was chosen in preference to the alternative single large diameter tube, because of the difficulty (and high cost) of manufacture a large diameter thick section carbon composite tube.

The system architecture is unusual in that there is no single "brain" or central computer which carries out the mission management functions on the vehicle. Rather there is a distributed network of 14 sensor control and actuation nodes distributed throughout the vehicle (McPhail and Pebody 1998). The nodes communicate over a single twisted pair network, using the LonTalk® network protocol, and the LonWorks ® network operating system (ANSI 1999), (LonTalk and Lon Works are registered trademarks of the Echelon Inc.). A single data logger records "black box" type of engineering, navigation, and science sensor data. Throughout the lifecycle of the project this modular approach was found to have distinct advantages compared to the more traditional centralised approach. The major advantage is that it splits the problem of developing the mission management system into manageable modules, none of which are excessively complex, and each of which can be individually and thoroughly tested.

In the science mission campaigns between 1999 and 2000, this modular approach and the use of LonWorks to integrate science sensors was a major factor in managing the rapid changes in sensor and system configuration on the vehicle. Post mission data processing is also simplified by the use of integrated sensors and a single data

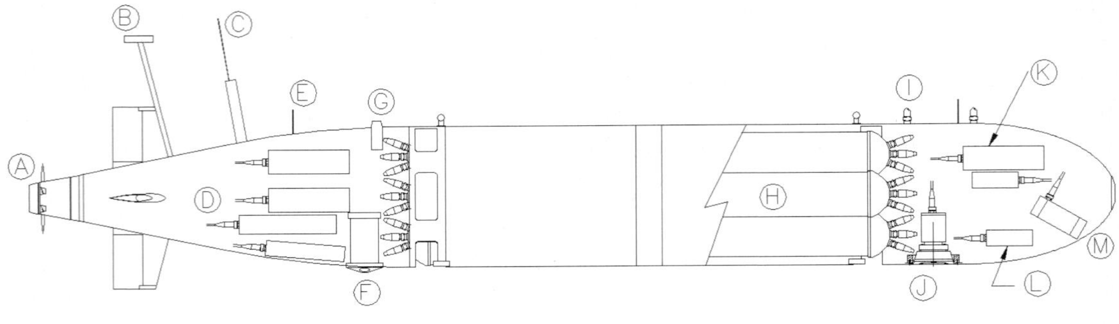

Fig. 6. Autosub mechanical arrangement. Longitudinal section through Autosub (6.8 m long). A: Brushless d.c. motor; B: GPS antenna; C: Orbcomm satellite antenna; D: Rear flooded volume (for science instruments); E: ARGOS satellite recovery beacon antenna; F: Doppler sonar; G: 12 kHz acoustic transponder; H: one of 7 carbon fibre composite tubes holding batteries and mission management system; I: Flashing lights (2); J: Emergency abort system and ballast weight; K: Seabird CTD electronics; L: Depth sensor; M: Forward collision avoidance sensor.

logger, as the data from such sensors are already synchronised with the vehicle navigation data. The sensors derive power from a distributed 24 volt and 48 volt power supply.

Early history

The first Autosub vehicle, Autosub-1 was constructed in 8 months, and following completion in April 1996 was launched within the confines of Empress Dock, Southampton for a series of engineering trials. In July and November 1996 two engineering trial campaigns were successfully completed at Portland Harbour, UK, an enclosed area of 5 km square, with water depths of 15 m. The vehicle acceptance trials took place near Oban, Scotland, in April 1998 with a campaign that included box surveys, depth and terrain-following undulating sections down to 195 m (Millard et al. 1998). During these trials Autosub demonstrated its ability to leave a quiet harbour autonomously using DGPS navigation (Griffiths et al. 1998a).

Autosub-1's first open ocean experience was off the east coast of Florida in December 1997, operating at the edge of the Florida Current (Griffiths et al. 1998b). The mission comprised terrain-following and "lawn-mower" pattern sections with the vehicle carrying ocean environment sensors including three CTDs, a pH probe and a Doppler current profiler. The longest mission covered 110 km in a lawn-mower pattern of 64 legs of 1.3 km east-west sections, with short north-south joining sections across the Florida shelf break. Autosub was operating from a basin within Port Everglades with access to the sea through the busy main entrance. The available support vessel was a 15 metre pleasure-diving boat (Lady-go-diver), too small to carry Autosub. Autosub was launched from the shore, towed out of the small basin with an outboard motor boat and the tow transferred to Lady-go-diver for the tow to the start of the mission. These missions also gave valuable experience of operation in strong currents (4 to 5 knots at the surface), and in strong winds.

In August-September 1998 Autosub-1 completed 11 missions in oceanic waters off Bermuda in the Autonomous Vehicle Validation Experiment (AVVEX), using the RV *Weatherbird II* as the support ship (Griffiths et al. 2000a). One aim was to demonstrate that it would be possible to fit Autosub with instruments suitable for making routine observations around a site known as hydrostation "S" in a cost effective manner, regardless of weather conditions. Nine missions were carried out to demonstrate and to test various aspects of Autosub's engineering and science data gathering ability, culminating in a 263 km mission with profiling down to a depth of 500 m. One problem that became all too apparent, was the difficulty in safely recovering the Autosub onto the *Weatherbird II* in the conditions experienced (4 m high long period swell, which is quite typical

Range, endurance	1000 km, 8 days at 1.4 m/s (calculated from empirical trails data)
Energy supply	Manganese alkaline primary cells
Length , Diameter	6.8 m , 0.9 m
Displacement	3.7 m^3
Energy cost	$10 per km. (at 1.4 m/s)
Maximum depth	1600 m at safety factor of 2:1
Payload space and power	500 Litres, 400 W
Acoustic tracking	ORE Trackpoint II system, range of 6 km
Mission plan upload	by UHF radio-link, range 1000 m
Control system architecture	LonWorks distributed network
Flight modes	Constant depth: Standard deviation of 5 cm Constant altitude: 2 m to 200 m altitude
Navigation	(D)GPS when surfaced. DVL based error of 1% of distance travelled (within 200 m of the seabed)
Relocation	Flashing lights (2), ARGOS satellite beacons (2), 12 kHz acoustic transponders (2)

Table 3. Specification summary of Autosub.

for the North Atlantic.) This experience was a spur to the development of a purpose made launch and recovery platform for Autosub.

Autosub science missions

In 1999, Autosub began an 18 month NERC funded thematic programme "Autosub Science Missions". So far there have been four separate science campaigns as part of this programme, with a total of 70 missions. A total of 2025 km have been run over 400 hours, and over 4 Gbyte of data have been collected. Some sixteen different science instruments have been integrated onto the vehicle. During the programme the operators became gradually more confident at allowing the vehicle to operate completely unattended. By May 2000 (for the Straits of Sicily campaign) this had become a routine way of operating the vehicle.

Under Sea Ice and Pelagic Surveys (USIPS) - (July 1999)

The campaign was part of a collaborative project between Dr. Paul Fernandes of the Marine Laboratory, Aberdeen, and Drs. Andrew Brierley and Mark Brandon of the British Antarctic Survey. The main aims of phase I, carried out on the Fisheries RV *Scotia* in the northern North Sea in July 1999 were: To demonstrate the use of an AUV for fisheries research; to test whether herrings avoid

the survey ship, and hence their quantity are underestimated; to estimate the occurrence of herring in the near-surface dead-zone of the *Scotia* (whose echosounder cannot provide information in the upper 10 m of water) by using an upward looking echosounder on Autosub.

Fish schools were detected on all missions and on one particular occasion a very large mid-water herring school was detected at less than 7 m range. This provides good evidence that fish are not sensitive to the vehicles' presence beyond this distance. In the missions where *Scotia* followed behind Autosub, the total mass of fish detected by both Autosub and Scotia were similar, showing that the herring do not significantly avoid a research ship with noise characteristics such as *Scotia's* (Fernandes et al. 2000).

This campaign provided the first opportunity to test the operation of Autosub in a completely autonomous mode, with no attempt to track it from the support ship. A total of eight unescorted missions were successfully carried out by Autosub, where the vehicle was deployed and directed away from the *Scotia* to rendezvous at a fixed location up to 16 hours later (Fig. 7) On no occasion did the vehicle deviate from its intended cruise track, however, on one occasion (Mission 186), the vehicle did fail to rendezvous with the research vessel at the planned location. The cause turned out to be a simple software error in the posi-

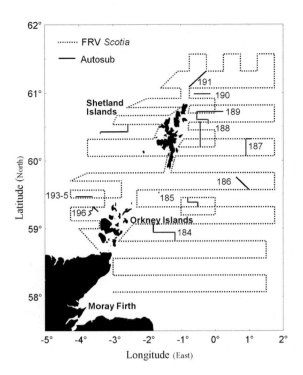

Fig. 7. Track of Autosub and FRV *Scotia* during fish survey in the North Sea. Cruise track of the *Scotia* (dotted lines) and the twelve mission tracks of the autonomous underwater vehicle Autosub-1 (solid lines) during the North Sea herring acoustic survey. The numbers 184 to 196 refer to the Autosub mission number. Missions 184 to 192 were unattended, missions 193 to 196 were escorted by FRV *Scotia* . The cruise started in the Moray Firth on 2 July 1999 and proceeded north, to the east of the Orkney and Shetland Islands, and then south, to the west of the Islands, ending at the north-east tip of mainland Scotland on 23 July 1999.

tion control system. Following backtracking by the *Scotia* along the intended track the AUV was detected using the acoustic homing system and later recovered.

The campaign also allowed the new launch and recovery gantry (Fig. 8) to be tested in winds of up to force 6. Operating from the starboard side of *Scotia*, the gantry worked well, proving to be a remedy for the deficiency noted on the AVVEX campaign. A new system for deploying a recovery line from the Autosub was successfully tested during this campaign, which together with the use of the gantry, removed any requirement for a small boat to be used in any part of the operations,

thereby increasing the weather window for deployment and recovery. This was another step in demonstrating the viability of AUVs in real-world conditions. In early 2001, this programme will be concluded with operations under sea ice in the Southern Ocean.

Spatial variability of bottom turbulence over sandbanks (August 1999)

Only three weeks after the *Scotia* campaign, Autosub was reconfigured, from a fisheries survey vehicle, to one capable of making turbulence measurements close to the seabed on a sandbank in the North Sea. The objectives of the campaign were to map bottom stresses in tidal flows over a linear ridge and to use the measurements to understand the role of bottom stress in the sediment transport processes that maintain the sandbank. Stationary, ship and AUV-borne observations were collected during a 10-day period over Broken Bank, a tidally dominated sand ridge situated 100 km offshore of the East Anglian coast (Griffiths et al. 2000b).

This campaign tested the robustness of the sensor and control functions for a prolonged period, with Autosub flying at only 4 m off the seabed for up to 36 hours, and with tidal speeds of up to 60 cm/s. Instruments on the vehicle included a 1.75 MHz Acoustic Doppler Velocimeter (ADV), a 1200 kHz Broadband Acoustic Doppler Current Profiler (ADCP), and a single axis pulse-coherent sonar, which were used to measure the near bed (0.4 to 4 m) flow.

Sonar and turbulence studies in the upper ocean (February 2000)

This campaign operated in the Scottish Inner Western Hebrides between the Islands of Mull and Colonsay. Its scientific objectives were to measure the horizontal structure of turbulence dissipation, its fine scale temperature variability, its variation with depth in the upper 10 m and its relation to breaking waves and wind speed; to measure the structure of dissipation within bubble bands produced by Langmuir circulation; to examine the variation of turbulence in the mixed layer in re-

Fig. 8. Recovery of Autosub onto RV *Calanus* using gantry. Autosub being recovered onto the RV *Calanus*, following trials in a Scottish sea loch. The recover gantry is essential for safe recovery conditions up to sea state 6. The gantry arm extends 5 m beyond the ship, and two recovery lines lift the vehicle, fore and aft, into the cradles, which can be rotated through 360 degrees. The cradle rotation, gantry extension, and two winches are separately controllable.

sponse to internal wave straining; and to measure turbulence in the thermocline and to relate it to local features such as internal waves. The campaign made use of Autosub's good depth keeping ability between 2 and 20 m, and the suitability of Autosub as a platform for turbulence measurements due to its low self induced noise and vibration. Sensors were a turbulence probe (courtesy of Tom Osbourne of John Hopkins University), upward looking 248 kHz sidescan sonar for imaging bubble bands, and a bubble resonator to determine the size of bubbles.

During this campaign the operators became even more confident at running the vehicle completely unattended. Part of this confidence was due to the installation of an Orbcomm satellite transceiver on the vehicle, which when surfaced transmitted relevant vehicle engineering data via a Low Earth Orbit (LEO) satellite and the internet to the operators, who could in principle be anywhere in the world. In practice, following deployment of the AUV, the operators found convenient sheltered anchorages around the Islands of Iona, Colonsay, and Jura. Steady improvements had been made to the navigation system during the previous campaign period, such that a mission leg of 24

hours was possible, without the need to surface to get a GPS position fix.

Exchange flow through the Strait of Sicily (May 2000)

Between the sea loch trials and the next campaign, Autosub underwent a major structural change. The complete central section, comprising a single glass-fibre composite pressure case was replaced with a new centre section with seven carbon fibre pressure cases. This modification was necessary to increase the depth rating from 500 m to 1600 m, and to increase the endurance from 300 km to 800 km.

The aims of the campaign were to make high-resolution measurements of the spatial distribution of the sea water overflow at the sills in the Straits of Sicily and to quantify the turbulent mixing and dissipation in the overflow. The Autosub carried dual Seabird CTDs, O_2 sensors, a turbulence probe de-ployed through the nose, and an upward looking as well as the usual downward looking 300 kHz ADCP (used for navigation).

A 170 km mission was successfully carried out, gathering upward as well as downward looking

ADCP data of good quality. But on the next mission the vehicle was unable to cope with a 150 m uncharted cliff rising almost perpendicularly from 450 m, at a point in the mission where Autosub was flying at 30 m above the seabed. The AUV hit the cliff wall, and became stuck beneath an cliff overhang at a depth of 320 m.

The Italian research support ship *Urania* had launched Autosub in the normal manner, and then left to carry out a complementary survey using lowered ADCP and CTD systems. The ship arrived at the rendevous point 2 hours early. When 4 hours had past and there was still no sign of Autosub, a search procedure was initiated using acoustic, radio, and satellite relocation aids. The *Urania* slowly moved back along the vehicle's track. After a further 29 hours contact was made with Autosub using two acoustic transponders (much to the relief of the scientists and engineers on board!). It was fortunate that the vehicle navigation was good (the vehicle was eventually found off track by 400 m after a 60 km run), and that the acoustic transponders on the vehicle were working, otherwise the vehicle may never have been found. Three weeks later Autosub was rescued using an ROV deployed from the Sonsub survey ship *Polar Prince*. For the ROV the recovery was simple, its manipulator arm grabbed the tail of the Autosub, dragged the vehicle clear of the overhang and then let go. The buoyant Autosub rose to the sea surface where it was recovered onto *Polar Prince*.

These events, and subsequent analysis of the data (equivalent to "black box" flight data) showed that Autosub could have cleared the cliff if it had used maximum control effort (Stern planes fully up) when the cliff was detected by the forward looking sonar. The crash data will be valuable for developing a better collision avoidance behaviour for the vehicle.

Autosub under ice

In 2002 to 2004 Autosub will embark on its most ambitious campaigns to date - into the ice cavities under glaciers in the Arctic and the Antarctic as part of the NERC programme "Autosub Under Polar Ice Shelves".

Polar oceanography will continue to increase its already high profile in climate research, but the little we know of the conditions in the ice cavities are based on seismic studies, bore holes (very expense), or measurements of oceanographic conditions at the ice front. The only way of obtaining a detailed information of the conditions below the ice shelves is to use an AUV such as Autosub. These missions will present several severe challenges to Autosub, and several new capabilities will be developed:

• Emergency abort - the option of rising to the surface in response to an emergency will no longer be an option. The mission management system will have to execute more sophisticated emergency strategies, such as commanding the vehicle to retrace its steps to the recovery point.

• Navigation - an Inertial Navigation System, integrated with the Doppler Velocity Log will be needed, improving navigation accuracy from 1% to 0.1% of distance travelled. The vehicle will need at some stages in the mission to Doppler navigate relative to the underside of the stationary ice sheet.

• The convergence zone, where the ice and the sea bed meet, the Autosub will need to detect when the "head room" falls below a minimum safe level, and react according.

Discussion and areas for improvement

AUV technology has developed greatly during the past 10 years, and there are now several operators with a good deal of experience, and who are able to carry out useful scientific as well as commercial missions, at very competitive costs compared to the alternative towed or ROV platform options. Using AUV technology, science data can be obtained which are of better quality and from areas (such as under ice, or near to boundaries) which are inaccessible by other means. This technology is very relevant to the exploration and exploitation of the ocean margin regions. So why are AUVs not being used extensively for research in the oceans margins? There are several reasons. Mostly are related to perceptions that are not necessarily correct.

• Operating costs are perceived as being too high. Although, when the full cost of running research ships are taken into account, AUVs appear already to be competitive with towed instruments (particularly in deep water), there needs to be a further quantum leap improvement in operating efficiency before scientist will be convinced to change technologies. Attention to the propulsion and hydrodynamic efficiency issues mentioned in this paper, could half energy costs per km, as well as double operating range.

• AUVs are too expensive. AUVs such as Autosub, MARTIN, HUGIN, very roughly cost around $1 million (US). This limits their availability to science. Costs must be reduced. Some of the key components (such as computer systems or INS navigation systems), are reducing in price. Some reduction will result from making the vehicles smaller, but the biggest reduction will be in economy scale of production of larger quantities, and when the manufactures have recouped the large development costs for expensive items such as pressure vessels.

• AUVs are considered to be at high risk of loss. This is a perception, which is not born out of any statistics. Although under-ice operations are clearly risky, other operations in the open ocean have not proved to be risky, with few AUVs reported missing. Between the main operators there have been over a thousand missions without loss. Because AUVs are generally run positively buoyant, and have additional emergency ballast drop systems, together with satellite (usually ARGOS) relocation systems, and are, barring the two exceptions of catastrophic leaks, or getting stuck in a moored net, at very low risk of total loss.

But perhaps one of the most important reasons that AUVs have not be widely adopted for scientific research can be blamed on the engineers and academics who have been developing AUVs in the past two decades. AUVs are perceived as "high tech", and "complex". Expressions such as "Artificial Intelligence" (AI), and "Advanced Robotics" are used in connection with AUVs. This perception has been propagated by some of the community developing AUVs, who naturally want their work to be seen as exciting and at the leading edge. Naturally the potential science users (who are not interested in technology for technologies sake) are scared away by this image. They perceive that AUVs are difficult to programme, need a team of highly expert operators, and are probably unreliable.

The reality is different. All operating AUVs use a mission control system which essentially operates exactly like a washing machine controller; with event and action pairs, e.g. an event "waypoint has been reached", is followed by an action such as "go to next waypoint" (drum is full of water → begin wash cycle). AUVs are not difficult to programme. The challenge is to further de-skill the operation of AUVs so that fewer, and lower cost operatives are needed. In the early days of Autosub six engineers were needed on field trails. This has been reduced to three engineers.

The other major challenge is in the integration of the science sensors and their data onto the vehicle, and the subsequent data processing. This is going to remain an expensive and time consuming process until the sensor manufactures agree on a standard (probably network based) electrical and communication interface for the sensors, so that the system integration of any new sensors on to a platform does not involve a lot of wasted effort in providing special bespoke interface software and hardware. This development is of course relevant to the integration of sensors on to any platform. The standards already exist (e.g. LonWorks, CAN), but we need to see them implemented in the marine sensor world. Then we as engineers can devote ourselves to more interesting and productive tasks than solving the same interfacing problems, again and again.

References

ANSI/EIA 709.1 (1999) Control Network Protocol Specification. ANSI/EIA 709.1-A-1999

Bjerrum A, Ishoy A (1995) AUV for surveys in coastal waters. Sea Techn 36(2):19-22

Bradley AM, Duester AR, Liberatore SP, Yoerger DR (2000) Extending the endurance of an operational scientific AUV using Lithium Ion batteries. Proc Unmanned Underwater Vehicle Showcase 2000, PGI Spearhead Ltd, pp 149 -158

Dhanak RD, An E, Holappa K, Hartmut P, Shay LK (2000) The role of AUVs as complimentary measurement platforms in physical oceanography experiments. Proc Unmanned Underwater Vehicle Showcase 2000, PGI Spearhead Ltd, pp 159 -167

Dunn P (2000) The navy unmanned undersea vehicle (UUV) master plan. Proc Unmanned Underwater Vehicle Showcase 2000, PGI Spearhead Ltd, pp 159 -167

Fernandes PG, Brierley AS, Simmonds EJ, Millard NW, McPhail SD, Armstrong F, Stevenson P, Squires M (2000) Fish do not avoid survey vessels. Nature 404: 35-36

Griffiths G (1999) Technology Needs for AUVs. EURO-MAR Conference of Technologies for Ocean and Coastal Survey, Brussels, November 1999

Griffiths G, McPhail SD, Rogers R, Meldrum DT (1998a) Leaving and returning to harbour with an autonomous underwater vehicle. Proc Oceanology International 98, Brighton, Spearhead Exhibitions Ltd, Kingston upon Thames

Griffiths G, Millard NW, McPhail SD, Stevenson P, Perrett JR, Pebody M, Webb AT, Meldrum DT and Russell D (1998b) Towards environmental monitoring with the Autosub autonomous underwater vehicle. Proc IEEE Underwater Technology 98, Tokyo April 98, pp 121-125

Griffiths G, Millard NW, McPhail SD, Stevenson P, Perrett JR, Pebody M, Webb A (1999) Open ocean operational experience with the Autosub-1 AUV. Proc Unmanned Untethered Submersible Technology Symposium, New Hampshire, USA. 23-25 August 1999

Griffiths G, Knap AH, Dickey T (2000a) The autonomous vehicle validation experiment. Sea Techn 41(2): 35-45

Griffiths G, Fernandes PG, Brierley AS, Voulgaris G, Millard NW, McPhail SD, Stevenson P, Perrett JR, Pebody M, Webb AT, Harris A (2000b) Unescorted science missions with the Autosub AUV in the North Sea. Proc International UUV Symposium, Newport, Rhode Island, April 2000

Hoerner SF (1965) Fluid-dynamic drag. 2nd ed. Brick Town, NJ: Hoerner Fluid Dynamics

Kukowski et al. (2002) Subsurface fluid flow and material transport. In: Wefer et al. (eds) Ocean Margin Systems. Springer, Berlin pp 295-306

McCartney BS, Collar PG (1990) Autonomous submersibles - instrument platforms of the future. Underwater Technology, 15(4), pp 19-25

McFarlane JR (1997) The AUV revolution: Tomorrow is today! Proc Underwater Technology International,

Aberdeen. Society for Underwater Technology, London, UK ISBN 0 906940 30 3, pp 323-336

McNeill A (2000) Survey requirements for oil & gas construction support. Proc Unmanned Underwater Vehicle Showcase 2000, PGI Spearhead Ltd, 2000. pp 11-18

McPhail SD (1993) Development of a simple navigation system for the Autosub Autonomous Underwater Vehicle. Proc Oceans 93: Engineering in Harmony with the Ocean, 1993, Victoria, British Columbia, Canada, New York: Institute of Electrical and Electronics. Vol.II pp 504-509

McPhail SD, Pebody M (1998) Navigation and control of an autonomous underwater vehicle using a distributed, networked, control architecture. Underwater Techn 23(1):19-30

Meldrum DT, Haddrell T (1994) GPS in autonomous underwater vehicles. Proc Sixth International Conference on Electronic Engineering in Oceanography, 19-21 July 1994, Cambridge, UK. Institution of Electrical Engineers, London, UK. pp 11-17

Millard NW, Griffiths G, Finnegan G, McPhail SD, Meldrum DT, Pebody M, Perrett JR, Stevenson P, Webb AT (1998) Versatile autonomous submersibles - the realising and testing of a practical vehicle. Underwater Techn 23(1): 7-17

Pind J (1999) Autonomous survey offshore Namibia. Proc. PING '99, Technical University of Denmark, September 1999. Unpaginated

Rogers et al. (2002) Life at the edge: Achieving prediction from environmental variability and biological variety. In: Wefer et al. (eds) Ocean Margin Systems. Springer, Berlin pp 387-404

SARA (1999). See http://www.tecnomare.it

Smith SM, Dunn SE, An E (1996) Data collection with multiple AUVs for coastal oceanography. Proc Oceanology International 96, Vol. 1. pp 263-280

Smith KL Jr, Glatts RC, Baldwin RJ, Uhlman AH, Horn RC, Reimers CE, Beaulieu SE (1997) An autonomous, bottom-transecting vehicle for making long time-series measurements of sediment community oxygen consumption to abyssal depths. Limnol. Oceanogr 42(7):1601-1612

Storkersen N, Indreeide A (1997) HUGIN - an untethered underwater vehicle system for cost-effective seabed surveying in deep waters. Proc Underwater Technology International, April 1997, Aberdeen, Society for Underwater Technology, London, ISBN 0 906940 30 3. pp 337-348

Thomas H, Petit E (1997) From Autonomous Underwater Vehicles (AUV's) to Supervised Underwater Vehicles (SUV's). OCEANS '97, MTS/IEEE, October 97, Halifax (Canada). pp 875-887

Thomsen et al. (2001) Margin building - regulating processes. In: Wefer et al. (eds) Ocean Margin Systems. Springer, Berlin pp 195-203

Vincent CA (1999) Lithium batteries. IEE Review 45(2). pp 65-68

Von Alt C (2000) News from the front - Why some UUVs are in demand. Proc Unmanned Underwater Vehicle Showcase 2000, PGI Spearhead Ltd. pp 133 -142

Physical Processes and Modelling at Ocean Margins

R. Neves[1*], H. Coelho[2], R. Taborda[3] and P. Pina[4]

[1] *MARETEC, Instituto Superiror Técnico, Lisbon, Portugal*
[2] *University of Algarve, Faro, Portugal*
[3] *Faculty of Sciences, University of Lisbon, Lisbon, Portugal*
[4] *MARETEC, Instituto Superior Técnico, Lisbon, Portugal*
* *corresponding author (e-mail): ramiro.neves@ist.utl.pt*

Abstract: Ocean margins are very productive areas and consequently they are interesting both for scientific and socio-economic reasons. Their economical importance was the main reason to support integrated projects to understand and quantify the processes responsible for high biological productivities, in order to create the scientific knowledge required for its management. For long time it was believed that high productivity of ocean margin areas was a consequence of the discharge of nutrients form the continents. As the scientific knowledge of the processes taking place on those areas increased, it was shown that biological productivity of the ocean margin is mainly a consequence of the complexity of the physical processes taking place on those areas and only in semi-enclosed areas (e.g. estuaries) a consequence of continental discharge. This conclusion has enhanced the importance of the development of integrated studies involving fieldwork and modelling. The complexity of physical problems taking place on ocean margins is a consequence of local depth gradients (e.g. continental slope and submarine canyons), but also of the wide range of forcing mechanisms driving the flow - wind, density and tides. The combination of these forcing mechanisms lead to an even more wide range of phenomena like, upwelling, fronts, internal waves, surface gravity waves, etc. To understand processes going on, process-oriented models can be used. However the final product for modelling processes in coastal areas must be an integrated model based on the primitive equations for mass and momentum. For management purposes this model has to couple physical and biological processes. In this paper a general modelling framework is described. This tool is developed to accommodate models for physics, biology and sediment transport. Numerical solutions, processes and results for the Iberian margin and for the Tagus Estuary (Portugal) are described.

Introduction

A typical ocean margin has a continental shelf about 200 m deep, and a steep slope (5 to 10%) down to the abyssal plain with depths of the order of 4 to 5 km. Although the continental shelf is the transition zone from the continent to the deep ocean, this does not mean that biological production on the shelf is directly related to the discharges from the continent. Exchanges between the shelf and the deep ocean due to the circulation pattern (horizontal and vertical) are the main source of nutrients for shelf areas.

There are several mechanisms that are candidates to promote this exchange. Coastal upwelling typical of subtropical eastern ocean boundaries generates cross slope exchange due to Ekman off-shore transport. Equatorwards directed wind forcing together with Coriolis force push surface water seaward, resulting in a lower sea level along the coast. This surface depression, together with vertical shear and with coriolis force generate a vertical distribution of velocity that creating an onshore subsurface transport bringing nutrients from the deep ocean into the shelf region, to the photic zone. Associated to coastal upwelling there are equatorward jets that often become unstable via barotropic and/or baroclinic instability generating eddies and

From WEFER G, BILLETT D, HEBBELN D, JØRGENSEN BB, SCHLÜTER M, VAN WEERING T (eds), 2002, *Ocean Margin Systems.* Springer-Verlag Berlin Heidelberg, pp 99-124

filaments that can transport considerable amounts of material across the shelf break. Other typical features of the ocean margins are along shore currents generated by pressure gradients. Some examples are the typical slope currents of the eastern ocean boundaries driven by the meridional pressure gradients (Huthnance 1984) and the currents originated by river plumes of fresh water. In both cases, the currents are topographically trapped but in the presence of bottom irregularities such as submarine canyons or capes they can produce cross slope exchange.

Tides acting on a stratified fluid generate internal tides, which are amplified on the slope, which can propagate across the shelf. Internal tides and the shorter internal waves propagating on the shelf generate vertical movement and mixing, bringing nutrients to the surface layers and enhancing primary production.

The importance of surface gravity waves generated by wind depends on the ratio between their height and local depth and on their frequency. Along the coast they are always important, generating coastal currents, which play a crucial role on beach sediment transport. The importance of currents generated by the waves decreases with depth. However even in deep areas they can induce instantaneous high frequency velocities that, added to lower local low frequency velocity, can create conditions for resuspending bottom sediments, which become available for transportation. This process is essential for geological studies, as it is important e.g. for reoxygenation of the upper layer of bottom sediments, increasing the mineralisation rate of sedimentary organic matter.

Classical ocean models use finite-difference methods and rigid laws to perform vertical discretisation. In finite-difference methods, partial differential equations are transformed into algebraic equations, replacing derivatives by differences between state variable values calculated at points of a space (or temporal) grid. On a Cartesian reference, the most convenient horizontal grid is rectangular. For large size models the most convenient coordinates are geographical and grid lines coincide with meridians and parallels. Finer grids generate in general the most accurate results.

In ocean systems the horizontal dimension is orders of magnitude larger than the vertical one and strong depth gradients can be found, which are generally at maximum along the ocean margins. Persistent features of ocean circulation are in general density gradients associated to vertical distributions of temperature and salinity. In traditional models the vertical coordinate is chosen according to the importance given to each of those aspects (topography or density). A common drawback of those models is that they cannot shift from one coordinate to another.

In sigma coordinates each layer occupies a constant percentage of the water column and the number of layers becomes independent of the local depth. This type of coordinate is adequate when the topography plays a major role in the circulation. In isopycnic coordinates, layers are coincident with isopycnic levels (levels of equal density). This type of coordinate is the most convenient when the flow follows isopycnic levels, which is the case when density is the major forcing mechanism. If topography plays a major role or vertical transport destroys the vertical gradients, this type of coordinate becomes inadequate. Cartesian coordinates do not attribute the major role to any of those mechanisms, being in fact a compromise between different types of coordinates, in models that do not allow more than one type. Some of the most emblematic examples for each category are: the Miami Isopycnic Coordinate Ocean Model - MICOM (Bleck et al. 1992); the cartesian Modular Ocean Model – MOM (Cox 1984); the sigma coordinate model Semi-spectral Primitive Equation ocean circulation Model - SPEM (Haidvogel et al. 1991); and the sigma coordinate Princeton Ocean Model - POM (Blumberg and Mellor 1987).

That there is no ideal solution for the vertical coordinate is clear from the results of the DYnamics of North Atlantic MOdels - DYNAMO - project (Barnard et al. 1997). The most satisfactory grid should always be oriented with the flow and so must be a compromise between the various available possibilities, depending on the physical processes that determine the flow in any particular region. Sigma models reveal strong topographically-determined currents, making these models the best choice whenever flow is constrained by depth contours. However, if the flow follows surfaces of constant density, as may be the case near the seasonal thermocline in periods of low turbulence intensity, sigma models can, numerically speaking, erode these surfaces, and in such instances isopycnal models are a better choice, in spite of the numerical difficulties associated with them. The shortcomings of sigma models in stratified regions can be reduced by a slight change in the conceptual formulation. The computational model may be divided vertically into two sigma models, separated by an interface placed at a level of nearly horizontal motion (Deleersnijder and Beckers 1992; Santos 1995). This is a compromise between Cartesian and sigma coordinates, often called *a double sigma coordinate model*. Generalising this concept to equal the number of sigma domains to the number of layers, a Cartesian model is achieved.

Finite-volume integral approaches introduced recently give more flexibility for the choice of the vertical coordinate (Martins et al. 1998; Mellor et al. subm.). Using finite-volumes, equations are solved in their integral form (the rate of accumulation of a property inside a volume is equal to the integral of the fluxes across its boundaries plus the production inside the volume). In this case, the vertical discretisation is limited only by the complexity of the integrals involved in the calculation. A great achievement of this method is the possibility of combining different classical discretisations in one simulation (e.g. Cartesian below the thermocline and sigma above).

The choice of the vertical coordinate is essentially a problem of the physics. It is very important to simulate the circulation and is also essential for the simulation of advection-diffusion on biogeochemical models. The vertical coordinate is irrelevant for the simulation of the biochemical reactions involved on marine environment processes, which depend on the local properties of the water only (concentrations, temperature, light, etc.). Most of these processes are site independent, occurring in coastal areas, on the shelf or in the deep areas. The complexity of the biochemical processes requires a large number of developers, which is easily achieved if a common model is used on all those areas. More than a model, the simulation of the marine environment needs a modelling framework

based on a modular approach, where modules are easily coupled.

This paper describes a modelling framework, a circulation module, the turbulence closure, a sediment transport model and some results.

Modelling framework

The need to understand the processes in the ocean soon conduced to the development of models. The starting points for physicists and for other disciplines were very different. Physicists knew the general equations for fluid dynamics, their problem being the incapacity to solve them without simplifications; other disciplines still had to look for empirical equations.

Before numerical calculation became possible, both physicists and ecologists had to use analytical procedures. Physicists started to develop models considering a subset of processes and simplified boundary conditions. Ecologists developed simple models of the type predator-prey and investigated on the factors affecting the rates of production and destruction of relevant properties. The advent of computers allowed the development of more complex models resulting in an increasing knowledge of the processes going on in the ocean environment.

Actual computers are powerful enough to develop integrated models coupling physics to other disciplines and are available to everyone wishing to develop modelling. This means that a new philosophy of modelling is being created. Modelling is becoming more and more a group task. It is becoming more and more difficult to a single modeller to know all the system features and all the parts of the code. Data to be entered into a model becomes more and more complex as well as the results of the models. Graphical interfaces are becoming essential accessories of models, for managing input/output complexity. Input and output modules also need to be sufficiently intelligent to allow the input file to hold only data required for the processes under simulation. As the complexity of models increases, the need of a modelling work-bench becomes clear.

A modelling workbench must include separated modules for tasks common to several modelling activities. Examples are input and output tools, grid processing tools, advection-diffusion. A graphical interface to enter data and a post-processing tool able to visualize results can save a lot of time to new users of the system. MOHID (Neves 1985; Santos 1995; Martins et al. 1998) and TELEMAC (Hervouet and van Haren 1996) were developed following this philosophy. After the development of a structured tool, it was very simple to develop a model for ground water flow, replacing the hydrodynamic module solving the shallow water equations by another one solving the Darcy equation.

MOHID was initially developed as a 2D depth integrated hydrodynamical model for tidal flows in coastal areas (Neves 1985). This model was afterwards extended to simulate free-surface waves using Boussinesq equations, 3D baroclinic flow (Santos 1995), sediment transport (Cancino and Neves 1998), ecology (Portela 1996) and lagrangian transport, and was applied in a variety of conditions (Cancino and Neves 1998; Taboada et al. 1998; Coelho et al. 1999; Martins et al. 1999; Miranda et al. 1999). From these applications, the need for a structured code and a versatile vertical discretisation became obvious. MOHID2000 was the answer found for this problem.

MOHID2000 is programmed using an object-oriented approach. Each task is performed by a different module, which manages its own data and processes. Objects to enter data into the model and to generate the output were developed using standard formats - ASCII for input and HDF (Hierarchical Data Format) for output. Major modules included in the modelling system process are: bathymetry, discretisation, advection-diffusion (Eulerian and Lagrangian), settling/erosion/deposition of particulate matter, discharges and ecology. Graphical interfaces to process input and output were also developed. Such a system is very flexible and allows quick development of models for other purposes. That was the case of a model for groundwater flow obtained replacing the very complex hydrodynamic module by a simpler module solving the Darcy equation. The rationale used to design the module structure was based on individuality and multi-purpose. Individuality is important to allow the system to be developed by a group, making the work of each member of the group as in-

dependent as possible. Multi-purpose is important for development and maintenance efficiency. Input, output, bathymetry, advection-diffusion and geometry are examples of modules providing basic services to higher-level modules.

Hydrodynamical model

Equations of a baroclinic 3D model assuming hydrostatic pressure and Cartesian coordinates can be written as (Leendertsee and Liu 1978; Santos 1995):

$$\frac{\partial u}{\partial t} + u\frac{\partial u}{\partial x} + v\frac{\partial u}{\partial y} + w\frac{\partial u}{\partial z} - fv =$$

$$-\frac{1}{\rho_r}\frac{\partial p}{\partial x} + \frac{\partial}{\partial x}(A_H\frac{\partial u}{\partial x}) + \frac{\partial}{\partial y}(A_H\frac{\partial u}{\partial y}) + \frac{\partial}{\partial z}(A_V\frac{\partial u}{\partial z})$$

$$\frac{\partial v}{\partial t} + u\frac{\partial v}{\partial x} + v\frac{\partial v}{\partial y} + w\frac{\partial v}{\partial z} + fu =$$

$$-\frac{1}{\rho_r}\frac{\partial p}{\partial y} + \frac{\partial}{\partial x}(A_H\frac{\partial v}{\partial x}) + \frac{\partial}{\partial y}(A_H\frac{\partial v}{\partial y}) + \frac{\partial}{\partial z}(A_V\frac{\partial v}{\partial z})$$

$$\frac{\partial p}{\partial z} + \rho g = 0$$

$$\frac{\partial u}{\partial x} + \frac{\partial v}{\partial y} + \frac{\partial w}{\partial z} = 0$$

$$\frac{\partial(S)}{\partial t} + \frac{\partial(uS)}{\partial x} + \frac{\partial(vS)}{\partial y} + \frac{\partial(wS)}{\partial z} =$$

$$\frac{\partial}{\partial x}(K_H\frac{\partial S}{\partial x}) + \frac{\partial}{\partial y}(K_H\frac{\partial S}{\partial y}) + \frac{\partial}{\partial z}(K_V\frac{\partial S}{\partial z})$$

$$\frac{\partial(T)}{\partial t} + \frac{\partial(uT)}{\partial x} + \frac{\partial(vT)}{\partial y} + \frac{\partial(wT)}{\partial z} =$$

$$\frac{\partial}{\partial x}(K_H\frac{\partial T}{\partial x}) + \frac{\partial}{\partial y}(K_H\frac{\partial T}{\partial y}) + \frac{\partial}{\partial z}(K_V\frac{\partial T}{\partial z})$$

Where u, v and w are the velocity components on space directions x, y and z respectively and p, S and T are the pressure, salinity and temperature. A_H and K_H are horizontal diffusivities and A_V and K_V vertical diffusivities and ρ is the water density related to the temperature (°C) and salinity by:

$\rho = (5890 + 38T - 0.375T^2 + 3S)/(1779.5 + 11.25T - 0.0745T^2 - (3.8 + 0.01T)S + 0.698(5890 + 38T - 0.375T^2 + 3S))$

A prognostic equation for sea surface elevation is obtained by vertical integration of the continuity equation over the entire water column:

$$\frac{\partial \eta}{\partial t} = -\frac{\partial}{\partial x}\int_{-h}^{\eta} u dz - \frac{\partial}{\partial y}\int_{-h}^{\eta} v dz = -\frac{\partial \overline{U}}{\partial x} - \frac{\partial \overline{V}}{\partial y}$$

Vertical integration of the hydrostatic pressure equation yields for pressure:

$$p(z) = p_{atm} + g\rho_r(\eta - z) + g\int_{z}^{\eta}\rho' dz$$

Where $\rho' = \rho - \rho_r$

This equation relates the pressure at depth z to the atmospheric pressure, the surface level and the vertical integral of density anomaly between that level and free surface. Two first terms are the barotropic and the last one is baroclinic.

The vertical coordinate

The choice of the vertical coordinate is put into different terms for finite-difference and finite-volume approaches. Finite-difference approaches look for approximations for spatial derivatives, which must be written on a predefined spatial reference. On the contrary, finite-volumes solve integral forms of the conservation principles. In this case fluxes across the faces of a finite-volume are computed knowing the values on both sides. In this case the shape of the volumes is limited only by the ability to compute the fluxes.

The finite-difference approach

The choice of the vertical coordinate in a circulation model is yet a matter of discussion in the marine modelling community. The ideal mesh should always be oriented with the flow in order to minimise numerical diffusion. The sigma (σ) type coordinates (Phillips 1957) transform the model domain into a constant depth domain and resolve the equations in that transformed domain, allowing the same number of grid points whatever is the local depth being adequate to solve problems where topography plays a major role.

In stratified flows isopycnals are nearly horizontal. Using a σ coordinate, along the shelf break, they are represented by sloping lines that can cross several layers. Another difficulty of the σ coordinates to deal with stratified flows is that the vertical resolution is linked to the local depth and thus can be too poor in areas where density gradients are strong. A double sigma coordinate does not completely solve these problems but can minimise them. In this type of coordinates (Deleersnijder and Beckers 1992; Santos 1995) the water column is split into two domains and in each of them a σ transformation is applied. The grid used in the upper ocean is linked to the local depth only if the bottom is above the plan splitting the two domains. Locating the splitting plan under the ther-mocline, layers in the upper domain - which are nearly horizontal - represent better the flow along isopycnals.

Cartesian and isopycnic coordinates are alternatives to the σ type coordinates. In the former grid lines are horizontal, while in the latter they are co-incident with isopycnals. Isopycnic coordinates are suitable to simulate flows where the density plays the major role, while Cartesian coordinates are a compromise between the 3 types, since they are not optimised for any process existing in the ocean. In general they need a big number of vertical layers and can become computationally expensive.

The finite-volume approach

The finite-volume approach solves the equations in their integral form:

$$\frac{\partial}{\partial t}\iiint_{CV}\beta dV = -\iint_{surface}\left(\beta \vec{v}.\vec{n} - v(\vec{\nabla}\beta)\cdot \vec{n}\right)dA + S$$

where β is the volumetric value of the property being calculated, v is the diffusivity and S represents the sources and sinks of the property. The geometry of the volume is limited for the complexity of the calculation of the surface integral.

Flexible implementations of this approach are getting computing distances in the real space (this is not the case of the coordinate transformations in finite-difference methods). Discretisations equivalent to Cartesian, σ or isopycnic coordinates are easily obtained if initial shapes of the volumes are drawn on those grids and if they are deformed following the rules intrinsic to those transformations. Other discretisations can also be used.

The big advantage of the finite-volume approach is that a single computer program is used for all the discretisations considered and conveniently organised, different coordinates can be used in different parts of the domain, according to the local conditions. This is the case for MOHID 2000. In estuarine applications sigma coordinates are mostly used, while in ocean applications Cartesian coordinates are generally more convenient.

The numerical algorithm

The numerical algorithm is independent of the technique used for spatial discretisation. In oceanic areas, stability limitations arise mainly from vertical transport and from the propagation of gravity waves. Time splitting methods are the most convenient to handle those limitations. In these methods to perform the calculation in a time step one or more intermediate time levels are considered. A set of processes (e.g. advection/diffusion) modifies the property values known at the beginning of the time step and then other processes present in the equation correct these estimated values to conclude time iteration. These methods are generally more stable than the explicit methods and allow the calculation of the different terms of the equation using different numerical schemes. Benqué et al. (1982) uses time splitting to allow the calculation of advection with the method of characteristics and the pressure with a classical Eulerian implicit algorithm.

In MOHID, a major goal of the splitting method is to obtain the solution using only tridiagonal matrixes with a time-centred Coriolis term (*fu* and *fv* in hydrodynamic equations) in order to increase the accuracy of its calculation. Several methods are used in vertical integrated models that can be easily extended to a 3D calculation (Leendertsee 1967, 1970; Abbot et al. 1968). The first method uses 6 finite-difference equations in each time step, while others use 4. The former can be more adequate to simulate intertidal areas, but the latter are more efficient in deeper zones if the coriolis term is time-centred. The Coriolis term is a non-derivative term and then its relative importance increases as the size of the modelling area increases.

In MOHID unknown velocities in the free surface finite-difference equation are eliminated using the corresponding momentum equation leading to tridiagonal matrixes. Knowing the new elevation in each ½ time step the corresponding momentum equation can be resolved with an implicit calculation for vertical transport, inverting again a tridiagonal matrix. In this way the most limiting stability factors, gravity wave propagation and vertical diffusion, are resolved implicitly. Courant numbers of 5 are typically used in MOHID applications.

Spatial discretisation is based on Arakawa C grid. In this grid, scalars are computed in the centre of the basic finite-volume, while velocities are computed on their faces. To compute velocities, secondary finite-volumes are defined based on scalar property volumes. Advective fluxes are calculated considering upstream values. Diffusive fluxes are computed considering central differences referenced to each volume face. For simplicity and accuracy reasons, the pressure is computed using the traditional barotropic and baroclinic components. The baroclinic term is integrated between the level of the velocity being computed and the free surface and the barotropic pressure is computed directly from the surface slope (per unit of volume).

Turbulence modelling

Previous work, simulating the diurnal cycle of temperature observed during the Long Term Upper Ocean Study (LOTUS) in the Sargasso Sea and the seasonal cycle of temperature off the Iberian coast, showed similar results produced by two different models, based on the Gaspar et al. (1990) one-equation turbulence closure and on the quasi-equilibrium version of the level 2.5 Mellor and Yamada closure scheme (Galperin et al. 1988), respectively. For simplicity, therefore, the one-equation closure scheme was adopted in MOHID 2000.

The vertical turbulence fluxes are parameterised using the turbulent viscosity/diffusivity concept:

Viscosity and diffusivities are related to length and velocity scales according to: $K_m = c_k \lambda_k E^{1/2}$ and $K_s = K_h = K_m/P_{rt}$ where c_k is a constant to be determined, λ_k is the mixing length, E is the turbulence kinetic energy (TKE), $E = 0.5(u'^2 + v'^2 + w'^2)$, and P_{rt} is the turbulent Prandtl number, assumed to be 1. To close the system TKE is determined from its balance equation:

$$\frac{\partial \overline{E}}{\partial t} = -\frac{\partial}{\partial z}\left(\overline{E'w'} + \frac{\overline{p'w'}}{\rho_0}\right) - \overline{u_H'w'}\frac{\partial \overline{U}_H}{\partial z} + \overline{b'w'} - \epsilon$$

where p being the pressure; e is the dissipation rate of TKE; b is the buoyancy, $b = g(\rho_0 - \rho)/\rho_0$, where g is gravity. The density ρ is determined by a state equation: $\rho = \rho_0[1 - \alpha(T - T_o) + \beta(S - S_o)]$ where 0 refers to a reference state and α, β are respectively the coefficients of thermal expansion and haline contraction calculated according to Bryan and Cox (1972). $\overline{X}$ denotes mean quantities and X' denotes fluctuations around the mean.

For the diffusivity of density, $K_\rho = K_m/P_{rt}$. The turbulent diffusivity concept is also used to parameterise the vertical flux of turbulent kinetic energy:

$$-\left(\overline{E'w'} + \frac{\overline{p'w'}}{\rho_0}\right) = K_e \frac{\partial \overline{E}}{\partial z}$$

with the usual assumption $K_e = K_m$. The dissipation rate is parameterised as follows: $\epsilon = c_\epsilon E^{3/2}/\lambda_\epsilon$, c_ϵ being a constant to be determined and λ_ϵ the length scale for dissipation.

A difficulty of models that parameterise the turbulent viscosity based on the velocity and length scales is the determination of such scales, especially the length scale. In this model, very simple definitions of the length scales are used, avoiding a large number of coefficients and leading to very reasonable results as were obtained by Bougeault and

Lacarrère (1989). The mixing length definitions are $\lambda_k = \min(l_u, l_d)$ and $\lambda_\epsilon = (l_u l_d)^{1/2}$, λ_k and λ_ϵ being the length scales for mixing and dissipation respectively; l_u (upward) and l_d (downward) are obtained according to Bougeault and André (1986).

$$\frac{g}{\rho_0}\int_z^{z+l_u}[\overline{\rho}(z) - \overline{\rho}(z')]dz' = E(z),$$

$$\frac{g}{\rho_0}\int_{z-l_d}^z[\overline{\rho}(z) - \overline{\rho}(z')]dz' = E(z)$$

Two constants are to be determined, c_k and c_ϵ. The determination of the constants is part of the model calibration. However, based on laboratory experiments, Bougeault and Lacarrère (1989) deduced that $c_\epsilon = 0.7$ is an adequate value for simulations. The choice of c_k is more difficult to justify from observations. Based on the definition of the mixing efficiency coefficient, $\gamma = R_f/(1-R_f)$, where

$$R_f \equiv \overline{b'w'}/(\overline{u'w'}\partial \overline{U}_H/\partial z)$$

is the flux Richardson number, it is possible to deduce that $c_k = 0.15\,c_\epsilon$ (for details see Gaspar et al. 1990).

To avoid unrealistically small diffusion and dissipation rates in the pycnocline, Gaspar et al. (1990) suggested that a minimal value E_{min} for TKE should be imposed. To match the results of Gargett (1984) E_{min} is set equal to 10^{-6} m^{2}s^{-2}. This represents only a simple solution to obtain realistic diffusion rates in the thermocline. Gaspar et al. (1990) suggested that better results could probably be obtained by parameterising E_{min} as a function of internal wave activity and surface forcing.

Sediment transport modelling

The continental shelf is a highly dynamic environment, where the flow induced by waves and currents extends to the bottom inducing marine sediment transport. The phenomena of sediment transport in combined wave-current conditions are complex, and not yet fully understood, largely due to the nonlinear interaction between the flow, bed micromorphology and the moving sediment. For-tunately, substantially progress has been made dur-ing the last years concerning the understanding of all the aforementioned aspects of the problem. Excellent recent reviews are available, including those by Fredsoe and Deigaard (1992), Nielsen (1992), Van Rijn (1993), Nittrouer and Wright (1994) and Soulsby (1997).

Two different approaches have been used for predicting sediment transport rate, the so-called energetic method, and the 'process' models. The energetic models, which relate the transport rate to the turbulent energy dissipation (e.g. Bailard 1981), are very popular amongst morphological modelers due to their simplicity and ease of use. However, in this type of models the whole physics are incorporated in the coefficient of proportion-ality between dissipation rate and sediment trans-port, which can seriously affect their range of ap-plication, so they will not be further discussed in this paper.

Process models involve solving transport equa-tions for momentum and sediment concentration are subject to appropriate boundary conditions. Differences between these models lie in the as-sumptions made for eddy viscosity and bottom sediment concentration, which will be discussed in more detail.

Wave-current boundary layers

On the continental shelf, wind generated waves are generally responsible for the existence of an os-cillatory boundary layer of centimetre scale em-bedded in a much thicker (usually of metre scale) boundary layer driven by winds or tidal currents. These two boundary layers interact non-linearly, enhancing both mean and oscillatory bottom shear stresses.

There are numerous models to describe the bottom boundary layer in combined flows. For ex-ample, in the scope of the MAST G6M Group (Soulsby et al. 1993) a list with 21 models has been compiled, and after this some more have been developed increasing the available list. These models can be divided into five major groups de-pending on the turbulent closure scheme used: time-invariant eddy viscosity models (e.g. Smith

1977; Grant and Madsen 1979, 1986), vertically integrated models (e.g. Fredsoe 1984), mixing length models, one- and two- equation turbulence models (e.g. Davies et al. 1988; Justesen 1988) and Reynolds stress equation models. In spite of the widely differing formulations, Soulsby et al. (1993) in an intercomparison of eight typical wave-current boundary layer models, showed that the general forms of their prediction of mean and maximum bed shear-stress were broadly similar. These results justify the use of the simpler mod-els, like the time-invariant eddy viscosity model of Grant and Madsen (1979, 1986) and the vertically integrated model of Fredsoe (1984), to predict large-scale sediment transport on the con-tinental shelf.

Bed micromorphology

One of the most remarkable features of sediment transport over a non-cohesive sediment bottom is the development of bed geometric shapes in a much larger scale than the sediment particles in an amazing variety of shapes and patterns. These bedforms have a decisive influence on the struc-ture of the bottom boundary layer, the near bed turbulence and, consequently, on sediment trans-port. However, in spite of the unquestionable im-portance or these features, especially in a wave dominated shelf where their presence is almost ubiquitous, their generation mechanism is still poorly understood, as can be seen from the differ-ent approximations for predicting their geometry. The most widely used models for predicting ripple geometry due to waves are those of Nielsen (1981), Grant and Madsen (1982) and Wiberg and Harris (1994). While the Grant and Madsen (1982) meth-ods tends to strongly overpredict field ripple roughness (Li et al. 1996), those of Nielsen (1991) and Wiberg and Harris (1994) generally give more realistic results, although the average error is commonly greater than 100% (Taborda and Dias 2000).

Under combined waves and currents, most authors usually use the wave-ripples predictors solution in a wave-dominated case and a current alone solution for current dominated cases. Re-cently, Li and Amos (1998) have found that this methodology can lead to large errors and have proposed a new ripple predictor for combined flows.

Ongoing investigations, many of which are being carried out in the scope of the EU MAST program, are focusing on better predictions and parameterizations of bed micromorphology.

Sediment resuspension and transport

Sediment transport within the bottom boundary layer takes place in two modes: suspended load and bed load.

Bedload

Bed load, which involves rolling, sliding and jumping (saltation) of grains along the bed, is the dominant mode of transport for low flow rates and/or large grains. In this mode of transport the particles are supported by intergranular forces as opposed to suspended load where particles are supported by the upward fluid motion.

Several empirical formulae have been proposed to compute bed load transport, being most of them expressed in the form $\Phi = f(\theta, \theta_{cr})$, where Φ is the dimensionless bedload transport rate, θ is the Shields parameter and θ_{cr} the critical value required for sediment movement. One of the most used formulae is that of Meyer-Peter and Muller, originally developed from data obtained in rivers and channels, given by:

$$\Phi = 8(\theta - \theta_{cr})^{3/2}$$

Madsen (1991), using a conceptual mechanics-based model for sediment transport processes in steady and unsteady turbulent boundary layer flows, supported the use of a generalized Meyer-Peter and Muller bed load transport relationship for the combined action of waves and currents in the coastal environment.

Suspended load

In a combined flow, sediment is suspended within the wave boundary layer, and diffused further up into the flow by the turbulence associated with the current. Typical wave-current concentration profiles can be divided in two main parts: very close to the bed, in the wave boundary layer, turbulence is mainly originated by wave oscillatory motion and the concentration profile is similar to the pure wave case, while further up the concentration profile is dominated by current related processes. Following Glenn and Grant (1987), the vertical distribution of time-average suspended sediment under combined waves and currents can be computed by the following Rouse–type equations, were the suspended sediment concentration $(c(z))$ is predicted in terms of the reference concentration (c_R) at elevation (z_R) by:

$$c(z) = c_R \left(\frac{z}{z_R} \right)^{-b_m} \quad for \ z < \delta_{cw}$$

$$c(z) = c_{\delta_w} \left(\frac{z}{\delta_w} \right)^{-b_c} \quad for \ z > \delta_{cw}$$

where b (the Rouse number) is:

$$b_m = \frac{\beta w_s}{\kappa u_{*m}} \quad for \ z < \delta_{cw}$$

$$b_c = \frac{\beta w_s}{\kappa u_{*c}} \quad for \ z > \delta_{cw}$$

and u_{*m} and u_{*c} are the maximum and mean shear velocity in a wave cycle, respectively, c_{δ_w} is the concentration at the top of the wave boundary layer (δ_w), computed from the first equation, and w_s is the particle-settling velocity. The value of the β coefficient, the ratio of sediment diffusivity to eddy viscosity, is object of some controversy. While some authors argue that in a turbulent flow the particles can not fully follow the turbulent motion, which implies a value less than one, others claim that β should be greater than one due to the centrifugal effects in the eddies. In the literature values between 0.1 and 10 can be found. The observed discrepancies might be related to the attempt to describe all the concentration profiles exclusively by diffusive processes neglecting the convective terms (see Nielsen 1991 and Deigaard 1991 for a discussion). As the behaviour of β is still very poorly understood, a value of $\beta = 1$ is probably the safest for many purposes (Dyer and Soulsby 1988), thus, assuming that the eddy viscosity and sediment diffusivity can be freely interchanged.

With the knowledge of the particle-settling velocity the only missing parameter to solve this equation is the reference concentration. In the literature there is a large variety of relationships to compute the reference concentration from the characteristics of flow and sediment properties, being most part of the form (Garcia and Parker 1991):

$$\overline{c}_R \propto \tau'^P \propto u_*'^{2P}$$

where τ' is the skin friction shear stress and μ_*' is the related shear velocity. For the P exponent values between 1 and 15 have been suggested, which express the high uncertainty related with the determination of the near-bottom reference concentration. A major problem concerning the reference concentration is the specification of a reference height. Smith and McLean (1977) have identified z_r with the top of the saltation layer, some authors assumed that the reference level is proportional to the grain diameter (Engelund and Fredsoe 1976; Madsen 1991) while others have used a constant height above the bottom (Vincent and Green 1990). This last approach is generally used in the field experiments due to practical constrains. In the continental shelf scope most authors have used a form of the expression originally developed by Smith and McLean (1977), given by:

$$C_0 = C_b \frac{\gamma_0 \theta'}{1 + \gamma_0 \theta'}$$

where C_b is the volume sediment concentration in the bed, θ' a normalized excess shear stress defined as $(\tau'-\tau_{cr})/\tau_{cr}$ and γ_0 is the resuspension parameter, representing the relative efficiency of sand resuspension. For low values of θ', this expression can be reduced to $C_0 = C_b \gamma_0 \theta'$. In their original work, Smith and McLean (1977) have proposed a value of 24×10^{-4} for the resuspension parameter, and since then this value has been involved in some controversy. For example, in a tidal flow over a rippled sand bed Dyer (1980) found a value of $\gamma_0 = 0.78 \times 10^{-4}$, while the flume experiments of Hill et al. (1988) suggested a value of $\gamma_0 = 1.3 \times 10^{-4}$. More

recent studies by Drake and Cacchione (1992) and Vincent and Downing (1994) conflicts with the previous concept of constant γ_0, indicating that ripple roughness, bed armoring and down-core increase of sediment cohesion can significantly affect the sediment resuspension coefficient. The values of γ_0 from these studies differ more than one order of magnitude, though they both show a systematic decrease in γ_0 with the increase of excess of shear stress. Another source of uncertainty is related with real shelf sediment characteristics. In fact, bottom sediments are generally composed of particles with different grain sizes and different compositions, including cohesive fine ones, and are frequently modified by biologic activity, which may deviate the observed values considerably from the theoretical ones. To overcome this problem, a wide suite of tools for monitoring sediment transport *in situ* has been developed in the last years, which have proven very successful. This new instrumentation has given new insights into sediment transport processes enabling the development of a new generation of process-based models.

Integrated modelling at estuarine scale

The Tagus estuary is one of the widest estuaries at the west coast of Europe and the largest in Portugal, covering almost 320 km^2 (Fig. 1). The Portuguese capital, Lisbon, is the most important city built on its banks.

The metropolitan area has nowadays around 2 million inhabitants, an important harbour and big industrial complexes around the estuary.

The estuary is a mixing place of river and oceanic waters. The salinity distribution depends mostly on the river flow and on the mixing imposed by the tidal regime, which is the main mechanism controlling the distribution of aquatic organisms and suspended particulate matter in the estuary. In ecological terms, it works as a nursery for several species.

Hydrodynamics can be seen as the first driving mechanisms of a cascade of complex processes. The water flow is responsible for transporting chemical (e.g. ammonia), biological (e.g. phytoplankton) and geological (e.g. sediments) materials in the water column. It is also responsible for the sediment fluxes between the bottom and the water column. The hydrodynamic model was forced only with tide because the main goal is the study of salt marshes and intertidal areas where tide is the main forcing mechanism. The cohesive-sediment model uses shear stress computed by the hydrodynamic model to quantify bottom fluxes. The sediment concentration strongly interacts with water quality related processes. The light extinction factor that regulates the amount of light that primary producers receive, is sensible to sediment concentration, causing low production rates in high turbidity areas (Portela 1996).

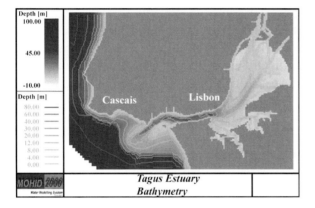

Fig. 1. Tagus bathymetry.

Hydrodynamic processes

Residual velocities presented in Fig. 2 (surface values) were obtained through time integration of transient velocities. Residual velocities do not usually provide much direct information but they can be helpful to understand long-term phenomena with time scales much larger than the tidal period. There is a jet outward the estuary associated with a strong anticyclonic eddy off Cascais; a cyclone and an anticyclone inside the channel reveal a very complex hydrodynamic system coupled with the topography.

This figure shows the Cascais' bay periodic anticyclone (it appears during ebb time) and the outward jet, the maximum velocity occurs in the channel. These features have a strong influence in

the bathing of the coastal area of Cascais; because of this gyre the estuarine ebb water weakly affects the area. Model results (and other field studies) strongly suggest that water quality in this area depends first of all on the proper control of local pollution sources.

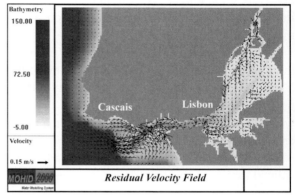

Fig. 2. Tagus Estuary surface residual velocity field.

Cohesive sediments processes

Cohesive sediment transport is simulated solving the 3D-advection-diffusion equation in the same sigma-grid used by the 3D hydrodynamic model using finite volumes for spatial discretization. Horizontal transport is solved explicitly, while vertical transport (including settling) is solved implicitly for numerical stability reasons.

The erosion algorithm is based on the classical approach of Partheniades (1965). Vertical sediment transport between layers is due to vertical diffusion, vertical advection and sediment settling. The hydrodynamic model computes diffusivity and vertical velocity. Settling velocity depends on flocculation processes and is calculated as a function of the concentration (Dyer 1986). Deposition is modeled as proposed by Krone (1962) and modified by Odd (1986).

The sediment properties used by the model are those of fine (or cohesive) sediments (particle diameter less than 64 μm) found in the literature for the Tagus estuary. The total mass of suspended sediments can change only due to fluxes across the estuarine boundaries (open boundaries and bottom) and a zero flux condition is used at the free surface. The fluxes across the river boundaries are imposed using field data. In the ocean boundary a constant value is imposed.

Sediment transport plays an important role in water quality. Firstly, the crucial role that suspended sediments impart to the attenuation of the available photosynthetically useful radiant energy. Secondly, contaminants and nutrients are generally transported along with the sediments upon which they are adsorbed.

Two example cases are presented to illustrate the kind of results obtained: the evaluation of the importance of the seasonal river variability and the potential consequences of sea-level rise. To do so we have computed the difference between a reference situation and the two scenarios.

In what concerns the first case, it must be considered that the river input depends essentially on the policy of management of the river basin. An increase of the agriculture activity, without any modification of the agricultural techniques, increases, in general, soil erosion and, therefore, sediment input to the estuary. Climate changes are expected to increase storm strength and, consequently, erosion. On the contrary, a forestry increase is expected to reduce the sediment discharge.

In Figure 3a, one can see strong modifications of the residual fluxes and sediment concentration, mainly in the upper part of the estuary due to a strong reduction of the river input of sediments. These results confirm the observations made by Vale and Sundby (1987) and Vale et al. (1993), about the importance of the river input in the dynamical process of sediment transport at the Tagus estuary. Those kinds of results may help the local authorities to better manage the system, at least in what concerns the parameters that depend on the human activity.

The other aspect presented is the effect of sea level rise. This problem is being object of a great concern mainly during the last decade. According to most climate change models, a rise of mean sea level is expected in the future. Some predictions point to differences of one meter in certain locations. This value is probably too pessimistic, but it was chosen for our simulation. Being an extreme value it also gives a clear insight of its importance.

Results show that the effects in the estuary will be different according to the regions but, for instance, one of the consequences will be an increase of the erosion processes with direct impacts in the salt marsh areas (e.g. Fig. 3b).

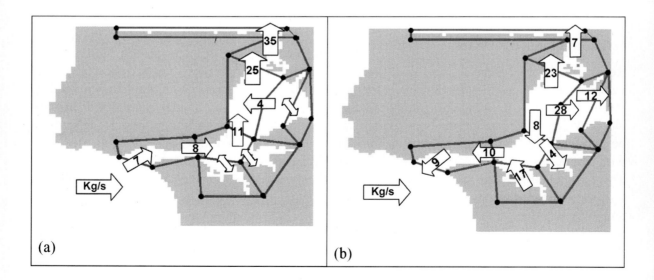

Fig. 3. Residual sediment fluxes. Differences between the reference situation and a situation with no river input **a)** and a scenario considering a sea level rise of 1 m **b)**.

Water quality processes

The water quality module has been developed in terms of sinks and sources. Such an approach is convenient to give these models the desired flexibility, providing it with the capability of being coupled to either a Lagrangian or an Eulerian resolution method. Because of the properties' interdependency, a linear equation system is computed for each control volume and this system can be computed forward or backward in time.

The simulation of the water quality processes is developed with the following considerations: Autotrophic producers consume inorganic nutrients and depend on both their availability and sunlight as a source of energy for photosynthesis. Nitrate and ammonia are the inorganic nitrogen forms that primary producers consume. The primary and secondary producer's excretions are considered to fuel the nitrogen cycle. Primary producers are consumed by secondary producers, which in turn are consumed by higher trophic levels.

The following results show time series comparisons between model and field data from the Tagus Field Station 3.5 (Fig. 4) for four consecutive years: 1980, 1981, 1982, 1983 (Silva et al.1986).

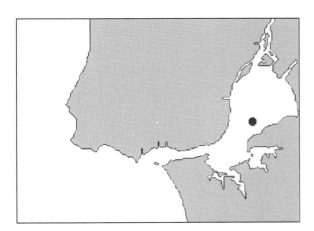

Fig. 4. Field station 3.5 location in the Tagus estuary.

The model results show a higher phytoplankton production in June (Fig. 5), caused by nutrient availability and increased sun radiation. After the bloom the phytoplankton concentration is controlled essentially by strong zooplankton growth (not represented). Ammonia (Fig. 6) and nitrate (Fig. 7) are

consumed during the phytoplankton peak, afterwards ammonia increases duo to zoo- and phytoplankton respiration and excretion loses and nitrate increases duo to nitrification processes.

The next results show the spatial distribution of phytoplankton, nitrate and ammonia during the summer period (7, June 1999). Fig. 8 shows a high concentration of phytoplankton in the upper part of the estuary especially in the salt marsh region. Due to the low water level (more light available)

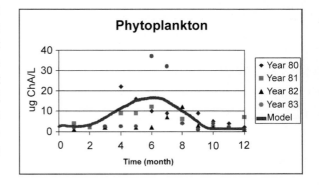

Fig. 5. Phytoplankton variation over a year.

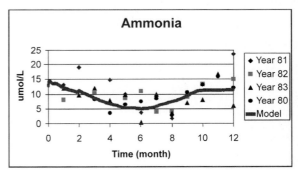

Fig. 6. Ammonia variation over a year.

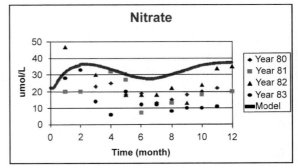

Fig. 7. Nitrate variation over a year.

and high nutrient concentration this region will have an intense production. The assimilation by phytoplankton preferably towards ammonia causes a strong depletion of nitrogen especially in the higher production areas (Fig. 9 and Fig. 10).

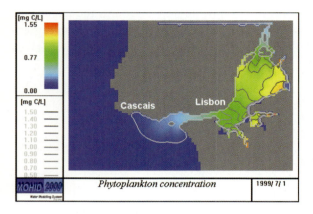

Fig. 8. Phytoplankton distribution at the Tagus Estuary.

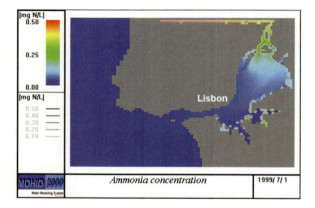

Fig. 9. Ammonia distribution at the Tagus Estuary.

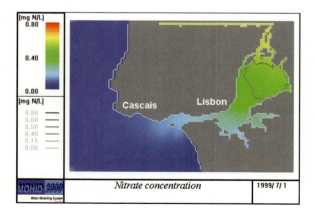

Fig. 10. Nitrate distribution at the Tagus Estuary.

The next pictures show the time and spatial integrated fluxes of phytoplankton, nitrate and ammonia over a year in the Tagus estuary. In every case the estuary is exporting to the ocean. The phytoplankton fluxes (Fig. 11) show small river input, all the production occurs inside the estuary and is afterwards exported to the ocean. The ammonia fluxes (Fig. 12) show the estuary exporting less then it receives from river input. This can be explained by the fact that respiration and excretion loses being smaller then consumption by phytoplankton. The estuary exports more nitrate to the ocean then it receives from river input (Fig. 13). This means that the source term of nitrate, nitrification, is higher then the sink terms, denitrification and assimilation by phytoplankton. These results are influenced by the fact that the pelagic mineralization process was increased because benthic mineralization was neglected for simplicity. This fact gives an unrealistic mobility to remineralized nitrogen that could explain a higher output flux of nitrate.

Circulation at the Iberian margin

Many authors have provided evidence for a poleward flow along the West European slopes (Ambar 1985; Ambar et al. 1986; Huthnance 1986; Frouin et al. 1990; Haynes and Barton 1990; Pingree and LeCann 1989). Very similar poleward flows have been described in other eastern boundary regions such as the California current system (Lynn and Simpson 1987) and the Leeuwin current at the West Coast of Australia. These flows, mainly concentrated along the upper continental slope and outer continental shelf, appear as undercurrents in the upwelling season and sometimes as surface currents in the non-upwelling season. Barton (1989) suggested that the poleward flow is continuous along the entire eastern boundary and attributed a key role to the Iberian poleward flow in the transport of Mediterranean water ultimately into the Norwegian Sea.

Frouin et. al. (1990) described a flow 200 meters deep with geostrophic velocities ranging from 0.2 to 0.3 m s^{-1} and associated transports varying from 300 x 10^3 m^3 s^{-1} at about 38°N to 500 – 700 x 10^3 m^3 s^{-1} at about 41°N. They concluded that the

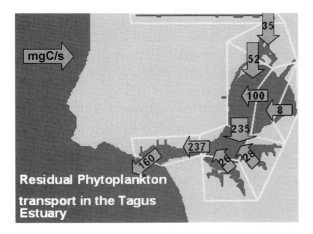

Fig. 11. Residual phytoplankton transport.

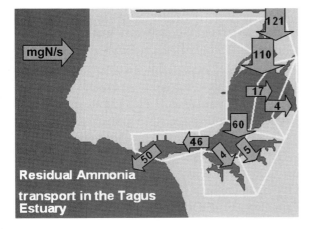

Fig. 12. Residual ammonia transport.

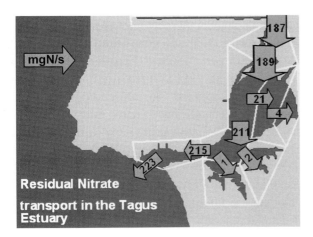

Fig. 13. Residual nitrate transport.

poleward current off the Iberian peninsula runs for about 1500 km along the upper continental slope of western Portugal, north-west Spain, northern France and south-west France and that it is about 25 to 40 km wide. Ambar (1984) suggested that the current extends from 1600 m depth to the bottom of the surface layer during the upwelling season and to the surface during the non-upwelling season. Pingree and Le Cann (1989, 1990) summarised current meter data collected from the Bay of Biscay and presented a residual circulation pattern. Further mention of their work can be found in the section where the model results are discussed. During the last twenty years, several driving models to explain slope currents have been put forward. Of these, the most-frequently-studied have been wind-stress (Ambar et al. 1986), wind-stress curl (McCreary et al. 1987) and thermohaline forcing (Huthnance 1984; Haynes and Barton 1990; Frouin et al. 1990; Lynn and Simpson 1987; McCreary et al. 1986; Weaver and Middleton 1990). In fact, off the Iberian coast, onshore Ekman convergence induced by south-south-westerly winds forces a poleward surface flow. The shelfward transport induced by these winds causes a rising of sea level near the coast. The geostrophic adjustment to this sea level distribution will then generate a poleward current. In this case, the longshore acceleration is given by $\delta V/\delta t = \tau_y/\rho H$, where H is the depth of the frictional layer, V the longshore velocity averaged over the depth H, τ_y the meridional component of wind-stress and ρ the seawater density. Using $\tau = 0.03$ Nm^{-2}, $\rho = 1027$ kgm^{-3} and $H = 200$ m, Frouin et al. (1990) found a longshore acceleration of 0.013 m s^{-1} d^{-1}, which gives $V = 0.4$ m s^{-1} after 30 days. However, the authors argued that other effects, particularly friction, retarded the flow. Assuming a steady state, reached when the bottom stress balances the wind stress ($C_d V^2 = \tau^y/\rho$, being C_d the bottom drag coefficient taken equal to 0.001), they obtained $V \cong 0.17$ m s^{-1} which is in agreement with observations. This current should decay seaward from the shelf break and the associated spatial scale is the internal radius of deformation ($\cong 15$ km off Iberia). This is what is generally observed both from satellite images and from *in situ* observations. Evaluation of the Ekman volume transports based on windstress measured at Cabo Carvoeiro (west of the Portuguese Coast, 39° N) revealed that only 1/5 of the estimated transport could be explained by the wind, which therefore could not be regarded as the main mechanism driving the poleward current. On the other hand, estimates of large-scale geostrophic eastward transport gives $1.0 \ m^2 s^{-1}$ per meter of meridional coastline. A value of the same order as that estimated from hydrographic sections can be calculated by integrating along the western Iberian coast and adding wind-driven transport. The poleward cooling of the sea surface leads to a meridional increase of surface density causing the dynamic height to drop towards the pole. The large scale eastward flow is generated by this meridional pressure gradient and occurs in the upper 200-300 m. Near the eastern ocean boundary, this flow forces coastal downwelling and a surface poleward current, as confirmed by model results obtained by McCreary et al. (1986) and Weaver and Middleton (1990) for the Leeuwin Current. Huthnance (1984) showed that a combination of shelf-slope bathymetry with a northward density gradient provides a local mechanism that can drive a current towards the pole, as can be expressed by the relation $\rho \delta \eta / \delta y = -h \delta \rho / \delta y$ where η is the sea surface elevation and h the water depth (see also Pingree and Le Cann 1989). This relation states that sea level decline is proportional to depth h. Therefore sea level declines faster in deep water than in shallow water, so implying a cross-slope sea level gradient. The existence of this gradient leads to a poleward flow over the slope. The cross-slope sea level gradient increases northward and so consequently does the along-slope transport, but this is not a situation that can continue, since friction acts to balance the forcing mechanism (Pingree and Le Cann 1990). Huthnance also showed that if the cross-shelf density diffusion is large, the along-slope current is given by:

$$v = \frac{1}{2} \frac{g}{\rho} \frac{\partial \rho}{\partial y} \frac{H}{k} h \left(1 - \frac{h}{H}\right)$$

where H is the oceanic thermal depth and k the bottom friction coefficient. According to this equation maximum velocity must be expected over the slope.

The model domain encompasses the west coasts of Iberia and Morocco, extending from 32°N to 46°N and from 6° to 16°W. The horizontal grid spacing is 8.5 km in both directions. Bottom topog-

raphy was derived from ETOPO5 by means of an interpolation for the model grid followed by a smoothing using a five point Laplacian filter. The bottom depth is then determined, using shaved cells (Visbeck et al. 1997). The model uses 18 vertical layers centred at constant z-levels at depths of 5,20, 45, 80, 130, 200, 290, 400, 530, 680, 850, 1040, 1250, 1480, 1750, 2200, 3000, 4250 m. The western, southern and northern boundaries are open while the eastern boundary is open only at the Strait of Gibraltar.

Lateral heat and momentum diffusion coefficients are 50 and $300 \ m^2 s^{-1}$, respectively. On the open boundaries we use the previously referred conditions except at the Strait of Gibraltar where salinity, temperature and transports are imposed.

The model is initialised with climatological temperature and salinity fields, horizontal sea level and zero velocity. The climatological temperature and salinity fields are extracted from Levitus and Boyer (1994) and Levitus et al. (1994) and are interpolated to the model grid and then smoothed using a simple cubic spline algorithm. In the upper 500 m, objectively analysed mean monthly temperature and salinity fields were used, estimated from CTD/XBT data supplied by the British Oceanographic Data Centre (BODC). This procedure provides a more detailed density field very useful to describe the distribution of the meridional density gradient. The spin-up phase consists of a 6 month run using surface climatological momentum fluxes derived from the near surface analyses of the European Centre for Medium-Range Weather Forecasts ECMWF (Trenberth et al. 1990). Surface temperature and salinity are relaxed to climatological data during the spin-up phase. After this period the model is run for 1 year using daily heat, mass and momentum fluxes from the ECWMF large scale forecast model for the year of 1994. The spatial resolution of the ECWMF fluxes was 0.5° by 0.5°. The data are interpolated spatially for the model grid and temporally for the model time step.

The results were compared with available data and previous work concerned with the circulation in the area. The model was able to reproduce the general patterns of the circulation as well as the seasonal variability (Figs. 14 and 15 show the velocity fields in winter and summer for the OMEX

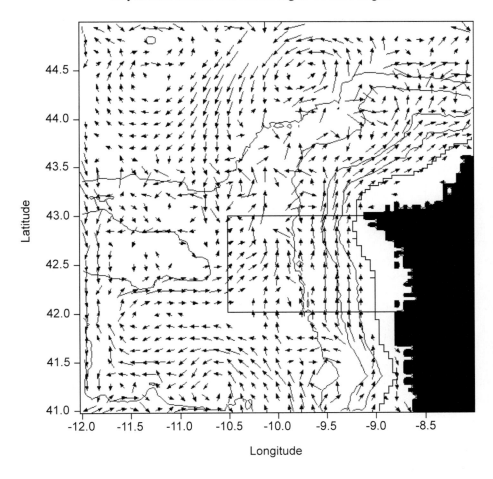

Fig. 14. Velocity field at layer 5 (130 m) on 15 January 1994. Maximum velocity plotted is 30 cm/s and minimum is 2 cm/s. Depth contours represented are for 200 m, 500 m, 1000 m, 2000 m and 3000 m. The overlaid box represents the OMEX box.

(Ocean Margin Exchange is a MASTII/III project funded by EU DG Research) study area.

The transports were predominantly along slope especially in the OMEX box (OMEX box is the area off Galicia between 41°30'N and 43°30'N and 11°W). However the total amount of water exported from the shelf/slope to the deep ocean along the west coast of Iberia was relatively high (2 to 4 Sv). The exchange seems to have preferential locations since most of the cross-slope transport occurred between 38°N and 40°N (the role of the canyons of Nazaré and Setubal is not focused here but it is probably very important and should be subject of detailed studies in the future). This emphasizes the need to look at the fluxes for the whole Iberian margin, instead of considering only the OMEX box. Finally we should point out that fila-

ments of cold water during the upwelling season might contribute significantly to cross slope exchange. However, to generate filaments in the model we need a very high resolution – 2 km. We also need to cover a very large area to obtain a good description of the large-scale circulation and this is not compatible with the resolution needed to simulate filaments.

Two other major findings of this modelling work were: 1) The integrated transport in the upper 1500 m between 10.5°W and the coast was always poleward for the forcing conditions considered in this study. 2) The transport decreased to the north and the decline seemed very well correlated with topography. This finding seems to be supported by current meter data and by previous studies (Mazé et al. 1997). Our results indicate

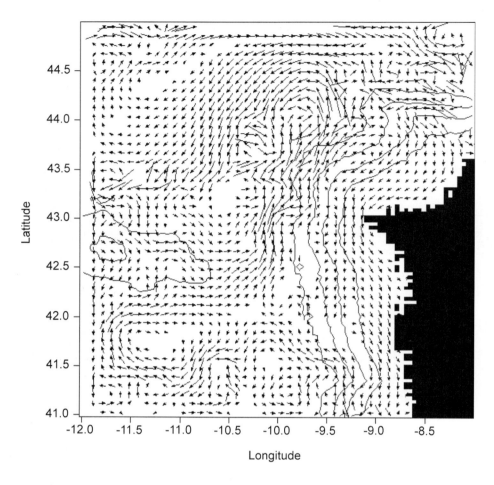

Fig. 15. Velocity field at layer 2 (20 m) on 15 July 1994. Maximum velocity plotted is 20 cm/s and minimum is 2 cm/s. Depth contours represented are for 200 m, 500 m, 1000 m, 2000 m and 3000 m. Note the equatorward jet over the shelf associated with upwelling favourable winds and the poleward flow further offshore.

that the surface poleward current over the slope was a permanent feature at least for 1994. However, this was a year with a low upwelling index in summer months and may be viewed as an anomalous year. It is tempting to state that: 1) the poleward surface current is always there as long as the meridional component of windstress is not strong enough to revert the flow and 2) The reversal of the flow occurs first over the shelf and may not occur over the slope.

To have a more realistic description of the circulation and fluxes off the Iberian peninsula, we need to improve the horizontal resolution to simulate filaments. Other possible improvements are: 1) to consider river runoff; 2) to use biharmonic diffusion that acts predominantly on submesoscales

allowing mesoscale eddy generation and 3) to consider temporal variability in conditions at the Strait of Gibraltar that might be important to have a better description of the variability of MW levels.

Results of sediment transport modelling along the Portuguese coast

The sediment transport model was used to study the northern Portuguese continental shelf sediment dynamics using one of the most extensive data sets available, acquired by the Portuguese Hydrographic Institute. Although the data set is still sparse both in terms of temporal and spatial coverage, model predictions capture the main processes related to shelf sedimentary dynamics. Coupling this model to a circulation and wave propagation model is certainly the best strategy for future developments.

Currents

Current meter data were acquired in the mid and inner shelf, off Cabo Mondego at depths of 27 m, 37 m and 83 m (see Fig. 16 and Table 1 for details). Current meter time series were filtered with a low-pass Butterworth filter of order 7 and a cut-off period of two hours. Since current meter re-cords were contaminated by surface wave energy, it was subtracted from mean orbital wave velocities computed from the knowledge of the surface wave field. The rectified current velocities are plotted in Fig. 17. These velocities are typical of shelf velocities, with magnitudes around 10 to 15 cm/s, without any significant difference between the three records. Spectral analysis has shown that tidal motion contains most of the flow variability.

Waves

Analysed wave data were recorded by a wave buoy located at a depth of 83 m (Fig. 16), maintained by the Portuguese Hydrographic Institute (Fig. 18). In the studied time period, it is possible to perceive that "summer" conditions are very well represented while the more energetic conditions, typical of "winter" conditions, are much worse represented. Nevertheless, during the studied period, three storms with waves heights greater than 5.0 m were registered.

Sedimentary cover

The sediment-grain-size used as input into the model represents the median of the two main deposit types identified on the inner and mid shelf

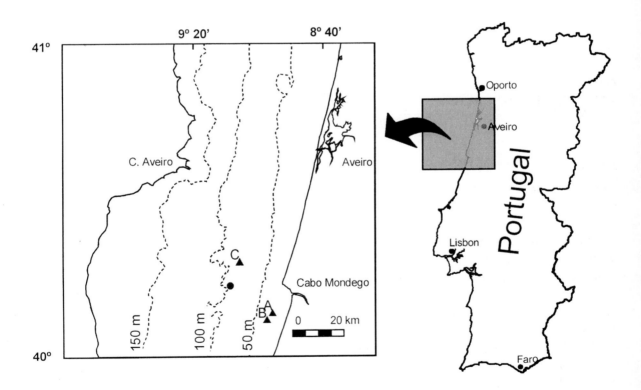

Fig. 16. Current meter deployment sites (solid triangles) and wave-buoy location (solid circle). Dashed lines represent depth contours.

Location	Deployment code (figure 16)	Depth (m)	Height from bottom (m)	Sampling interval (min)	Start	End
40°05.1'N 08°56.6'W	A	27	1.0	20	19:00 03/06/82	23:00 30/06/82
40°03.7'N 08°58.5'W	B	37	1.0	20	19:40 02/06/82	10:20 11/08/82
40°13.5'N 09°06.0'W	C	83	1.5	10	15:50 13/04/83	19:40 09/06/83
40°13.5'N 09°06.0'W	C	83	1.5	10	16:20 22/06/83	12:10 30/08/83
40°13.5'N 09°06.0'W	C	83	1.5	20	21:00 30/08/83	03:40 14/09/83

Table 1. Synthesis of current meter data.

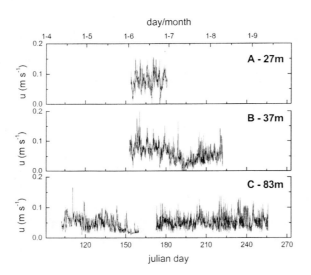

Fig. 17. Current time series.

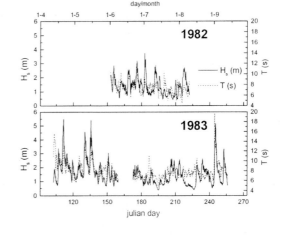

Fig. 18. Time series of significant wave height (solid) and wave period (dashed).

(Magalhães et al. 2000): littoral and mid shelf deposits consist of fine sand deposits, while on the mid shelf coarse sedimentary deposits dominate (Fig. 19).

Results of the sediment transport model

Results confirmed that the Portuguese shelf (at least the mid and inner parts) is clearly wave dominated, with the majority of sediment transport occurring during time of energetic long period waves (Fig. 20). During the studied time intervals, the waves revealed to be the only mechanism capable of remobilising the sedimentary particles, while the current presented always a reduced intensity, functioning only as a transport mechanism to particles put in suspension by the waves (Fig. 20).

Moreover, while the major current component is tide related, the resulting current related transport has a very weak magnitude. This fact explains the apparent contradiction between the existences of high energetic levels at bottom, with frequent remobilisation occurrences, and the apparent im-

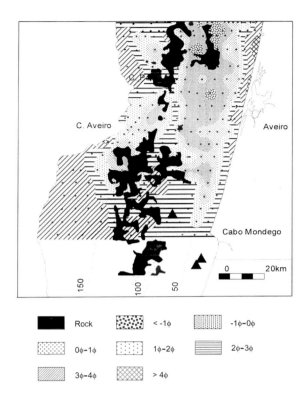

Fig. 19. The superficial sediment distribution (contours in phi) on the northern Portuguese continental shelf (Magalhães 1999). Dashed lines represent depth contours in meters.

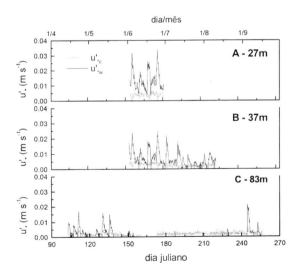

Fig. 20. Comparison between current (u'_{*c} - dashed) and wave (u'_{*w} - solid) related shear velocity computed for a fine sand bottom.

mobility of the shelf deposits observed by several authors (Dias 1987; Magalhães 1999). In fact, as supported by the present observations, as the frequent mobilization is not associated with strong currents, the effective sediment transport is always of reduced intensity, enabling the conservation of deposit identity.

In the middle and outer shelf the modern sedimentation is composed by fine particles (very fine sand or smaller) transported exclusively in suspension, and there is no exchange between median and coarse sand with the inner shelf. The sedimentary transport of the coarse fractions of sand is low and essentially related with low frequent and high energetic events. In the middle shelf, the presence of high energetic conditions (wave related) associated with the frequent presence of large bed forms (associated with relict deposits, Fig. 21) turns this area in a temporary deposition zone. The definitive deposition of fine particles is made in the outer shelf or at greater depths.

Conclusions

Common physical and biogeochemical processes take place in the continental shelf and in estuaries, although the relative importance varies a lot from place to place. In this paper a revision of those processes was done and results were presented for the Portuguese shelf and Tagus estuary using the same computer code. Some basic conditions to obtain a generic model were described.

To achieve a general modelling tool, the model must be able to use the most common vertical coordinates and need to be interdisciplinary. To do so, the model need to be organised in a modular way and be able to accommodate alternative modules, developed by different teams, for each discipline. These features allow its use in areas where the relative importance of physical processes is different and for different purposes (research, management or coastal engineering). This type of tool can generate the critical number of user required to develop supporting software (pre- and post-processing) required to minimize the time necessary to obtain results.

High biological activity taking place on the shelf and slope is directly related to the supply of nutri-

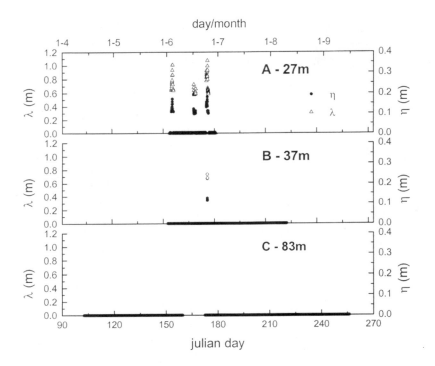

Fig. 21. Computed bedform length and height variation for coarse sand deposits.

ents from the deep ocean by vertical movements induced by the slope current, internal tides and coastal upwelling. In semi-enclosed basins continental discharges and recycling of nutrients are the main support of primary production.

In very shallow areas topography plays a major role for circulation and sigma type coordinates are the most adequate to simulate hydrodynamics. On the contrary, in deeper areas, baroclinic effects can be as important as topographic and a Cartesian type of coordinate is most adequate. An integrated model of the ocean margin must allow more than one type of coordinates.

A model must have a grid as fine as possible to increase accuracy and the number of processes simulated. However, increasing the resolution increases the difficulty to analyse results. Time and space integration tools must be developed together with other post-processing tools. Using these tools it was possible to identify the contribution of each part of Tagus estuary for the overall estuary budget of sediments, nutrients and primary producers.

A sediment transport model has shown that the vertical distribution of sediments in the water col-

umn over the Portuguese shelf is determined mainly by the local physical forcing. To apply this model to the whole shelf lots of data are required. Coupling of this model to a circulation and wave propagation model will fill gaps of experimental data and will allow the calculation of lateral transport and consequently the calculation of budgets.

Acknowledgements

Results on the circulation along the Iberian coast and in the Tagus estuary were obtained in the framework projects funded by DG Research of the European Union. The former by OMEX II – II, contract number MAS3-CT97-0076 and the latter by EUROSAM, contract number ENV4-CT97-0436.

References

Abbott MB, McCowan A, Warren IR (1968) Numerical modelling of free surface flows that are two dimensional in plan. In: Fischer HB (ed) Proceedings of Predictive Ability of Transport Models for Inland and Coastal Waters. Academic Press

Ambar IJ (1984) Seis meses de medições de correntes, temperaturas e salinidades na vertente continental ao largo da costa alentejana (in portuguese). Grupo de Oceanografia, Universidade de Lisboa. Technical Repport pp 1/84:47

Ambar IJ (1985) Seis meses de medições de correntes, temperaturas e salinidades na vertente continental Portuguesa a 40° N (in portuguese). Grupo de Oceanografia. Universidade de Lisboa. Technical Repport pp 1/85: 40

Ambar I, Fiúza A, Boyd T, Frouin R (1986) Observations of a warm oceanic current flowing northward along the coasts of Portugal and Spain during Nov-Dec 1983. Eos 67(144), 1054

Bailard JA (1981) An energetics total load sediment transport model for a plane sloping beach. J Geophys Res 86:10938-10954

Barnard S, Barnier B, Beckman A, Boening C, Coulibaty M, DeCuevas D, Dengg J, Dieterich C, Ernst U, Herrmann P, Jia Y, Killworth P, Kroeger J, Lee M, Le Provost C, Molines J-M, New A, Oschlies A, Reynauld T, West L , Willebrand J (1997) DYNAMO: Dynamics of North Atlantic Models: Simulation and assimilation with high resolution models. Berichte aus dem Institut fuer Meereskunde an der Christian-Albrechts-Universitat Kiel, n° 294 334 p

Barton ED (1989) The poleward undercurrent on the eastern boundary of the subtropical North Atlantic. In: Neshyba et al. (eds) Poleward Flows Along Eastern Ocean Boundaries, Coast Estuar Stud 34:82-92

Benqué JP, Cunge J, Feuillet J, Hauguel A (1982) A new method for tidal current computation. EDF, Rapports HE41/81.26, HE42/81.17

Bleck R, Rooth C, Hu D, Smith LT (1992) Salinity-driven thermocline transients in a wind- and thermohaline-forced isopycnic coordinate model of the North Atlantic. J Phys Oceanogr 22:1486-1505

Blumberg A F, Mellor GL (1987) A description of a three-dimensional coastal ocean model. In: Heaps N (ed) Three-Dimensional Coastal Ocean Models. Coast Estuar Sci 4:1-16

Bougeault P, André JC (1986) On the stability of the third-order turbulence closure for the modeling of the stratocumulus-topped boundary. J Atmosph Sci 43:1574-1581

Bougeault P, Lacarrère P (1989) Parameterization of orography-induced turbulence in a meso-beta scale model. Monthly Weather Rev 117:1872-1890

Bryan K, Cox MD (1972) An approximate equation of state for numerical models of ocean circulation. J Phys Oceanogr 2:510-514

Cancino L, Neves RJ (1998) Hydrodynamic and sediment suspension modelling in estuarine systems. Part II: Application in the Scheldt and Gironde Estuaries. J Mar Syst 22:117-131

Coelho HS, Neves RJ, Leitão PC, Martins H, Santos A (1999) The slope current along the Western European Margin: A numerical investigation. Bol Inst Esp Oceanogr 15:61-72

Cox MD (1984) A primitive equation, 3-dimensional model of the ocean. Technical Report number 1, Ocean Group, Geophysical Fluid Dynamics Laboratory, Princeton, NJ, USA 143 p

Davies AG, Soulsby RL, King HL (1988) A numerical model of the combined wave and current bottom boundary layer. J Geophys Res 93:491-508

Deigaard R (1991) On the turbulent diffusion coeficient for suspended sediment. Progress Report n° 73, Inst of Hydrodynamics and Hydraulic Engineering, ISVA, Techn Univ Denmark pp 55-66

Deleersnijder E, Beckers JM (1992) On the use of σ-coordinate system in regions of large bathymetric variations. J Mar Syst 3:381-390

Dias JA (1997) Historical aspects of marine geology in Portugal. In: Saldanha L, Ré P (eds) One Hundred Years of Portuguese Oceanography: In the Footsteps of King Carlos de Bragança. Publicações Avulsas do Museu Bocage, Lisboa, 2ª Série, n° 2, 173-226 pp

Drake DE, Cacchione DA (1992) Wave-current interaction in the bottom boundary layer during storm and non-storm conditions: Observations and model predictions. Cont Shelf Res 12:1331-1352

Dyer KR (1980) Velocity profiles over a rippled bed and the threshold of movement of sand. Estuar Coast Mar Sci 10:181-199

Dyer KR (1986) Coastal and estuarine dynamics. J Wiley & Sons, Chichester 342p

Dyer KR, Soulsby RL (1988) Sand transport on the continental shelf. Ann Rev Fluid Mech 20:295-324

Engelund F, Fredsøe J (1976) A sediment transport model for straight alluvial channels. Nordic Hydrol 7:293-306

Fredsøe J (1984) Turbulent boundary layer in wave-current motion. J Hydraul Engin, ASCE, 110:1103-1120

Fredsøe J, Deigaard R (1992) Mechanics of Coastal Sediment Transport. Advanced Series on Ocean Engineering, vol. 3, World Scientific, 366 p

Frouin R, Fiúza A, Ambar I, Boyd TJ (1990) Observations of a poleward surface current off the coasts of Portugal and Spain during the winter. J Geophys Res 95:679-691

Galperin B, Kantha LH, Hassid S, Rosati A (1988) A quasi-equilibrium turbulent energy model for geophysical flows. J Atmosph Sci 45:55-62

Garcia M, Parker G (1991) Entrainment of bed sediment into suspension. J Hydraul Engin 117:414-435

Gargett AE (1984) Vertical eddy diffusivity in the ocean interior. J Mar Res 42:359-393

Gaspar PG, Grégoris Y, Lefevre J-M (1990) A simple eddy kinetic energy model for simulations of the oceanic vertical mixing: Tests at station Papa and Long-Term Upper Ocean Study site. J Geophys Res 95:16179-16193

Glenn DA, Grant WD (1987) A suspended sediment correction for combined wave and current flows. J Geophys Res 85:1797-1808

Grant WD, Madsen OS (1979) Combined wave and current interaction with a rough bottom. J Geophys Res 84:1797-1808

Grant WD, Madsen OS (1982) Moveable bed roughness in unsteady oscillatory flow. J Geophys Res 87:469-481

Grant WD, Madsen OS (1986) The continental shelf bottom boundary layer. Ann Rev Fluid Mech 18:265-305

Haidvogel DB, Wilkin JL, Young R (1991) A semi-spectral primitive equation ocean circulation model using vertical sigma and orthogonal curvilinear horizontal coordinates. J Comput Phys 94:151-185

Haynes R, Barton ED (1990) A poleward flow along the Atlantic coast of the Iberian peninsula. J Geophys Res 95:11425-11141

Hervouet J-M, van Haren L (1996) Recent advances in numerical methods for fluid flows. In: Anderson MG, Walling DE, Bates PD (eds) Floodplain Processes. J Wiley & sons, Chichester pp 183-214

Hill P, Nowell RM, Jumars PA (1988) Flume evaluation of the relationship between suspended sediment concentration and excess boundary shear stress. J Geophys Res 92:12499-12509

Huthnance JM (1984) Slope Currents and "JEBAR". J Phys Oceanog 14:795-810

Huthnance JM (1986) The Rockall slope current and shelf edge processes. Proc R Soc Edinburgh Sect B 88:83-101

Justesen P (1988) Prediction of turbulent oscillatory flow over rough beds. Coast Engin 12:257-284

Krone RB (1962) Flume studies of the transport +in estuarine shoaling processes. Hydr Eng Lab,Univ of Berkeley, California, USA

Leendertsee JJ (1967) Aspects of a Computational Model for Long Water Wave Propagation. Rand Corporation, Memorandum RH-5299-RR, Santa Monica

Leendertsee JJ (1970) A water qualtiy simulation model for well mixed estuaries and coastal seas. Rand Cor-

poration, Memorandum RM-6230-RC, Santa Monica

Landertsee JJ, Liu SK (1978) A three-dimensional turbulent energy model for non-homogeneous estuaries and coastal sea systems. In: Nihoul JCJ (ed) Hydrodynamics of Estuaries ans Fjords. Elsevier, Amsterdam 387-405

Levitus S, Boyer TP (1994) World Ocean Atlas 1994. Volume 4: NOAA Atlas NESDIS 4, 117p

Levitus S, Burgett R, Boyer TP (1994) World Ocean Atlas 1994. Volumes 1 and 2: NOAA Atlas NESDIS 3, 99p

Li MZ, Amos CL (1998) Predicting ripple geometry and bed roughness under combined waves and currents in a continental shelf environment. Cont Shelf Res 18:941-970

Li MZ, Wright LD, Amos CL (1996) Predicting ripple roughness and resuspension under combined flows in a shoreface environment. Mar Geol 130:139-161

Lynn RJ, Simpson JJ (1987) The California current system: The seasonal variability of its physical characteristics. J Geophys Res 92:12947-12966

Madsen OS (1991) Mechanics of cohesionless sediment transport in coastal waters, Proceedings of Coastal Sediments '91, ASCE, 15-27

Magalhães F (1999) Os sedimentos da plataforma continental portuguesa: Contrastes espaciais, perspectiva temporal, potencialidades económicas (in portuguese). Ph D Dissertation, Univ of Lisbon, 289p

Magalhães F, Cascalho J, Dias JA, Matos M (2000) Surface sediments of the portuguese continental shelf north of Espinho / Sedimentos superficias da plataforma continental portuguesa a norte de Espinho. 3° Simpósio sobre a Margem Continental Ibérica Atlântica, Faro, 263-264

Martins H, Santos A, Coelho EF, Neves R, Rosa TL (1999) Numerical Simulation of Internal Tides. J Mechan Engin Sci 214C:867-872

Martins FA, Neves RJ, Leitão PC (1998) A three-dimensional hydrodynamic model with generic vertical coordinate.In: Babovic V and Larsen LC (eds) Proceedings of Hidroinformatics98 (Copenhague, Denmark, August 1998). Balkema, Rotterdam 1403-1410

Martins FA, Leitão PC, Silva A, Neves R (in press) 3D modelling of the Sado Estuary using a new generic vertical discretization approach. Oceanolo Acta

Mazé JP, Ahran M, Mercier H (1997) Volume budget of the eastern boundary layer off the Iberian Peninsula. Deep-Sea Res 44:1543-1574

McCreary JP, Shetye SR, Kundu P (1986) Thermohaline forcing of eastern boundary currents: With application to the circulation off west coast of Australia. J Mar Res 44:71-92

McCreary JP, Kundu P, Chao SY (1987) On the dynamics of the California current system. J Mar Res 45:1-32

Mellor GL, Hakkinen S, Ezer T, Patchen R (subm.) A generalization of a sigma coordinate ocean model and an intercomparison of model vertical grids. In: Pinardi N (ed) Ocean Forecasting: Theory and Practice. Springer, Berlin

Miranda R, Neves R, Coelho H, Martins H, Leitão PC, Santos A (1999) Transport and mixing simulation along the continental shelf edge using a Lagrangian approach. Bol Inst Esp Oceanogr 15:39-60

Neves RJJ (1985) Étude Expérimentale et Modélisation Mathematique de l'Hydrodynamique de l'Estuaire du Sado. PhD Thesis, Université de Liège, Belgium

Nielsen P (1981) Dynamics and geometry of wave generated ripples. J Geophys Res 86:6467-6472

Nielsen P (1991) Combined convection and diffusion: A new framework for suspended sediment modelling. Proceedings of Coastal Sediments '91, ASCE, 418-431

Nielsen P (1992) Coastal bottom boundary layer and sediment transport. Advanced Series on Ocean Engineering, vol. 4, World Scientific, 324 p

Nittrouer C, Wright D (1994) Transport of particles across continental shelves. Rev Geophys 31:85-113

Odd NVM (1986) Mathematical modelling of mud transport in estuaries. Int Symp Physical Processes in Estuaries, 9-12 September 1986

Partheniades E (1965) Erosion and deposition of cohesive soils. J Hydr Div, ASCE, 91, No. HY1:105-139

Phillips NA (1957) A coordinate system having some special advantages for numerical forecasting. J Meteorol 14:184-185

Pingree R, Le Cann B (1989) Celtic and Armorican slope and shelf residual currents. Prog Oceanogr 23:303-338

Pingree R, Le Cann B (1990) Structure, strength and seasonality of the slope currents in the Bay of Biscay region. J Mar Biol 70:857-885

Portela LI (1996) Mathematical modelling of hydro-dynamic processes and water quality in Tagus estuary, PhD thesis, Instituto Sup Técnico, Tech Univ of Lisbon, (in portuguese)

Santos AJP (1995) Modelo hidrodinâmico de circulação oceânica e estuarina (in portuguese). PhD Thesis, IST Lisbon 273 p

Silva MC, Moita T, Figueiredo H (1986) Controlo da qualidade da água. Resultados referentes às observações realizadas em 1982 e 1983. Estudo Ambiental do Estuário do Tejo (3°série) n°7. Secretaria de Estado do Ambiente e Recursos Naturais, Lisboa pp 1-139

Smith JD (1977) Modeling of sediment transport on continental shelves. In: The Sea, 6. J Wiley & sons, New York 539-577

Smith JD, McLean SR (1977) Spatially averaged flow over a wavy surface. J Geophys Res 82:1735-1746

Soulsby RL (1997) Dynamics of marine sands. Thomas Telford Publ, London, UK, 249p

Soulsby RL, Hamm L, Klopman G, Myrhaug D, Simons RR, Thomas GP (1993) Wave-current interaction within and outside the bottom boundary layer. Coast Engin 21:41-69

Taborda R, Dias JA (2000) Prediction of wave related bedform geometry. 3° Simpósio sobre a Margem Continental Ibérica Atlântica, Faro 257-258

Taboada JJ, Prego R, Ruiz-Villarreal M, Gómez-Gesteira M, Montero P, Santos AP, Pérez-Villar V (1998) Evaluation of the seasonal variations in the residual circulation in the Ria of Vigo (NW Spain) by means of a 3D baroclinic model. Est Coast Shelf Sci 47:661-670

Trenberth KE, Large WG, Olsen JG (1990) The mean annual cycle in global wind stress. J Phys Oceanogr 20:1742-1760

Vale C, Sundby B (1987) Suspended sediment fluctuations in the Tagus estuary on semidiurnal and fortnightly time scales. Est Coast Shelf Sci 27:495-508

Vale C, Cortesão C, Castro O, Ferreira AM (1993) Suspended sediment response to pulses in river flow and semidiurnal and fortnightly tidal variations in a mesotidal estuary. Mar Chem 43:21-31

Van Rijn LC (1993) Principles of sediment transport in rivers, estuaries and coastal seas. Aqua Publications, Amsterdam

Vincent C, Downing A (1994) Variability of suspended sand concentrations, transport and eddy diffusity under non-breaking waves on the shore-face. Cont Shelf Res 14:223-250

Vincent C, Green MO (1990) Field measurements of the suspended sand concentration profiles and fluxes and of the resuspension coefficient γ_0 over a rippled bed. J Geophys Res 95:11591-11601

Visbeck M, Marshall J, Haine T, Spall M (1997) Representation of topography by sahved cells in a height coordinate ocean model. Mon Wea Rev 125:2293-2315

Weaver AJ, Middleton JH (1990) An analytic model for the Leeuwin Current off western Australia. Cont Shelf Res 10:105-122

Wiberg PL, Harris CK (1994) Ripple geometry in wave-dominated environments. J Geophys Res 99:775-789

Seabed Classification at Ocean Margins

Ph. Blondel

Department of Physics, University of Bath, Bath BA2 7AY, UK
corresponding author (e-mail): pyspb@bath.ac.uk

Abstract: Ocean margins have become the focus of numerous geophysical and environmental surveys, because of their economic, scientific and oceanographic significance. These surveys deliver increasingly larger volumes of data, acquired by many types of techniques and sensors. Despite their importance, most of these data are still interpreted visually and qualitatively by skilled interpreters. Human interpretation is time-consuming and difficult to standardise; and, in certain conditions, it can be error-prone. Current research in data processing is shifting toward computer-based interpretation techniques, and in particular seafloor classification. After a brief review of the main characteristics of ocean margins, the different notions of classification will be presented, along with the desired aims. These will be followed by a review of non-acoustic and acoustic (mainly sonar) classification techniques, supplemented with actual examples when applicable. Seabed classification, in general and at ocean margins, is fast becoming a major tool in seafloor surveying and monitoring. The last section will assess the latest tendencies and the technical developments that can be expected in the near future.

Introduction

Scientific advances of the last decades have revealed the extreme importance of ocean margins in all areas of human endeavours. Seafloor exploration has revealed their sheer size and variety. Oceanographic studies have revealed their importance for global ocean circulation. Marine and terrestrial geology have revealed their role in the generation and propagation of often catastrophic tsunamis. And, in general, ocean margins have proved important in all spheres of human activity: political (Exclusive Economic Zones), industrial (oil and gas exploration, communications), ecological (fisheries, environmental assessments), military (potential or actual theatres of operation), etc.

The ocean margins are very varied and complex environments, harbouring a multitude of variations in geology (sedimentation, volcanism, tectonics), and acting as markers of chemical, biological and human activity (coral reefs, pollution, etc.). Many portions of ocean margins around the world have not been mapped yet, either in general or in detail, and more areas still need to be surveyed and monitored. Technological progress ensures that the number of instruments available for exploring, surveying and monitoring increases steadily. The same is true of their resolution, and in general of the amount of (often digital) data that these instruments generate. Traditional, qualitative interpretation by skilled specialists is time-consuming and difficult to standardise. The presence of knowledgeable interpreters with the appropriate expertise is also not guaranteed in all marine operations. This is why current research efforts are turning toward computer-based interpretation techniques. These classification techniques must supplement the raw data with quantitative results going as far as possible beyond what is achievable through human interpretation.

Seabed measurements are actually made at scales varying from the millimetre to the kilometre, with varying degrees of penetration below the seafloor. These data are acquired at sea in often complex and limiting conditions, with a bewildering variety of sensors, ranging from acoustics to radioactivity. Each will require a different type of classification technique, which most often cannot be transferred to another domain or needs to be developed specifically. The present article aims at providing an overview of these techniques, how they work and what they achieve.

From WEFER G, BILLETT D, HEBBELN D, JØRGENSEN BB, SCHLÜTER M, VAN WEERING T (eds), 2002,
Ocean Margin Systems. Springer-Verlag Berlin Heidelberg, pp 125-141

The initial description of the different types of seafloor structures at the ocean margins will be followed by the definition of seabed classification, its aims and its means. This will be followed by a review of non-acoustic and acoustic classification techniques. Each will be supplemented with actual examples, and pointers to the appropriate references detailing these techniques and their achievements.

This article does not pretend to be an exhaustive list of all possible seabed classification techniques, using all possible instruments. Some techniques have been left out, either voluntarily (untested or inconclusive techniques) or involuntarily. A complete treatment of past, present and future seabed classification techniques would require a complete book, and only the main domains will be presented here.

Seabed classification, in general and at ocean margins in particular, is fast becoming a major tool in seafloor surveying and monitoring. The last section will thus assess the latest advances, the technical developments that can be expected in the near future, and their likely impacts.

Ocean margins and seabed classification

General setting of continental margins

Ocean margins are the site of, and influence many aspects of our socio-economic activities. In addition, they form a major source of biodiversity and are a key habitat of the world's ecology. The Exclusive Economic Zones of most industrialised countries occupy continental margins, from which they derive an important portion of their wealth. Hydrocarbon exploitation is presently concentrated on continental shelves (for example in the North Sea or the Gulf of Mexico), but is steadily moving over the shelf edge into deeper waters. The world's largest fish reserves are also located on the continental margins, and their management often proves politically and economically difficult. Therefore, a detailed knowledge of the ocean margins is of eminent importance, as for example slope failures and the resultant mass-movement flows, are capable of destroying marine installations and submarine cables, and even of generating tsunamis.

Recent environmental studies have also shown the importance of a more complete and detailed knowledge of continental margins, such as for global climate studies and modelling of the carbon cycle, for exchange processes at the shelf edge, as well as for studies of anthropogenic inputs. These may provide a direct link to coastal pollution, and may reflect the off-shore dumping of harmful chemical products in poorly known areas that were wrongly assumed deep and stable enough.

The continental margins can be broadly divided into the continental shelf close to the coast (down to about 200 m on average), the adjacent and somewhat steeper continental slope (down to 3000 m), and the continental rise at the limit with the abyssal plains (Fig. 1). The continental shelf is a relatively flat area (slopes less than 1:1000, low local relief). Underlain by continental crust, it is also remarkably shallow; generally less than 250 m. Depending on the regional geological setting, the shelf will extend from a few (tens of) kilometres (near subduction zones, for example) to several hundreds of kilometres (near passive margins) away from the coastline. The continental shelves are areas of input of (sometimes very large) quantities of sediments by the rivers, sometimes even at large distances from land (e.g. the Indus, the Amazon or the Yangtze rivers). Also on the shelves, pelagic settling takes place, and depending on local tidal, wave and current conditions, these particles may settle either on the shelf or be transported over the shelf edge. During this transport, and quantified over time, large volumes of sediments ultimately settle on the continental slope. This rapidly descending area ranges in depth between ca. 200 m and 3000 m. Another means of transport of particles from the shelf over the slope and into the deep-sea is through channels and canyons, bringing the sediment far out into the abyssal plains. In the Western Atlantic, canyons extend for several hundreds of kilometres as tightly sinuous channels with no tributaries, whereas in the Southeastern Pacific, they meander on longer distances and leave behind abandoned channel segments (see Blondel and Murton 1997, for examples). Locally, sediments are contained within contourite bodies (sediment drifts) that have formed by sedimentation under influence from cur-

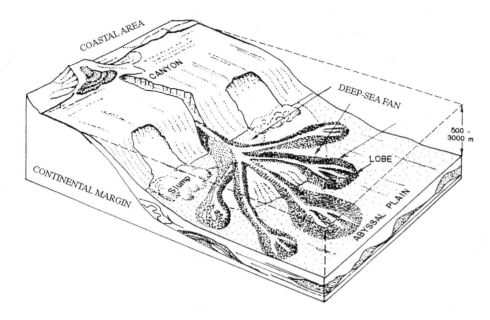

Fig. 1. Morphological diversity of continental margins.

rents following the bathymetric contours, sometimes with sediment or sand waves on their flanks (e.g. Blondel and Murton 1997).

Because of the occasionally steep slopes, and in combination with for example fast sedimentation, the slopes may become unstable, and earthquakes or minor sea-level changes can trigger their collapse, creating large-scale submarine slides and slumps (the larger ones producing internal waves or tsunamis). In combination with particles settling from the pelagic zone, sediments are carried over the slope to the continental rise. This physiographic province marks the limit between the continental margins and the abyssal plains. Deeper than 3000 m, the continental rise can go down to 6000 m (as in the North Atlantic). The more prominent structures at the continental rise are the distal deep-sea fans, lobes and sediment channels marking the outer limits of canyon-influenced sedimentation. In case of large-scale slope failures (e.g. Storegga slide or Grand Banks earthquake), this may also lead to transport of large amounts of slope sediments forming local bulges.

Complex and varied environments

Continental margins are the geological regions with the largest depth range, and with a great diversity of morphological structures. Figure 1 shows only a couple of them; from the sediment input by rivers, caused by erosion of the continent, and biogenic carbonate deposits in the coastal area and at the shelf and shelf edge (in tropical areas), to the junction with the flat, comparatively featureless abyssal plains. The inventory of all possible structures goes beyond the remit of this article, and much of it is already covered in reference books (e.g. Blondel and Murton 1997; Stoker et al. 1998 and references therein).

To grossly simplify what is truly a complex environment, the sediments coming from the continents themselves are deposited in the coastal area (Fig. 1), forming large-scale sedimentary structures such as estuaries, deltas and fans. A large portion of the accumulated sediment moves downslope, transported through large slides, often called debris slides, or along narrow channels merging into submarine canyons (Fig. 1). Not all submarine canyons look alike. Some canyons just cut into the continental slope, while others may extend far onto the continental rise.

Large sediment accumulations are usually unstable, and can move downslope as slumps, slides, and debris flows, leaving scars on the sea-floor and the marks of downslope transportation and deposition of sediments. Factors known to influence these catastrophical events include loading or over-steepening of slopes, underconsolidation of sediments due to rapid sedimentation, earthquakes, gas build-up and sea-level changes. Mass-movement deposits are often called debris, and include mixed lithologies of muds, clays and sands, with inclusions of pieces of hardrock, pebbles and boulders, with sizes ranging from decimetres to several hundreds of metres (e.g. Blondel and Murton 1997; Stoker et al. 1998).

Tectonic structures (such as faults and folds) and volcanic structures below the seabed or emerging above, may also be found. When still intact, these volcanoes look very much like their counterparts at spreading centres. Other structures, discovered in the last decade, are actively forming mud volcanoes, or mud mounds, often in association with local gas seeps, or gas hydrate below the seabed, and often in tectonically active regions. These are interesting structures, as potential traps for hydrocarbons, and because they can create important modifications of the seafloor, which are potential hazards to telecommunication cables and pipelines.

Other structures visible on continental shelves and margins are pockmarks, associated with past or present seepage of gases and fluids. Most pockmarks are approximately circular in plan, but there is a considerable variety in shapes and sizes, as they are active features (e.g. Blondel and Murton 1997). Because of their origin and of their locations (often in close relationship with hydrocarbon reservoirs in the deeper seabed underneath), pockmarks generate a considerable interest in offshore industries. This interest is enhanced by the consequences pockmarks may have on the stability of offshore platforms and the security of drilling exploration. Catastrophic gas escapes can form large, deep pockmarks in very short periods of time. In one occurrence, the gas blowout at a depth of 240 m reached to the surface, and instabilities led to the abandonment of a neighbouring drilling platform.

Structures related to local brine accumulation can be found in various environments, and manifest themselves as bottom depressions filled with hypersaline water. Brine pools have been particularly well documented on the Mediterranean Ridge and in the Gulf of Mexico; the Mediterranean ones are thought to result from the rapid dissolution of evaporites of the Late Miocene. Brine lakes and pools are uncommon features on the seafloor, and have been often discovered in the proximity of mud volcanoes or of important faults (e.g. Blondel and Murton 1997). Brine pools are also of biological interest, since the recent discovery of deep-sea chemosynthetic communities and faunal assemblages.

Biological activity at continental margins can manifest itself in many ways. For seabed classification, the organisation of biota in large-scale structures, such as slumps of large algae, local accumulations of dead and living plants, or formation of carbonate structures, reefs and deposits, are all of considerable importance. In shallower waters, coral structures are also visible. Halimeda mounds (bioherms), composed of clustered dead and living plants, have for example been documented on the western Indian continental shelf, in water depths close to 90 metres. They can be tens of metres high and several hundred metres long. Bioherms on the shelf edge appear more linear, whereas deeper bioherms generally have mound-like morphologies (cf. Blondel and Murton 1997). Linear algal ridges and cold-water coral carbonate mounds, sometimes aligned along tectonic faults and fault structures, have also been documented along the Eastern Atlantic margin, from Spain to Norway, in water depths ranging from 300 to 1200 m.

Seabed classification

Continental margins appear therefore as a strongly variable environment. Confronted with many and diverse types of (sometimes new) structures, often arranged in complex patterns, the interpreter of marine data needs classification techniques. But what is seabed classification? A rapid search through the scientific and technical literature shows many different definitions of the word "classification". These definitions are often blurred by commercial hype, and by the need to communicate between different disciplines (e.g. image processing, marine geology, underwater acoustics) with different vocabularies. These semantic differences can be quite confusing. One usually distinguishes three different stages: segmentation, classification and characterisation.

Segmentation comes from the image processing world, and defines the partition of an image into several regions with different characteristics. The image may come from any source (satellite image, medical image etc.), and the partitions of the image are not definitely interpreted. They just share the same numerical characteristics, and are generally labelled only with letters or numbers.

Classification goes one step further. Classification recognises these different regions as distinct physical entities, even if a physical feature can correspond to several partitions (e.g. "mud = partition 1 & partition 2"), or a partition can correspond to several physical features (e.g. "corals and aggregated pebbles").

Characterisation is the next step, where the regions recognised in the images or in the maps correspond to definite characteristics, physical (e.g. particular shear wave attenuation), chemical (e.g. oil slick or metallic object), geological (e.g. fine-grained silt), or biological (e.g. coral mound).

Notwithstanding confusions between scientific disciplines with different jargons, one can see how a new system can quickly be marketed as achieving "seafloor characterisation", even if it only performs "seafloor segmentation", as this makes it more likely to attract potential customers. For the sake of simplicity, the definition of seabed classification adopted here will be broader, and encompass both classification per se, and characterisation. The techniques and systems presented in the next two sections have been selected because they could achieve at least a basic recognition of the type of seafloor surveyed, or because they could identify a set of seabed intrinsic characteristics (e.g. grain size, porosity, density) from which the type of seafloor could be recognised. The capability and spatial coverage offered by the different tools will vary, and they have been regrouped into two categories: non-acoustic systems and acoustic systems (by far the most present in the literature).

Non-acoustic seabed classification

In situ seabed analysis

The most simple and traditional way of characterising the seabed is through the taking of samples and their on-board analysis by skilled interpreters. Historically, it was by using line soundings that the first samples of the deep ocean floor were collected for scientific purposes. Dredging has not evolved much since the nineteenth century: a strong welded steel frame, holding open a sturdy bag of steel chain, is attached through hinged attachment arms to the heavy steel wire that lowers the dredge to the seafloor. The dredge is towed directly on the bottom, and the steel frame of this rudimentary device pulls out portions of the seafloor. Its positioning on the seafloor is much more precise now than it was in the nineteenth century, owing to GPS and transponder technology. There are different types of dredges (e.g. Juteau and Maury, 1999), but even the most adapted dredge might bounce off obstacles (e.g. rocky outcrops) or scrape the seafloor without sampling it. Furthermore the samples collected are in complete disorder and their precise location along the dredge track is impossible to know.

Coring methods are commonly used for sampling still-soft and water-saturated oceanic sediments. There are many types of coring techniques (Hebbeln this volume), but all enable the sampling of a complete cross-section of sediments, and its laboratory analysis for different properties, using geoacoustics, X-ray tomography, microscopic analysis etc. (e.g. Orsi et al. in Pace et al. 1997; Balsam et al. 1999). Cores can be very precisely located on the seafloor, and recent use of video data adds to the final interpretation, in particular by showing how each core is representative of the area. The problem is that cores are very localised (a few centimetres or tens of centimetres in diameter), and usually very time-consuming and/or expensive to acquire. This becomes even more true as one moves to the deeper parts of the continental margins.

Free-fall instruments have also been designed for rapid assessment of specific geoacoustic and geotechnical parameters. Penetrometers dropping through the water column penetrate the seabed to different depths and at different rates depending on the geometry, mass, and impact velocity of the probe, and on the shearing strength of the sediment. If the physical dimensions and terminal velocity of the probe are fixed, it is possible to infer the shearing strength and certain other properties based on an analysis of the deceleration signature, compared with a database compiled from tests on different types of sediments. This is the approach used by Stoll and Akal (1999) for the XBP (eXpendable Bottom Probe), now available commercially. Rosenberger et al. (1999) developed a similar instrument, measuring the resistivity of the seafloor. Longer stations on the seafloor are possible with complete instrumentation platforms, which can take more time-demanding measurements of the seabed properties. This is for example the case of ISSAMS, the In Situ Sediment geoAcoustic Measurement System developed by Barbagelata et al. (1991), which measures compressional and shear speed and compressional attenuation. Like cores, the use of these tools is very localised and time-consuming.

Direct sampling by manned submersibles or Remotely-Operated Vehicles (ROVs) is more adaptable and selective. Recent technological and engineering advances allow to use many instruments simultaneously, even at large depths. For example, the German Ministry for Research and Technology is developing a ROV equipped with instruments for localizing and analysing chemical pollutants in the water column and on the seafloor. These instruments include a lidar (laser and fluorometre), an acoustic sensor for measuring the acoustic impedance of the seabed, a membrane induction/gas chromatograph/mass spectrometer, and a quartz microbalance sensor array (Harsdorf et al. 1998). But such sampling is even more time-consuming and expensive, and submersible constraints limit greatly the amount of samples that can be taken.

Optical seabed classification

Very often now, manned and unmanned dives are constantly recording video and still imagery. These images are interpreted and quantified with traditional image processing techniques (e.g. Lebart et al. 2000), as can be used for satellite imagery (e.g. feature-space classification). The general type of seafloor may be identified, even from complex images (Fig. 2), but the other characteristics of the seabed remain inaccessible. The conditions in which the images are acquired are often causing problems: inadequate and variable lighting (Fig. 2 a); unaccounted variations in the water column inside the field of view (e.g. shimmering), etc. Standard video cameras are also often unsuitable for seafloor imaging in case of high turbidity. Furthermore, pools of chemicals on sediment can hardly be seen on video images taken from above. Especially on coarse-grained sediments, their contours follow the ground, so that the low gradient of the index of refraction does not provide sufficient contrast.

Laser-based techniques offer the advantage of constant, quantifiable illumination, and independence from all but the most severe disturbances in the water column. Laser line-scan systems, such as the Raytheon Model LS4042, sweep the seafloor with a pencil-thin laser beam, using a mechanism similar to the "push-broom" device of Earth-orbiting satellites. The reflected light is analysed, and the use of different filters yields multi-spectral images (and access to all the relevant classification techniques developed for satellite imagery). The narrow laser beam allows images to be acquired at distances three to four times that possible with conventional imaging systems. The emergence of high repetition solid state blue-green lasers and the latest developments in fast electronics provide new possibilities, although the operational range remains determined by the red channel (more attenuated in water). Originally designed to identify mines and mine-like objects in the shallow-water littoral environment, this and other similar systems are now used for the identification of anthropogenic materials and biological targets on the seafloor (Coles et al. 1998). Polarisation signature aids in material discrimination and identification, but the applications presented have not broached geological objects yet. Compared to classic, optical imagery, these laser systems can sweep further from the ROV or submersible, and cover sub-

a)   b)

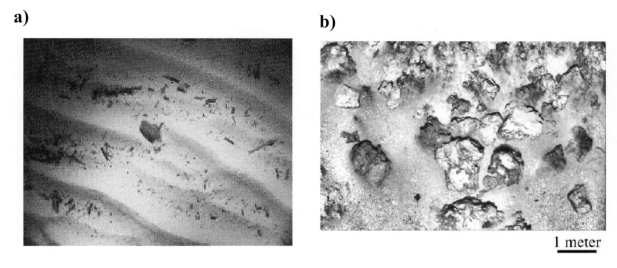

1 meter

Fig. 2. Comparison of a normal image **a)** of sandy ripples (© NATO/SACLANTCEN), and a high-definition digital still image of **b)** sulphide rubble, taken from directly above (© WHOI).

stantially wider portions of the seabed. Airborne laser bathymetry, with systems like SHOALS (Scanning Hydrographic Operational Airborne Lidar), cover even wider strips, much faster. But these tools are limited to the shallowest parts of the continental shelf (Irish and Lillycrop 1999), and their applications for seabed characterisation have not been assessed yet.

Using the seabed radioactivity

An interesting approach has been tested at sea by Noakes et al. (1999) and de Meijer et al. (2000), using the measurement of naturally occurring radionuclides to distinguish different types of seabeds. All natural materials incorporate some radioactive substances. The elements ^{40}K, and the decay products of ^{232}Th and ^{238}U emit γ radiation. This γ radiation is characteristic of the radionuclides and should therefore be detectable with scintillation detectors. The concentration of radionuclides in a sediment depends on its size, its origin, and its mineral composition. These concentrations are usually very low, but still measurable with state-of-the-art equipment. According to de Meijer et al. (2000), mud shows more radioactivity per unit mass than sand or gravel. Sediments originating from granites are more radioactive than the weathering products of basalt. In addition, sedi-

ments occasionally also contain radionuclides released by anthropogenic processes (e.g. fertiliser production). Each type of seafloor, sedimentary or not, should therefore exhibit a distinctive radioactive signature.

The implementation by de Meijer et al. (2000) led to the Multi-Element Detector system for Underwater Sediment Activity (MEDUSA), based around a highly sensitive scintillation crystal encased in a watertight casing. MEDUSA is towed over the bottom at speeds of about 2 m/s (ca. 4 knots), and the γ rays emitted by the seabed are measured at 10-second intervals. Radioactivity concentrations are analysed in real time, using calibration points from a couple of samples taken at the beginning of the survey. The applications presented by de Meijer et al. (2000) include the mapping of mud concentrations in a dumping site 20 km northwest of Rotterdam Harbour (for an area of 15 × 10 km), and a high-resolution sand/mud map of the Hollandsch Diep (for an area of 20 × 3 km).

Radioactive characterisation of the different types of seabeds is accurate and relatively fast. However, the method has been tried only in shallow waters (up to 100 m with the GIMS/CS³ system of Noakes et al. (1999)), and both technological and scientific developments may be needed for use in deeper waters such as on continental margins. And this approach, like many others, provides information only for the small portion of seafloor directly below the instrument. This is not the case of acoustic techniques, which either penetrate the seafloor or image wide portions of the seafloor each time.

Acoustic seabed classification

Seismic seabed classification

Seismic surveying is used to get information about the region immediately below the seafloor. Various acoustic energy sources can be used: air or water guns, boomers, sparkers, dynamite or other explosives (e.g. Telford et al. 1990; Juteau and Maury 1999). These sources usually transmit at frequencies between 10 Hz and several hundred Hz. After propagating through the water column, the acoustic wavefronts penetrate the sedimentary layers or the hard rocks below the seabed. They are reflected or refracted by geological discontinuities, and some of the energy returns back towards the surface. They are recorded by one or, more often, several receivers towed behind the survey vessel (Fig. 3). Mono-channel seismic reflection provides a good resolution for shallow penetration, whereas multi-channel seismic reflection provides a variable resolution, but deeper penetration. Seismic refraction techniques are used to study the deepest structures, usually with two ships car-rying out long seismic profiles, or using one ship and a buoy. Because the present article focuses on the seabed itself, only seismic reflection techniques will be considered.

Figure 4 shows a typical seismic profile, collected on the outer edge of a continental shelf. This profile shows the bathymetry along the survey line, expressed as the time taken for the sound waves to reach back to the receivers. It also shows the reflections from the first sediment layers below the seabed. The profile covers approximately 5 kilometres of seafloor, imaged to a depth of approximately 30 metres (with a vertical exaggeration of 28). Traditional analysis includes the picking of the "seismic horizons", i.e. the different layers or targets, and their interpretation as different geological features (e.g. Telford et al.

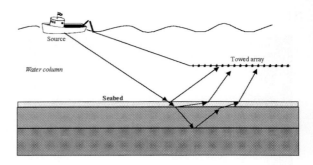

Fig. 3. Principle of seismic surveys. The acoustic signal transmitted by the source is reflected by the different sediment layers, and recorded on the hydrophones positioned on the array towed at some distance behind the ship.

1990). This method is fraught with difficulties, especially as some of the seismic structures might be poorly contrasted, or buried in acoustic noise. It is also inherently qualitative, and does not provide much more than basic information about the sea-floor characteristics. It works best when used in conjunction with information from direct sampling (coring, drilling).

Werby et al. (1986) addressed this problem using beamforming of the signals received by each hydrophone in the array, and direct inversion techniques. They use the fact that the propagation of seismo-acoustic modes at low frequencies is governed by the geophysical properties of the seabed (including P- and S-wave velocity and attenuation, density, and layer thickness). An appropriate measure of the signals reflected from the seafloor (or just beneath) might then indicate its general acoustic character. As presented, their method is restricted to shallow water depths (ca. 100 m), and a spatial resolution of the order of a few wavelengths (for a frequency of 50 Hz, the acoustic wavelength is 30.8 m). It will therefore provide only a rough idea of the overall characteristics of the seafloor.

Davis et al. (in Pace et al. 1997) use a different technique to construct maps of the relative seabed reflectivity strength and shot-to-shot reflectivity variance. The acoustic source is a surface-towed boomer. The raw data from each hydrophone in the towed array is processed by detrending and correcting for geometrical spreading. A "swell"

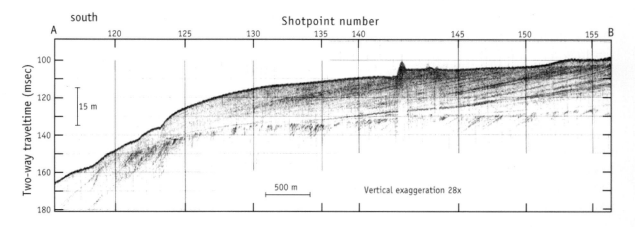

Fig. 4. Portion of a seismic profile on the outer edge of the Mississippi continental shelf. From QuarterDeck Online, (www-ocean.tamu.edu/Quarterdeck/1998/3/sager-profile.html) (© Department of Oceanography; Texas A&M University).

filter is then applied to remove the effects of water-wave motion. This enhances the lateral continuity of the subsurface layering. In the next stage, the amplitude envelope trace is calculated, and used for the detection of the seafloor and direct wave. The amplitude envelope trace is computed by convolving the absolute values of the original data trace with a rectangular window as long as the acoustic pulse. Bottom detection is then achieved with a simple time-delay threshold, robust and computationally efficient. The first peak after bottom detection is taken as the relative bottom reflectivity strength. The bottom reflectivity variance is calculated over windows of 40 shots. The maps of these two parameters allow more quantitative descriptions of the structures detected. Calibration data (from cores analysed on-board) allows correlating relative seabed reflectivity values to absolute bottom reflection coefficients, using empirical relationships from the literature. These coefficients can in turn be inverted to produce maps of the bulk density, porosity and mean grain size of the seabed sediments. Davis et al. (in Pace et al. 1997) were thus able to recognise different types of seabed, including coarse-grained ripples, glacial till or outcropping bedrock. These results were validated with independent ground-truthing from cores and grab samples.

Some other approaches (e.g. Liu and Fu 1982; Simaan 1998) consider the two-dimensional seismic sections as images in their own right, and apply traditional image processing techniques to charac-terise the seismic units. These techniques are similar to the ones used for optical images (cf. above) or sidescan images (cf. below). They can use a priori information about what the seismic structures should look like (e.g. Liu and Fu 1982), or local statistical descriptors of the seismic record (e.g. Simaan 1998). Although they do not seem to have been incorporated into the major seismic analysis software, these approaches are interesting for the sub-surface applications. But they are of limited interest for the water-seabed interface, about which they give little or no information.

The extraction of physical/geotechnical information from the digital seismic reflection response is a domain of on-going research. But seismic data are limited spatially, as they give information about the seabed directly below the survey lines only, and no information athwartship. The solution is to survey along closer lines, creating a denser grid. The reflections can be treated as a volume, rather than a profile, so that the analyst can look at the data from many angles in three dimensions (e.g. Kidd 1999). Many commercial software exist for this purpose, such as EarthVision (Durham 1999) or VoxelGeo (Kidd 1999), but their classification routines are mostly semi-quantitative. The other problem is that three-dimensional seismic data are very expensive to acquire, and estimated to cost around

$1,000,000 for an area of 30 km², which leaves 3-D seismics out of reach of most academic institutions.

Echosounder-based techniques

Echo-sounders are the most simple acoustic systems to use. They transmit a single beam, oriented toward the ship's nadir (Fig. 5). They generally use low-frequency signals (< 20 kHz) transmitted in short pulses (< 2 ms). The first return from the seabed corresponds to the point closest to the ship, and the next returns correspond to points further away, as the sound cone spreads. If the seabed is rough, the echo will continue longer, returning from wider angles. The echo begins with a steep edge, due to normal backscattering enhanced by the maximal response of the transducer around its beam axis. The echo then follows with an overall decrease, due to increasing incident angles and to the lower portions of the directivity pattern. Secondary lobes may sometimes be observed, and, if the conditions are adequate, second returns (after reflection on the sea surface) might be recorded as well.

The size of the echosounder footprint is a function of the beamwidth and the water depth. As with all acoustic signals, the return from the seabed will be affected, in decreasing order of importance, by the local geometry of the sensor-target system (angle of incidence of the beam, local slope, etc.), the morphological characteristics of the surface (e.g. micro-scale roughness) and its intrinsic nature (composition, density, relative importance of volume vs. surface diffusion for the selected frequency) (Fig. 6) (Blondel and Murton 1997).

The information contained in the echo can be compared to theoretical curves corresponding to different types of seafloor, and tempered by such factors as the echosounder directivity pattern and the depth. This is the approach chosen by Pouliquen and Lurton (1992). They integrate the return signal to a certain time, and normalise it by its total value. Their system (patented) recognises seven types of seabed commonly found on the continental shelf: rock, gravel, sand, fine sand, muddy sand, mud and soft mud. Sea trials showed good accuracy, and the maximum depth to which the system has been used is 200 m. Compared to other classification techniques (see below), this approach does not require systematic calibration and/or seafloor sampling before each survey.

The RoxAnn system was developed in the late 1980s, and uses two values derived by the analogue integration of the tail of the first echo and the full extent of the second echo (known as E1 and E2). E1 corresponds to the direct reflection from the seabed, and can be related to its roughness. E2 corresponds to the signal reflected from the sea surface and the seabed again, and can be related to its "hardness" (e.g. Heald and Pace 1996). Two-dimensional displays of E1 and E2 allow the recognition of the different types of seafloor, whose echoes are clustered together. Commercially available, the RoxAnn system has been used in many surveys in diverse regions. A thorough review of its performance with varying sonar parameters and seabed types can be found, *inter alia*, in Dyer (2000). The RoxAnn system requires calibration each time it is installed on a new sonar system, even if it has been used in a similar region and with a similar echosounder, but on a different ship. Finally, RoxAnn works by "gating" the incoming signal, i.e. applying a lower and upper threshold to the signal. If these thresholds are inappropriately chosen, the signal loss is irreversible.

Another famous commercial system is QTC-View (e.g. Preston et al., in Zakharia et al. 2000). Developed at the same time than ISAH-S, this system quantifies the shape of each echo with 5

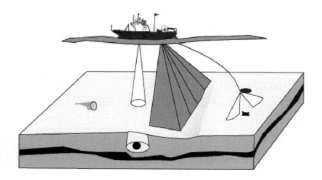

Fig. 5. Schematic view of the different acoustic tools: hull-mounted echosounder and multibeam system, deep-towed sidescan sonar. From the European Community ISACS project home page (http://www.isacs.ntnu.no/ISACS).

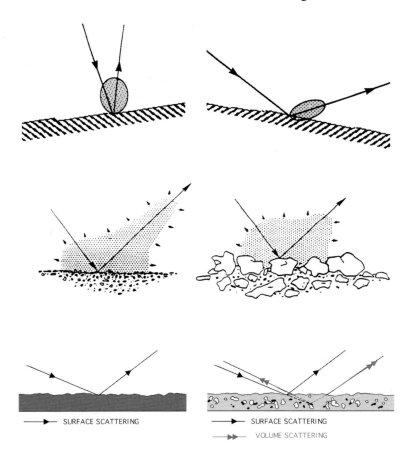

Fig. 6. Acoustic scattering is influenced by the geometry of seafloor imaging, and by the intrinsic character of the seafloor (composition, roughness, volume properties), at scales comparable to the acoustic wavelength. From Blondel and Murton 1997.

algorithms (histogram, quantile, integrated energy slope, wavelet pack coefficients, spectral coefficients), yielding 166 descriptors of each echo. Principal-Component Analysis reduces this to 3 descriptors, which prove enough to recognise the different types of seafloor. Whenever prior information is available, PCA need not be performed. The recognition of seabed types from the 3 remaining descriptors is achieved through a proprietary algorithm (not disclosed). System calibration needs to be performed before each survey, by collecting a few hundred echoes at sites representative of each bottom type to be mapped (known from *in situ* sampling). New seabed classes can be recognised during post-processing. QTC-View is also able to recognise subtle differences between sediments, including gas inclusions.

Similar techniques are being developed around the world. One interesting approach includes fractal description of the echo, combined with backscatter strength and the duration of the echo, and input into a neural network (Tegowski and Lubniewski 2000). Recent developments of "chirp" sub-bottom imaging have been used in basic classification schemes using the sediment impedance profile and attenuation estimation (e.g. LeBlanc et al. 1992), and in later refinements using Biot-Stoll-Gassmann models and time-domain inversion algorithms (e.g. Turgut and Wolf 1998).

Overall, the classification results can be heavily influenced by the beamwidth and frequency of the echosounder, as well as the sea state (making the echosounder point toward a different seabed patch). The information provided by these echosounders can be quite accurate, but the actual penetration inside the seafloor is not always known, and the footprint covered at each measurement can be quite important.

Classification with sidescan sonars

Sidescan sonars cover a much larger portion of the seabed away from the ship: from a few hundred metres to 60 km or more (Blondel and Murton 1997). This coverage is attained by transmitting two broad beams, one on each side (Fig. 5). The new sidescan systems can acquire bathymetry as well. Using different frequencies, they can reach resolutions on the seabed of 60 m down to 1 cm. The images produced can be processed with the arsenal of image classification techniques already developed in the field of satellite remote sensing, some of which are already applied for optical image classification (cf. above). But these techniques must be adapted to the acoustic processes which created these pictures, because the actual backscatter is heavily influenced by the physical processes on the seafloor (Fig. 6).

The tonal information (i.e. from the backscatter alone) is a combination of too many factors to be unambiguously interpreted (e.g. Blondel 1996; Blondel and Murton 1997). A few studies have endeavoured to use the tonal information alone, in some cases with co-registered bathymetry and/or with neural networks, but they were limited to geologically simple terrains, or were of very limited accuracy.

Looking closely at a sonar image (like the one in Fig. 7, left), we realise that the information about the type of seabed is in fact mainly conveyed by the texture of the image. This has been confirmed by the theoretical and experimental studies of the last 20 years, presented in Blondel (1996) and Blondel and Murton (1997). Textures can be intuitively described as smooth or rough, small-scale or large-scale, random or organised. The textural properties correspond to the spatial organisation of the grey levels within a neighbourhood. They are best quantified with stochastic methods, such as Grey-Level Co-occurrence Matrices (GLCMs). This forms the basis of the classification software TexAn (e.g. Blondel 1996; Blondel 1999).

GLCMs address the average spatial relationships between pixels of a small region, by quantifying the relative frequency of occurrence of two grey levels at a specified distance and angle from each other. More than 25 textural indices have been developed to describe GLCMs. Their usefulness for sonar images has been assessed in detail (e.g. Blondel 1996; Blondel 1999), and only two indices have been retained: entropy and homogeneity. Entropy measures the lack of spatial organization inside the region where the GLCM is computed. Entropy is high when all co-occurrence frequencies are equal, i.e. very low. This marks rougher textures. Conversely, entropy is low when the texture is smoother and more homogeneous. Sedimentary facies will show low entropies, increasing with the number of heterogeneities (gravel patches, ripples, rocky outcrops). More complex structures such as slumps or turbidity channels will present higher entropies. All geological features visible on sonar images are characterised by specific entropy signatures. Local-scale variations will also be enhanced by changes in entropy. The other textural parameter, homogeneity, is directly proportional to the amount of local similarities inside the computation window. This parameter was specially modified to ensure its invariance through linear transformations of the grey levels (Blondel 1996). If the Angle-Varying Gain or Time-Varying Gain are changed during sonar data acquisition, or if the image contrast is changed but the seafloor remains the same, then the modified homogeneity parameter will remain the same. Homogeneity will be higher in regions of homogeneous backscatter or in regions with a few grey levels organised along at the scale of the computation window. Homogeneity is able to quantify the differences between smooth sediments and faulted or deformed areas (including ripples or slumps).

Like the parameters used in RoxAnn or QTC, these two parameters are sufficient to completely classify seabed types, with accuracies ranging between 60% and 100% (as measured through intensive and extensive comparison with existing ground truth). The classification algorithm used is called Measurement-Space Guided Clustering (Blondel 1999), and is designed to make maxi-

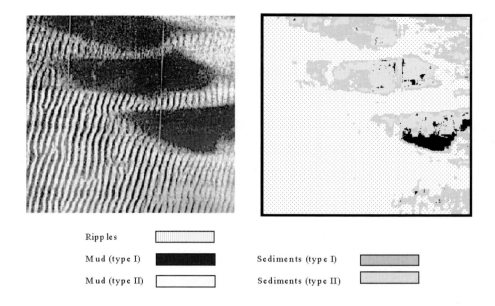

Ripples	▨
Mud (type I)	■
Mud (type II)	☐

| Sediments (type I) | ▧ |
| Sediments (type II) | ▨ |

Fig. 7. Example of seabed classification with TexAn (Blondel 1999). The image is ensonified from the right, at an altitude of 6 m above the seafloor, and covers an area approximately 30 m wide. Lighter tones in the original sonar image (left) correspond to higher backscatter levels.

mum use of the interpreter's knowledge, if available. Figure 7 shows a typical application of TexAn, in a region chosen, on purpose, as simple, because of the possibility that finer details might get lost in reproduction. The image on the left was acquired with a 100-kHz sonar (Klein 590), in a water depth of 18 m in the North Sea. Only very limited processing was performed after the acquisition. The image can be divided into two major elements: dark homogeneous patches of low backscatter and linear parallel backscatter variations. On the basis of their characteristics, and with subsequent ground-truthing (Hühnerbach, pers. comm.), these areas are respectively interpreted as patches of mud, and wave-driven ripples in medium-grained sand, with both bifurcated and regular crests (Milkert and Hühnerbach, in Blondel and Murton 1997). TexAn maps out exactly the different regions of mud and sand ripples, but goes further by detecting other textural variations. Some areas of the sand ripples are less distinct: the ripples may be smaller, or made from a different material. They are located mainly at the proximity of the mud patches, supporting the idea of ripples decreasing in height as the geological setting changes. Similarly, the dark

mud patches are in fact composed of distinct regions. The darker tones delimit homogeneous areas where the textures are smoother. Logically, they are located away from the regions where ripples are destroyed or reduced. Finally, the artefacts visible on the image (sub-vertical lines) appear only as slight variations in the textures, but do not influence the final results, nor the geological interpretation.

TexAn has been validated by comparison with ground-truthing at all ocean depths, and for frequencies ranging from 6.5 kHz to 657 kHz (e.g. Blondel 1996; Blondel 1999). It was also shown to detect details invisible to the human eye, however trained, because of its physiological limitations (Blondel 1996). TexAn is hardware-independent, and can be used at any stage of processing (or non-processing). Contrary to some seafloor characterisation techniques, TexAn does not lose the original information. It is always possible to come back to the original sonar data, and reprocess/reinterpret it. This technique makes most of the user's experience, and with Artificial Intelligence techniques, can build self-learning, self-adapting systems.

Another processing package, called SAPS, is cited for completeness as it includes some sediment classification capabilities based on similar textural analyses. But its applicability to complex terrains or the types of seabeds found in deeper waters of the continental margins has never been tested. From the scarce marketing literature available on the subject, it seems restricted to frequencies of a few tens of kilohertz.

The performance of any classification technique relies heavily on the quality and amount of data processing performed (Blondel and Murton 1997). One cannot stress enough the progress that could be achieved if all surveying sidescan sonars were properly calibrated, and if the final images were expressing the backscatter in dB and not relative grey levels. Such progress was seen in satellite imagery, where each pixel in the image corresponds to a well-calibrated reflectance, which can be used in turn to refine the classification accuracy. More sidescan instruments are now delivering high-resolution bathymetry, or collecting some kind of underway data (chemical, optical, etc.). This new information should also be added to refined or new classification schemes. The current trend toward standardisation of the proc-essing techniques (see Blondel and Murton 1997, for details), the systematic calibration of instru-ments, and the incorporation of sidescan sonar and other data all promise that this field of research will remain open and active in the next decades.

Multibeam seabed classification

Multibeam systems are the latest addition to deep-sea surveying tools. They transmit several beams (up to 120 for some instruments), covering a wide swath (Fig. 5). These systems acquire bathymetry for each beam, and backscatter strengths can be derived from the individual measurements. Multibeam systems have proved particularly attractive for the mapping of Exclusive Economic Zones (in particular for their extension beyond the 200 nm-limit, for which bathymetry is paramount), and for repeat surveys of near-shore areas. Addressing the comments made at the end of the previous section, multi-

beam systems are always calibrated, and the data processing steps are practically standardised.

Knowledge of the local bathymetry, at each point where backscatter has been acquired, can be used to correct the imagery and represent it using the exact local incidence angles. Analysis of the imagery itself proves different from the side-scan sonar case, partly because the footprint on the seabed is generally much larger (100 m compared to a few metres), partly because of the difference in frequency and beam widths. Textural analysis techniques, for example, have not been very successful. Markov Random Fields have met with more success. They are used to partition the backscatter image into regions as close as possible to the geological partition (e.g. Lurton et al. in Pace et al. 1997). Markov fields provide a simple mathematical model of local interactions at the pixel scale. Many types of optimisation algorithms are available in the literature, such as Iterated Conditional Mode (ICM) or simulated annealing. Some of these techniques are now incorporated into IFREMER proprietary software.

A complete suite of different software tools for multibeam data classification has been developed by the Ocean Mapping Group of the University of New Brunswick (Canada). Mayer et al. (in Pace et al. 1997) use waveform characterisation of the vertical incidence measurements, and association with a particular seabed type, either provided by the user or from a waveform dictionary. Alternatively, more refined analysis can be performed on the waveform, and analysed in multivariate space with a graphical tool called Lassoo. This allows to mimic waveform descriptors such as the E1 and E2 parameters used by RoxAnn, or to perform more complex analyses. Hughes-Clarke et al. (in Pace et al. 1997) combine the angular response of the backscattered intensity with information from the local bathymetry. The angular response is characterised based on its mean level, on the slope over pre-defined angular sectors, and on the presence or absence of abrupt changes in the slope. Because the angular response is derived from a finite area, the presence of sediment boundaries must be tested as well. Mitchell and Hughes-Clarke (1994) use a similar technique, adding measurements of topographic curvature to improve the precision. Classification maps generally show a good accuracy. Some of the examples presented in these articles were acquired at an impressive speed of 16 knots, with the Simrad EM-1000 multibeam system.

Conclusion

This article aimed at showing the different tools of seabed classification, and how they had been applied, or could be applied to ocean margins. After a reminder of how varied and complex the ocean margins are, the article proceeded to define seabed classification, its different meanings, and its objectives. The definition chosen here is that seabed classification techniques should achieve at least a basic recognition of the type of seafloor surveyed, or identify a set of intrinsic characteristics from which the type of seafloor could be identified. The more intuitive tools were presented first: *in situ* analysis, in particular with free-fall instruments; and optical seabed classification, from video or still imagery. Interesting techniques using the chemical characteristics of the seabed, and the differences in γ radioactivity, were presented next.

The main part of the article was the review of acoustic classification techniques, by far the most present in the literature. The particularities of acoustic scattering on the seafloor mean that physical processes are more important, and need to be accounted for. Seismic techniques provide some information on the seabed, but have comparatively lower spatial resolutions and do not offer as much field for computer-based seabed classification algorithms. Echosounder-based techniques are more promising, and this was seen with the flurry of commercial packages available, of which Roxann and QTC-View are but two examples. Classification of sidescan sonar images has been progressing much in the last decades, in part because of the increase in computer power which allowed to test new or more intensive approaches. The technique presented here (TexAn) has been the most intensively tested, covering the whole ocean depth at frequencies from 6.5 kHz to 675 kHz (spanning two orders of magnitude). Finally, multibeam classification techniques were also presented, with software packages from IFREMER and the University of New Brunswick.

The present article does not aim at being an exhaustive list of classification software and techniques, and their compared merits. Rather, it aims at showing the basic mechanisms by which seafloor classification can be attained, and how its accuracy can be assessed. The examples given throughout the article are far from being the only examples available, and the choice was dictated as much by the proven merits of these systems as by the space available. A whole book would instead be necessary.

The capability and spatial coverage offered by the different classification techniques will vary, depending on the tools and on the applications they are intended for. Some tools will prove very important for large-scale geological studies or EEZ mapping, while others will be more important for marine biology or environmental assessments. Some tools are very accurate, but require extensive calibration and hardware modifications; while others are maybe less accurate, but more user-friendly and simple to deploy.

Seafloor classification is an open and fast-moving field of research. Current trends are based as much on technological improvements as on scientific advances. Some of the developments most likely to impact on future seabed classification are:

• Increasing computer power and storage capabilities, for steadily decreasing costs;
•Addition of underway geophysical or chemical measurements, which might provide additional information about the seabed;
• Increasing use of Artificial Intelligence (such as belief and decision networks) and data mining techniques;
• Use of multi-frequency measurements, for example with parametric arrays (e.g. Caiti et al. in Pace et al. 1997). This creates the potential for multispectral classification as performed so successfully in satellite remote sensing;
• Move toward higher frequencies (1 MHz and beyond) and therefore higher resolutions on the ground;
• Use of multistatic geometries, where the sonar and one or several receivers are physically

decoupled (e.g. Blondel et al. in Zakharia et al. 2000);

• Better understanding of the physical processes behind high-frequency acoustic scattering on the seabed, through theoretical and experimental studies.

Computer-based seabed classification is steadily gaining a foothold in even the more traditional domains of marine surveying. Its reasoned use makes it an invaluable tool to help the human interpreter and the ocean-going scientist, whether they are specialists of the phenomena investigated or not. Seabed classification tools are not primarily intended as substitutes for human interpretation, and they should not be viewed as such. The main powers of decision/interpretation still rest with the scientist. This flexibility, supported by the latest technical advances, guarantees a promising future for seabed classification and its applications in multi-disciplinary studies of the ocean margins.

Acknowledgements

I am very grateful to Dr. D. Billett (SOC, UK) and Prof. Dr. G. Wefer (U. Bremen, Germany), for their invitation. I would also like to thank Dr. I. Mehser (Hanse Inst., Germany) and the Hanse 2000 organising committee. Discussions and thorough reviews of this manuscript by Prof. W. Berger (UCSD, USA) and Dr. T.C.E. van Weering (NIOZ, Netherlands) are gratefully acknowledged.

References

Balsam WL, Deaton BC, Damuth JE (1999) Evaluating optical lightness as a proxy for carbonate content in marine sediment cores. Mar Geol 161:141-153

Barbagelata A, Richardson MD, Miaschi B, Muzi E, Guerrini P, Troiano L, Akal T (1991) ISSAMS: An *in situ* sediment geoacoustic measurement system. In: Hovem JM, Richardson MD, Stoll RD (eds) Shear Waves in Marine Sediments. Kluwer, Dordrecht, pp 305-312

Blondel Ph (1996) Segmentation of the Mid-Atlantic Ridge south of the Azores, based on acoustic classification of TOBI data. In MacLeod CJ, Tyler P, Walker CL (eds) Tectonic, Magmatic and Biological Segmentation of Mid-Ocean Ridges. Geol Soc Spec Publ 118:17-28

Blondel Ph (1999) Textural analysis of sidescan sonar imagery and generic seafloor characterisation. J Acoust Soc Amer 105:1206

Blondel Ph, Murton BJ (1997) Handbook of Seafloor Sonar Imagery. PRAXIS, Chichester

Coles BW, Radzelovage W, Laurant JP, Reihani K (1998) Processing techniques for multi-spectral laser line scan images. Proc IEEE/MTS Oceans'98

Durham LS (1999) 3-D models are making cents. AAPG Explorer

Dyer CM (2000) Studies of an acoustic sediment discrimination system. MPhil thesis, University of Bath

Harsdorf S, Janssen M, Reuter R, Wachowicz B, Willkomm B (1998) Lidar as part of an ROV-based sensor network for the detection of chemical pollutants on the seafloor. Proceedings IEEE/MTS Oceans'98

Heald GJ, Pace NG (1996) An analysis of 1st and 2nd backscatter for seabed classification. ECUA-1996 Proceedings, Crete 649-654 p

Hebbeln D (2002) State of the art and future prospect of scientific coring and drilling of marine sediments. In: Wefer et al. (eds) Ocean Margin Systems. Springer, Berlin pp 57-66

Irish JL, Lillycrop WJ (1999) Scanning laser mapping of the coastal zone: The SHOALS system. ISPRS J Photogram Rem Sens 54:123-129

Juteau T, Maury R (1999) The oceanic crust, from accretion to mantle recycling. PRAXIS, Chichester

Kidd GD (1999) Fundamentals of 3-D seismic visualization. Sea Techn 40:19-23

Lebart K, Trucco E, Lane DM (2000) Real-time automatic sea-floor change detection from video. Proceedings IEEE/MTS Oceans'2000 337-343

LeBlanc LR, Mayer L, Rufino M, Schock SG, King J (1992) Marine sediment classification using the chirp sonar. J Acoust Soc Amer 91:107-115

Liu HH, Fu KS (1982) A syntactic pattern recognition approach to seismic discrimination. Geoexplor 20:183-196

de Meijer RJ, Venema LB, Limburg J (2000) Synoptical mapping of seafloor sediment. Sea Techn 41:21-24

Mitchell NC, Hughes-Clarke JE (1994) Classification of seafloor geology using multibeam sonar data from the Scotian Shelf. Mar Geol 121:143-160

Noakes JE, Noakes SE, Dvoracek DK, Bush PB (1999) Survey and mapping system for surficial marine sediments. Sea Techn 40:49-54

Pace NG, Pouliquen E, Bergem O, Lyons AP (eds) (1997) High-Frequency Acoustics in Shallow Water. Saclantcen Pub, Lerici, Italy

Pouliquen E, Lurton X (1992) Identification of the nature of the seabed using echo sounders. J Phys 4-2(C1):941-944

Rosenberger A, Weidelt P, Spindeldreher C, Heesemann B, Villinger H (1999) Design and application of a new free fall *in situ* resistivity probe for marine deep water sediments. Mar Geol 160:327-337

Simaan MA (1998) Texture-based techniques for interpretation of seismic images. Proceedings Oceans'98 MTS/IEEE

Stoker MS, Evans D, Cramp A (eds) (1998) Geological Processes on Continental Margins: Sedimentation, Mass-Wasting and Stability. Geol Soc Spec Publ 129

Stoll RD, T Akal (1999) XBP – Tool for rapid assessment of seabed sediment properties. Sea Techn 40:47-51

Tegowski J, Lubniewski Z (2000) The use of fractal properties of echo signals for acoustical classification of bottom sediments. Acta Acustica 86:276-282

Telford WM, Geldart LP, Sheriff RE (1990) Applied Geophysics. Cambridge University Press, Cambridge

Turgut A, Wolf SN (1998) Inversion of sediment properties from chirp sonar data using an extended Biot–Stoll–Gassmann model. J Acoust Soc Amer 104:1788

Werby MF, Tango GJ, Ioup GE (1986) Remote seismo-acoustic characterization of shallow ocean bottoms using spatial signal-processing of towed array measurements. Geophys 51:1517

Zakharia M, Chevret P, Dubail P (eds) (2000) Proceedings of the Fifth European Conference on Underwater Acoustics ECUA-2000. Lyon, France

The Benthic Boundary Layer

L. Thomsen

IUB, International University Bremen, P.O.Box 750561,
28725 Bremen, Germany
corresponding author (e-mail): l.thomsen@iu-bremen.de

Abstract: Processes in the benthic boundary layer (BBL) at different continental margins are described and the importance of lateral particle transport, particle aggregation, biological mediated particle deposition and resuspension within the BBL is discussed. Theoretical and methodical aspects of BBL research are demonstrated and examples of modern continental margin studies are given.

Introduction

Ocean margins and continental slope regimes are natural boundaries common to oceanic and shelf sea domains. They are areas of enhanced productivity, and biologically-mediated downslope transport of matter within the benthic boundary layer (BBL) may be large, providing strong links between the two realms. The burial of particle-associated organic matter in continental margin sediments is directly linked to the global cycle of carbon over geologic time scales (Hedges et al. 1999). Although continental margins account only for $\approx 15\%$ of total ocean area and 25 % of total ocean primary production, today more than 90 percent of all organic carbon burial occurs in sediments built up by particle deposition on continental shelves, slopes, and in deltas (Hartnett et al. 1998). This burial is intrinsically linked to the cycling of biogeochemically important elements. Lateral transport of particles at the continental margin is largely controlled by the hydrodynamics within the BBL, which is most important for the exchange between sediments and water column (McCave 1984); however, detailed observations of BBL characteristics at continental margins are rare.

Most people regard benthic boundary layer studies and particle transport as an interdisciplinary endeavour between physical oceanography and sedimentology, but the brand of fluid dynamics used is unfamiliar even to most physical oceanographers. For these reasons, a short summary of particle transport approaches is worthwhile to provide a context in which physical, sedimentological, geochemical and biological interactions can be evaluated before a review on BBL studies at continental margins is given. The term "benthic boundary layer" is the most commonly used term for the water layers above the sediments although in sedimentological/physical-oceanography terminology "bottom boundary layer" would be the right phrase to use. In the whole chapter, useful new references are predominantly cited which lead the interested reader to important previous work on the topic.

Methodologies

The velocity boundary layer is the region above the sea floor where flow is measurably slowed from a more vertically uniform mean velocity in overlying water. When both bottom microtopography is relatively uniform and the overlying flow is steady, flow conditions in the BBL can be easily described. Approaching the sea floor from above, the upper limit of the BBL is defined as the distance above bottom at which the mean flow velocity is $0.99\,u_\infty$, where u_∞ is the free-stream velocity. The BBL at continental margins is in the order of 5 to 50 m thick. More than the theoretical 10 % of the BBL thickness obeys what is known as the law of the wall:

$$\overline{u}(z) = \frac{u_*}{\kappa} \ln \frac{z}{z_0}$$

From WEFER G, BILLETT D, HEBBELN D, JØRGENSEN BB, SCHLÜTER M, VAN WEERING T (eds), 2002, *Ocean Margin Systems.* Springer-Verlag Berlin Heidelberg, pp 143-155

which is often used to describe the mean velocity profile in regions near the bed where the flow is fully turbulent and neutrally stable (i.e. not stratified). Convention in BBL work is to use z as a distance [L] upward from the bottom, κ is von Karman's constant (0.41, dimensionless), u_* is shear velocity [LT^{-1}] and z_0 is the roughness height. The z_0 of a natural sea floor is strongly affected by benthic organisms structuring the microtopography of the sediment surface. Under turbulent conditions $z_0 \approx d/30$, where d is the mean height of the bottom microtopography as long as it is closely spaced. With suitable time averaged measurements of the velocity at a series of measurements heights within the logarithmic layer (log-layer) the slope of the fitted line (Fig. 1) can be used to estimate u_*, and z_0 is defined as the intercept. Higher u_* corresponds with greater vertical shear and greater horizontal stresses.

For a reliable calculation of the bottom shear velocity (u_*) at least 2 flow measurements within the log-layer are required. The latest down-looking ADCP (acoustic doppler current profiler) technology is the most promising tool because it allows velocity profiles within the BBL of 1.5 - 5 cm sectors and offers the best, non disturbing solution for flow measurements, even in the deep-sea (van Weering et al. 2000).

The flow velocity measurements at distinct heights above sea floor within the lower 40 % of the logarithmic layer (the so-called "constant stress layer") are plotted against log height and used to calculate the shear velocity (u_*) from the von Karman-Prandtl log-profile relationship via

$$u_* = \frac{\overline{\Delta u}}{5.75 \Delta \log z}$$

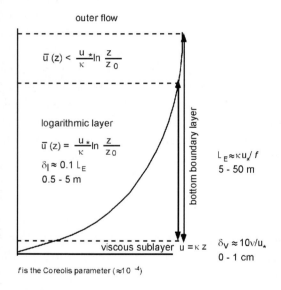

Fig. 1. Vertical structure of the bottom boundary layer. The thickness of each layer can be calculated via the u_*.

It is really a shear stress on the wall-in disguise:

$$\tau_0 = \rho u_*^2$$

where ρ is the density of the fluid, thus in water the square of the shear velocity approximates numerically the mean shear stress operating on the bed.

Within the log-layer turbulent flow is characterized by eddies that are carried downstream in the direction of the general flow. As turbulence is an irregular rotational motion, eddies are formed and their energy is dissipated by transfer to smaller and smaller-scaled eddies. The size of these smallest eddies (the so-called "Kolmogoroff-length", λ) transported depends on the kinematic viscosity of the fluid (ν) and the turbulent energy dissipation (ε) in the constant stress layer, where

$$\varepsilon = u_*^3/\kappa z$$

and

$$\lambda = (\nu^3/\varepsilon)^{0.25}$$

Both ε and λ are useful parameters for the determination of solute fluxes, microbial processes and particle encounter rates for aggregation and particle capture processes (McCave 1984; Ritzrau 1996; Thomsen and McCave 2000). The kinematic viscosity ν is strongly temperature dependent and has an enormous influence on the behavior of particles and small organisms of low Reynolds numbers. ν nearly doubles from warm surface waters (0.01 $cm^2 s^{-1}$ at 20 °C) to cold bottom waters (0.018 $cm^2 s^{-1}$ at 1 °C) at continental margins.

A criterion for the transition from laminar to turbulent flow is formulated in terms of the Reynolds number:

$$Re = \frac{ud}{v}$$

which relates the velocity of the fluid u (or instead the particle/organism velocity relative to the surrounding fluid) to the kinematic viscosity and a characteristic length scale d of the flowsystem (i.e. the depth of the flow-system, or for flow around particles, the particle/organism diameter or length). For Re < 24 viscous forces dominate, for Re > 100 turbulence becomes important. In general Re describes the ratio of the inertial forces to the viscous force acting on a body of interest, be an organism, particle or fluid.

Below the log-layer, within the lowermost millimeters above sea floor, flow velocity is so reduced that viscous forces dominate in the "viscous sublayer" (Fig. 1). The thickness is $\delta_v \approx 10v/u_*$ but when roughness is more than about $\delta_v/3$ it exists discontinuously, and when it is greater than $1\delta_v$ the layer does not exist at all and flow is fully turbulent right to the bed. Where the bottom is rough, turbulent eddies develop at the bottom that dissipate upwards into the main flow. Thus near the bottom, the turbulence is usually more intense, and of smaller scale than higher in the flow.

For mass flux calculations the settling velocity of a particle as another important parameter should be determined. As the particle transport represents intense exchange of suspended material with the bed, some of the relevant dynamics are made more transparent by looking at the reason for persistence of a vertical concentration gradient dC/dz, under turbulent diffusion. This dynamic balance is of gravity, in the form of particle settling velocity, moving particles toward the bed and acting to steepen the gradient and of turbulent diffusion moving particles away from the region of high concentration near the bed:

$$w_S \frac{\partial C}{\partial z} = K_S \frac{\partial^2 C}{\partial z^2}$$

where Ks is the turbulent (eddy) diffusion coefficient for particles and w_s is the settling velocity. For particles with Reynolds numbers < 1, Stokes law can be applied to determine the settling velocity:

$$w_S = \frac{d^2 (\Delta\rho)g}{18v\,\rho}$$

Where d is the particle diameter, $\Delta\rho$ is the excess density of the particle over seawater, g is the gravitational constant and v is the kinematic viscosity of the fluid. As Stokes law was originally applied to rigid, impermeable spherical particles of known density, it is difficult to apply to non-spherical aggregated particles which virtually represent most particulate material in the ocean. Therefore empirical particle-size/settling-velocity relationships have been developed (Fig. 2). However, Stokes law can then be used to back-calculate the particle density of the aggregates as dicussed by McCave (1984).

Further, it is expected that the main factor that determines the distribution of suspended sediments with height above the bed is the ratio of the settling

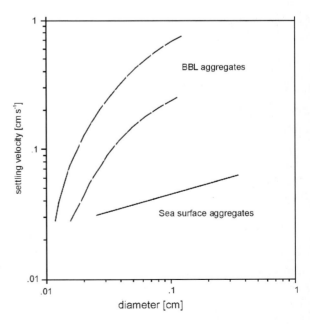

Fig. 2. Particle settling velocities/particle diameter relationships of aggregates.

velocity to the shear velocity (w_s/u_*). If the Rouse number [$p_s = w_s/\kappa u_*$] is of order 1, but the sediments are eroded, the material moves as bedload. If p_s is of order of 0.1 to 1, settling is still important with increasing particle concentrations towards the sea floor. If p_s is < 0.1, material will stay in suspension. Given p_s the particle concentration at any height above bottom within the constant stress layer can be determined via:

$$C_z = C_a(a/z)^{ps}$$

where C_a is the known concentration at height (a) above bed and C_z is the concentration to be determined at height (z) above bed. The lateral particle flux at a given distance (z) off the sea bed is:

$$J(z) = u(z) \times C(z)$$

where J is the flux ($L^2 T$), u is the mean flow velocity at height (z) above the bed and C is the particle concentration.

The methods of particle-size measurements may be divided into those which give information on the whole size distribution and those for which the information comes from a size window with upper and lower limits. Wholesale misunderstanding arises from the attempts to attribute significance to differences in parameters obtained for the same sediment by different methods. For the determination of the size/settling velocity relationship of particles > 100 μm in diameter, settling cylinders with cameras are now quite reasonable (synopsis by Eisma et al. 1996). Instrument systems with water samplers and cameras allow the study of potential cause and effect relationship associated with the morphological, chemical and biological characteristics of large particles. The principle here is to capture the suspended flocs very carefully so that the large particles are not destroyed. Next the suspended flocs settle in a vertical tube under still water conditions and from the video recordings both floc sizes and settling velocities can be determined. For smaller particles in the size range < 100 μm the Sedigraph method is the most reliable but for low concentrations the coulter-counter is best (McCave et al. 1995).

In situ particle transport measurements in the bottom boundary layer have increased significantly in recent years. Instrumented tripods with flow meters, transmissometers, optical backscatter sensors, *in situ* settling cylinders and programmable camera systems have often been used in marine environments. These instruments were deployed to study particle transport dynamics in the benthic boundary layer and were able to collect water samples (1-10 l) at given distances from the seafloor. The latest technology available includes the already mentioned ADCPs, arrays of miniature optical backscatter sensors, particle cameras, settling cylinders, *in situ* filtration and sonar systems. These techniques allow the exact study of sediment dynamics at continental margins, even on longer time scales of months to years (van Weering et al. 2000; Sternberg and Nowell 1999; Thomsen et al. in press).

Implication of the BBL for fluxes at continental margins

The distribution of sediment components as a function of grain size and of size parameters for the inference of current strength has been commonplace in marine geology for sand sized material, which is increasingly of biogenic origin with increasing water depth at continental margins. As hydrodynamic conditions at continental margins are enhanced (i.e. higher flow velocities), sediment sorting occurs principally during resuspension or deposition by processes of particle-breakup and – selection according to settling velocity and shear stress. Sorting takes place through differing rates of sediment transport, so that an originally unsorted mixture is converted downstream into narrower distributions.

Sedimentological implications

Several attempts have been made to provide grain size parameters that respond to changes in current strength for palaeoceanography. However these sediment parameters are not capable of resolving whether the currents responsible for sorting have a small mean and large variability or large mean and small variability (McCave et al. 1995). It is not

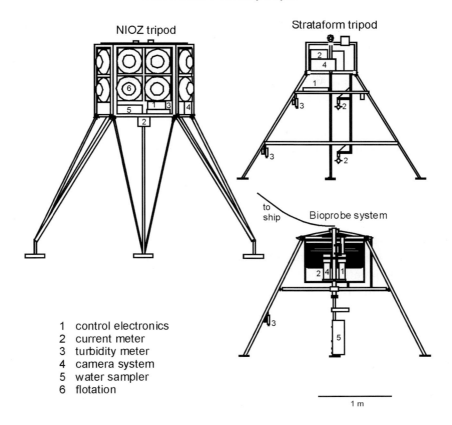

NIOZ tripod

Strataform tripod

to ship

Bioprobe system

1   control electronics
2   current meter
3   turbidity meter
4   camera system
5   water sampler
6   flotation

1 m

Fig. 3. Examples of latest BBL monitoring technology. Left: NIOZ BoBo long term lander for deep-sea studies. Lower right: The GEOMAR BIOPROBE II short term water sampler, with hydrodynamically shaped 5-40 cm sampling device. Upper right: The UW Washington STRATAFORM long term lander for shelf research.

yet possible to distinguish the nature of a current i.e. whether a change in size is due to an increase in the variability of speed. However by using actual data-sets on BBL dynamics one can identify those oceanic regions where strong mean flow and those where strong eddy variability occurs. The effects of eddy energy on the deep-sea floor have been recognized as "benthic storms". In these, long-period variability of the deep flow shows short episodes of intense sediment resuspension. This is in particular the case for continental margin environments, where boundary currents, internal waves and tides can create shear stresses exceeding the threshold needed to erode the sediments. Size distributions under modern ocean currents show a general relationship between coarser silt mean and the path of deep western boundary currents as defined by turbidity and hydrography but the data do not allow a relationship to be derived between

mean size and long-term mean current velocity. According to McCave et al. (1995) the best current-related parameters are probably modal or median size of the 10 - 63 µm fraction. Well-sorted, fine bed sediments respond as non-cohesive particles down to 30 - 10 µm diameter, in which case Shields-type diagrams provide links between flow, bed stability/behaviour and transport mode; below this size, particles bond electrochemically and become cohesive, either as clay minerals with their charge imbalances entering the composition spectrum or as other particles (such as quartz) with size-specific surface charges.

The two layer concept of the sediment interface

Cohesive sediments with biological interaction, found at mid- and lower-slope regions of continental margins, are part of a complex chain of events which are not fully understood in their effects on bed stability, flow structure, material fluxes, and ecosystem dynamics. To understand and quantify these transport processes, both laboratory and field studies are required to link hydrodynamics with particle and bed formation. Advective near-bed fluid flow imports both particles and solutes from sources upstream (in addition to those arriving from above) that can serve as food (and fuel) for benthic communities. Through the coupling to the local ecosystem (the benthic mill), more or less nutrient-rich, reprocessed or remoulded particles such as faecal wastes or detritus, and dissolved metabolites are subsequently exported (moved downstream). The bottom stress (τ), expressed as friction velocity (u_*), the turbulence intensities and the mean local horizontal speed together with the controlling variables of sediment transport: the critical erosion stress (τ_c), the critical deposition stress (τ_d), and the particle settling velocity (w_s). In order to obtain τ_c, erosion devices have to be used. Thomsen and Gust (2000) presented the latest data-set on critical bed shear stress for continental margin sediments (Fig. 4).

The bottom sediments (200 to 5000 m water depth) at the continental slopes studied consisted of a thin surface layer which resuspended as aggregates (mean diameter 125 - 2400 μm) under critical shear velocities (u_{*c}) of 0.4 to 1.2 cm s⁻¹. For the underlying sediments, eroded as primary particles, u_{*c} increased with water depth from 0.7 cm s⁻¹ (sandy upper slope sediments) to 2.1 cm s⁻¹ (cohesive clay sediments). The authors discuss the concept of a two-layer sediment interface, which distinguishes between an underlying sediment layer, and a more easily resuspendable surface aggregate layer. The surface layer consists mainly of BBL aggregates in the 140 – 450 μm median size range, 0.05 – 0.35 cm s⁻¹ settling velocity range and is resuspended at mean thresholds u_{*c} of 0.8±0.1 cm s⁻¹. The aggregates consist of ≤ 75 % of organic matter, which was mostly refractory with a carbon/nitrogen ratio exceeding 10, and the lithogenic material was embedded in the amorphous matrix of the organic. Aggregates contained remnants of faecal pellets, meiofauna organisms and shell debris of foraminifera. 35 - 65 % of the bacteria of the BBL were particle attached and covered the organic matrix of the aggregates and approximately 1 % of the organic fraction was labile bacterial organic carbon (Thomsen and Gust 2000).

These aggregates can subsequently be transported in tide-related resuspension-deposition loops over long distances (Fig. 5).

The onset of a cohesion effect in their studies started to show at about 30 μm. Thus, organo-mineral particles with average sizes < 30 μm behaved in the same way as clay (< 2 μm), and particles coarser than 30 μm should display size sorting behaviour. This result was different from the calculations of McCave et al. (1995), who propose 10 μm as threshold between non-cohesive and cohesive sediment behaviour. This difference seems due to particle stabilisation from microbial exudates. The erosion threshold data were mainly obtained in summer, when biological activity in surface sediments at the study site is high (Duineveld et al. 1997). Evidence for the importance of biological adhesion on critical stress for incipient transport has been demonstrated by various authors (e.g. Dade et al. 1990). Microbial exudates can in-

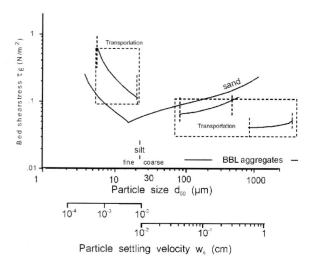

Fig. 4. Critical bed shear stress (τ_c) for erosion of continental margin particles, showing the onset of a cohesion effect at about 30 μm. The black curve represents the critical bed shear stress for erosion of quarz.

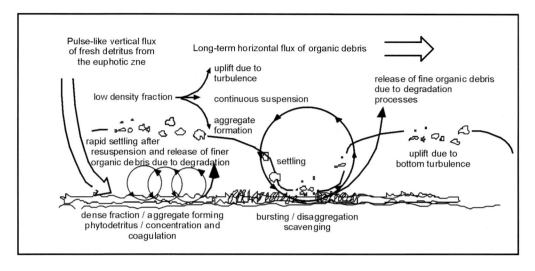

Fig. 5. Schematic model of the resuspension loop of aggregated particles at continental margins.

crease the critical bed stress by a factor up to 5, and the presence of animal burrows within sediments can also lead to the strengthening of the sediment bed. The biological exudates are also responsible for the formation and strength of large particles/aggregates transported within the resuspension loop and microbial processes play an important role in the formation of aggregates (Ritzrau 1996). Observation of the seabed accumulation of organic matter and the relevance of aggregates as a food source for the benthopelagic communities has been discussed by Thomsen et al. (in press).

The aggregation of suspended matter plays an important role in the transfer of particles through the water column. It has been increasingly recognized that in the deep ocean the vertical transport is dominated by large relatively rare particles that settle quickly. Unresolved, due mainly to lack of data, is the issue of whether particles of various sizes in deep BBLs are likely to be scavenged by large aggregates or transported to the bed as single grains. As this issue bears directly on interpretation of fine-grained size distributions in deep-sea sediments (McCave et al. 1995) and biodeposition rates of benthic communities (Graf and Rosenberg 1997), resolution is critical. Thomsen and McCave (2000) calculated small particle scavenging rates in the benthic boundary layer at the Celtic Sea continental margin. Particle scavenging of small particles ≤ 5 μm at the study site

took place under typical flow conditions of $u_* > 0.2$ cm s^{-1}. Scavenging of ≥ 10 μm particles could only take place under massive sediment erosion. It was expected that the BBL aggregates found at the continental margin at the mid and lower slope areas originate in sediment resuspension events i.e. under flow conditions high enough to erode the cohesive sediments below the surface aggregates. Under these hydrodynamic conditions dissolved organic carbon can be released into the BBL in large amounts, as pointed out by Otto and Balzer (1998) and can feed surface primary production after internal wave resuspension and subsequent vertical turbulent diffusion.

During times of enhanced flow conditions, aggregates are formed and compacted by shear, which accounts for the fact that they do not disaggregate when they enter the viscous sublayer at mid- and lower slope sediments. The formation of these BBL aggregates and the minerals incorporated into the particles might be responsible for the organic matter preservation on continental slopes. Total clearance of bottom nepheloid layers only occurs during times of transport of large aggregates > 1000 μm after the arrival of phytodetrital aggregates on the sea floor with number concentrations > 200 x 10^3 m^{-3}. This is comparable to the clearance of intermediate nepheloid layers during fall out of aggregated phytodetritus. Table 1 summarizes typical particle characteristics from continental margin BBLs.

$\overline{u}$	[cm s^{-1}]	2	-	50
u_{*c}	[cm s^{-1}]	0.1	-	2
τ_c	[Nm^{-2}]	0.01	-	0.4
Total particulate matter	[g m^{-3}]	0.01	-	0.4
Particulate organic carbon	[mg m^{-3}]	10	-	150
Chlorophyll equivalents	[mg m^{-3}]	0.01	-	0.3
BBL aggregate number	[n m^{-3}]	(10	-	1500) * 10^3
BBL aggregate diameter	d_{50}	140	-	2400

Table 1. Typical flow- and particle characteristics from continental margin BBLs.

The benthic-pelagic coupling

The pulse like seasonal input of aggregated phytodetritus at continental margins triggers an awakening response of benthic organisms (Graf 1992) within the top 10 cm sediment layer. Due to the hydrodynamic conditions at the margin, an active biogeochemical layer is produced which extends from 10 cm sediment depth to at least 30 cm height above the sea floor, where biodeposition, bioresuspension and bioentrainment alter particle and fluid fluxes within the BBL. The benthic organisms rely on advective processes for their food supply. They are dependent on the fluid medium for transport of this material and thus can in many cases influence these transport processes (Graf and Rosenberg 1997). Near-bed fluid flow imports both particles that provide new food and solutes that can fuel local microbial communities while it exports used particles (feacal wastes) and dissolved wastes. Thus benthic fauna plays an important role in determining the fate of organic material at the seabed. The majority (usually 80 % or more by numbers and taxa) of benthic animals are deposit feeders (Jumars 1993) which substantially modify sediment structure via bioturbation. Generally, the amount of deposit feeders increases, and the overall densities of macrofauna (i.e. animals not passing a sieve with 500 µm mesh size) decrease, with increasing water depth (Jumars 1993). However, while the macrobenthic community seems to be depth-related, further investigations have shown that other physical and chemical factors can also be important. Flach et al. (1998) showed that the hydrodynamic conditions at the Atlantic continental margin strongly influenced benthic community structure. A peak in density and biomass of interface feeders (i.e. those animals, which are able to switch between passive filter feeding and surface deposit feeding) between 1000 and 1500 m water depth coincided with strongest changes in sediment structure and bottom current regime. Studies on the feeding behaviour of many macrobenthic species have shown that biodeposition can be more important in providing food for the benthos than material deposited by sedimentation. Graf and Rosenberg (1997) summarized biodeposition rates of benthic communities, which varied from 0.002 gC m^{-2} d^{-1} for a deep-sea sponge community, via 0.004 – 0.039 gC m^{-2} d^{-1} for mid slope interface feeders to 70 gC m^{-2} d^{-1} for a shallow water mussel bed on the shelf. Thomsen et al. (in press) calculated a mean biodepositional flux of carbon at continental margins of 0.005 % of the horizontal carbon flux. This sounds little but repre-sents an important flux to the sea bed, even under moderate horizontal carbon fluxes within the BBL of $\approx$ 400 gC m^{-2} d^{-1} at u of 10 cm s^{-1} and an aver-age particulate organic carbon concentration of 50 mg m^{-3}. An important outcome of the various studies was that even under conditions of sediment erosion, the benthic community was still able to selectively biodeposit 0.03 gC m^{-2} d^{-1}, i.e. carbon, which would normally pass the highly energetic upper slope areas at continental margins was accumulated at the sediment. Work on bioresuspension is rare. Values range from 70 g (dry sediment) m^{-2} d^{-1} for shallow water communities to 0.5 g m^{-2} d^{-1} for mid slope interface feeders.

At continental margins a progressive change in macrofauna feeding types takes place. Suspension-feeding animals which rely on labile organic carbon dominate on the shelf and upper slope areas, interface feeders are found primarily at the more highly energetic mid-slope sites, and deposit feeders which feed on carbon of much lower nutritional value dominate the lower slope and continental rise (Flach et al. 1998). From the results of studies of flow conditions and particle mass fluxes at continental margins, it is evident that the lateral advection of organic particles is the most important food source for the benthos.

If sediment (or particle) movement is so frequent and intense that food limitation rarely occurs, benthic carbon demand should be balanced for all feeding types. At the Celtic Sea continental margin, the flux of organic carbon into near-bottom sediment traps (Wollast and Chou 2001) constituted 14 to 59 % of required total flow to the sediment surface. Heip et al. (2001) indicated that the reactivity of the organic matter in the sediment at the continental slope decreased with water depth by a factor of 5, whereby at the same time the total amount of mineralizable carbon increased by a factor of 9. However, the increased amount of carbon was of inferior nutritional value and its average residence time of 10 years was 36 times higher than on the shelf (Heip et al. 2001). That indicates that the flux of carbon in the BBL at mid and lower slope areas is high, but it mostly refers to less labile refractory carbon which has passed many resuspension loops. Studies on benthic carbon remineralization rates at the Celtic Sea continental margin suggest that the carbon arriving at the sea floor at the lower slope > 2000 m water depth is degraded slowly during most of the year, because no seasonality with enhanced oxygen uptake rates was observed (Duineveld et al. 1997). The authors also concluded that at upper and mid slope areas with clear seasonality, the carbon input mainly occurs laterally and that the continental rise in water depth > 4000 m was mainly supported by the vertical flux of organic carbon thus indicating a decoupling of these depth zones with respect to food supply. These findings are consistent with other studies. Smith and Kaufman (1999) measured a significant increase of carbon

on the sediment surface at abyssal depth of the Pacific after the arrival of detritus. However oxygen consumption was not enhanced and the seasonal arriving detritus accounted for less than 0.34 % of the total annual mineralization. This can be due to the fact that the labile fraction of the carbon has already been remineralized before analyses were done; Heip et al. (2001) discuss carbon degradation rates in the order of 0.03 to 0.16% d^{-1}. Or the hypotheses can be drawn, that for most of the year the benthos feeds on other carbon sources provided from an additional internal carbon cycling within the BBL.

Analyses of particulate matter in the BBL from all study sites showed that up to 4% of the transported particulate organic carbon is in the form of bacterial organic carbon. 35 to 65 % of these bacteria are attached to the aggregates and thus perform the same resuspension loops. The aggregates were highly organic ($\leq$ 75 % wt). However the average amount of fresh bacterial carbon on the aggregates, 1 - 1.3 % of the particulate organic carbon, is not high enough to alter either the age or the overall high C/N ratio of the particles.

Model calculations of microbial activity on the BBL particles are given by Ritzrau (1996) who shows that an increase of turbulence within the BBL stimulates microbial activity and hence most likely results in increased microbial biomass production. Further on he postulated that bacteria "use" the aggregates as transport vehicles to enhance the accessibility of nutrients. In contrast to diffusion-limited free living 0.2 to 1 µm long bacteria, particle associated bacteria benefit from turbulent nutrient transport towards 100 to 5000 µm large aggregates. Boetius et al. (2000) conclude that bacterial carbon turnover in the BBL of the Arabian Sea was substantially higher than the vertical flux of particulate organic carbon in this region. They pointed out that most likely the BBL aggregates and the enhanced dissolved organic carbon of the bottom water are the main source of energy for the bacteria of the BBL.

The BBL organo-mineral aggregates consist of particulate organic carbon which forms an "ocean background value" of permanently transported particles and the total amount of bioavailable carbon in the aggregates has not yet been determined over the full aggregate size spectrum. In any case, even under low flow conditions of < 5 cm s^{-1} the lateral fluxes of the particulate organic carbon and bacterial organic carbon are orders of magnitude higher than is the vertical flux, and it might be this carbon source that the benthos feeds on for most of the year.

Comparison of continental margins studies

As hydrodynamic processes are highly dynamic at the ocean margins (Huthnance et al. in press), the transport of both lithogenic sediments and organic material is enhanced and controlled by local topography and hydrography. As a result, particle input to the sediments is strongly related to sediment-transport processes across and along the margin; benthic biological communities covering the margin adapt to the extant hydrodynamic conditions and related carbon supply. During winter and the initial phase of annual spring phytoplankton bloom, a temporal delay in the development of heterotrophic biomass often occurs, causing an exponential increase of autotrophic biomass. During such periods, a considerable export of biomass to deeper waters can take place. This has raised questions regarding the modes of transport, final settling and burial of organic matter. Lateral export via the upper water layers appears to be very difficult due to oceanographic conditions (temperature, salinity front). As an alternative, export via lateral advection within bottom and intermediate nepheloid layers has been proposed.

Over the past decade a number of research programs related to these processes and fluxes have been implemented, notably along the north-eastern American Atlantic margin and Mid-Atlantic Bight (SEEP I, SEEP II, Biscaye et al. 1994). In Europe, the western Mediterranean margin was studied during the ECOMARGE project (Monaco et al. 1990), and more recently the Ocean Margin Exchange Project (OMEX I+II, Wollast and Chou 2001) and SFB 313 Project (Schäfer and Thiede 2001) addressed similar processes along the Atlantic margin. However, these studies showed that the processes to be investigated were much more complicated than expected, and that mass balances

were difficult, if not impossible, to establish. In addition to a recognition of the role of lateral transport of particulate matter parallel to the American coast, a carbon depocenter was identified at water depths of 1000 to 1500 m (Biscaye et al. 1994). However, the results of the SEEP II studies clearly indicate that the importance of the eastern American Mid-Atlantic Bight slope as a carbon depocenter is limited and the same holds true for the Celtic Sea continental margin (Wollast and Chou 2001). Sediment trap data provide further indications for the lateral transport of mainly refractory matter. The original hypothesis of a massive export of carbon produced on the shelf could not be confirmed, and < 5 % of the primary production was thought to be exported (SEEP II, Biscaye et al. 1994). The two possibilities for export, direct offshore dispersal through intensive water circulation at the shelf edge and/or cross-slope or slope-parallel transport via the bottom nepheloid layer, were under discussion. The importance of canyon-fed transport was noted in the Gulf of Lions during the ECOMARGE project (Monaco et al. 1990). Studies of the coupling between sinking fluxes of particulate organic matter and the activity of the benthic community (Graf 1992) reveal that the magnitude of vertical primary sedimentation at continental margins is not sufficient to balance benthic supply and the energy requirements of the benthic community. This paradox may originate from instrumental artifacts of sediment traps, which may either underestimate the vertical flux of the fine particle fraction or overestimate the flux of the aggregated fraction. However, also the lateral advection of organic matter within the bottom boundary layer may account for this discrepancy.

Measurements of the flow field on the northwestern European continental margin for example show that the upper slope is blanketed by a persistent northward flowing slope current. At the Celtic slope the near bottom currents (1 to 6 m above bottom) at mid depths are markedly directed downslope, reaching mean speeds of 15 cm s^{-1}. At the Celtic Sea continental margin, erosion of sediments, as well as along and downslope transport was postulated by Thomsen et al. (in press). Observations and experiments on particle characteristics permit linkage of *in situ* particle transport with local sources and sinks in view of the tidal forcing of the flow. Figure 6 demonstrates a typical time series station of BBL flow conditions at the Iberian continental margin at a water depth of 2200 m, measured with the long term NIOZ BoBo lander. The data set permits identification of the presently little-known downslope and net alongslope transport for a distinct period of the year. Here, tidally-driven erosion-transport-deposition loops exist for the benthic aggregates with different residence times at the sea bed and in the BBL near-bed water column, depending on the local flow field, sinking particle types, and particle composition of the sea bed. Strong tidal signals lead to friction velocities u_* varying between 0.5 and 2 cm s^{-1}.

The particle transport loop identified for periods of 1 – 3 month indicates that between 2000 - 3000 m depth particles remain in the BBL for long periods of time and are thus exposed to turbulent boundary-layer flow conditions. Consequently, enhanced mineralization rates can be expected of the particulate organic carbon they carry. This should also be the case for the upper continental margin, where little net accumulation of depositing matter (and thus carbon) occurs (van Weering et al. in press). Under the hydrodynamic conditions demonstrated the BBL exposure time of organic particles is up to 20 hours per day although the net transport of particles in one direction is small. The same holds true for particles from the BBL on the shelf, where long term alongslope transport and short term cross slope transport keep particles in suspen-

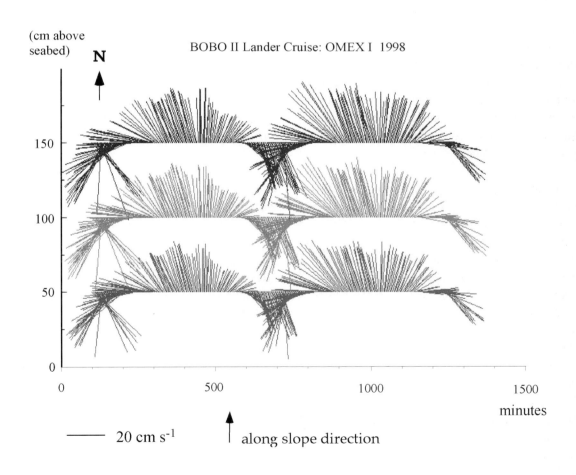

Fig. 6. Data on flow velocity of a deployment at the Iberian Margin at 2200 meters water depth in August 1998 (modified after van Weering et al. 2000).

sion (Huthnance et al. in press). As a consequence, particles of labile organic carbon contents are mostly remineralized within the water column of the BBL and not while settled on the sea floor during the 4 to 8 hours of sediment accumulation time per tidal cycle. That would result in low carbon accumulation rates (van Weering et al. in press), relatively slow first-order degradation rates (Epping and Helder in press), lower benthic biomass rates and a less intense seasonality in benthic carbon mineralization rates for this particular study site. Following the ideas of Hartnett et al. (1998), the results found here expand the concept that the extent of organic matter preservation in continental margin sediments is controlled by the average period that accumulating particles reside in oxic pore waters immediately beneath the water/sediment interface. It must include particle carbon remineralization in the BBL waters as well. This hypothesis is supported by the observation that the organic carbon/surface area ratios of particles decrease progressively offshore due to the oxic effect and could likewise explain the extremely low carbon loadings on the aluminosilicate surfaces of marine sediments (Mayer 1999). Thus, there is strong evidence for an additional mechanism of carbon mineralization at energetic continental margins.

Results from other deployments of the BoBo II lander suggest that lateral particle transport via aggregates can produce significant mass fluxes. Within the Rockall Trough at 2800 m water depth in May 1997, BBL aggregates of 1000 - 2000 μm in diameter and numbers of 80 - 210 x 10^3 m^{-3} were mainly transported to the north during flood tide. The resulting aggregate fluxes under moderate flow conditions of 4 - 10 cm s^{-1} at z_{150} (i.e. the height of the water layer above bottom focussed by the particle camera) were in the order of 1 - 10 x 10^7 m^{-2} h^{-1} (Fig. 7). Assuming an excess density for a 1000 μm aggregate of 4 x 10^{-3} g cm^{-3} (McCave 1984), 20 - 200g m^{-2} h^{-1} of particulate matter is transported in form of aggregates. Thus, a sig-nificant mass of material in the BBL is transported laterally in form of aggregates, which bypasses any common sampling technique and transports particles along the European continental margin to the north, following the residual along slope current in this region. These particles have passed many

resuspension loops and will feed lower slope benthic communities, which are adjusted to the nutritional value of these particles. These animals may also serve as tracer for palaeoceanographic studies, like benthic foraminifera, which might thus incorporate a signal from surface productivity located several hundreds of miles upstream.

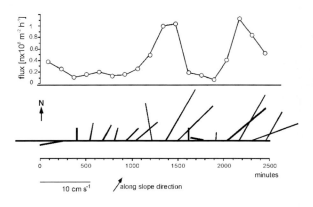

Fig. 7. Flow direction and aggregate flux in the BBL at z_{150} in the Rockall Trough at 2815 m water depth, measured during a 42 h deployment of the NIOZ BoBo II lander (modified after van Weering et al. 2000).

Conclusions

The review given here shows that open questions remain. As the studies indicate, multidisciplinary efforts only can give further insight into the processes mentioned above. The greatest needs are the ability to measure multiple chemical, physical and (bio)geochemical parameters or fluxes all at once and on the same scale. The carbon budget for continental margins still remains unsolved. The export of particulate matter, especially carbon might be much higher and in shorter timescales through transport within the BBL and subsequent dispersal at the continental rise. The idea that most of the mineralization takes place in the BBL and that bacteria in the BBL consume for most of the vertical flux of carbon leads to new discrepancies in the budget. The role of canyons and river input are still under discussion and the importance of particle fluxes during winter needs to be investigated.

References

Biscaye P, Flagg CN, Falkowski PG (1994) The SEEP II: An introduction to hypotheses, results, and conclusions. Deep-Sea Res 41:231-252

Boetius A, Springer B, Petry C (2000) Microbial activity and particulate matter in the benthic nepheloid layer (BNL) of the deep Arabian Sea. Deep-Sea Res 47: 2687-2706

Dade B, Davis JD, Nichols PD, Nowell ARM, Thistle D, Trexler MB, White DC (1990) Effects of bacterial exopolymer adhesion on the entrainment of sand. Geomicrobio J 8:1-16

Duineveld GC, A Lavaleye MS, S Berghuis EM, de Wilde PA, Weele J, Kok A, Batten SD, de Leeuw JW (1997) Patterns of benthic fauna and benthic respiration on the Celtic continental margin in relation to the distribution of phytodetritus. Int Revue Hydrobiol 82/3:395-424

Eisma D, Dyer KR, van Leussen W (1996) The *in situ* determination of the settling velocities of suspended fine-grained sediment - a review. J Sea Res 36:1-15

Epping E, Helder W (in press) Geochemical processes at the Iberian Sea continental margin. Progr Oceanogr

Flach E, Lavaye M, de Stigter H, Thomsen L (1998) Feeding types and particle transport at Goban Spur Celtic Sea. Progr Oceanogr 42:209-231

Graf G (1992) Benthic pelagic coupling: A benthic view. Ocean Mar Biol Annu Rev 30: 149-190

Graf G, Rosenberg R (1997) Bioresuspension and biodeposition - A Review. J Mar Syst 11:269-278

Hartnett HE, Keil RG, Hedges JI, Devol AH (1998) Influence of oxygen exposure time on organic carbon preservation in continental margin sediments. Nature 391:572-574

Hedges JI, Hu A, Devol A, Hartnett HE, Tsamakis E, Keil RG (1999) Oxygen exposure time as a controlling factor in the preservation of organic carbon in continental margin sediments. Amer J Sci 299:529-555

Heip C, Duineveld G, Flach E, Helder W, Herman P, Lavaleye M, Middelburg J, Pfannkuche O, Soetaert K, Soltwedel T, de Stigter H, Thomson L, Vanaverbeke J, Wilde P (2001) The role of the benthic biota in sedimentary metabolism and sediment-water exchange processes in the Goban Spur area NE Atlantic. Deep-Sea Res II 48:3223-3243

Huthnance J, Humphery J, Knight P, Chatwin P, Thomsen P, White M (in press) Near-bed turbulence measurements, stress estimates, and sediment mobility at the continental shelf edge. Progr Oceanogr

Jumars PA (1993) Concepts in Biological Oceanography. Oxford University Press, New York p 341

Mayer L (1999) Extent of coverage of mineral surfaces by organic matter in marine sediments. Geochimica et Cosmochimica Acta 63:207-215

McCave IN (1984) Size spectra and aggregation of suspended particles in the deep ocean. Deep-Sea Res 31:329-352

McCave IN, Manighetti B, Robinson S (1995) Sortable silt and fine sediment size/composition slicing: Parameters for palaeocurrent speed and palaeoceanography. Palaeoceanogr 10:593 - 610

Monaco AP, Biscaye J, Soyer R, Pocklington, Heusser P (1990) Particle fluxes and ecosystem response on a continental margin the 1985–1988 Mediterranean ECOMARGE experiment. Cont Shelf Res 10:809–839

Otto S, Balzer W (1998) Release of dissolved organic carbon DOC from sediments of the NW European continental margin Goban Spur and its significance for benthic carbon cycling. Prog Oceanogr 42:127-144

Ritzrau W (1996) Microbial activity in the Benthic Boundary Layer: Small scale distribution and its relationship to the hydrodynamic regime. J Sea Res 36:171-180

Schäfer P, Thiede J (eds) (2001) Climate change in the Northern North Atlantic. Springer, Berlin p 365

Smith KL, Kaufmann R (1999) Long-term discrepancy between food supply and demand in the deep Eastern North Pacific. Science 284:1174-1177

Sternberg R, Nowell ARM (1999) Continental shelf sedimentology: Scales of investigation define future research opportunities. J Sea Res 41:55-73

Thomsen L, Gust G (2000) Sediment stability and characteristics of resuspended aggregates of the western European continental margin. Deep-Sea Res 47:1881-1897

Thomsen L, McCave IN (2000) Aggregation processes in the benthic boundary layer at the Celtic Sea continental margin. Deep-Sea Res 47:1389-1404

Thomsen L, van Weering T, Gust G (in press) The particle resuspension loop at the Iberian continental margin. Progr Oceanogr

Van Weering TCE, Thomsen L, Heerwarden J, Koster B, Viergutz T (2000) A seabed lander and new techniques for long term *in situ* study of deep-sea near bed dynamics. Sea Techn 41:17-27

Van Weering TCE, de Stigter H, Boer W, de Haas H (in press) Recent sediment accumulation and carbon burial along the NW Iberian margin 39-44N. Progr Oceanogr

Wollast R, Chou L (2001) The carbon cycle at the ocean margin in the northern gulf of Biscay. Deep-Sea Res II 48:3265-3293

Ocean Margin Early Diagenetic Processes and Models

K. Soetaert[*], J. Middelburg, J. Wijsman, P. Herman and C. Heip

*Netherlands Institute of Ecology, Centre for Estuarine and Coastal Ecology,
PB 140, 4400 AC Yerseke, The Netherlands
* corresponding author (e-mail): soetaert@cemo.nioo.knaw.nl*

Abstract: The early diagenetic processes occuring in ocean margin sediments are discussed with an emphasis on the carbon, nitrogen and oxygen cycle. We first use simple mass-balance assumptions to derive a relationship between the fluxes at the sediment-water interface and the processes that occur in the sediments. This allows amongst others relating the sediment community oxygen consumption rate (SCOC) to total organic carbon deposition. It is shown, based on literature data that, overall, SCOC decreases significantly with depth, equivalent with a decline of organic carbon flux from about 35 g C m^{-2} yr^{-1} at 100 m to a flux of 4 g C m^{-2} yr^{-1} at 4000 m depth. A simple 0-dimensional and two more complex 1-dimensional models are then used to demonstrate the applicability of diagenetic modelling in interpreting the species distribution vertically in the sediment and quantifying the processes occurring in sediments. Finally, the different view of biochemists and biologists of the sediment is discussed. It is concluded that much can be gained from a combined biochemical and biological analysis of the sediments.

Introduction

During the last decades, quantifying the share of sediments in the ocean ecosystem or global bio-geochemical cycles has become an important objective of biological and geochemical oceano-graphy. There is a need to understand and to quantify the processes occurring in these systems, which is the subject of early diagenetic studies. Early diagenesis refers to the combination of biological, chemical, and physical processes that occur during burial in the upper layers of marine sediments (Berner 1980). Most of these early diagenetic processes are driven by biologically-mediated redox reactions, caused by the mineralisation of organic matter. The products of benthic mineralisation are then involved in a complex of secondary reactions including oxidation, sorption, dissolution and precipitation reactions. Many of these processes preferentially take place in sediments, and they affect both the amount and speciation of components released to the water-column. By releasing nutrients, essential for primary production, and by consuming oxidants such as oxygen, metal oxides and sulphate, the sediment organic

matter mineralisation plays a key role in the benthic-pelagic coupling of marine ecosystems (Graf 1992; Soetaert et al. 2000). Marine organisms incorporate the elements C, N, P, Si in relatively constant ratios, such that the preferential retention in the sediment of any of these constituents or the release of products that are unsuitable for uptake by primary producers, may strongly impact pelagic processes. Sediments also have a more subtle influence on water-column dynamics, by acting as temporary storage pools for nutrients, both on the seasonal or interannual time scale.

Because of the intricate complexity of the early diagenetic processes, kinetic models that express the transfer of matter in the sediment, have become an invaluable tool in their study. The theoretical basis of these diagenetic models goes back to the early sixties and was recently reviewed by Boudreau (1997). Initially, the models consisted of relatively simple, one-component descriptions, that were amenable to analytical solutions. By fitting sediment depth profiles, these models can be used to derive either the transport intensity or reaction

From WEFER G, BILLETT D, HEBBELN D, JØRGENSEN BB, SCHLÜTER M, VAN WEERING T (eds), 2002,
Ocean Margin Systems. Springer-Verlag Berlin Heidelberg, pp 157-177

rate of the constituent, provided that the other quantity is known. For instance, in the absence of important animal ventilation of burrows, the major transport mode of dissolved constituents is via molecular diffusion, corrected for the tortuous path induced by the sediment grains. As these diffusion coefficients are well known, fitting of solute profiles such as oxygen have proven to be extremely effective for calculating total mass fluxes (e.g. Bouldin 1967). For solid constituents, the main mode of transport in the sediment is spatial reallocation by organism activity. The intensity of this process is not *a priori* known, and therefore the derivation of this quantity is often one topic of diagenetic modelling. Most frequently used for this purpose are radionuclides, as they have well-known decay rates (Soetaert et al. 1996b), and chlorophyll (Sun et al. 1991).

Recently, a set of more complex numerical models was developed where the cycling of many elements is fully coupled (Boudreau 1996; Dhakar and Burdige 1996; Soetaert et al. 1996 a,b; Van Cappellen and Wang 1996; Hensen et al. 1997). These models not only have increased our insights in the diagenetic processes and how the cycling of the various elements are linked, but they can also be coupled to water-column models (Soetaert et al. 2000, 2001). In addition, they offer the potential to derive quantities that are not routinely estimated by the more simple models. Examples include the degradability (Soetaert et al. 1998) or the elemental composition (Hensen et al. 1997) of the organic matter deposited on the seafloor, or the vigour with which the animals rework the sediment (Soetaert et al. 1998) or impact solute profiles (Boudreau et al. 1998). Recent progress in diagenetic modelling has seen amongst others the inclusion of more statistical reliability in the application of diagenetic models (Berg et al 1998; Soetaert et al. 1996b, 1998), and the more realistic parameterisation of animal behavior and its impact on sediment mixing (François et al. 1998).

As it is far beyond the scope of this paper to give a complete overview of all diagenetic processes and how they can be represented in models, we will focus our discussion on those processes that relate to organic matter mineralisation, and more specifically on the C, N and O_2 cycles. We will explain the basic reactions that occur during early diagenesis and, based on mass balance assumptions, how these can be used to infer relationships between the sediment-water fluxes and the processes in the sediment. We will also shortly deal with the principles of mathematical modelling and the methods of solution. Then we will demonstrate the applicability of diagenetic models based on two different approaches. Using a simple 0-dimensional model of organic matter mineralisation, we will demonstrate how the quality of the organic matter impacts the response of the sediment. Then we will explain the basic principles of the more complex 1-D models that fully couple the production and consumption of several species that play part in the sediment mineralisation process. We will use such a model, a.o., to investigate how the diagenetic processes change along the ocean margin. Remark though that a 1-D approach is valid only when groundwater input and advective flows are unimportant. Wherever these processes are significant, more elaborate models are in order (see Hinkelmann and Helmig this volume). Finally, we will discuss the apparent discrepancy between the ecological and diagenetic approach to carbon flows in marine sediments.

Main diagenetic reactions

The main reactions directly involved in the mineralisation process are outlined in Table 1, while the secondary reactions are in Table 2, in the order in which they occur in the sediment. We will not discuss the dissolution of $CaCO_3$ resulting from respiration, although this reaction is important for the carbon cycle and affects pH and alkalinity of pore waters. This item is discussed in the paper by Zabel and Hensen (this volume).

The sequence of the mineralisation reactions is determined by the free energy yield of each, with the most energetically favourable occurring first. In addition, physiological constraints on the organisms may prohibit the consumption of one oxidant in the presence of other more reactive electron acceptors. In marine sediments, the organic matter is initially decomposed by oxygen (oxic or aerobic mineralisation pathway, reaction 1.1), after which follows the denitrification (consuming nitrate, re-

1.1. oxic mineralisation

$$(CH_2O)_x(NH_3)_y(H_3PO_4) + xO_2 + yH^+ \rightarrow xCO_2 + yNH_4^+ + HPO_4^{2-} + 2H^+ + xH_2O$$

1.2. Denitrification (Jorgensen, 1981)

$$(CH_2O)_x(NH_3)_y(H_3PO_4) + 0.8xNO_3^- + (0.8x + y)H^+ \rightarrow$$
$$xCO_2 + yNH_4^+ + 0.4xN_2 + HPO_4^{2-} + 1.4xH_2O$$

1.3. Mn oxide reduction:

$$(CH_2O)_x(NH_3)_y(H_3PO_4) + 2xMnO_2 + (4x + y)H^+ \rightarrow$$
$$xCO_2 + yNH_4^+ + 2xMn^{2+} + HPO_4^{2-} + 2H^+ + 3xH_2O$$

1.4. Iron oxide reduction:

$$(CH_2O)_x(NH_3)_y(H_3PO_4) + 4xFe(OH)_3 + (8x + y)H^+ \rightarrow$$
$$xCO_2 + yNH_4^+ + 4xFe^{2+} + HPO_4^{2-} + 2H^+ + 11xH_2O$$

1.5. Sulphate reduction

$$(CH_2O)_x(NH_3)_y(H_3PO_4) + 0.5xSO_4^{2-} + (0.5x + y)H^+ \rightarrow$$
$$xCO_2 + yNH_4^+ + 0.5x\Sigma H_2S + HPO_4^{2-} + H^+ + xH_2O$$

1.6. Methanogenesis

$$(CH_2O)_x(NH_3)_y(H_3PO_4) + yH^+ \rightarrow 0.5xCO_2 + yNH_4^+ + 0.2xCH_4 + HPO_4^{2-} + 2H^+$$

Table 1. The major oxidation reactions in the sediment, in order of occurrence. y=molar N/P ratio, x = molar C/P ratio (for Redfield stoichiometry, x=106, y=16).

action 1.2) and manganese reduction (MnO_2, reaction 1.3), iron reduction ($Fe(OH)_3$, reaction 1.4), sulfate reduction (anoxic mineralisation, reaction 1.5) and finally methanogenesis (reaction 1.6), until complete exhaustion of the metabolisable organic matter.

During this organic matter decomposition, a variety of reduced compounds are formed (e.g. ammonium, Mn^{2+}, HS^-..). These reduced substances diffuse upward where they may react with more oxidised species. All reduced forms are reoxidisable with oxygen. The more common reoxidation reactions in marine sediments are given in Table 2 and include the reactions consuming oxygen (reactions 2.1), such as nitrification, reactions using manganese or iron oxides (section 2.2), or using sulphate (2.3) as an oxidant. In addition, precipitation may occur (2.4), such as e.g. the formation of pyrite by reaction of sulfide with ferrous iron. A more complete description of the various reoxidation and precipitation reactions can be found in Van Cappellen and Wang (1996) and Boudreau (1996) and references therein.

Aerobic mineralisation occurs by benthic animals and micro-organisms alike, whereas the suboxic and anoxic mineralisation is an exclusively bacterially-mediated process. The oxic, denitrification and manganese pathway yield about eight times more energy than sulphate reduction, or methanogenesis (Froelich et al. 1979), where much of the energy, originally contained in the organic molecule, is transferred to reduced substances such as H_2S (Heip et al. 1995). Chemo-autotrophic bacteria can liberate that energy for their metabolism by re-oxidation of the reduced substances. Because these chemo-autotrophs require a mix of reduced substances and oxygen, they are usually found in quite narrow ranges in the sediment, where these two conditions meet.

From mass balance to models

Basically all process modelling, whether ecological, diagenetic or physical, starts with writing a balance equation, representing the inputs, losses and storage of mass or energy:

$$\text{storage} = \frac{dMass}{dt} = \text{sources} - \text{sinks}$$

This equation states that if sources are larger than sinks, the storage will be positive and the mass (or energy) will increase in time, and vice versa. Both the sources and sinks may be functions of time. This relationship is formalised in a more mathematical form as a differential equation, expressing the evolution of mass in time (the "rate of change", second term in the equation). Because differential calculus is such a well-developed field in mathematics, powerful techniques for solving these equations exist. Generally, a solution may either be estimated "analytically" or "numerically". Simple models often have an ana-

---

1. Re-oxidation reactions with oxygen as oxidant

$NH^{+4} + 2O_2 + 2HCO_3^- \rightarrow$
$\qquad\qquad NO_3^- + 2CO_2 + 3H_2O$

$Mn^{2+} + \frac{1}{2}O_2 + 2HCO_3^- \rightarrow$
$\qquad\qquad MnO_2 + 2CO_2 + H_2O$

$Fe^{2+} + \frac{1}{4}O_2 + 2HCO_3^- + \frac{1}{2}H_2O \rightarrow$
$\qquad\qquad Fe(OH)_3 + 2CO_2$

$\Sigma H_2S + 2O_2 + 2HCO_3^- \rightarrow$
$\qquad\qquad SO_4^{2-} + 2CO_2 + 2H_2O$

$FeS + 2O_2 \rightarrow Fe^{2+} + SO_4^{2-}$

$FeS_2 + 7/2\,O_2 + H_2O \rightarrow$
$\qquad\qquad Fe^{2+} + 2SO_4^{2-} + 2H^+$

$CH_4 + 2O_2 \rightarrow CO_2 + 2H_2O$

2. Re-oxidation with Mn-Fe oxides as oxidant

$MnO_2 + 2Fe^{2+} + HCO_3^- + 2H_2O \rightarrow$
$\qquad\qquad Mn^{2+} + 2Fe(OH)_3 + 2CO_2$

$MnO_2 + \Sigma H_2S + 2CO_2 \rightarrow$
$\qquad\qquad Mn^{2+} + S^0 + 2HCO_3^-$

$2Fe(OH)_3 + \Sigma H_2S + 4CO_2 \rightarrow$
$\qquad\qquad 2Fe^{2+} + S^0 + 4HCO_3^- + 2H_2O$

3. Re-oxidation reactions with sulfate as oxidant

$CH_4 + SO_4^{2-} + CO_2 \rightarrow 2HCO_3^- + \Sigma H_2S$

4. Precipitation reactions

$Fe^{2+} + HS^- + HCO_3^- \leftrightarrow FeS + CO_2 + H_2O$

$FeS + H_2S \rightarrow FeS_2 + H_2$

Table 2. Main secondary reactions in the different zones.

---

lytical solution, where the mass at a certain time t is expressed as a function of known quantities. This greatly reduces the burden of the calculations, but often requires the introduction of simplifying assumptions at the expense of realism. More complex models generally need to be solved by discretising the equations in time and/or space, so-called numerical approximations, which are solved by means of totally different techniques (e.g. Boudreau 1997). These techniques may be quite complicated and often require the use of high-level programming, but they have the advantage that there is virtually no limit to the complexity of processes that can be described in such models.

The models discussed in this paper are numerical models, implemented as a FORTRAN computer programme. The simplest model was solved using 4-th order Runge-Kutta integration, the more complex model was solved to steady-state using iteration by the Newton-Raphson method. Integration was done by an implicit method, VODE.

Mass balance of N and O_2 in sediments

Based on the reactions outlined in Tables 1 and 2, we can calculate nitrogen and oxygen mass budgets of sediments, assuming that temporary storage effects of the dissolved compounds are negligible, or the net production rates equal the net efflux, i.e. the system is at steady-state:

storage = 0 = sources - sinks
= production - consumption - efflux

For the nitrogen cycle, the impact of the processes related to mineralisation can be summarised as follows: during decomposition, ammonium is released from the organic nitrogen (mineralisation reactions 1.1-1.6, Table 1); part of this ammonium is re-oxidised to nitrate by the nitrification process (secondary reaction 1, Table 2). A small fraction of ammonium can be adsorbed to the sediment grains (we have neglected this term). Nitrate is directly consumed as an oxidant for organic matter mineralisation in the denitrification process, forming nitrogen gas (reaction 1.2, Table 1). The oxidation of one mole of carbon by denitrification requires 0.8 moles of nitrate.

Assuming steady-state, the mass balance equations for nitrate and ammonium can thus be written as:

Nitrate efflux = integrated nitrate production
    − integrated nitrate consumption
    = nitrification − denitrification * 0.8
    $$(1)$$
Ammonium efflux =
    integrated ammonium production
    − integrated ammonium consumption
    = carbon mineralisation * N:C ratio
    − nitrification
    $$(2)$$

where efflux is the flux out of the sediment, and where nitrification is expressed per unit ammonium oxidised, carbon mineralisation and denitrification are expressed per unit carbon, and N:C ratio is the molar nitrogen to carbon ratio of the decomposing organic matter.

This gives the following mass budget for total dissolved inorganic nitrogen (DIN):

TOTAL DIN efflux = carbon mineralisation
    * N:C ratio − denitrification * 0.8
    $$(3)$$

Only a fraction (PartDenitrified) of total carbon mineralisation is performed by denitrifying bacteria, such that the equation can also be written as:

TOTAL DIN efflux = carbon mineralisation
    * (N:C ratio -PartDenitritfied * 0.8)
    $$(4)$$

From this it is clear that, depending on the relative importance of denitrification, nitrate and ammonium will either flux into the sediment (i.e. when the part that is denitrified is larger than the N:C ratio/0.8) or out of the sediment. Thus, in areas where denitrification is important, the sediment can act as a sink for both organic nitrogen and dissolved inorganic nitrogen (Middelburg et al. 1996). Also, in highly anoxic sediments, where nitrification is relatively low, the ammonium efflux can be quite substantial. More oxic sediments have only limited ammonium efflux.

With respect to oxygen, the following applies: the various reduced compounds formed upon suboxic and anoxic mineralisation, with the exception of N_2, are all reoxidisable with oxygen. Generally, a substantial fraction of these reduced substances will be eventually reoxidised in the sediment. Some however escape re-oxidation either by burial below the bioturbated zone or by fluxing out of the sediment. In accordance with Soetaert et al. (1996a) we can express these effluxes of reduced substances by the amount of oxygen that is required to re-oxidise them, and calculate the total oxygen debt represented by these fluxes out of the sediment. We will call those effluxes "oxygen demand unit" (ODU) effluxes. The units of these effluxes must then be expressed in oxygen equivalents. For instance, based on the reaction outlined in Table 2 we learn that 0.5 mol of oxygen is required to reoxidise one mol of Mn^{2+}, so that half of the Mn^{2+} efflux contributes to the ODU efflux. Oxygen can thus either be directly consumed in the mineralisation process (oxic pathway) or indirectly, via re-oxidation of these reduced substances and ammonium. In the end, for every x moles of reduced substances (except N_2 and NH_3) formed, x moles of oxygen can be consumed. Adding to this the 2 moles of oxygen that are required to re-oxidise one mol of ammonium, and taking into account the escape of N_2, the oxygen influx can be calculated as:

Oxygen influx   = (carbon mineralisation
    − denitrification)*O:C ratio
    − burial of solid substances
    − ODU efflux
    + 2*nitrification

Or, rearranging, and based on equation (2)

Oxygen influx   = carbon mineralisation
    *(1-PartDenitrified)
    * O:C ratio-solidDeposition
    − ODU efflux
    + 2*(carbon mineralisation
    * N:C ratio–ammonium efflux)

Except for extremely eutrophic sites, both the sediment-water exchange of reduced substances (ODUefflux and ammonium) and the burial of olid substances are insignificant terms, and they can be safely ignored. Denitrification accounts for about 7 to 11% of the global organic carbon mineralisation (Middelburg et al. 1996), but in oligotrophic areas this may be less than 1%. Only in those regions where carbon loadings are intermediate and bottom water nitrate concentrations are elevated, denitrification may take an impressive share of total mineralisation, and the term becomes significant.

Ignoring all the terms that do not contribute significantly to the equation (burial of solid substances, ODUefflux, ammonium efflux), we then arrive at the following relationship expressing oxygen influx as a function of organic carbon mineralisation:

Oxygen influx = carbon mineralisation
                 * [(1 - Part Denitrification)
                 * O : C ratio + 2*N : C ratio]

Assuming Redfield remineralisation ratios and a 10% contribution of denitrification to the total mineralisation the following (but approximate) conversion between oxygen influx and carbon mineralisation is obtained:

Carbon mineralisation = 0.83*oxygen influx

Remark that oxygen is NOT a good proxy for the aerobic mineralisation process, as a rather large proportion of oxygen is consumed for the reoxidation of ammonium and the other reduced substances. The contribution of nitrification is usually relatively constant (~ 2*N:C ratio, i.e. about 30% for Redfield organic matter), but the amount of oxygen consumed for reoxidation of the other substances is extremely variable, and can be very important in more anaerobic sediments. Thus, the oxygen consumption cannot be simply related to the respiration of higher organisms and aerobic bacteria.

Because of this relatively simple relationship between total carbon mineralisation and oxygen influx, and because the sediment community oxygen consumption rate (SCOC) is relatively straightforward to measure, the latter has often been used to estimate total mineralisation rates in marine sediments.

In Figure 1, we have compiled 528 SCOC values from the literature, measured along the entire range from shelf to deep-sea (Wijsman et al. in prep.). The data consisted of *in situ* flux estimates using either benthic chamber incubations or were obtained via the modelling of *in situ* measured oxygen profiles. Overall, there is a significant decrease of SCOC with water depth, from an average of 0.35 mmol O_2 cm^{-2} yr^{-1} at 100 m water depth to about 10 times as low at 4000 m water depth (0.04 mmol O_2 cm^{-2} yr^{-1}). This amounts to an equivalent carbon deposition ranging from about 35 to 4 gC m^{-2} yr^{-1} at 100 and 4000 m depth, respectively.

A simple 0-dimensional model of organic matter mineralisation in deep-sea sediments

A common theme in biogeochemical modelling of early diagenetic processes is how to express the mineralisation of organic matter. The classical experiment of Westrich and Berner (1984) is typically taken as a starting point. These authors have recorded how the concentration of freshly produced organic matter changed in time. Their results are depicted in Figure (2a) and demonstrate that the organic matter decreased fast at the start of the experiment, but the rate of decrease declined as time proceeded.

Such a trend can be described by the sum of two exponential functions, expressing the organic matter concentration at time t as:

$$C(t) = C(t_0) \cdot \left[p \cdot e^{-k_1 \cdot t} + (1 - p) \cdot e^{-k_2 \cdot t} \right]$$

where $C(t_0)$ is the carbon concentration at the start of the experiment and p is a partitioning coefficient. This formula represents the analytical solution of a differential equation that considers two carbon fractions, each decaying at a fixed rate k_1 and k_2 respectively (see Figure 2a).

To use this description of organic matter decay in a model for a deep sediment system, we need to include the source of organic matter, which is deposition from the water column (Cflux; Fig. 2b). We assume that the contribution of carbon

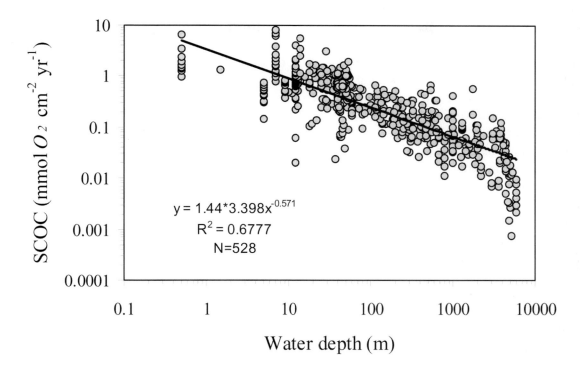

Fig. 1. Sediment Community Oxygen Consumption (SCOC) rates, as a function of water depth, and the best (exponential) regression equation. Data are from the literature.

component 1 consists of a fixed fraction of this flux (p).

The full equations of this simple diagenetic model then read:

$$\frac{dC_1}{dt} = -k_1 \cdot C_1 + Cflux \cdot p$$

$$\frac{dC_2}{dt} = -k_2 \cdot C_2 + Cflux \cdot (1-p)$$

where the first terms are the sink terms (mineralisation), the second term represents the source term (deposition).

These equations can be solved analytically when Cflux is a constant, but we will consider the more realistic case where carbon deposition varies in time.

In the first example (Fig. 2c), we imposed a time series of deposition fluxes, consisting of one year of data (Fig. 2c, thin line). As a good starting condition, we consider the steady-state condition

with respect to the yearly-averaged flux. The model was run for ten years, but only the results of the last year will be used and presented.

We ran the model for two contrasting cases, one in which the flux consisted of a mixture of reactive (k_1=0.07 d^{-1}, 26 yr^{-1}) and refractory organic matter (k_2=0.0007 d^{-1}, 0.26 yr^{-1}), the former comprising 75% of the total (values from fitting the experiments from Westrich and Berner 1984). In the second run all the deposited organic carbon consisted of the more refractory components (with decay rate 0.0007 d^{-1}).

These simulations demonstrate how the quality of the organic matter impacts the response of the sediment: when the organic matter is of high quality (1st run), the sediment responds vigorously, but the response is smoothed with respect to the input flux. The peak mineralisation rates lag about 14 days behind the peak deposition. In addition, the amplitude of the sediment response has decreased with respect to the input flux: the maximum/minimum value was 4.3 for mineralisation, compared to 10 for the deposition flux. In the second run, where the degradability of the organic matter was about 100 times lower, the sediment response is hardly discernible, with only a 7% difference between maximum and minimum mineralisation rate, and the optimal response now lags 47 days behind peak deposition. Remark that this is a well-known property of a box model with a fluctuating input. The box acts as a "filter" which decreases the rate and introduces a time lag.

To unravel the relationship between organic matter quality and sedimentary response, we ran a Monte Carlo analysis. The three parameters determining organic matter quality (p, k1, k2) were varied randomly within broad ranges and the time lag between peak deposition and peak mineralisation was recorded (Fig. 2d right), as well as the total range in mineralisation rates (Fig. 2d left). It is clear that the vigour of the sediment response (i.e. the difference between minimum and maximum response) increases at higher mean degradabilities. Organic matter that does not decay (first-order rate = 0) displays no response at all (Max/Min = 1), whereas at the highest reactivities, the Max/Min ratio of the response becomes similar to the ratio in the deposition flux (which was 10 in the model run). Secondly, the time lag between peak sediment deposition and peak response can

be as high as 30 to 40 days for extremely refractory detritus, but reduces to less than 10 days for highly reactive organic matter. Essential though is that, even when the organic matter is exceedingly fresh, a time lag does exist between peak depositon and mineralisation or respiration rates, such that establishing a direct day-to-day relationship between sedimentation pulses and sediment activity may not be feasible.

Because of this dependence of the sediment response on the reactivity of the organic matter, it becomes possible to estimate the quality of organic matter by using a model as simple as the one described above. Such a model has been applied by Sayles et al. (1994) to two deep-sea sites, one near Bermuda, and one in the North East Pacific. They showed that the latter site was characterised by large temporal variability in sediment oxygen consumption, as a response to the deposited organic matter, which therefore had to be of relatively high quality. Near Bermuda on the other hand, the seasonal variability in deposition flux was not met by significant variations in the oxygen influx and the quality of the organic matter was therefore low.

In Fig. 2e, the simple 2-box model was used to reconcile the response of the sediment at the same North East Pacific station, but now spanning a larger period of time (Smith and Kaufmann 1999).

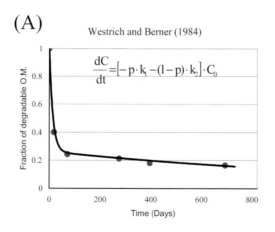

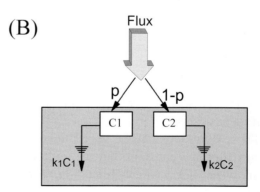

Fig. 2. Simple 0-dimensional model of organic matter mineralisation. **a)** Experiments performed by Westrich and Berner (1984) outlining the fraction of organic matter as a function of time, and the differential equation representing the dynamics (solid line). **b)** Schematic of the model derived from the experiments in a); p, k1 and k2 are model parameters; C1 and C2 are state variables.

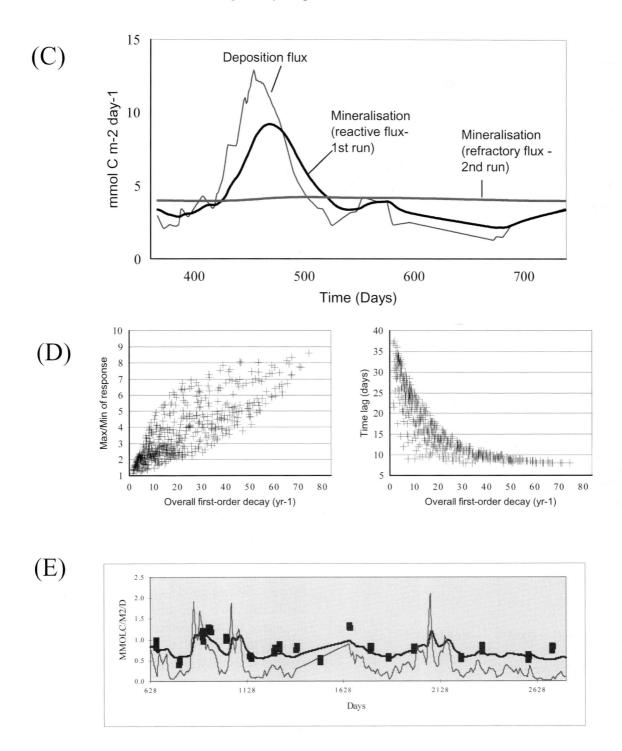

Fig. 2 cont. c) Two runs performed with the model as schematised in b). Deposition flux is the forcing function; the first run assumes the flux to consist of reactive material, second run assumes a refractory flux. **d)** Results of a monte carlo analysis, evidencing the impact of first-order decay rate of deposition flux on the amplitude of the sediment response (left), and on the time lag between deposition and response (right). Maximal amplitude (maximum value / minimum value) in forcing flux was 10. **e)** Model simulation reconciling the sediment trap fluxes and SCOC measures as obtained from Pacific station N. Data from Smith and Kauffman (1999).

The model performs well reproducing the trend observed in SCOC (thick line), except for one period where the deposition flux at the lower trap was not recorded (around day 1628). Remark that, although the POC flux was decreasing over the period, the SCOC did not exhibit the same decrease, mainly because of the large stock of the more refractory organic compounds.

Coupled 1-dimensional models of early diagenetic processes

Most diagenetic models do not consider the upper part of the sediments as a whole but look at the way the constituents change vertically in the sediment. All these vertically-resolved diagenetic models are basically similar and start from the general diagenetic equation proposed by Berner (1980) for solutes and solids. This equation includes advective transport (sediment deposition and compaction) and bioturbation for solid substances, molecular diffusion and if appropriate bio-irrigation for solute substances.

Simplified, the equations for solutes and solids are:

$$\frac{\partial \phi C_i}{\partial t} =$$

$$-\frac{\partial}{\partial x}\left[-\phi D_s \frac{\partial C_i}{\partial x} + \omega \phi C_i\right] + \phi \sum REAC_i$$

$$\frac{\partial (1-\phi) C_i}{\partial t} =$$

$$-\frac{\partial}{\partial x}\left[-(1-\phi) Db \frac{\partial C_i}{\partial x} + \omega(1-\phi) C_i\right] +$$

$$(1-\phi)\sum REAC_i$$

where the first bracketed term is the diffusive and advective transport; x is sediment depth (cm), t is time (day), ϕ is porosity. ϕC_i is the concentration of a solute species, $(1-\phi)C_i$ is the concentration of a solid species (both in units of nmol cm^{-3} of bulk sediment). D_s is the sediment diffusion coefficient (cm^2 day^{-1}), D_b the bioturbation coefficient (cm^2 day^{-1}), ω is the advection rate (cm day^{-1}) and

$\sum REAC_i$ is the sum of all production/consumption terms of species i in biogeochemical reactions.

The main mode of transport of solutes is by molecular diffusion, and sediment diffusion coefficients (Ds) are lower than in water, because of the tortuous paths that the molecules must follow around sediment grains. The actual diffusion coefficient can be estimated based on its free-solution value and the sediment porosity (Boudreau 1997). Bio-irrigation is the enhancement of sediment-water exchange due to animal ventilation of burrows. In some models, this process is included as an apparent increase of the diffusion coefficient (Soetaert et al. 1996), whereas other models describe the process as an advective process (Boudreau 1996; Van Cappellen and Wang 1996). The degree of bio-irrigation is not well known, and is especially important in shallow areas where animal activity is high. Its magnitude can be assessed by comparing solute fluxes estimated via benthic chambers (which include irrigation) and via diagenetic modelling of the involved solutes.

Bioturbation is the degree to which the benthic animals rework the sediment. In most models, this complex process is described as if it were a random diffusion-like process, assuming a so-called bioturbation coefficient (Db). The value of this bioturbation coefficient is highly variable and not known a priori. It is usually derived by diagenetic curve-fitting sediment depth profiles of specific radionuclides such as ^{210}Pb or ^{234}Th (Middelburg et al. 1997) or chlorophyll (Sun et al. 1991). Alternatively, finding good values of this bioturbation coefficient can be one goal of applying a coupled diagenetic model (Soetaert et al. 1998). Animals stir only the upper part of the sediment, so that this process is usually restricted to the upper layers (~ 10 cm) of the model.

Sediment advection is the process of accretion, due to the deposition of new material on top of the sediment. As a result, the sediment will be compacted which is shown by an exponential decrease in porosity with increasing depth into the sediment. In most ocean margin systems, advection is relatively unimportant compared to bioturbation (solids) or molecular diffusion (solutes) (Middelburg et al. 1997).

The standard two-type of organic matter consumption models as described above often forms the basis of the coupled biogeochemical reactions that now include the consumption and/or production of oxygen, nitrate, ammonium and other reduced substances. The production and consumption processes are regulated by the stoichiometry of the reactions as outlined in Tables (1) and (2). Because of that, the cycles of various elements are linked, and one set of parameters is used to describe the profiles of all the constituents that are being modeled.

The kinetics of utilisation of the different electron acceptors during the biodegradation of the organic matter is generally described by a Monod-type hyperbolic formulation (Fig. 3a top). Inhibition of a metabolic pathway by the presence of a substance, and hence the amount of overlap between different diagenetic zones is described by the reciprocal of a Monod formulation (Fig. 3a bottom). For the secondary reactions, the equations used are either of Monod-type, or bi-molecular kinetics. In the latter case, reaction rates have first-order dependence on both the electron acceptor and donor (e.g. VanCappellen and Wang 1996).

As an example, consider the model description of denitrification, which is inhibited by oxygen (constant kin), and limited by nitrate (constant ks).

$Denitrification =$

$$\left(\sum_{i=1}^{2} k_i \cdot C_i\right) \cdot \left[\frac{NO_3}{NO_3 + ks} \cdot (1 - \frac{O_2}{O_2 + kin})\right]$$

The first term represents the total organic matter mineralisation, whereas the second term (in []) denotes the fraction of total mineralisation performed by the denitrifying bacteria. Remark that the denitrification rate, because of its linear (first-order) dependence on the carbon concentration, is expressed in nmol C cm^{-3} solid d^{-1}. Therefore, in order to convert this rate to actual nitrate consumption rates (in nmol C cm^{-3} liquid d^{-1}), porosity must be taken into account. The denitrification rate then impacts the equations for each carbon fraction, nitrate and ammonium; for nitrate this leads to the following rate of change:

$$\frac{dNO3}{dt} = ... - Denitrification * 0.8 * \frac{(1-\phi_i)}{\phi_i}$$

where "..." denotes all the other processes affecting nitrate (transport, nitrification,...). Because of the complexity of reactions, and due to the inclusion of porosity gradients, these coupled diagenetic models can only be solved numerically. This is done by subdividing the modelled sediment column in a number of thin layers (~ 1mm), typically around 100 or more. In each layer, the governing equations are solved; transport is approximated by finite differences.

A typical outcome of a vertically-resolved diagenetic model is a set of depth profiles, such as exemplified in Fig. 3c. These profiles were generated by application of the model OMEXDIA (Fig. 3b) (Soetaert et al. 1996a), using a setting characteristic for a North-East Atlantic station at 1000 m depth. OMEXDIA is the simplest of the recently developed numerical models because it has lumped all the reduced components produced upon anoxic mineralisation into one "oxygen demand unit". This model then does not resolve the fine details of manganese, iron or sulphur cycling. However, it is very economical in terms of computational effort: for a model consisting of 100 sediment layers, and running on a 600 MHz computer, 10 to 50 steady-state calculations are performed per second, running a 1-year cycle takes about 30 seconds.

Fig. 3c illustrates the rapid consumption of oxygen in the upper first centimeter, the buildup of nitrate in this region due to the nitrification process, and the consumption of nitrate and increase in the ammonium concentration below the oxic zone. By integrating each process over the sediment layers, one then obtains depth-integrated rates. For instance, in the model simulation carbon mineralisation occurred for 68% by aerobic respiration, 9% by denitrification and 23% by anoxic processes. Some 26% of total oxygen consumption is due to nitrification, 17% due to reoxidation of other reduced substances, leaving only 56% for aerobic respiration. Because these models are numerical models, they are ideally suited to run

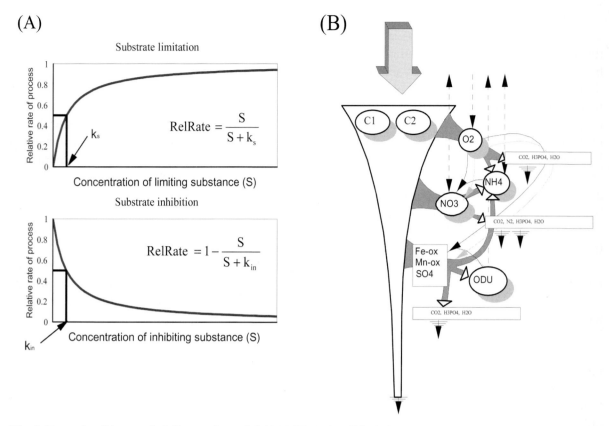

Fig. 3. Example of the coupled diagenetic model OMEXDIA (modified after Soetaert et al. 1996a). **a)** Formulations used to represent the limitation and inhibition of substances on the maximal rates. **b)** Schematic of the OMEXDIA model; ovals are state variables.

dynamically and simulate the changes of the various constituents through time. In addition, they can be coupled to water-column models in order to study the mutual interaction between the pelagic and benthic system. An example of such coupling is represented in Fig. 3d for a shelf system (from Soetaert et al. 2001). Note the upward migration of the nitrate peak after the deposition of the spring bloom (around day 500) and the gradual increase in water-column nitrate concentration, partly due to benthic nutrient generation. Similarly, ammonium displays rather pronounced seasonal dynamics, moving upward after deposition of the bloom, and retreating afterwards as oxygen penetrates deeper into the sediment.

In order to investigate the effect of increased carbon loading on the relative importance of the various mineralisation pathways, we applied another, similar model. This model describes the cycling of O_2, C, N, Fe, Mn and S (Wijsman et al. in press; Fig. 4a). In this relatively simple model, pH is imposed

rather than resolved. The equations used in this model are outlined in Tables (1) and (2).

We ran the model 300 times to steady-state, with the total deposition flux of organic matter varying in between 0 and 0.6 mmol C cm^{-2} yr^{-1}, representing the gradient observed from the shelf to the deep-sea. Details about the model settings can be found in Wijsman et al. (in press), results of the Monte Carlo simulation are in Figure 4b. Superimposed on this figure (white line) is the empirical depth- relationship of total mineralisation rate as derived above (Fig. 1). The main trends observed with increased depth along the ocean margin (e.g. Lohse et al. 1998) are reproduced: the oxic mineralisation is most prominent at low carbon fluxes, typical for deep areas; sulphate reduction is most important at high loadings, found generally at shallow water depths. At intermediate carbon loadings and water depth, manganese and iron cycling play an important role in the sediment mineralisation. There is a rather abrupt succession

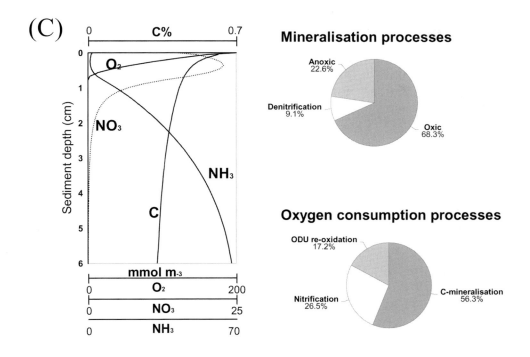

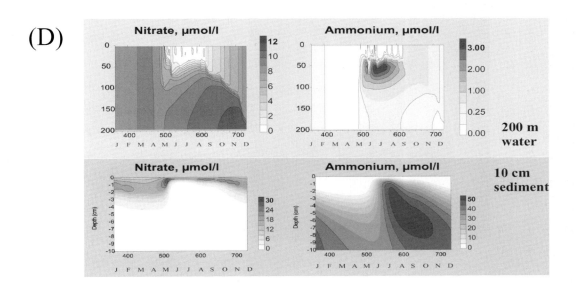

Fig. 3 cont. c) Example of output generated by virtue of a coupled diagenetic model. Left: sediment depth profiles of various substances (oxygen, nitrate, ammonium, organic carbon) right: share of the various mineralisation pathways in total mineralisation (upper) and of oxidation processes in total oxygen consumption. **d)** Example of a diagenetic model, coupled to a water column model (excerpt from Soetaert et al. 2000).

(A)

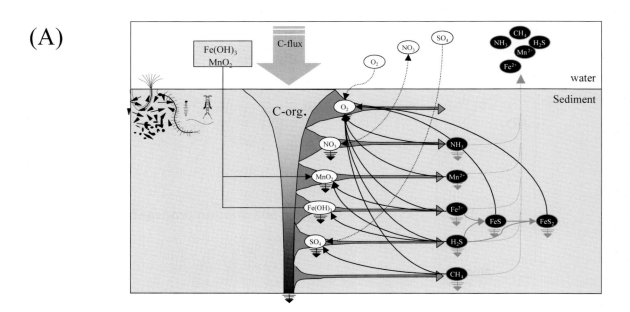

(B)

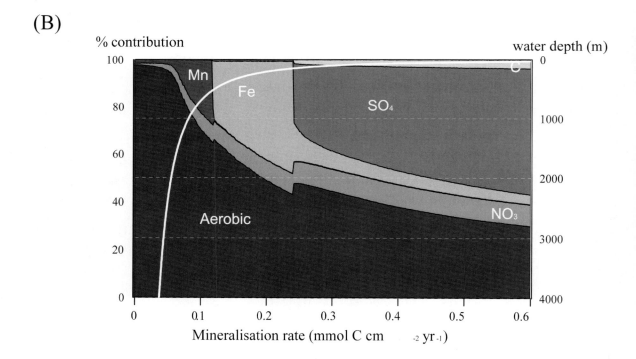

Fig. 4 a) Schematic representation of the coupled diagenetic model BSDIA (Wijsman et al. in press). **b)** Contribution of the various mineralisation pathways with increased loading, as resulting from a monte carlo analysis (200 runs). Redrawn from Wijsman et al (in press).

from the manganese pathway towards iron reduction. This positive feedback in the model occurs when reduced iron compounds are reoxidised by manganese oxide, so that the latter becomes unavailable for organic matter respiration. A similar positive feedback occurs when sulphide reacts with ferric oxides and competes with organic matter.

Discussion

The characteristics of early diagenetic processes along the ocean margin.

The functioning of marine sediments is driven both by the flux or *in situ* production of organic matter and the conditions of the overlying bottom water. The former fuels benthic activity, whereas the latter mediates the magnitude and direction of the sediment-water exchanges of various substances. For instance, low oxygen, high nitrate bottom water reduces the importance of aerobic mineralisation at the benefit of denitrification and/or anoxic mineralisation processes (Middelburg et al. 1996).

Before organic matter is deposited on the sediment floor, it has undergone decomposition in the water column, and the compounds that are more easily assimilated will be taken up first (Westrich and Berner 1984). The longer its residence time in the water, the more the organic matter will have been degraded, so that both the quantity and the quality of the organic matter will decrease with water depth. As total sediment community oxygen consumption rates provide a valid proxy for total sediment mineralisation rates, this depth-dependence of total carbon mineralisation rates is nicely illustrated by the order-of-magnitude decrease of SCOC from the shelf to the deep-sea (Fig. 1).

The property that most distinguishes the sediments from the water column is the predominance of particles, which reduce exchange rates, magnify solid-based consumption of dissolved substances and directly interfere with the biogeochemical cycles by adsorption (Soetaert et al. 2000). Because of that, significantly sharper gradients exist in the sediments, and the biogeochemical diversity is much higher here than in

the water column. Some reactions are almost exclusively restricted to the sediments, because the necessary biogeochemical conditions are never met in the water column. As a consequence, there exists within the sediment a typical depth succession of oxidants that are being consumed in the mineralisation process (Froelich et al. 1979; Table 1). Oxygen is the first oxidant to be reduced (oxic mineralisation), followed by nitrate (denitrification) and manganese oxide, iron oxide, sulphate (anoxic mineralisation) and eventually organic matter itself (fermentation) (Table 1). Only when organic matter mineralisation has exhausted the oxidants providing the highest energy yield, will the next energetically favorable be consumed. Because of this succession, the presence of a certain mineralisation pathway in the sediment is strongly linked to the amount of organic matter produced or deposited to the sediment. The highest biogeochemical diversity is usually observed in intermediate rich organic sediments, with oligotrophic sediments being oxic till great depths, and strongly eutrophic sediments being dominated by sulphate reduction or methanogenesis (Fig. 4b).

Because of the clear-cut decrease in mineralisation with water depth, the typical depth succession of mineralisation reactions that is observed in the sedimentary sequence can also be discerned along the ocean margin (Fig. 4). Deep-sea sediments, where total mineralisation rates are low are oxic till great sediment depths and oxic mineralisation is by far the dominant process. With increased organic loading (decreased water depth), the denitrification and dissimilatory manganese reduction become more important, after which iron and finally sulphate reduction take over. It has been amply demonstrated that, near the ocean margin, where organic matter deposition is moderate, mineralisation pathways other than aerobic respiration and sulphate reduction may be prominent (Lohse et al. 1998). Because of this high biogeochemical diversity at the ocean margin, these systems play a very important role in many of the ocean's elemental cycles. For instance, the denitrification, which releases reactive N as N_2 gas to the atmosphere, is mainly a sedimentary process and occurs preferentially on the shelf and upper slope (Middelburg et al. 1997). The phosphorus cycle is often tightly linked to the iron cycle, and iron-rich sediments may possess an enormous capacity of binding P. Recent estimates suggest that continental margins may constitute a very important oceanic sink of reactive P (Delaney 1998). Diagenetic processes also affect the long-term preservation of organic matter in sediments, by reducing the degradation under anoxic conditions. Added to that the high rate of deposition of organic material on the margins, it has been estimated that today, over 90% of all organic carbon burial in the ocean occurs in continental margin sediments (Middelburg et al. 1997).

The link with the water column

Because of the impact of water-column processes on the quality of the deposited organic matter, the latter can provide a good indication about the main source of the food. For instance, whether organic matter is derived by lateral advection from a distant source or by rapid sinking will have important consequences for its quality as it settles on top of the sediment (Herman et al. 2001). This means that the sediments may provide strong constraints with respect to the importance of lateral advection to fuel the benthic ecosystem.

Unfortunately, it is not easy to find a good measure for the degradability of the deposited organic matter. The most down-to-earth method would be to look at the organic matter in the sediment itself, and, say, calculate the ratio of total mineralisation versus sediment carbon content. However, this merely estimates the reactivity of the organic matter in the sediments, which can be several orders of magnitude different from what originally settled on top (Soetaert et al. 1996). This is because the benthos will rapidly consume only the most palatable organic matter, not the more recalcitrant fraction, so that the bulk of organic matter in the sediments represent that part of the detritus that the organisms do not eat.

Modelling can provide help here, as it is possible to estimate the quality of organic matter deposits via dynamic modelling of sediment deposition/ response events (Sayles et al. 1994; Fig. 2e). Similarly, because of the constraints imposed by linking the cycles of different components in the more elaborate diagenetic models, it is possible to arrive at that quantity by fitting a combination of carbon, oxygen and nitrate profiles (Soetaert et al. 1998).

The geochemical approach versus the biological approach

During the past decades, biological studies have convincingly demonstrated the existence of a diverse, abundant and active benthic community extending from the coastal areas till the deep-sea floor (e.g. Heip et al. 2001). Similarly, biogeochemists have documented the diversity in biogeochemical conditions that exist in these environments (e.g. Lohse et al. 1998) and have tried to evaluate the impact of the sediment on the ocean's biogeochemical cycles (Middelburg et al. 1996). Most of the early diagenetic processes studied by biogeochemists are driven by the mineralisation of organic matter and are performed by the bacteria and animals inhabiting the sediment. Nevertheless, the diagenetic processes in the sediment are generally studied without consideration of its inhabitants. It may seem surprising that one can study a system that is set into motion by animal and bacterial activity with complete disregard of the actors themselves. To make clearer the difference in approach between the biologists and the geochemists, and the reason for this discrepancy, we must take a closer look at the physiology of marine animals and bacteria and how this is dealt with in the two disciplines (Fig. 5).

Food ingested by marine heterotrophs is used for growth, reproduction and respiration, with a certain amount being undigested, and expelled as faeces (animals only). The digested (assimilated) food is used partly to provide the building blocks for growth, the other part is respired to provide the energy necessary to maintain basal metabolism, to form new biomass and for other activities such as reproduction and locomotion. The end-products of respiration depend on the nature of the substrate: carbon dioxide is produced, and if excess nitrogen is present in the substrate, it will be excreted in the form of ammonium or urea. Animals and many bacteria are obligate aerobes and require oxygen

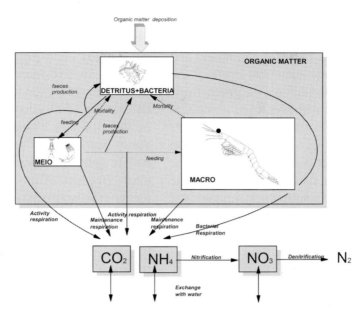

Ecological model

dMEIO = MEIO Feeding - MEIO Faeces - MEIO Respiration - MACRO Feeding
   - MEIO Mortality

dMACRO = MACRO Feeding - MACRO Faeces - MACRO Respiration
   - MACRO Mortality

dDETRITUS and BACTERIA = MACRO Faeces + MEIO Faeces
   - BACTERIA Respiration + MEIO Mortality + MACRO Mortality + DEPOSITION

Losses = MEIO Respiration + MACRO Respiration + BACTERIA Respiration

Diagenetic model

dORGANIC MATTER = DEPOSITION - ORGANIC MATTER Respiration

dCO2 = ORGANIC MATTER Respiration + Net exchange

dNH4 = ORGANICNITROGEN Respiration - Nitrification + Net exchange

dNO3 = Nitrification - Denitrification + Net exchange

Fig. 5. Sediment functioning as from the point of view of the ecologist (white boxes) and the biogeochemist (grey boxes) with a simplified budget of the main components.

for their respiration. The consumption of oxygen by the biota has major consequences. First of all, because of the limited exchange rates with the overlying water, and when food is abundant, the oxygen becomes exhausted at short distances below the sediment-water interface. Nitrate, manganese oxides, iron oxides and ultimately sulfate are now used by microbial organisms for their respiration, so that part of the food becomes inaccessible to the obligate aerobic animals. These suboxic and anoxic metabolic pathways equally produce CO_2 and ammonium, but in addition other reduced substances such as Mn^{2+}, Fe^{2+} or H_2S are formed (Table 1). These are all energetically-rich species, and, upon reoxidation, provide enough energy for the metabolism of so-called chemo-

autotrophic bacteria. These reoxidation processes however consume a considerable amount of oxygen, and therefore compete with the oxic respiration chain (higher organisms). It is not uncommon to find, in more eutrophic sediments that a significant fraction (>50%) of total oxygen uptake is linked to the reoxidation of reduced substances.

When considering the elemental cycles of carbon, nitrogen and phosphorus (the interest of the geochemist), the processes of grazing, faeces production and assimilation (secondary production) are nothing more than a conversion from one pool of labile organic matter to another pool of labile organic matter. More interesting in this respect are respiration and excretion, which cause a transfer from the particulate (or for bacteria, dissolved), organic phase to the dissolved, inorganic pool. It is only after excretion and respiration that the dissolved substances such as nitrate, ammonium, and phosphate can flux out of the sediment and impact the global biogeochemical cycle of the elements in the ocean. In addition, the geochemists argue that, when integrated over time, respiration measurements are constrained by organic matter import, whereas grazing, secondary production and defaecation are not constrained. Indeed, the respiration measurements can never exceed the net import of organic matter to the system and the changes in organic stock. However, this is not true for the other terms, in which recycling can be quite important (one molecule of carbon can be incorporated several times in animal tissue before it is finally respired). Remark that this difference in interest does not imply negative judgement over the biological approach. It is extremely interesting to study secondary production, as this is the fraction of organic matter that is available for grazing by predatory organisms. If one is interested in the amount of "harvestable resources" or the part of the organic matter that can be exported from the system, secondary production is the process to look at, not respiration. Geochemists generally show more interest in the bacterially-mediated suboxic and anoxic processes, because these are the ones that have the largest impact on the recycling of nutrients. Examples are denitrification, which represents a loss of nitrogen from the system, or the coupling between phosphate and iron cycling in suboxic environments, resulting in either preferential phosphate burial or release relative to organic material (Berner 1980). These processes have direct impact on the global cycles, yet they are very difficult to study from a biological point of view. In contrast to the more bacterially-minded bias from the geochemists, the biologists tend to focus attention towards the higher organisms.

Now that the motivation for the geochemical interest is clear, it still needs to be resolved what is the rationale behind their approach. To do that, it is instructive to consider what happens to organic matter in sediments after it has been deposited on the sea floor (Graf 1992). The food that eventually reaches the sediment is a complex mixture of biochemical compounds, some of which can be directly assimilated by the biota and other fractions that are more resistant to biotic attack. After a deposition pulse, part of the food will be rapidly taken up by the organisms, and their respiration rate will start to increase, concurrent with the assimilation of the food into new tissue. A substantial fraction of this deposited organic matter will be stored in new biomass. These organisms may die or become food for other larger organisms, which again require energy to make new body mass. Because of this sequestration of organic matter in organism tissue, and the cascading effect of incorporation from prey towards predatory animals, there does not exist an instantaneous response between deposition and respiration or release of nutrients. Instead, organismal activity, and consequently the return flux of nutrients and uptake of oxygen, will peak with a certain time lag after the deposition event. This biotic response of a sediment ecosystem to an episodic supply of organic matter, although it is the correct sequence of things, is extremely complex as it involves the time scales of growth of all the biological components, from bacteria to megafauna, and the complex interactions between them. Our degree of understanding of the sediment food web has not yet reached the point at which we can predict the response of a sedimentary system based on its biotic components. Due to methodological uncertainties and limitations, it is not trivial to constrain the role of each biotic component in determining the fate of organic matter in the system (but see Heip et al. 2001 for a different view). In addition, there is no easy way to quantify each functional group, from mega- to microfauna, from aerobic bacteria to the sulphate reducers.

To overcome this difficulty, biogeochemists have taken the pragmatic approach of looking at the effect that the organism have on the sedimentary system rather than to the organisms themselves. The first step is to ignore the biological complexity and group together the organisms with the detritus into a lumped sum of "organic matter". The respiration of organisms is now called "mineralisation" or "decomposition" of organic matter, which is usually considered to proceed at a constant rate relative to organic matter concentration. To allow for the sequential uptake of the various components, the so-called multi-G approach has been introduced, where several fractions of "organic matter" are considered, each decaying with a constant rate, which differ several orders of magnitude (Westrich and Berner 1984; see Fig. 2).

The different view of essentially the same system is schematised in Fig. 5. It may be clear that whereas in geochemistry one lumps or ignores the biological species, biologists tend to do the same with the chemical species.

Nevertheless, there is a lot to be gained by a combined approach. First of all, the geochemical rates can be used to set an upper bound to the functional role of the benthic animals on the fate of organic matter (Heip et al. 2001). As animals can only gain energy via aerobic respiration, the contribution of the aerobic pathway to total mineralisation is the result of either the animals or the aerobic bacteria. Suboxic and anoxic mineralisation is an exclusive bacterially-mediated process and therefore represents the fraction of organic matter unavailable to the benthic animals. Secondly, the geochemical models are able to estimate the degree and the vertical zone where the chemo-autotrophic re-oxidation of reduced substances occurs and, theoretically, this can be used to crudely estimate bacterial chemo-autotrophic production. Admittedly, the biomass production will be much less than the amount of detritus that originally settled on top of the sediment, but it consists of a very nutritious food source (bacterial biomass) and is located in relatively narrow zones of the sediment. In a food-diluted environment, such micro-niches of high-quality organic matter may be important for the benthic organisms. The representation of the effect of animal activity on the redistribution of the "organic matter" and the solutes remains one of the least constrained and most difficult process to parameterise in terms of models. For instance, increased animal activity may cause on the one hand more rapid translocation of organic matter below the oxic zone, thus favoring microbial anoxic processes. On the other hand, the ventilation of burrows folds the oxic layer into the sediment and may therefore significantly increase the importance of the oxic mineralisation pathways (Herman et al. 1999). The burrow wall then can be considered an extension of the sediment-water

interface, with similar gradients, now directed radially from the internal axis of the burrow (Boudreau 1997). Which way the system will be affected is not often clear, and requires the expertise from biologists. In addition, certain animals may bridge the oxic/anoxic interface with their body and feed in the anoxic zone of the sediment, yet respire aerobically. Others may migrate from one zone to the other periodically. Finally, organisms may upgrade the quality of the organic matter, e.g. by increasing the nitrogen content (bacteria). None of these processes have been adequately quantified such that they can be included in the geochemist's models.

Conclusion

The processes related to organic matter mineralisation in sediments produce a rich variety of chemical species, the diversity of which peaks at the ocean margin. The distributions of these species in the sediment are useful indicators about the processes that generated them, and undoubtedly, diagenetic modelling has helped a great deal in interpreting them and quantifying the magnitude of the processes. Because of the coupled nature of the more recently developed models, it is now even possible to derive quantities such as the degree of sediment reworking or the quality of the deposited food that can be related to biological processes studied in the sediment or to water-column processes. The impact of animal behaviour on diagenetic processes is complex and incompletely understood. The classical approach to modelling biogeochemical sedimentary cycles is to ignore the biological complexity and interactions, but rather study the organic matter as a whole. This keeps the equations simple and has led to the wide application of diagenetic modelling. Similarly, the biological approach is to keep the biogeochemical complexity of the sediment simple. However, these simplifications prevent the simulations from being complete descriptions of the sedimentary system. As the biota essentially control the biogeochemical cycles, but are in turn strongly impacted by the biogeochemical processes they set in motion, we believe that more complete understanding of the functioning of sediment systems will come by the concurrent studying and modelling of geochemical and biological processes.

Acknowledgements

We thank Dr. K. Smith and X. Kauffman for providing data shown in Fig. 2E. This research has been supported by the European Union (OMEX-I, MAS 2-CT93-0069, OMEX-II MAS3-CT96-0056, EROS, nr. ENV4-CT96-0286, the "glamor" part of the METROMED project, MAS 3-CT96-0049). This is publication no 2814 from the NIOO-CEMO and ELOISE publication no 232.

References

Berg P, Risgaard-Petersen N, Rysgaard S (1998) Interpretation of measured concentration profiles in sediment pore water. Limnol Oceanogr 43:1500-1510

Berner RA (1980) Early Diagenesis - A Theoretical Approach. Princeton University Press, Princeton

Boudreau BP (1996) A method-of-lines code for carbon and nutrient diagenesis in aquatic sediments. Computers Geosci 22:479-196

Boudreau BP (1997) Diagenetic Models and Their Implementation - Modelling transport and Reactions in Aquatic Sediments. Springer, Berlin

Boudreau BP, Mucci A, Sundby B, Luther GW, Silverberg N (1998) Comparative diagenesis at three sites on the Canadian continental margin. J Mar Res 56:1259-1284

Bouldin DR (1967) Models for describing the diffusion of oxygen and other mobile constituents across the mud-water interface. J Ecol 56:77-87

Delaney ML (1998) Phosphorus accumulation in marine sediments and the oceanic phosphorus cycle. Glob Biogeochem Cycl 12:563-572

Dhakar SP, Burdige DJ (1996) A coupled, non-linear, steady state model for early diagenetic processes in pelagic sediments. Amer J Sci 296:296-330

Francois F, Poggiale JC, Durbec JP, Stora G (1997) A new approach for the modelling of sediment reworking induced by a macrobenthic community. Acta Biotheoretica 45:295-319

Froelich PN, Klinkhammer GP, Bender ML, Luedtke NA, Heath GR, Cullen D, Dauphin P, Hammond D, Hartman B and Maynard V (1979) Early oxidation of organic matter in pelagic sediments of the eastern equatorial Atlantic: Suboxic diagenesis. Geochim Cosmochim Acta 43:1075-1090

Graf G (1992) Benthic-pelagic coupling: A benthic view. Oceanogr Marine Biol: An Annual Review 30:149-190

Heip CHR, Goossen NK, Herman PMJ, Kromkamp J, Middelburg JJ, Soetaert K (1995) Production and consumption of biological particles in temperate tidal estuaries. Oceanogr and Marine Biol: An Annual Review 33:1-149

Heip CHR, Duineveld GCA, Flach E, Graf G, Helder W, Herman PMJ, Lavaleye MSS, Middelburg JJ, Pfannkuche O, Soetaert K, Soltwedel T, De Stigter HS, Thomsen L, Vanaverbeke J and De Wilde PAWJ (2001) The role of benthic biota in sedimentary metabolism and sediment-water exchange processes in the Goban Spur area (NE Atlantic). Deep-Sea Res II 48:3223-3243

Hensen C, Landenberger H, Zabel M, Gundersen JK, Glud RN, Schulz HD (1997) Simulation of early diagenetic processes in continental slope sediments off southwest Africa: The computer model CoTAM tested. Mar Geol 144:191-210

Herman PMJ, Middelburg JJ, Van de Koppel J, Heip CHR (1999) Ecology of estuarine macrobenthos. Adv Ecol Res 29:195-240

Herman PMJ, Soetaert K, Middelburg JJ, Heip CHR, Lohse L, Epping EHG, Helder W, Antia AN and Peinert R (2001) The seafloor as the ultimate sediment trap - using sediment properties to constrain benthic-pelagic exchange processes at the Goban Spur. Deep-Sea Res II 48:3245-3264

Hinkelmann R, Helmig R (2002) Numerical modelling of transport processes in the subsurface. In: Wefer et al. (eds) Ocean Margin Systems. Springer, Berlin pp 269-294

Lohse L, Helder W, Epping EHG, Balzer W (1998) Recycling of organic matter along a shelf-slope transect across the NW European Continental Margin (Goban Spur). Progr Oceanogr 42:77-110

Middelburg JJ, Soetaert K, Herman PMJ, Heip C (1996) Denitrification in marine sediments: A model study. Glob Biogeochem Cycl 10(4):661-673

Middelburg JJ, Soetaert K, Herman PMJ (1997) Empirical relationships for use in global diagenetic models. Deep-Sea Res 44:327-344

Sayles FL, Martin WR, Deuser WG (1994) Response of benthic oxygen demand to particulate organic carbon supply in the deep-sea near Bermuda. Nature 371:686-689

Smith KL and Kaufmann S (1999). Long-term discrepancy between food supply and demand in the deep eastern North Pacific. Science 284:1174-1177

Soetaert K, Herman PMJ and Middelburg JJ (1996a) A model of early diagenetic processes from the shelf to abyssal depths. Geochim Cosmochim Acta 60: 1019-1040

Soetaert K, Herman PMJ, Middelburg JJ, Heip CHR, Stigter HSD, Van Weering TCE, Epping EHG, Helder W (1996b) Modeling ^{210}Pb-derived mixing activity in ocean margin sediments: Diffusive versus nonlocal mixing. J Mar Res 54:1207-1227

Soetaert K, Herman PMJ, Middelburg JJ, Heip C (1998) Assessing organic matter mineralisation, degradability and mixing rate in an ocean margin sediment (Northeast Atlantic) by diagenetic modeling. J Mar Res 56:519-534

Soetaert K, Middelburg JJ, Herman PMJ, Buis K (2000) On the coupling of benthic and pelagic biogeochemical models. Earth Sci Rev 51:173-201

Soetaert K, Herman PMJ, Middelburg JJ, Heip C, Smith CL, Tett P, Wild-Allen K (2001) Numerical modelling the shelf break ecosystem: Reproducing benthic and pelagic measurements. Deep-Sea Res II 48:3141-3177

Sun M, Aller RC, Lee C (1991) Early diagenesis of chlorophyll-a in Long Island Sound sediments: A measure of carbon flux and particle reworking. J Mar Res 49:379-401

Van Cappellen P, Wang Y (1996) Cycling of iron and manganese in surface sediments: A general theory for the coupled transport and reaction of carbon, oxygen, nitrogen, sulfur, iron and manganese. Amer J Sci 296:197-243

Westrich JT, Berner RA (1984) The role of sedimentary organic matter in bacterial sulfate reduction: The G-model tested. Limnol Oceanogr 29:236-249

Wijsman JWM, Herman PMJ, Middelburg JJ, Soetaert K (in press) A model for early diagenetic processes in sediments of the continental shelf of the black sea. Est Coast Shelf Sci

Zabel M, Hensen C (2001) The importance of mineralisation processes in surface sediments at continental margins. In: Wefer et al. (eds) Ocean Margin Systems. Springer, Berlin pp 253-267

Slope Instability of Continental Margins

J. Mienert[*], C. Berndt, J.S. Laberg and T.O. Vorren

Department of Geology, University of Tromsø, Dramsveien 201, 9037 Tromsø, Norway
**corresponding author (e-mail): Juergen.Mienert@ibg.uit.no*

Abstract: Giant submarine landslides occur on almost every contintental margin. Individual slides involve up to 20,000 km^3 of slope material and cover an area of up to 113,000 km^2. Their wide spread distribution and their large dimensions make them important geological features, particularly as many of them are located within hydrocarbon exploration areas. The factors that are controlling slope stability are still poorly understood in spite of significant research efforts, and there are only few landslides for which the trigger is known with certainty. It appears that ground motion due to earthquakes, rapid sedimentation, and slope destabilization by gas hydrates are among the most important factors, whereas slope angles seem to be less important.

Introduction

Submarine landslides are a global phenomenon. They occur in the sedimentary successions of continental margins and within the basaltic edifices of volcanic ocean islands. In the scope of this book we concentrate on landslides on continental margins. Submarine landslides have been reported from passive, active and sheared margins (Fig. 1), and they occur on all scales. Factors such as sedimentation, earthquake loading, and gas hydrates have proven to be of variable importance for submarine slope stability. The goal of this paper is to review the recent development and current understanding of slope stability based on examples from active and passive continental margins.

During the last decade research on submarine mass wasting has increased significantly for a number of reasons: (1) hydrocarbon exploration moves into deep-water areas where slides occur (Fig. 1), and where even small slides can have the potential to endanger installations on the seafloor, (2) public awareness of climate change demands understanding of the interrelationship of slope failure and gas hydrate stability, (3) regional side-scan sonar and seabeam bathymetry acoustic images have revealed the importance of erratic down-slope transport of sediments for under- standing ocean margin systems, and (4) submarine mass wasting provides the means to transport large amounts of sand into the ocean basin. Recent discoveries of hydrocarbon in ancient sand deposits of deep-water areas makes an understanding of its distribution important for hydrocarbon reservoir assessment in the ocean deep-water domain.

There are three major types of gravity driven processes: generation of transport by slides or slumps, transport of sediments in laminar motion by debris flow, and transport of sediment in turbulent motion by turbidites. Because it is often difficult to determine from the acoustic images whether a mass movement is a slump or debris flow the terms "blocky" or "cohesive" are used. Here "blocky" describes failures that leave rubble at the base of the eroded sea floor. Another term is the "disintegrative" landslide, which has a distinct scar at the upper source area of gravity-driven transport, but lacks the failure evidence at the base of the failure. Traditionally rotational and translational slide deposits, debris flow deposits, and turbidites have been associated with distinct slope failure processes. However, this might not be true as submarine slides evolve dynamically, for instance slumps frequently turn into debris flows and these into turbulent flows during the downward passage of the involved material (Fig. 2).

From WEFER G, BILLETT D, HEBBELN D, JØRGENSEN BB, SCHLÜTER M, VAN WEERING T (eds), 2002, *Ocean Margin Systems.* Springer-Verlag Berlin Heidelberg, pp 179-193

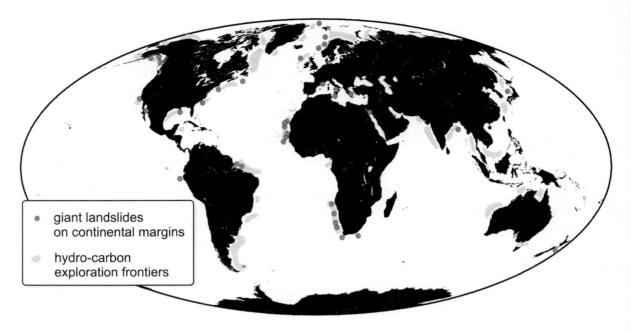

Fig. 1. Global distribution of major submarine landslides on continental margins and location of frontier areas of hydrocarbon exploration (modified after Stow and Mayall 2000). Note, that landslides are found in all well investigated margins, and that they coincide with today's exploration activity.

Modeling of submarine slope failure has improved significantly during the last decade. We will touch on the modeling results, but a thorough review would be beyond the scope of this paper, and we refer the reader to recent publications (Mohrig et al. 1999; Huang and García 1999; Mello and Pratson 1999; Dimakis et al. 2000) and references therein.

Giant landslides on continental margins

Global distribution

Compilation of major landslides (> 1000 km²) reported in the literature (Fig. 1 and Table 1) shows that landslides occur on virtually every continental margin. However, the large number of known small landslides (< 1000 km²) and the expected high number of still undiscovered small landslides, which escape the resolution of acoustic mapping systems, may be equally important in area and volume. Gaps are rather caused by poor data coverage than by the absence of landslides. Evaluation of side-scan sonar data (McAdoo et al. 2000) shows that approximately 10 % of the U.S. conti-

nental margins (in the Gulf of Mexico up to 27%) are influenced by submarine slope instabilities. Two inferences can be directly drawn from these observations. First, landslides are an important phenomenon for understanding sediment transport rates from the upper continental slope to deep-sea basins as well as for maintaining the slope angle. Secondly, the geological processes that are responsible for causing landslides cannot be deducted from their spatial distribution alone. Therefore, we focus in the following on areas with data coverage dense enough to address the individual processes. These areas are the N-American, the European and the NW-African margin. They include active and passive margins, fluvial dominated and non-fluvial dominated, and glaciated and non-glaciated margin segments.

Geomorphology of submarine landslides - an example

Submarine landslides are downslope movement of sediments above a basal shear surface and can result in little-deformed to intensely folded, faulted and brecciated masses that have translated downslope from the original site of deposition.

Location, Name	Area (km²)	Volume (km³)	Run-out (km)	Reference
North America				
Grand Banks, Canadian east coast	27500	150 -200		Piper et al. 1999
Continental slope off Maryland	2000			Embley and Jacobi 1977
Gulf of Mexico	5509			McAdoo et al. 2000
Gulf of Mexico	2913			McAdoo et al. 2000
Gulf of Mexico	2460			McAdoo et al. 2000
Gulf of Mexico	1394			McAdoo et al. 2000
Gulf of Mexico	1098			McAdoo et al. 2000
Icy Bay, Alaska	1080			Schwab and Lee 1986
Sur submarine landslide, California	~1000	35	70	Gutmacher and Normark 1993
Middle and South America				
Amazon Fan	10000	1000		Piper et al. 1997
Peruvian Margin	1000	250		Duperret et al. 1995
Africa				
Agulhas, S Africa	79500	20300		Dingle 1977
Chamais, SW Africa	69000	17400		Dingle 1980
Cape Town, S Africa	48000	10000		Dingle 1980
Childs Bank, SW Africa	28500	3800		Dingle 1980
Spanish Sahara	18000	1100	700	Embley 1975
Dakar, W Africa	6577	395		Jacobi 1976
Dakar, W Africa	4952	495		Jacobi 1976
Walvis Bay, SW Africa	3500	90		Summerhayes et al. 1979
Concepcion Bay, SW Africa	2500	150		Summerhayes et al. 1979
Dakar, W Africa	1102	66		Jacobi 1976
Europe				
Western Mediterranean	60000	500		Rothwell et al. 1998
Storegga Slide, Norway	112500	5600	800	Bugge et al. 1988
Trænadjupet Slide, Norway	14100			Laberg and Vorren 2000
Andøya Slide, Norway	9700			Laberg et al. 2000
Gela Slide, Mediterranean	1500			Trincardi and Argnani 1990
Rockall Bank	1100	300		Faugeres et al. 1981
Asia				
Bassein Slide, Bay of Bengal	3940	960		Moore et al. 1976

Table 1. Reported submarine landslides involving areas larger than 1000 km².

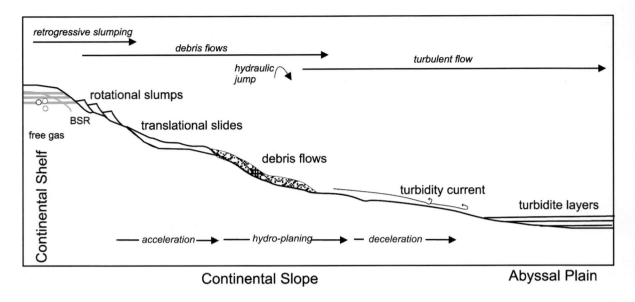

Fig. 2. Schematic diagram for downslope mass movements on continental margins.

Once initiated, the shear surface will propagate upslope from its nucleation point leading to a scoop-shaped, concave-downslope slide scar, often with an irregular outline (Martinsen 1994).

The Towed Ocean Bottom Instrument (TOBI) of the Southampton Oceanography Centre, UK, is a high-resolution side-scan sonar which produces sea-floor images of much higher resolution than previously (Fig. 3). It is now possible to study submarine slides almost at the same level of detail as onshore landslides. Based on TOBI data part of the Trænadjupet Slide offshore Norway was studied. The slide probably occurred during the mid-Holocene, sometime prior to 4 ka BP (Laberg and Vorren 2000).

The headwall of the Trænadjupet Slide is up to 150 m high and 20 km long and controls the nature and location of the shelf break in front of a large transverse shelf trough. Immediately downslope from the headwall initial sediment disintegration produced detached sediment ridges. Near the headwall the ridge spacing is relatively dense, but the ridges increase in spacing and decrease in size downslope. The downslope decrease in size involves breakdown of the ridges into sediment blocks. The sediment ridges moved by back-tilting or through basal deformation (Fig. 3). Transition to sediment streams comprising more-or-less dis-

integrated sediments occurred over some kilometres. Areas dominated by sediment streams are imaged as marked downslope lineations which represent an interplay of erosion by the downslope-flowing sediments and the formation of smaller escarpments delineating lobes of sediments (Fig. 3).

The slide deposits are characterised by what are interpreted to be four prominent debris flow lobes terminating in the southern Lofoten Basin. The lobes have a maximum height of about 150 m above the surrounding sea-floor. GLORIA images indicate small and large blocks. Thus some of the failed mass characterised by the longest run-out distance, probably the most-consolidated sediments, were not remoulded completely during the downslope flow.

Processes influencing landslides

Slope stability depends on the shear strength of the slope material and the applied forces, i.e. primarily gravity. Slope failure occurs if the applied forces exceed the shear strength. The forces applied to a slope will rarely change in nature. Although sea-floor ground motions due to earthquakes are difficult to quantify (Spudich and Orcutt 1982), it

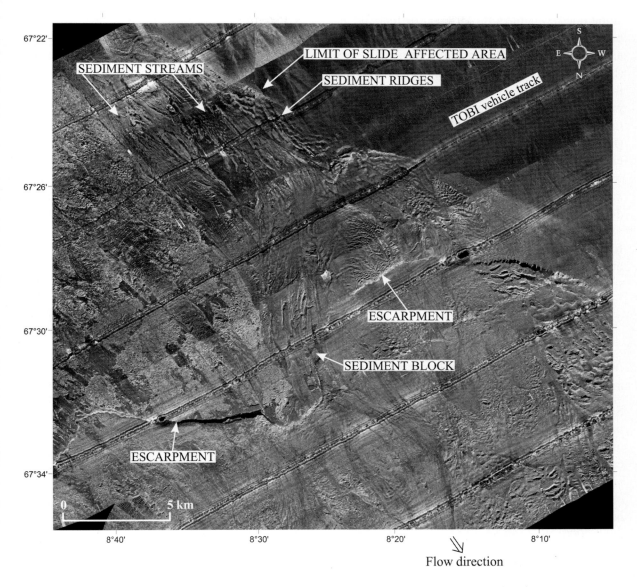

Fig. 3. Sea-floor image of Towed Ocean Bottom Instrument (TOBI) side-scan sonar covering the upper, western part of the Trænadjupet Slide. A prominent escarpment up to 100 m high can be followed from the lower left part of the image across the slide. Sediment ridges near the sidewall, areas of sediment stream and a large sediment block are also visible. Sediment flow direct into the lower right and TOBI vehicle tracks are indicated.

may be possible that they can generate forces strong enough to start landslides (Keefer 1993).

It is widely accepted that landslides initiate when the shear strength of the slope material decreases in a short time. The most efficient way to decrease the shear strength is the increase of pore pressure because the effective stress is the applied stress, i.e. gravity, minus the pore pressure (Scheidegger 1982). Mechanisms that increase the pore pressure include sedimentation

rates that are high enough to trap fluids, wave loading, earthquake loading, and localized transport and accumulation of gas and fluids. Schwab and Lee (1988) demonstrated for the Gulf of Alaska that wave loading primarily occurs at shallow water depth on the shelf and leads to rotational slumps, whereas earthquake loading dominates at water depth greater than approximately 50-100 m and is responsible for debris flows. Dissociation of gas hydrates on the continental slope due to changes in sea level or increase in bottom water temperature may produce large amounts of free gas within sediment layers (Mienert et al. 1998; Bouriak et al. 2000). This increase of free gas within the sediment column will decrease the bulk shear strength of the slope material and can potentially lead to slope failure (Paull et al. 1996; Mienert et al. 1998). Finally, erosion of the slope foot by bottom water currents will decrease the stability of the slope, e.g. in deltaic environments. However, so far it has not been demonstrated that the latter process can lead to major landslides on continental margins. In the following we will refer to and discuss the most quoted triggering mechanisms earthquakes, (rapid) sedimentation, and gas hydrates on landslides.

Seismicity and landslides

North American Atlantic and Pacific margins

Seismicity on North American margins is distinctly more frequent and stronger at the active margins on the west coast than at the passive margins of the east coast and in the Gulf of Mexico (Fig. 4). Gloria side-scan sonar and bathymetric data provide evidence of wide spread slope failures on both the active and the passive margins (McAdoo et al. 2000). It is important to note that the active Oregon margin generally has a lower seismic activity than for example the active Japanese margin. Therefore it may not be ideally suited as a type location for an active margin. However, the detailed studies on the Oregon margin allow a first comparison of landslides on passive and active margins using similar data sets. The seismically active margins like the sheared margin of southern California and the margin off Oregon with its subduction zone show landslides only on 7.1 and 3 % of their area, respectively. The passive margin of the Gulf of Mexico has 27 % and the passive New Jersey Margin 9.5 % covered by landslides. This observation indicates that earthquakes, which not only have a high frequency but also a high magnitude may not be the sole triggers for landslides (Fig. 4a).

On the other hand, the North American margins also include the most widely accepted example for a landslide that was caused by earthquake-related ground motion. In 1929 a major submarine landslide occurred south of the Grand Banks immediately after a 7.2 magnitude earthquake had happened (Piper et al. 1999). Temporal proximity and the seismic quiescence of the area at other times, which excludes earthquake loading as a possible release mechanism, make it likely that the slope instability was a direct result of a single earthquake-related ground acceleration.

Norwegian margin

Most of the earthquake foci on the passive Norwegian Margin are at depths less than 25 km and thus generally much shallower than earthquakes with high magnitudes on the active N-American Margins. The locations of epicenters of earthquakes with magnitudes greater than 4 correspond approximately to the outer boundary of postglacial rebound (Fjeldskaar et al. 2000; Byrkjeland et al. 2000) (Fig. 4b).

It is not likely that the pattern of seismic activity was distinctly different during the Early Holocene when the slides occurred, because most workers believe that post-glacial rebound is still the dominating reason for seismic activity on the Norwegian Margin. Comparison of the epicenters of earthquakes (Ms > 4) along the margin with the location of large slides shows that the main center of seismic activity during the 20th century is located south of the Storegga Slide off the Norwegian coast. Additional seismic activity is scattered over the entire margin with a small concentration east of the Trænadjupet Slide. The relation of earthquakes and slides is ambiguous and earthquakes may not be the most important reason for the initiation of slides in this area.

Mediterranean margins

Tectonic activity causes earthquakes with magnitudes greater than 4 in the Mediterranean. A locally destructive earthquake of magnitude 6.1 in the eastern Mediterranean, the Gulf of Corinth, Greece, caused small sized submarine landslides

(Papatheodorou and Ferentinos 1997) in fan delta deposits. The sediment movements occurred on bedding planes. The failures are located about 9 km from the former epicentre. The three dominant instability mechanisms considered in association with this earthquake are liquefaction of a shallow sub-surface horizon (Papatheodorou and Ferentinos 1997). Secondly, a multi-block rotational slide is considered to be caused by remoulding and/or liquefication, and thirdly an elongated slide happened by a combination of shear stress changes of the unconsolidated sediments. The slope failure conditions in the eastern and northeastern Mediterranean Sea most likely have been generated by cyclic loading resulting from earthquakes. A combination of earthquake loading, undercutting, and increases in pore pressure is considered for past sea floor failures on the continental slope of the western Mediterranean. In general, the western Mediterranean Sea landslides and possibly the Adriatic Sea landslides took place during low sea level stands (Rothwell et al. 1998) in regions of high sediment input and prograding wedges. There is little evidence for

seismically triggered slope instabilities in the western Mediterranean Sea.

Sedimentation and landslides

NW-African margin

The trade wind driven upwelling belts of the NW - African Margin receive a considerable amount of biogenic material from ocean productivity and high inputs of aeolian sediments, which in turn cause high sedimentation rates, between 5-10cm/ kyr (Ruddiman et al. 1987). There is a very low seismicity on this passive margin. Dating of turbidites with biostratigraphic methods (Weaver et al. 1992) shows that mass-movement events occurred during times of climate change, and not at times of low or high sea level stands. While the Senegalese margin receives fluvial input to the head of one of the major turbidity current pathways, the Mauritainean Margin receives seasonal fluvial input. Here also lies one of the major turbidity current pathways. The giant slides and slumps observed on the NW-African margin have occurred

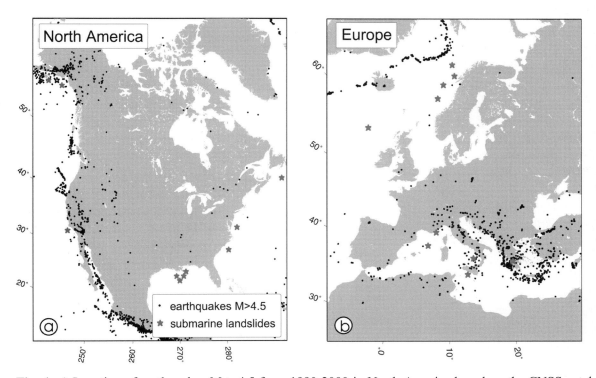

Fig. 4. a) Location of earthquakes M > 4.5 from 1990-2000 in North America based on the CNSS catalog. **b)** Location of earthquakes M > 4 from 1990-2000 in Europe based on the CNSS catalog. Location of landslides based on the literature cited in Table 1 and references therein.

in water depths greater than 2000 m on slopes less than 1.5 ° (Embley and Jacobi 1977). The slides occurred both in pelagic sediments rich in organic matter, and in terrigeneous sediments. No consistent relationship has yet been established between the landslides and the sedimentary systems in which they were initiated.

Norwegian margin

The formerly glaciated margin of Norway experienced sedimentation rates exceeding 10 cm/kyr during advances of the Fennoscandian Continental Ice Sheet to the shelf edge. The late Cenozoic depocenters comprise eight trough-mouth fans (TMF) varying in size between 2700 and 215,000 km² (Vorren and Laberg 1997). These fans are dominated by debris flows. It is suggested that each stacked debris flow unit documents an ice-sheet advance to the shelf break (Vorren et al.

1998). Of the four large late Quaternary landslides on the Norwegian margin (Bjørnøyrenna-, Andøya-, Trænadjupet- and Storegga slides) (Fig. 5), one is located on a fan (Bear Island TMF), one is located on the flank of a fan (North Sea TMF), and two are located in the inter-fan areas. AMS[14]C datings indicate that the acoustically imaged Storegga, Trænadjupet, Bjørnøyrenna, and Andøya slides all occurred during the last 10,000 years in Holocene times (Haflidason et al. 2001; Laberg et al. 2000; Laberg and Vorren 2000). Also, much older slides are observed in seismic reflection profiles in various locations pointing towards prolonged margin instability (Evans et al. 1996).

The two largest fans (Bear Island TMF and North Sea TMF) also contain several large buried slides (King et al. 1996). Thus, it is quite evident that the high sedimentation rate areas, represented by the TMFs, are sites of frequent slide events. The

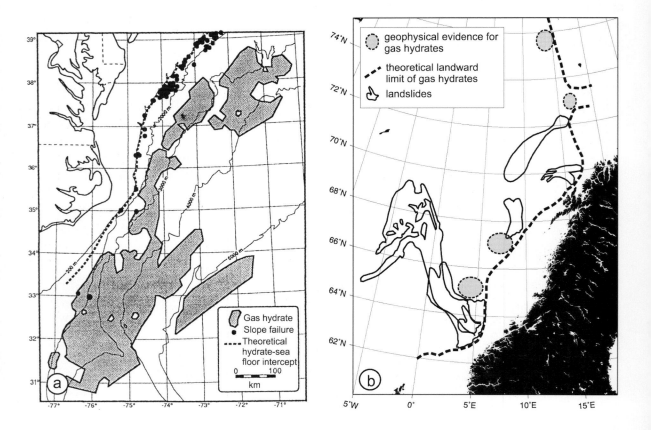

Fig. 5. a) Location of mapped gas hydrates and landslides on the US east coast (after Booth et al. 2000) and **b)** location of gas hydrates and major landslides on the Norwegian margin.

two late Quaternary slides in the inter-fan areas are situated in front of transversal troughs. The troughs were important drainage routes for active ice streams during the late Cenozoic glaciations. It is not clear why TMFs did not develop in these areas. Possibly it relates to the steeper slopes in these areas, or possibly it is due to frequent sliding activity that evacuated the material delivered by the ice-streams.

Gulf of Mexico margin

The sedimentary system of the Gulf of Mexico Margin is characterized by salt tectonics along the entire margin and high sedimentation rates at the Mississippi Delta. Salt diapirs are distinctive tectonic features that create vertical movements including uplift over the diapirs and subsidence between them. These movements change locally the dip of the slope (McAdoo et al. 2000). The most important factors determining location and rate of sediment deposition are tectonics, subsidence, slope channel switching and sea level changes (Bouma et al. 1992). The channel and delta switching occurs over periods from 1000 to 2000 years. It has a substantial influence on the rapid transport of enormous amounts of sediments from the shelf to the deep-sea, the deposition of sediments and slope instability. Sediment supply increases during a relative lowering of the sea level, when the river mouth is more closely located at the shelf edge. This interactive process of sea-level fall, river mouth location, and subsidence dictates whether the sediment will be stored on the shelf as deltaic sediments or if it will bypass the shelf and contribute substantially to a prograding continental slope. The accumulation may be so rapid that the normal compaction of fine grained sediments cannot follow up, which may easily result in slope failure (Coleman et al. 1983). Rather steep headscarps and retrogressive slumping may be formed, which in turn cause the classical processes resulting in debris flows and muddy and sandy turbidites (Bouma et al. 1992). This describes the frequently proposed process of instability of continental margin sediments during sea level lowering. However, we still cannot quantify the relative importance of the different proc-

esses during the different phases of a complete sea level cycle. The largest landslides occur in the vicinity of the Mississipi Canyon where sedimentation is high. This is an interesting candidate for comparable studies between the N-American and European Margin such as the Rhone Canyon.

Gas hydrates and landslides

Blake Outer Ridge

Circumstantial evidence indicates landslide initiation by gas hydrate decomposition on the east coast of the United States (Booth et al. 2000). On this continental margin gas hydrates and submarine landslides are abundant (Fig. 5). The landslides are exclusively located within the gas hydrate stability zone, although gas hydrates have not been observed in the direct vicinity of many of the landslides. However, it is possible that gas hydrates have existed at the head of the observed landslides before the last glaciation during which most of the landslides occurred, because a glaciation and corresponding sea level lowering might have resulted in dissociation of much of the gas hydrate.

One of the best localities to observe the effects of gas hydrate dissociation exists on the Blake Outer Ridge gas hydrate field (Paull et al. 1996). A depression of the Outer Ridge crest is clearly seen in a seismic profile (Fig. 6) including faulting in the upper 400 to 500 m of the sediment column. The faults develop near the base of the gas hydrate stability zone and extend from there to the seafloor (Fig. 6). The base of the hydrate stability zone is marked by a bottom simulating reflector drilled during Leg 164 (Paull et al. 1996). It marks the boundary between the hydrate-bearing sediments above and gas-bearing sediments below. Gas hydrates may also exist without a bottom simulating reflector if no free gas is trapped beneath it (Paull et al. 1996). As the ridge has not been deformed by extension it is interpreted that a downdrop and a rotation of blocks caused the observed features (Dillon et al. 1998), where also mobilisation of sediments is indicated. Pockmark-like depressions exist at the location of the major faults, which are potential gas escape

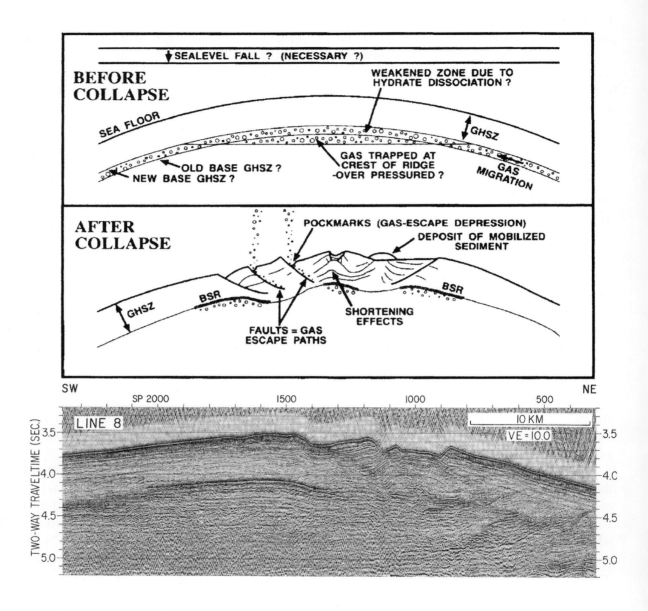

Fig. 6. a) + b) Development of the gas hydrate related collapse structure on the Blake Outer Ridge. **c)** Seismic line of the same structure at a different location (after Dillon et al. 1998).

paths (Fig. 6), documented for other sea floor environments (Hovland and Judd 1988). The Outer Ridge collapse structure was likely caused by the conversion of sediments at the base of the gas hydrate stability zone into a liquefied layer. It seems likely that gas hydrate-induced sediment mobilisation of this order of magnitude can cause submarine landslides in an area more prone to slope failure than the relatively flat top of the Blake Outer Ridge.

Norwegian margin: Storegga

The Norwegian-Barents Sea Margin presents good examples of gas hydrates, and venting-type processes such as diapirs, pockmarks and mud volcanism (Mienert et al. 1998; Bouriak et al. 2000). This margin has experienced some of the world's largest known slides such as the Storegga, Trænadjupet and Andøya Slides (Bugge et al. 1988; Laberg et al. 2000). Laberg et al. (2000) have suggested that the triggering mechanism for the Andøya Slide is earthquake loading. Bugge et al. (1988) discuss the possibility that the Storegga Slide is related to gas and gas hydrates, because a well-defined bottom simulating reflector occurs on seismic profiles from the north-eastern flanks of the current slide scar (Posewang and Mienert 1999). Seismic reflection profiling and ocean bottom hydrophone wide angle seismic experiments suggest that the bottom simulating reflector in this area represents the gas hydrate to free gas interface (Mienert et al. 1998). Mienert et al. (2000) have shown that the gas hydrate stability zone on the Norwegian Margin shifted by about 200 m

between glacial and interglacial times. Also, a collapse structure similar to the one on the Blake Outer Ridge is located directly at the northern fault scarp (Fig. 7). The associated sediment mobilisation might have triggered the slide. A main factor for causing a weakening of sediment strength in hydrated areas is the decomposition of gas hydrates. The decomposition of hydrates is causing a solid hydrate to transform into a fluid and free gas that increases the pore fluid pressure. It is important to note that the distribution of present-day gas hydrates on the Norwegian Margin has been recognised by seismic investigations (Posewang and Mienert 1999) but only in a few in-

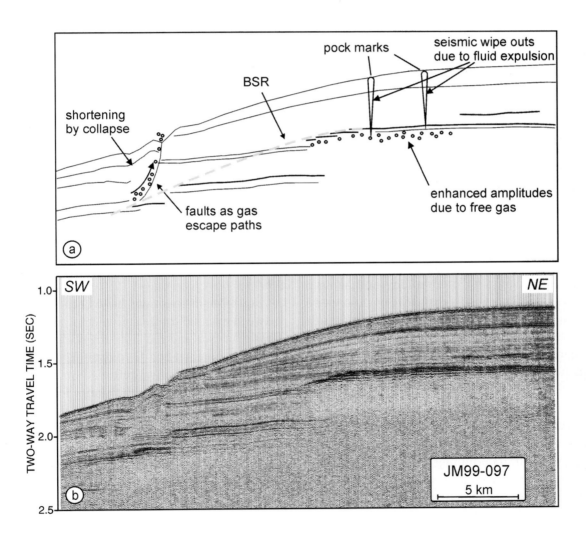

Fig. 7. a) Interpretation of **b)** seismic line JM99-097 on the northern rim of the Storegga Slide, off Norway.

stances by gas hydrate sampling. Past distributions of hydrates are inferred from modeling results (Mienert et al. 2000).

Modeling of the hydrate stabiltity zone (HSZ) as a function of temperature and pressure shows a distinct decrease of the HSZ at the Norwegian margin from the Last Glacial Maximum (LGM) to the present time (Mienert et al. 2000). Important is that the inflow of warm Atlantic Water at the end of the deglaciation 13.000 - 9600 yrs BP caused a thermal wave from the ocean into the seafloor, which has since strongly affected the hydrate reservoirs of shallow and moderate water depths (less than 1500 m). The proposed dissociation of gas hydrate fields within several thousand or even less years depending on the position of the gas hydrate stability base beneath the seafloor places it near or well within the observed slide frequencies. During this period, some of the major slides of the Norwegian margin have taken place. The southern Norwegian margin sites where water mass temperatures increased more distinctly in the upper 800 meters indicate not only a larger change in gas hydrate stability fields (Mienert et al. 2000) but also a much more unstable margin (Vorren et al. 1998).

Discussion and outlook

Acoustic images of morphology and acoustic backscatter have greatly expanded our knowledge about submarine landslides in the last decade. However, the information is biased towards more recent failures, and generally the slide triggers cannot be decided with certainty. Information is even sparser for buried sediment failures observed in seismic reflection profiles.

Recent work by McAdoo et al. (2000) documents that the slope angle is not the most important factor in determining where a slope failure will occur. Sedimentation, erosion, and local geology affect the location of a landslide much more. Moreover, sediment rheology appears to affect the size of landslides. Although most of the commonly called triggers for submarine landslides are earthquakes, it is becoming increasingly evident that there are large segments of the world's continental

margins which show slope failure and are far away from seismically active regions.

In the past, many of the failure mechanisms along passive margins have been assumed to have the greatest probability of triggering slope failure during sea level lowstands (Piper et al. 1997). However, it is now becoming clear that the largest acoustically imaged slope failure events on the Norwegian Margin have all occurred during the last 10000 years, in a period of rapid sea level rise. This weakens the hypothesis that slope failure is directly related to slope erosion or decompression of gas hydrates, that have been proposed for low sea levels.

Despite substantial mapping efforts in the frame of various research programmes on the North American (McAdoo et al. 2000; Hampton et al. 1996), North West African (Wynn et al. 2000) and European Margin (Mienert et al. 1998; Vorren et al. 1998), there are very few examples of submarine landslides on continental margins for which we can name with confidence the release mechanism (Piper et al. 1999). The geomorphic products of submarine landslides are well described, but due to the dynamics of landslides evidence of the release mechanisms can rarely be obtained directly from the observed landslide deposits. Therefore, it is necessary to find methods that can constrain these processes. Numerical modeling can be one way to exclude impossible scenarios (Mello and Pratson 1999).

Equilibrium equations have been established between external forces and internal forces at the failure surface and the slide interfaces based on both experimental and field data (Elverhøi et al. 1997). However, software based on this theory takes only into account the static 2-D case and does not have the ability to model the dynamic evolution of slides and retrogressive sliding mechanism. Thus further developments of dynamic modeling techniques for retrogressive sliding mechanism in sensitive and gassy sediments are required. Also, the initiation of slides may be closely related to the presence of weak layers, which we need to identify and investigate in upcoming studies. The commonly used undrained shear strength and standard slope analysis may thus have underestimated the risk of sliding. Improvements and major advances can be achieved through a combination of field stability analysis, laboratory experiments and finally modeling of observations.

Risk assessment needs to incorporate all available information for a potential slide area such as the geologic evolution, the seismicity, the abundance and potential for gas hydrates, glacier loading, weak layers, and bottom currents. The overall goal should be to develop a procedure for the quantification of the probability (yearly return period) for a specific area to be affected directly by slope instabilities or impacted by the long run out distance of individual slides. This will reduce statements such as "the above observations suggest that triggering of large slope failures identified on the margin may have been caused by a large earthquake".

Conclusions

The large number and the extensive amount of involved slide material make submarine mass wasting one of the most important geological processes shaping the continental margin architecture. Today's knowledge of acoustically mapped continental margin areas is not sufficient to draw firm conclusions on the stability or risk neither on passive nor on active margin areas. Integrated 3-D seismic interpretation and visualisation methods are therefore requested for future stability analysis.

The complex relationship between tectonic and sedimentary processes that are active on different time scales, i.e. tectonic evolution of the margin to almost instantaneous earthquakes, high sedimentation rates to low sedimentation rates due to rapid climatic changes, make it extremely difficult to assess the ultimate reason for an individual slide. Observations without improved modeling of mass movement mechanics and release mechanism only lead to more speculation about the importance of individual or a combination of different processes.

The dynamic evolution from rotational slumps, to translational slides, to eventually turbidity currents destroys the initial structure of the involved sediments, and often obscures evidence of the release mechanisms. Therefore, it is difficult to deduct the release mechanisms just from the geomorphic products of slides. We need new methods to improve "backstripping" of the sea floor observations to improve our understanding of the old slope failures.

Some of the most important questions await more firm answers: (1) What are the specific triggers that decide where slope failure will occur? (2) What are the specific triggers for slope failures on continental margins? (3) Why will one region of seafloor fail while neighbouring regions remain undisturbed? (4) What determines the location of the slip planes? (5) What is the role of gas hydrates in slope stability?

Acknowledgement

This study is a contribution to the European COSTA (Continental Slope Stability) project financed by the European Commission under contract EVK-CT-1999-00006 and the Ocean Margin Deep-Water Research Consortium (OMARC) cluster. The Norsk Hydro AS funding for cooperative COSTA - SEABED research is kindly acknowledged. The University of Tromsø also acknowledges support of this research by Landmark Graphics Corporation via the Landmark University Grant Program.

References

Booth JS, Winters WJ, Dillon WP (2000) Circumstantial evidence of gas hydrate and slope failure as-sociations on the United States Atlantic continental margin. Ann NY Ac Sci 714:487-489

Bouma AH, Roberts HH, Coleman JM (1992) Late Neogene Louisiana continental margin construction timed by sea-level uctuations. In: Watkins JS, Zhiqiang F, McMillen KJ (eds) Geology and Geophysics of Continental Margins. AAPG Memoir 53:333-341

Bouriak S, Vanneste M, Saoutkine A (2000) Inferred gas hydrates and clay diapirs near the Storegga Slide on the southern edge of the Vøring Plateau, offshore Norway. Mar Geol 163:125-148

Bugge T, Belderson RH, Kenyon NH (1988) The Storegga Slide. Philos Trans R Soc London 325: 357-388

Byrkjeland U, Bungum H, Eldholm O (2000) Seismo-tectonics of the Norwegian continental margin.

J Geophys Res 105 (B3):6221-6236

Coleman JM, Prior DB, Lindsay J (1983) Deltaic influences on shelf edge instability processes. In: Stanley DJ, Moore GT (eds) The Shelf Break: Critical Interface on Continental Margins. SEPM Spec Publ 33:121-137

Dillon WP, Danforth WW, Hutchinson DR, Drury RM, Taylor MH, Booth JS (1998) Evidence for faulting related to dissociation of gas hydrate and release of methane on the southeastern United States. In: Henriet JP, Mienert J (eds) Gas Hydrates: Relevance to World Margin Stability and Climate Change. Geol Soc London Spec Publ 137:293-302

Dimakis P, Elverhøi A, Høeg K, Solheim A, Harbitz C, Laberg JS, Vorren TO, Marr J (2000) Submarine slope stability on high-latitude glaciated Svalbard-Barents Sea margin. Mar Geol 162:303-316

Dingle RV (1977) The anatomy of a large submarine slump on a sheared continental margin (SE Africa). J Geol Soc London 134:293-310

Dingle RV (1980) Sedimentary basins on the continental margins of southern Africa; an assessment of their hydrocarbon potential. Erdöl Kohle Erdgas Petrochem 33 (10):447-463

Dupperet A, Bourgois J, Lagabrielle Y, Suess E (1995) Slope instabilities at an active continental margin: Large-scale polyphase submarine slides along the northern Peruvian margin between 5° S and 6° S. Mar Geol 122:303-328

Elverhøi A et al. (1997) On the origin and flow behavior of submarine slides on deep-sea fans along the Norwegian-Barents Sea continental margin. Geo-Mar Lett 17:119-125

Embley RW (1975) Studies of the deep-sea sedimentation processes using high frequency seismic data. Thesis, Columbia University, New York

Embley RW, Jacobi R (1977) Distribution and morphology of large sediment slides and slumps on Atlantic continental margins. Mar Geotech 2:205-227

Evans D, King EL, Kenyon NH, Brett C, Wallis D (1996) Evidence for long-term instability in the Storegga Slide region of western Norway. Mar Geol 130:281-292

Faugeres JC, Gonthier E, Grousset F, Poutier J (1981) The Feni Drift; the importance and meaning of slump deposits on the eastern slope of the Rockall Bank. Mar Geol 40 (3-4):49-57

Fjeldskaar W, Lindholm C, Dehls JF, Fjeldskaar I (2000) Postglacial uplift, neotectonics and seismicity in Fennoscandia. Quat Sci Rev 19:1413-1422

Gutmacher C, Normark W (1993) Sur submarine slide, a deep-water sediment slope failure. US Geol Surv Bull 2002:158-166

Haflidason H, Sejrup HP, Bryn P, Lien R (2001) The Storegga Slide, Chronology and flow mechanism, XI Europ U Geosci meeting, 8-12 April, Strasbourg, France, Journal of conference, Abstract 6 (1), p 740

Hampton M, Lee H, Locat J (1996) Submarine landslides. Rev Geophys 34 (1):33-59

Hovland M, Judd AG (1988) Seabed Pockmarks and Seepages. Impact on Biology, Geology and the Environment. Graham and Trotman, London

Huang X, Garcìa MH (1999) Modeling of non-hydroplaning mud flows on continental slopes. Mar Geol 154 (1-4):132-142

Jacobi RD (1976) Sediment slides on the northwestern continental margin of Africa. Mar Geol 22 (3):157-173

Keefer DK (1993) The susceptibility of rock slopes to earthquake-induced failure. Bull Int Assoc Eng Geol 30 (3):353-361

King EL, Sejrup HP, Haflidason H, Elverhøi A, Aarseth I (1996) Quaternary seismic stratigraphy of the North Sea Fan: Glacially fed gravity flow aprons, hemipelagic sediments, and large submarine slides. Mar Geol 130:293-315

Laberg JS, Vorren TO (2000) The Trænadjupet Slide, offshore Norway - morphology, evacuation and triggering mechanisms. Mar Geol 171:95-114

Laberg JS, Vorren TO, Dowdeswell JA, Kenyon NH, Taylor J (2000) The Andøya Slide and the Andøya Canyon, north-eastern Norwegian-Greenland Sea. Mar Geol 162:259-275

Martinsen O (1994) Mass movements. In: Maltman A (ed) The Geological Deformation of Sediments. Chapman and Hall, London, pp 127-165

McAdoo B, Pratson LF, Orange D (2000) Submarine landslide geomorphology, US continental slope. Mar Geol 169:103-136

Mello UT, Pratson LF (1999) Regional slope stability and slope-failure mechanics from two-dimensional state of stress in an infinite slope. Mar Geol 154: 339-356

Mienert J, Posewang J, Baumann M (1998) Gas hydrates along the northeastern Atlantic margin: Possible hydrate-bound margin instabilities and possible release of methane. In: Henriet JP, Mienert J (eds) Gas Hydrates: Relevance to World Margin Stability and Climate Change. Geol Soc London Spec Publ 137: 275-291

Mienert J, Andreassen K, Posewang J, Lukas D (2000) Changes of the hydrate stability zone of the Norwegian Margin from glacial to interglacial times. Ann NY Ac Sci 912:200-210

Mohrig D, Elverhøi A, Parker G (1999) Experiments on the relative mobility of muddy subaqueous and subaerial debris flows, and their capacity to remobilize antecedent deposits. Mar Geol 154 (1-4):117-129

Moore GA, Curray JR, Emmel FJ, Yount JC (1976) Dynamic processes of upper Bengal Fan and Swatch of No Ground Canyon, Northeast Indian Ocean. AAPG Bull 60 (4):699

Papantheodorou G, Ferentinos G (1997) Submarine and coastal sediment failure triggered by the 1995, Ms = 6.1 R Aegion earthquake, Gulf of Corinth, Greece. Mar Geol 137:287-304

Paull CK, Buelow WJ, Ussler III. W, Borowski WS (1996) Increased continental-margin slumping frequency during sea-level lowstands above gas hydrate-bearing sediments. Geology 24 (2):143-146

Piper DJW, Pirmez C, Manley PL, Long D, Flood RD, Normark WR, Showers W (1997) Mass-transport deposits of the Amazon Fan. Proc. Ocean Drill Program Sci Results 155:109-146

Piper DJW, Cochonat P, Morrison M (1999) The sequence of events around the epicentre of the 1929 Grand Banks earthquake: Initiation of debris flows and turbidity current inferred from sidescan sonar. Sedimentol 46:79-97

Posewang J, Mienert J (1999) The enigma of double BSRs: Indicators for changes in the hydrate stability field? Geo-Mar Lett 19:157-163

Rothwell RG, Thomson J, Kähler G (1998) Low-sealevel emplacement of a very large Late Pleistocene "mega-trubidite" in the western Mediterranean Sea. Nature 392 (26):377-380

Ruddiman WF et al (1987) Proc Ocean Drill Program Initial Rep, vol 94

Scheidegger A (1982) On the tectonic setting of submarine landslides. In: Saxov S (ed) Marine Slides and other Mass Movements. Plenum Press, New York, pp 11-20

Schwab WC, Lee HJ (1986) Causes of varied slope failure types in clayey silt, Northeast Gulf of Alaska continental shelf, in Abstracts presented at the SEPM Midyear Meeting, vol. 3, pp 99-100

Schwab WC, Lee HJ (1988) Causes of two slope-failure types in continental-shelf sediment, northeastern Gulf of Alaska. J Sediment Petrol 58 (1):1-11

Spudich P, Orcutt J (1982) Estimation of earthquake ground motions relevant to the triggering of marine mass movements. In: Saxov S (ed) Marine Slides and other Mass Movements. Plenum Press, New York, pp 219-231

Stow DAV, Mayall M (2000) Deep-water sedimentary systems: New models for the 21st century. Mar Petrol Geol 17:125-135

Summerhayes CP, Bornhold BD, Embley RW (1979) Surficial slides and slumps on the continental slope and rise of South West Africa; a reconnaissance study. Mar Geol 31 (3-4):265-277

Trincardi F, Argnani A (1990) Gela submarine slide: A major basin-wide event in the Plio-Quaternary foredeep of Sicily. Geo-Mar Lett 10:13-21

Vorren TO, Laberg JS (1997) Trough mouth fans - palaeoclimate and icesheet monitors. Quat Sci Rev 16:865-881

Vorren TO, Laberg JS, Blaume F, Dowdeswell JA, Kenyon NH, Mienert J, Rumohr J, Werner F (1998) The Norwegian-Greenland Sea continental margins: Morphology and late Quarternary sedimentary processes and environment. Quat Sci Rev 17:273-302

Weaver PPE, Rothwell RG, Ebbing J, Gunn D, Hunter PM (1992) The geochemistry of North Atlantic abyssal plains. Mar Geol 109 (1-2):1-20

Wynn R, Masson D, Stow D, Weaver PPE (2000) The northwest African slope apron: A modern analogue for deep-water systems with complex seafloor topography. Mar Geol 17:253-265

Margin Building - Regulating Processes

L. Thomsen[1*], T. van Weering[2], P. Blondel[3], R. Lampitt[4], F. Lamy[5],
N. McCave[6], S. McPhail[4], J. Mienert[7], R. Neves[8], L. d'Ozouville[9],
D. Ristow[10], C. Waldmann[5] and R. Wollast[11]

[1]IUB, International University Bremen, P.O.Box 750561,
28725 Bremen, Germany
[2]NIOZ, P.O.Box 59, 1790 AB Den Burg, Texel, The Netherlands
[3]University of Bath, Department of Physics, Bath BA2 7AY, UK
[4]Southampton Oceanography Center, European Way, Southampton SO14 3ZH, UK
[5]Universität Bremen, Fachbereich Geowissenschaften, 28359 Bremen, Germany
[6]University of Cambridge, Dept. of Earth Science, Downing Street,
Cambridge CB2 3EQ, UK
[7]University of Tromsø, Dept. of Geology, Dramsveien 201, 9037 Tromsø, Norway
[8]MARETEC, Instituto Superior Técnico, Av. Rovisco Pais, 1049-001 Lisbon, Portugal
[9]ESF Marine Board, European Science Foundation, 1 quai Lezay-Marnésia,
67080 Strasbourg Cedex, France
[10]GEOMAR, Forschungszentrum für Marine Geowissenschaften,
Wischhofstr. 1-3, 24148 Kiel, Germany
[11]Université Libre de Bruxelles, Laboratoire d'Océanographie Chimique,
Campus de la Plaine, CP 208, Boulevard du Triomphe, 1050 Brussels, Belgium
* corresponding author (e-mail): l.thomsen@iu-bremen.de

Abstract: Processes at the ocean margins are described and the importance of sedimentary settings, the carbon cycle, the benthic boundary layer, canyons and mass movements is discussed. Mayor outstanding problems are formulated and new enabling technologies are demonstrated.

Introduction

The principal boundary between any continent and ocean basin is the submerged area called the continental margin, a fundamental feature of the planet. The margin of the ocean can be defined as the region between the upper limit of the tidal range and the base of the continental slope. A wide variety of processes are involved in the accumulation and destruction of margins and these may be considered as those which are primarily controlled by biological processes and those which are physically driven. The time scales of the variability of the different processes vary considerably from, for example, seasonal changes in downward flux of biogenic material to the millennial variability associated with sea level change and the destruction of margins due to slumps and slides.

It is important to recognise that in the past 1 million years sea-level has only been as high as at present for less than 5% of the time, and it has been high for only the past 7,000 years. Much of the form of the continental shelves was developed subaerially by fluvial processes at lower sea-level (as low as -130 m) and smoothed off by wave reworking during subsequent sea-level rise. For much of the last 1 Ma sediment was discharged much more directly to the ocean basins and shelf-

From WEFER G, BILLETT D, HEBBELN D, JØRGENSEN BB, SCHLÜTER M, VAN WEERING T (eds), 2002,
Ocean Margin Systems. Springer-Verlag Berlin Heidelberg, pp 195-203

margins were built out of sediment deposition. Thus, the present day situation is anomalous (geologically), but exciting (scientifically).

The continental margins have received recently an increased interest in the attempts made to gain a better understanding of the global carbon cycle in the oceans and to evaluate the importance of the marine system in scavenging anthropogenic CO_2. Most of the models developed so far to answer these questions were restricted to the open ocean, because of the complexity and the variability of the hydrodynamic properties and of the biogeochemical processes occurring at the boundary between the open ocean and the continents. It has, however, been recognized that the role of this boundary in the global carbon cycle can no longer be ignored.

Sedimentary settings

Continental margins constitute the most significant interface on Earth for sediment distribution. It has been estimated that as much as 90% of the sediment generated by erosion on land is deposited on continental margins, particularly in major deltaic deposits. The controls of this deposition are many,

ranging from the tectonics and climate of the hinterland where sediment is produced, to tectonics and sea-level changes on the margin itself which define the rate at which space is made available for the deposition of sediment. More recently, man-induced effects are becoming of increased importance, including effects of increased land use, deforestation, inland waterways construction and coastal civil engineering constructions.

Carbon cycle

The ocean margins including the shelf, slope and rise represent only about 20% of the total area of the ocean, but compared with the open ocean, it is the center of an enhanced activity of physical and biogeochemical processes which cannot be neglected in a global description of the marine system. Even if the coastal zone is directly under the influence of nutrient input of continental origin by rivers and the atmosphere, these sources have only a negligible or a very local influence on the primary production when compared with nutrients input from the deep ocean transported by local vertical processes (upwelling, internal waves and secondary flows associated to the slope current).

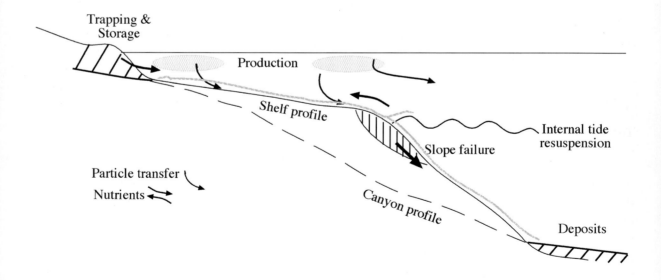

Fig. 1. Conceptual diagram of processes at the continental margin.

The coastal zone receives significant amounts of dissolved or particulate organic and inorganic carbon (about 1 GTC yr[-1]), and little is known about the fate of this material. Compared with the open ocean, the coastal zone is also characterized by enhanced biological productivity due to the large input of nutrients, not only from the continents via rivers. Further offshore at the shelf edge, the transfer of nutrient-rich deep oceanic water across the shelf break is of importance. Highest primary production, export of organic carbon and carbon burial typically occur on the continental shelf, although relatively little of this material is thought to be permanently buried there but will be transported by a variety of processes to the deeper ocean. In addition, the shelf and shelf edge form a favorable environment for the production of biogenic calcium carbonate, that is accumulated locally, or becomes exported as debris of benthic skeletons and dissolved in deep waters, thus contributing to the carbonate pump.

Because of the close correlation between sediment accumulation and organic carbon burial, continental margin sediments, and especially the relatively fine-grained sediments on the continental shelf and slope, must be considered as major potential reservoirs of buried organic carbon. It thus appears evident that continental shelves and slopes are of crucial importance for assessment and understanding of continental margin development and of their non-living resources potential.

The local wave and tidal regimes cause dispersal of sediments across the shelf, and many factors including sediment dynamics, calibre of sediment load and tectonics control sedimentation systems on continental slopes. Especially at and along the shelf edge, the influence and effects of dissipation of internal wave energy may induce resuspension of sediments making them available for transport in a narrow zone along the shelf edge. In combination with along-slope directed currents of variable intensity and on variable time scales, this strongly affects particle (re)-distribution, settling and composition patterns. Inherently, these factors also affect the biological productivity and organic carbon distribution and settling in the water column, and thus have a direct and strong influence on the carbon cycle and on pelagic and benthic community development in the coastal ocean.

The Benthic Boundary Layer

Lateral transport of particles at and over the continental margin is largely controlled by the hydrodynamic conditions within the benthic boundary layer (BBL), which is most important for the exchange between sediments and water column. However detailed observations of BBL characteristics are rare. The resuspension of sediment at the continental margin leads to dispersal of the material offshore over the shelf edge in relatively thin turbid layers (intermediate nepheloid layers, INL). As these spread over the adjacent slope and ocean basin, the coarser components sink out and are subsequently deposited on the lower slope and continental rise. In the INLs normally only the fine residue of what was originally resuspended, without its very fine sand and coarser mud aggregates, can be detected (mainly by optical methods). The bottom nepheloid layer (BNL) on the continental slope can have a strong downslope component and contains large organo-mineral aggregates, which demonstrate rapid transport of both labile and refractory organic carbon over the slope and margin down to the adjacent abyssal plain within a few tens of days. The present day processes acting on the outermost shelf and upper slope therefore lead to the seaward dispersal of fine grained, organic bound particles that were originally mainly deposited during periods of low sea-level stand. This reworked material then is distributed over the continental slope and rise and becomes entrained in the deep-sea circulation system.

Canyons and mass movements

Rapidly deposited fine-grained sediment on continental slopes is potentially unstable. Almost all continental slopes display abundant evidence of slope failure and mass movement in debris flows. It is likely that such slope failure leads to formation of sediment flows, which become turbidity currents and which in turn erode gullies which may grow into canyons. Some of these incipient canyons grow headward and ultimately form canyons which cut into the shelf edge. Other canyons start lower down the slope. Canyons are often located at the low sea-level stand terminus of major river systems.

Canyon systems form an extreme environment, seen in the fact that an actual transfer of components in the water column and of sediment particles from the shelf and shelf edge most likely takes place during short duration events and may have a catastrophic impact on seabed structure, stability and benthic communities in the direct vicinity of and in the canyon. On the other hand, the canyons contribute towards locally enhanced productivity and carbon cycling, because of the special focussed effect of internal tidal waves and currents in the canyons creating upwelling conditions in the canyon head at the shelf edge and on the shelf. Here upwelled waters spill over the nearby continental shelf and upper slope, causing improved conditions for near coastal and shallow shelf fisheries. Numerous, poorly studied canyons, dissect the European upper slope and shelf edge, mainly in the central and southern, formerly non-glaciated areas and in the Mediterranean region. The NE Atlantic and Mediterranean margins are deeply incised and dissected by a large suite of gullies and canyons that extend from a depth of 200 meters or less at the shelf edge down to the abyssal plain and contribute to the formation of turbidite and fan deposits.

Submarine landslides

Giant submarine landslides occur on almost every continental margin. However, the large number of known small landslides and the expected high number of still undiscovered small landslides, which escape the resolution of acoustic mapping systems may be equally important in both area and volume. Individual slides involve up to 80,000 km^2. Their widespread distribution and their large dimensions make them important geological features, particularly as many of them are located within hydrocarbon exploration areas. The factors controlling slope stability are poorly understood in spite of significant research efforts, and there are only few landslides for which the trigger is known with certainty. It appears that ground motion due to earthquakes, rapid sedimentation,

and slope destabilization by gas hydrates are among the most important factors, whereas slope angles seem of less importance. Rapid decompression of the underlying sediments in areas with BSR's (indicating methane hydrate) could lead to catastrophic release of methane. There is strong social benefit to be gained by improving the understanding of slope failure and possible tsunami formation and the modeling of submarine slope failure has improved significantly during the last decade.

Major Outstanding problems

Climate change

The fact that the margins are characterised by high biological productivity has several consequences regarding climate change. First climatic change will induce modifications in ocean circulation patterns and thus change the supply of nutrients to the surface waters. Warming of the surface may increase the degree of thermal stratification and thus potentially reduces vertical mixing. Vertical mixing may thus be reduced at the shelf break where internal waves are the main turbulent mixing process. However, it has been argued that the warming, greater over continents than over the adjacent ocean, would strengthen alongshore winds and thus enhance the upwelling along eastern boundaries of oceanic gyres. The time scale involved for these changes at the margins and the associated biological response might be decades rather than centuries for the modification of the global thermohaline circulation in the open ocean. Changes in the biological pump in the most productive marine system in thus likely to induce significant feedback mechanisms at short time scales. However, in the heterogeneous continental margins, the physical and biogeochemical processes vary strongly on a local scale and must be understood on those scales in order to determine possible global effects.

Ocean Margin Processes

Despite considerable recent work on compiling data sources, the supply of particles from the land to the oceans at present is poorly known, particularly, its changes due to sea level change and human impact. Although river input is known, the amount of sediments trapped in deltas and flood plains is not, and thus the continental supply to the ocean is not very well constrained. Given the importance of assessing coastal stability and the susceptibility of low-land areas to flooding under a regime of rising sea level, it is important that the coastal input sediment budget is assessed. A further major unknown is the particle size distribution of the material delivered to the sea. It needs to be known as e.g. damming selectively removes sand from a system but allows a flux of fine mud to the coast. Thus, natural replenishment of beach sands can be significantly reduced. Sedimentation under storm surges is of considerable interest but has not been investigated in satisfactory depth. The solution of these boundary problems is not simply one for marine sciences. It will require interdisciplinary research on the hydrology of rivers and deltas, and on coastal lagoons, tidal flats and beaches. Many shelves are covered with shell material produced *in situ* in both subtropical areas (e.g. West Florida shelf, Saharan shelf) and temperate to cold water shelves (e.g. South Australian shelf, Hebridean shelf). The amount of carbon and carbonate stored in "cool water carbonates", including the recently discovered cold water carbonate mounds and reef provinces fringing the NE Atlantic Margin need to be established.

The fate of the organic matter produced on the shelf, however, is a question of debate. It is generally admitted that the food web in the coastal zone is short and thus that a large fraction of the primary production is available for export either by burial in the continental shelf sediments or by transport to the open ocean. For mass balance requirements over long time scales, this export of organic carbon and nutrients must correspond to the new production sustained by nutrients of continental or oceanic origin. Furthermore, the tides generate high bottom stresses which are responsible for resuspension and transport of sediments and possibly for the export of freshly deposited material to the slope and rise. Several recent flux measurements of organic material collected in sediment traps deployed along the continental slope indicate a significant lateral transport from the shelf break to the open ocean, often probably linked to an initial phase of bottom transport along the slope.

Generally fine lithogenic and organic particles are transported in form of organo-mineral aggregates towards the shelf edge and then transported across-slope. These aggregates are easily resuspended and microbial activity on these particles is increased with increasing water depth at the slope.

The storage of organic and inorganic carbon at the slope is one order of magnitude higher than in the open ocean. Although rivers only contribute small amounts of nitrogen to the new production, they can export large quantities of particulate organic carbon to the shelf and adjacent slope. The great deltas store huge amounts of organic carbon, which can then be further dispersed by the coastal currents.

Modeling

The quantification of the aforementioned processes needs detailed descriptions and must be achieved through integrated modeling. Accurate models for each compartment (physics, biochemistry, sedimentology and geophysics/geology) are required. These models must cover coastal areas, the shelf and the slope. The complex relationship between tectonic and sedimentary processes that are active on different time scales make it extremely difficult to assess the ultimate reason for individual land slides. Observations without improved physical and/or numerical modeling of mass movement mechanics and release mechanisms only lead to more speculation about the importance of different processes. New numerical methods are needed to improve the "backstripping" of the sea floor observations to improve the understanding of past slope failures.

Physical processes associated with different forcing and topographic conditions are mainly responsible for different behaviors of continental margins. Physical models are general enough to be applicable in all those areas. Accuracy of the results is mainly conditioned by the quality of data used to specify boundary conditions. Since development of integrated models requires accurate physical inputs, studies must be carried on in places where boundary conditions are well defined.

Integrated modeling requires detailed description of physics (advective and diffusive transport) and its coupling to primary production and benthic boundary layer models. Major limitations to the development of integrated models including biology and physics are due to the grid size of physical models (both vertical and horizontal), which is too large to capture the details of the physical processes responsible for biological variability. Another major limitation in the development of coupled models is the time required for calculations, which is not compatible with the number of test runs necessary to biological model calibration. In a smaller extent the same difficulties apply to physical and sedimentological processes.

As computing power increases, obstacles to run integrated models will decrease. Integrated models must then be developed in sites where spatial and temporal scales are not limiting factors. Canyons provide those conditions and there is an urgent need to understand the importance of canyons to trap and channel along slope particle (pollutant) transport and its subsequent downslope export. Generally there still exists a severe lack of the understanding of the different boundary conditions acting at the continental margin.

Key Questions

What is the balance between frequent flux events of small magnitude and rare but major events such as slides?

River input, coastal erosion, delta storage and particle export

• How much sediment and organic carbon (marine and terrestrial) is trapped in deltas and flood plains and how much leaves the coastal zone and enters the coastal current systems?
• What are the particle size distributions of the material leaving the terrestrial domain?

Carbon cycling

• What is the importance of the coastal zone with its tightly coupled benthic/pelagic processes for the global carbon cycling?
• How much terrestrial organic carbon is stored at the continental shelf and slope?
• Do frequent carbon transfer events occur, which transport labile organic carbon to the lower continental margin where it is mineralized within short periods of time and where it can so far not be detected with common sampling techniques?

Benthic boundary layer

• How fast is organic carbon degraded during cross shelf export? Better estimates of the vertical versus lateral flux of organic carbon will be needed when newly developed sediment traps are used.
• How often does cross-slope transport of resuspended particles within the benthic boundary layer occur, and does near bottom transport cause fast downslope transport of labile organic carbon to the continental rise?

Canyon systems

• Do canyons and gullies act as fast transport pathways for particulate matter to greater depth and if so, how often do these processes occur?
• Do canyons trap particles (pollutants) during the along slope transport and enhance shelf/slope transport?
• What is the importance of canyons for the carbon cycle?
• What is the magnitude, frequency and potential impact of seasonal transport events?

Slides

• What are the specific triggers that decide where slope failure will occur?
• Why will one region fail while neighboring regions remain undisturbed?
• What is the likely impact of slides on gas hydrates destabilization and vice versa?
• Where are the weak sediment layers with higher porosity and water content which are responsible for the initiation of slides and how can these processes be modeled (and predicted)?

Key areas for studies

For the study of input of sediments, the coast of Europe offers a huge range of coastal environments, tectonic uplift and subsidence, tidal flats, lagoons, fjords, deltas, beaches, dunes, cliffs, rias and volcanic islands. Inputs include rivers, glaciers, aeolian transport and coastal erosion.

Reasonable carbon budgets are available now for the western European continental margins from the western Barents Sea down to the Iberian Margin and the western Mediterranean areas. There is little data available from the eastern Mediterranean continental margin. Here an additional effort could be undertaken to develop a mass balance for the carbon transfer into the ocean, with new sampling and monitoring techniques.

To address the importance of canyons and cross-shelf edge transport, a canyon site at the Iberian continental margin could be selected. There, both, canyon transport and regular cross slope transport could be studied in close vicinity.

To evaluate the risk of massive sliding, further detailed studies can be carried out along the NE - European margin, where both science and industry are interested in cooperating on this topic.

In general, areas at the continental margin should be determined which are free of any commercial activity and thus allow long-term stations to work under undisturbed conditions.

Enabling technologies

To achieve the envisaged goals of observational programs within the next ten years, we need to consider what technology is already available (and can be refined) and what new technological developments should be initiated. Most of our current understanding of the complex processes in ocean margin systems is based on shipboard measurements that are limited to a sequence of measurements made at single locations. We need new tools, more cost-effective and more productive, to take measurements at a variety of spatial and temporal scales.

The scientific needs are:
• To measure multiple chemical, (geo)physical and (bio)geochemical parameters or fluxes over a longer time scale. Within the last few years, new sensor technology has evolved that will enable long-term observations, of, for example: zooplankton abundance, pCO_2, O_2 and pH. Improved flux measurements are possible by refining sediment trap systems and combining them with upcoming optical observational techniques like multi-frequency laser systems.

• To improve our capability in the determination of the spatial and temporal variability of different process parameters. To this effect, we need to deploy observatories that encompass the use of autonomous, mobile platforms. Autonomous Underwater Vehicles (AUVs) can be used for survey and mapping at a variety of scales, from detailed optical and sidescan imaging to bathymetric mapping. AUVs are also useful for specialised purposes such as measuring benthic boundary layer turbulence and for survey under ice cover. Profiling platforms for studying processes in the water column are cost efficient, as only a single sensor suite is needed to cover a significant depth range. To address the spatial variability of the seafloor, moving lander systems are under development that will allow intense sampling in areas of up to 1-km radius. Deployment or exchange of instruments can be accomplished by using Remotely Operated Vehicles (ROVs). ROVs equipped with high-resolution digital video can precisely sample in the water column or on the seafloor.

• Further development of instruments for mapping the seafloor and detecting seismic activity. Due to the uncertainties in determining the topography and stratigraphy of the ocean bottom, processes such as slope instabilities and submarine landslides are not well understood. The involvement of gas hydrate in slides is an area of intense study, which requires new methods. The use of high-frequency sonar together with stable platforms like AUVs offers completely new perspectives in this field. Envisaged collecting of 3-D and 4-D acoustic observations of the ocean bottom will improve the knowledge of boundary layer processes. Developments on the hardware side need to be leveraged off with developments in data processing and analysis software (e.g. seafloor classification, acoustic scattering and propagation).

• The need for real-time access to ocean data as a unique opportunity for science, for process monitoring and for advanced educational activities. Today, real-time access is achieved through two different approaches: satellite transmission and fixed undersea cables. Satellite links appear cheaper and easier to deploy, particularly far offshore. Conversely, submarine cables give a much higher data bandwidth, and the option of providing power to instruments. At narrow continental margins, the cable approach seems to be advantageous, especially if existing infrastructure or support from activities of offshore companies can be used. The technology of real-time data transmission also enables the presentation of ongoing scientific investigations to the public.

A prerequisite for the successful development and widespread use of new observational tools is to involve commercial interests. They can bring in their experience and, in some cases, take over the running of the infrastructure, reducing the cost to academic scientists. Commercial and other interests will also be able to access the data for their own goals, e.g. when assessing risks to planned underwater structures.

Most of the tools presented here are already accessible, or in the last development phases, and many of the proposed modifications and improvements are within reach. Building on the existing synergy between instrument developers and end-users will benefit both the ocean margin community and the society at large.

Societal relevance

Within the next 50 years, up to 70% of the European population is expected to be concentrated in the coastal zone (within 50 km) and coastal and sea-floor management will become increasingly important. The anthropogenic impact on continental margin systems is increasing. Oil and gas exploration extend to greater water depth and monitoring systems are required to assess the risks connected with these activities. Cross-shelf particle export may act as a pathway for pollutants, which are either dispersed offshore in the upper water column and/or are transported downslope to the continental rise. Canyons are expected to trap and channel pollutants to great depth and are already used as illegal waste disposal sites. The fast export of atmospheric CO_2 to continental shelf sediments and possibly downslope could be enhanced via fertilization. CO_2 disposal into the sea floor may create fluid and gas expulsions at the upper slope with a possible danger of slope failure. Increasing deep-sea fishery activities may destroy the benthic communities and create enhanced resuspension of continental slope sediments. Further on, there is an increasing degree of human intervention in the sediment cycle. Dam construction has resulted in the reduction to almost zero of sediment yield from several major rivers (e.g. Nile, Indus, Colorado and perhaps soon the Yangtze) and many smaller rivers. Dredging of small rivers for sand and gravel frequently takes out more sand and gravel than is supplied from upstream, meaning that the long term depletion of sand in the beds of these rivers leads to accelerated deposition and an effective cessation of their contribution to the coastal zone. (This has led to severe problems of beach erosion, for example along the Italian Adriatic Coast).

Educational outreach

The idea to connect underwater technology to the Internet via fiber-optic cable and/or satellites offer rich educational opportunities for students of all ages. Because the Internet capabilities will provide real-time output from some of the most dynamic Earth-ocean systems, continental margin study products will be well suited for use in classrooms, laboratories, and even the living rooms of interested learners. The excitement of viewing life on the deep-sea floor is an optimal mechanism for capturing the interest of the public. With the increasing use of the continental margin this part of the ocean is also central to social issues, especially for environmental protection. Coupling these facts with the inherent fascination the marine world holds for students and the general public reveals a great educational potential. By using online data transfer and instrument manipulation via the Internet both educational strength and scientific motivations are brought together because the same internet technology that

will offer scientists continuous, long-term access
to the study area will also allow citizens of all ages
to explore worlds on, above and below the seafloor.

Fluid Flow in Continental Margin Sediments

M. Schlüter

Alfred-Wegener-Institut für Polar- und Meeresforschung, Am Handelshafen 12,
27570 Bremerhaven, Germany
corresponding author (e-mail): mschlueter@awi-bremerhaven.de

Abstract: The transfer of solutes and gases across the sediment-water interface through fluid flow at vents and seeps are complementing the inputs of nutrients, methane and xenobiotics through riverine and atmospheric input and diffusive fluxes from the seafloor to the water column of the continental margin. Fluid advection through sediments provide an efficient mechanism for the upward transport of reactive components and trace gases like methane and carbon dioxide, otherwise remineralised or precipitated within the sediment without impact on the chemistry of the ocean or the biota at the seafloor. Studies on submarine groundwater discharge and fluid venting along accretionay ridges emphasise the importance of fluid flow for hydrological budgets, biogeochemical cycles, and physical properties of sediments. Offshore plankton blooms, methane and trace element plumes, and release of large amounts of fresh water reveal the ecological and economic significance of submarine fluid discharge in the coastal zone. Along accretionary ridges, vent sites are characterised by the unique consortium of benthic organisms, formations of massive authigenic carbonates, methane plumes, and occurrences of gas hydrate. These features are associated with fluid flow in sediments. Techniques and tracers for localisation of discharge sites and quantification of discharge rates are introduced. The similarities between these two environmental settings characterised by fluid flow are considered and potential needs for future studies are discussed.

Introduction

Offshore springs and emanations of fluids from coastal sediments were reported by seamen and natural scientists since at least the days of the Romans. Nevertheless, the importance of fluid flow from the seafloor for hydrological cycles, biogeochemical fluxes, distribution of benthic organisms and habitats, or release of methane from the seafloor was essentially overlooked until the beginning of the 1980's. Especially within the last decade a number of geochemical, geological, seismic, and biological investigations provided striking evidence for the widespread occurrence of different environments along continental margins affected by fluid flow (Johannes 1980; Suess et al. 1985, 1998, 1999; Le Pichon et al. 1990; Vogt et al. 1999; Moore 1996, 1999).

For continental margin sediments of the world ocean, including the coastal zone, different modes of fluid flow have to be distinguished. In shallow water, fluid flow through permeable sediments is caused by bottom currents and wave or tidal pumping (Vanderbourgh et al.1977; Hüttel et al.1996). For submarine karstic systems and muddy sediments fluid flow due to submarine groundwater discharge and recirculated flow driven by hydraulic or density gradients were investigated by Kohout (1966), Johannes (1980), Valiela and D'Elia (1990), Cable et al. (1996), Moore (1996), Corbett et al. (1999). At several convergent continental margins (i.e. off Oregon or Peru) tectonic processes cause the formation of accretionary prisms. Lateral compression, sediment compaction, and dehydration of clay minerals cause an upward flow of porewater and discharge of fluids from the seafloor at vent and seep sites (Suess et al. 1985, 1999; Le Pichon et al. 1990; Moore and Vrolijk 1992). Furthermore fluid flow is associated with geological features such as mud volcanoes, submarine slides, subbottom recirculation of seawater, and gas bubble release (ebullition) from sediments (e.g. Hovland

From WEFER G, BILLETT D, HEBBELN D, JØRGENSEN BB, SCHLÜTER M, VAN WEERING T (eds), 2002,
Ocean Margin Systems. Springer-Verlag Berlin Heidelberg, pp 205-217

and Judd 1988). Consequently, fluid flow through continental margin sediments can be observed from nearshore locations down to seep and vent sites at accretionay prisms in several hundred meters of water depth.

These different types of flow, related to geological settings and driving processes, have in common the fast transfer of dissolved and gaseous constituents from the sediment into the bottom water and enhanced exchange between different sedimentological and geochemical domains in the sediment itself. The considerable impact of fluid flow on physical properties of the sediments and the development of specific benthic habitats provide first evidence for location of sites at the seafloor affected by emanating fluids.

One of the most prominent observations associated with fluid flow are vent sites off Oregon, characterised by a unique consortium of organisms nourishing from emanating fluids in water depths of more than 600 m (e.g. Suess et al. 1985, 1999). Associated with fluid flow in accretionary prisms is the transport of methane from deeper sources within the sediment to the seafloor and formation of gas hydrates, massive carbonate and barite precipitates in the sediment (Carson et al. 1994; Bohrmann et al. 1998). Furthermore, porewater-overpressure (associate to fluid flow) affects the physical properties of sediments and has to be considered in the context of sediment slides and slope stability along accretionary prisms and some continental slopes around the world.

In contrast to those vent locations and their specific habitats, submarine groundwater discharge (SGD) from the seabed is essentially "invisible" (Burnett 1998). Due to widespread occurrence and often considerable discharge rates SGD is of importance for the influx of nutrients and formation of "off-shore" plankton blooms (Johannes 1980; Valiela and D'Elia 1990), for the release of methane plumes in the bottom water, and from a hydrological perspective. For example, for Mediterranean countries like Greece it is assumed that a considerable amount of the annual groundwater renewal is lost to the sea by submarine groundwater discharge. The Mid-Atlantic Bight off the US coast is another example for the importance of SGD. For this region Moore (1996) estimated that ~40 % of

the freshwater input is not caused by river run-off but by SGD. High discharge rates were also assumed for the north-eastern Gulf of Mexico (Cable et al. 1996).

The importance of fluid flow for gas hydrate formation and its spatial distribution or the development of unique habitats is generally accepted and scientists and water authorities are becoming aware of the significance of submarine groundwater discharge for considerations of land-ocean interactions – from a scientific and an economic perspective. However, still little is known about the spatial distribution of fluid flow in the different study areas, discharge rates, and especially the impact on biogeochemical cycles and the processes controlling fluid flow. The present manuscript provides an overview concentrating on fluid flow associated with sediment compaction in accretionary prisms and with submarine groundwater discharge from karstic and muddy sediments. Tracers and techniques for localisation and quantification of discharge are considered. The main focus is set to processes controlling fluid flow and implications of advective transport for sediment-water exchange along continental margin sediments.

Indications and tracers for fluid discharge

Whether fluid flow from the seafloor is due to sediment compaction at accretionary prisms, sub-bottom recirculation of seawater, or submarine groundwater discharge in coastal regions, a set of different tracers and techniques are necessary to detect discharge sites and to assess discharge rates. These methods include visual inspection (e.g. via submersibles or towed video systems), seismic techniques, direct measurements by chamber systems or seepage meters, use of naturally occurring radio-nuclides as ^{226}Ra or ^{222}Rn, and investigations of tracers dissolved in the porewater system.

For localisation of discharge sites a combination of visual observation via towed video systems and seismic techniques is typically applied. Vent sites along accretionary prisms off Oregon and Peru, the Nankai through, the Aleutian subduction zone, and the Barbados mud volcanoes are often characterised by colonies of large shells and poly-cheates nourished by emanating fluids (e.g. Suess et al. 1985, 1999; Barry et al. 1996; Olu et al. 1997). Direct observation of these often cluster-like colonies by submersibles, Remotely Operated Vehicles (ROV) or towed video systems allows localisation and mapping of vent sites.

In coastal regions, the capability of visual techniques to trace submarine groundwater discharge is currently limited to karstic systems where SGD occurs via fractured flow and where submarine springs can be observed (e.g. the submarine springs off the Mellisi Coast, Greece, or off Florida). In contrast, visual inspection by bottom photography or video systems provides generally no clear indications for dispersed SGD from sandy or muddy sediments. In such settings bacterial mats or other hints observable at discharge sites are no unique indication.

Echo-sounder and seismic techniques, including boomer systems, side scan sonar, and reflection seismic are efficient tools providing information about fluid flow and conduits in deeper strata of the sediment (von Huene and Scholl 1991; Carson et al. 1994; Wever et al. 1998; Vogt et al. 1999; Mienert et al. 2001). One well studied example for

the capability of these techniques is the accretionary prism off Oregon (water depth > 600 m), where fluid flow along fractures and the dislocation of the bottom simulating reflector (BSR), a seismic reflector in several tens to hundred meter sediment depth running almost parallel to the seafloor, was observed by Westbrook et al. (1994), Suess et al. (1999) and others (Fig. 1). Furthermore, seismic investigations of accretionary prisms allow estimates about porosity reduction during pilling up of sediments. Based on such information fluid flow over large areas can be derived by volumetrically budgeting (e.g. von Huene and Scholl 1991).

Even for investigations of submarine groundwater discharge, shallow seismic techniques provide first and valuable information about discharge sites. For example, studies by echosounder, side scan sonar, and shallow-seismic revealed pockmark like structures in the Irish Sea or in muddy sediments of the Baltic Sea (Fig. 2). Such structures are reported from different locations of the world ocean (Hovland and Judd 1988) and are assumed to be caused by events of large gas eruption, groundwater discharge, or human impacts. In combination with site specific studies like geochemical investigations the driving process can be

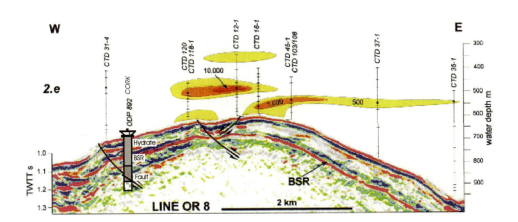

Fig. 1. Fluid flow at the accretionary prism off Oregon. *In situ* flow measurements, observation of vent organisms, deployment of the CORK system (see text), location of the bottom simulating reflector (BSR), and methane plumes in the water column indicate fluid flow from the seafloor (modified from Suess et al. 1999).

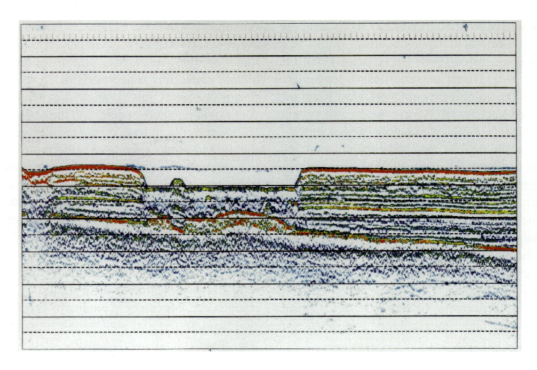

Fig. 2. Pockmark location (water depth ~26 m) in the Eckernförde Bay (W-Baltic Sea) detected by sediment echosounder (IOW, Warnemüde, Germany). The width of the trough like pockmark structure is ~75-90 m. Geochemical investigations of porewater composition, ^{222}Rn measurements in the water column, flow measurements at the seafloor, and deployment of a submarine well revealed the discharge of freshwater (unpubl. results: Schlüter, Sauter, Jensen, Kuijper, Dahlgaard, and Dando).

deciphered and, as in the case for Eckernförde Bay (W-Baltic Sea), submarine discharge of groundwater at such pockmarks was proven. Consequently, visual inspection and seismic surveys provide first information for the localisation of vents and discharge sites.

Currently, neither of the former mentioned visual nor seismic techniques allow estimations of discharge rates. For this purpose, geochemical tracers or direct measurements by seepage meters and chamber systems are necessary. Lee (1977) was one of the first measuring volumetrically seepage fluxes in lakes by a seepage meter lowered to the seafloor and collecting emanating fluids. Since then, several improved seepage meter and chamber systems, equipped with flow sensors and water samplers, were used for direct measurement of discharge rates and chemical composition of fluids (e.g. Sayles and Dickinson 1991; Linke et al. 1994). To record low discharge rates, typical for dispersed outflow, thermistor based flow sensors or heat-pulse techniques are applied. Especially in the deep water of accretionary prisms the positioning of flow chambers on the vent site (the diameter of such sites is often only a few meter to tens of meters) is crucial and requires video guided systems and specific launcher techniques. Although chamber systems allow direct measurements of discharge rates, exact localisation of vent sites and deployment of the device is rather time consuming. Therefore, only a limited data set of direct measurements is available.

Estimates about the importance of fluid flow from continental margin sediments, with regard to nutrient influx, release of methane, and hydrological budgets, requires information about discharge rates and chemical composition of emanating fluids for large areas. Geochemical tracers are alternatives to direct measurements by flow meters and provide the opportunity to derive discharge rates and to calculate spatial budgets for the release of dissolved and gaseous components from the seafloor. Such tracers are naturally occurring radio-nuclides (i. e. ^{222}Rn, ^{226}Ra), chloride, ^{4}He, CH_4, and barium, largely enriched or depleted in emanating fluids if compared to the ambient seawater composition (e.g. Moore 1996; Shaw et al. 1998; Suess et al. 1999; Sauter et al. in press).

Tracer studies in the water column, exemplary ^{222}Rn and ^{226}Ra tracer studies, and in the porewater of the sediment are considered subsequently. The former integrate discharge or recirculation of fluids over larger areas whereas porewater studies provide direct information about the composition of fluids entering the bottom water and enable the calculation of discharge rates by diagenetic modelling.

Related to tracer studies carried out for the water column, especially Cable et al. (1996) and Moore (1996) showed the suitability of radio-nuclides like ^{222}Rn and ^{226,224}Ra for studies on SGD or recirculation of seawater along continental shelves. The radium isotope ^{226}Ra (half life: 1600 years) is a relatively soluble decay product in the uranium decay series, and is thus abundant in groundwaters as well as in sea water. It decays to the inert gas ^{222}Rn (half life 3.82 days). One difference between seawater and groundwater is that groundwater often has a much higher level of radon than seawater. High levels of radon in groundwater result from radium in the geological structure through which the groundwater flows. In contrast, radon in seawater is supported only by the (low) content of radium in the water itself and from a thin layer of sand and sediments at the seafloor (radon diffusion). Likewise, groundwater also tends to have a larger content of radium than seawater. For this reason, radon and radium can be used to trace groundwater supply to the sea.

To derive discharge rates based on ^{226}Ra or ^{222}Rn the distribution of the tracer in the study area has to be mapped by water column sampling. The calculated mass budget, including source/sink terms and exchange rates across the boundaries of the study area such as (1) influx by e.g. molecular diffusion from sediments, (2) seawater exchange between the study area and the open ocean, and (3) air-sea exchange (in case of ^{222}Rn), provides an estimate for the influx of submarine groundwater or recirculated seawater along con-

tinental shelves (Moore 1999). Therefore, these tracers integrate the effect of fluid input over the entire study area. Basically the "radio-nuclide attempt" can be applied to estimate discharge rates for vent areas at accretionary prisms as well as for focused and disperse SGD from sediments in shelf and coastal environments.

Investigating geochemical tracers in the porewater allows the calculation of discharge rates and to investigate the chemical composition of fluids emanating from the seafloor without an imprint due to intense admixture with ambient seawater. As in terrestrial groundwater research non-conservative tracers, undergoing chemical or physical interactions, and conservative tracer like chloride have to be distinguished.

Studies on SGD can take advantage of the chloride composition of porewater to trace fresh-water flow. Figure 3 shows a Cl profile measured in sediments sampled at a SGD site in Eckernförde Bay (W-Baltic Sea). The considerable impact of freshwater is obvious by the strong downward Cl gradient. Diffusion-advection modelling of porewater profiles enables computation of discharge

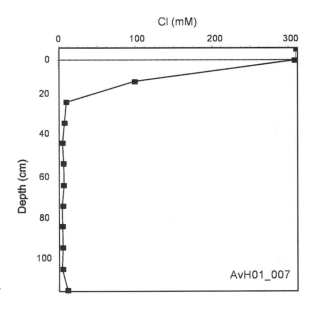

Fig. 3. Chloride concentration of porewater measured at a submarine groundwater discharge site of Eckernförde Bay (26 m water depth). The steep chloride gradient reveals the strong impact of freshwater on the sediment.

rates (Hinkelmann and Helmig this volume). Especially for investigations of disperse SGD from soft sediments the relative ease of sampling, compared to seepage-meter deployments, make reliable spatial coverage of the study area possible and presents information about regional distributions of fluid discharge and release of nutrients or methane by inflow of freshwater into the bottom water.

Although it is here not possible to consider the pros and contras of each technique and tracer in detail, it seems obvious that a set of different attempts has to be combined for estimates of discharge rates and spatial distributions. For example, "point-related" techniques as chamber-based measurements should be combined with water column investigations or seismic studies about the spatial distribution of vent locations which integrate the release of gaseous or dissolved components over large areas.

Fluid flow due to tectonic processes and large scale compaction of sediments at continental margins

Along continental margins like the accretionary prisms off Oregon or Peru, the Aleutian subduction zone, the Nankai trough, and at mud volcanoes like the Barbados and Haakon Mosby mud volcano fluid flow is observed for vent sites located in several hundred meters of water depth (e.g. Suess et al. 1985, 1998, 1999; Le Pichon et al. 1990; Olu et al. 1997; Vogt et al. 1999). Studies carried out by submersibles or towed video systems revealed that colonies of large benthic organisms (e.g. *Calyptogena*, *Pogonophora*, and *Cladorhizidae*) living in symbiosis with bacteria are indicators for fluid flow from the seafloor. These cluster-like colonies of benthic macroorganisms around vent sites derive their energy chemosynthetically by oxidation of dissolved constituents as methane, hydrogen sulphide, or ammonia delivered from below by emanating fluids (Suess et al. 1985; Barry et al. 1996; Olu et al. 1997). Furthermore, massive barite and calcareous precipitates (e.g. Carson et al. 1994; Bohrmann et al. 1998; Suess et al. 1999) were observed for regions affected by fluid discharge.

Geochemical and geophysical investigations at convergent continental margins established the global distribution of tectonically induced fluid flow and its importance for hydrological and geochemical budgets. Tectonically induced dewatering is caused by gravitational and tectonic stress, associated with the formation of an accretionary wedge by subduction processes, which reduces considerably the porosity and affects fluid advection or formation of mud volcanoes (e.g. Carson et al. 1994; von Huene and Scholl 1991; Le Pichon et al. 1990).

Additionally, fluids derived from diagenetic reactions, dehydration of smectites and external sources like meteoric water may contribute to fluid expulsion (Moore and Vrolijk 1992). These processes cause a substantial release of fluids and dissolved components (e.g. calcium, ammonium, hydrogen sulphide, or methane) through the sediment/water interface.

Another important feature of accretionary prisms are gas hydrates (clathrates). These ice-like, methane-bearing components form in methane-rich environments under low-temperature and high-pressure conditions. The occurrence of hydrates in continental margins is often inferred by prominent bottom simulating reflectors (BSR's) observed by seismic investigations. These gas hydrates are abundant in accretionary prisms and passive continental slope sediments and in permafrost regions. They are suggested as a potential energy resource as well as an important greenhouse gas reservoir (Kvenvolden et al. 1993).

Fluid flow from accretionary prisms and mud volcanoes is often directly related to large inventories of methane within the sediment, stemming from thermogenic or biogenic sources. In the sediment, methane occurs as dissolved component as well as free gas and as solid gas hydrate. It is assumed that the BSR reflects the transition zone from dissolved to free gas and gas hydrate, respectively. Nevertheless, an unique relationship between the occurrence of BSR and gas hydrate is not supported by drilling operations (e.g. Westbrook et al. 1994).

Although the occurrence of tectonic dewatering is widely acknowledged, measurements of fluid flow rates are restricted to a few *in situ* measurements, as at vent sites on the Cascadia margin (Fig. 1) and off Peru (e.g. Linke et al. 1994) and indirect estimations based on calculated porosity reductions (e.g. von Huene and Scholl 1991), thermal anomalies (Henry et al. 1992), and hydrogeological and geochemical investigations (Screaton et al. 1995; Schlüter et al. 1998). Flow rates determined by these different means deviate significantly. For example, *in situ* measurements are up to 3 orders of magnitude higher than those derived by computed porosity reductions (Moore and Vrolijk 1992). In this context, different fluid sources (e.g. breakdown of hydrous minerals, meteoric waters, or expulsion of pore water) and modes of fluid transport (dispersed flow through the sediment or focused flow along fault zones) have been suggested for tectonic dewatering of accretionary prisms.

The stability of gas hydrates depends for example on the temperature, pressure, salinity, and content of trace gases as H_2S or higher hydrocarbons. For gas hydrates originating from deep thermogenic CH_4 sources, fluid flow is an efficient transport mechanism for formation of hydrate layers. Furthermore, dissolved constituents as barium are transported from anoxic, SO_4 depleted regions upward into redox conditions suitable for precipitation of barite or authigenic carbonates (e.g. Carson et al. 1994; Bohrmann et al. 1998). Often precipitation causes formation of massive chemoherms at the seafloor. Fluids leaving the sediments into the bottom water cause CH_4 plumes which can be traced along transects crossing active areas (Suess et al. 1999).

Gas hydrates occur along several accretionary ridges, where fluid flow has been observed. Nevertheless, gas hydrates are not an unambiguous indicator for fluid flow, since gas hydrates are coupled to the availability of thermogenic or biogenic methane. The latter requires considerable carbon fluxes reaching the seafloor, which might be restricted to regions of the ocean characterised by high primary production and/or low O_2 concentration of bottom waters.

It has to be mentioned that gas hydrates in the marine environment are not limited to accretionary prisms, but are also common in regions where fluids are transported along fractures (e.g. formed by salt tectonics or associated to other tectonic forces) to the seafloor. One example is the long and well known gas hydrate occurrence in shallow subsurface cores from continental slope sediments of the Gulf of Mexico, where percolation of methane bubbles and trace gases from gas hydrates were observed (e.g. MacDonald et al. 1994). Mud volcanoes like the Barbados mud volcano (e.g. Le Pichon et al. 1990) or the Haakon Mosby mud volcano at the Norwegian-Barents-Svalbard continental margin (e.g. Vogt et al. 1999) are additional examples for fluid flow, formation of gas hydrates, and release of methane to the bottom water.

Common to all these different regions is that fluid flow through sediments provides an efficient mechanism for upward transport of nutrients (e.g. NH_4) and highly concentrated trace gases like methane, otherwise remineralised or precipitated within the sediment without impact for the chemistry of the ocean or the biota at the seafloor. Characterisation of vent sites with their unique biological consortium, escape of trace gases with relevance to climatic change, and fixation and potential release of CH_4 from gas hydrates necessitate a detailed understanding of processes controlling fluid flow and especially about spatial distributions of discharge rates.

Submarine groundwater discharge and sub-bottom recirculation of fluids

Submarine springs, discharge of freshwater from sediments, and recirculated sub-bottom flow is reported for numerous coastal and continental shelf regions of the world ocean, including benthic environments such as muddy and sandy sediments, carbonatic platforms, volcanic and karstic rocks. Analog to fluid flow in tectonically active regions, flow occurs as focused flow through fractures or caves or as dispersed flow through porous media. According to the mode of flow, flow velocities, residence times of fluids, and the area of the seafloor directly affected by subsurface fluid flow differ.

Moore (1996) was one of the first who emphasised the significance of submarine groundwater discharge by investigations of [226]Ra inventories. Considering [226]Ra enrichment, Moore (1996) concluded for the Mid-Atlantic Bight (off the east coast of the US) that roughly 40% of the ^{226}Ra inventory can not be attributed to freshwater runoff by rivers or exchange with the open ocean but are probably due to submarine groundwater discharge or subbottom recirculation of seawater.

Shaw et al. (1998) and Moore and Shaw (1998) investigated subsurface flow along the continental shelf off the South Carolina coast. Considerable enrichment of ^{226}Ra, ^{228}Ra, and Ba by factors of 3-4 above ambient ocean concentrations were observed approximately 40-80 km from shore near the continental shelf break. This enrichment is ascribed to submarine fluid discharge. The salinity of water in the broad area affected by discharge is diluted by 0.2-0.6 relative to water at the same depth farther offshore, suggesting that the fluid flux is similar to the summertime flow of major rivers in the area (Moore and Shaw 1998). Additionally, high concentrations of chlorophyll in the water was observed for this area, probably due to nutrient supply via subsurface flow stimulating productivity.

Focused flow from the seabed is observed, for example, for large areas of the Mediterranean Sea, the Gulf of Mexico, the continental margin off Florida, and parts of the Mid-Atlantic Bight. For some of these areas direct observations and salinity measurements proofed discharge of freshwater whereas for other regions recirculation of sea water through the subbottom is assumed. For the Gulf of Mexico enhanced activities of ^{222}Rn, ^{226}Ra, and barium were observed by Cable et al. (1996), Bugna et al. (1996), and Corbett et al. (1999). By investigation of the ^{222}Rn tracer distribution, Cable et al. (1996) and Corbett et al. (1999) were able to localise discharge sites and to measure discharge rates for the eastern Gulf of Mexico. A considerable part of the submarine discharge, driven by the hydraulic gradient between the landside recharge area and discharge sites, occurs at well defined submarine springs. At such sites ^{222}Rn fluxes of >2420 dpm m^{-2} d^{-1} were measured by benthic chamber experiments (Cable et al. 1996). In contrast, ^{222}Rn fluxes by molecular diffusion from sediments unaffected by fluid flow are an order of magnitude lower. The derived ^{222}Rn inventory in combination with the modelling of the data revealed a regional subsurface flow

ranging from 180-710 m^3 sec^{-1} into the study area of 620 km^2. According to Cable et al. (1996) this is equivalent to 20 first magnitude wells (US classification for water supply wells on land). For the same region Corbett et al. (1999) investigated nutrient fluxes associated with subsurface flow, which revealed the significance of submarine discharge as a nutrient source to surface waters within the eastern area of Florida Bay. This supports the ecological significance of submarine discharge of groundwater already deduced by Johannes (1980) for continental margin areas off Australia or by Laroche et al. (1997) for Long Island's coastal waters.

For regions as the Gulf of Mexico groundwater induced discharge from the seafloor not necessarily means release of freshwater to the marine environment. For example, in the Florida Keys most subsurface fluids are saline to hypersaline. Furthermore, for karstic, hardrock, and coarse grained continental margin sediments the source of emanating fluids might be recycled seawater or, due to chemical interactions with the marine subbottom aquifer, considerably altered freshwater. For such settings, estimates about the volume of groundwater transferred from land to ocean are more complex since the freshwater input has to be deduced from the mixed freshwater-seawater signature including chemical interaction within the sediment. Despite inherent problems to distinguish between the sources of fluid flow from the sediment (groundwater vs. recirculated seawater), it has to be emphasised that the fluid flow finally introduces large volumes of water with a composition considerably different from ambient seawater composition to the coastal environment.

Discharge of freshwater from the seafloor occurs in several regions around the world. Obvious but still scarcely studied examples are submarine springs along the Melissi Coast (Greece), off Sicily, or off Crete. Freshwater discharge rates of several 1000 L m^{-2} hr^{-1} are reported for such locations in karstic environments.

Dispersed flow of freshwater from soft sediments was reported for example for areas of the Baltic Sea or Irish Sea (Bussmann and Suess 1998; Wever et al. 1998). For example, several fresh water discharge sites were observed in Eckern-

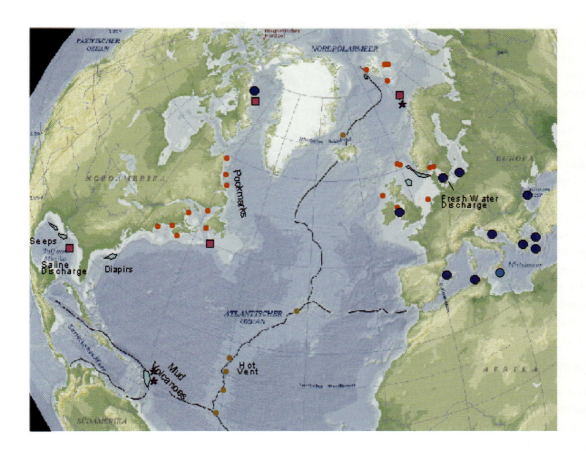

Fig. 4. Selected sites and regions along continental margins of the North Atlantic and adjacent seas where discharge and subbottom recirculation of freshwater are reported. Furthermore, sites along mid ocean ridges or possibly associated with diapirs or mud volcanoes affected by fluid flow are indicated (modified from Vogt et al. 1999).

förde Bay (W-Baltic Sea). Around such discharge sites, strong gradients of chloride concentration with depth in the sediment, enhanced ^{222}Rn inventories in the water column, low salinity in the bottom water, and high CH_4 concentration in the lower water column indicates fluid flow of fresh water. Within the bay such sites where observed in muddy sediments, where pockmark locations where discovered by seismic studies (Fig. 2), and in sandy deposits. Although pockmarks were not observed in the sandy sediment facies, porewater and water column data, as well as visual observations by divers revealed the efflux of freshwater. A dense grid of more than 200 sites of sediment and porewater sampling revealed that submarine groundwater discharge is widespread and of con-

siderable importance for the distribution of methane in the sediment and for the release of methane to the bottom water.

Especially the impact of fluid flow on the methane distribution, including the release of bubbles from the seafloor, reveals the close correspondence between submarine groundwater discharge in coastal areas and the impact of fluid advection along accretionary prisms of the outer continental margin. Subsequently, these similarities are considered by a process oriented approach. Despite different processes forcing fluid flow, comparing submarine groundwater discharge with fluid flow through sediments of accretionary prisms or mud volcanoes, numerous similarities related to biogeochemical cycles are obvious. In both cases

Schlüter

subsurface flow is an efficient transport mechanism for dissolved constituents from land to ocean or from deep sediment strata to the seafloor. Submarine groundwater discharge or recirculated flow has considerable impact on the chemistry of trace elements and gases of the benthic and water column environments of coastal zones and continental margins.

For areas of low river-run off the high significance of SGD from the seafloor for nutrient supply and "offshore plankton blooms" in some coastal areas has been well documented. Increasing requirements of drinking water and protection of pollution by subsurface fluid discharge in coastal zones are only two reasons for the considerable demand for efficient ways to locate discharge sites and for a better understanding of processes controlling fluid discharge and its chemical composition.

Summary and conclusion

The amount of evidence on the widespread occurrence and importance of fluid flow and discharge at continental margins, from a biological, geological and biogeochemical point of view are obvious. Land-ocean interaction, release of groundwater, trace element efflux, nutrient budgets, unique vent biota, flushing of potential greenhouse gases as methane, and formation and stability of gas hydrates are some of the keywords linked to fluid flow from accretionary prism, mud volcanoes, and by submarine groundwater discharge. Furthermore, fluid flow, essentially associated with porewater overpressure, affects slope stability and might alter sediment fabric by precipitation or dissociation of solid phases (e.g. carbonate crusts, gas hydrates).

Still little is known on small scale spatial distribution of flow fields (essential for considerations of biogeochemical and biological interactions or admixture of emanating fluids into the bottom water), spatial distribution of discharge sites, temporal variation of fluid discharge, and processes driving fluid flow including tides, internal waves, and physical properties of sediments. Due to close similarities between fluid flow processes operating in different environmental settings this applies to

studies on accretionary prisms, mud volcanoes, and submarine groundwater discharge. Fluid flow through e.g. beach sediments, salt marsh banks, or hydrothermal circulation along continental margins (e.g. Dando et al. 1995; Jahnke et al. 2000) has not been considered here.

Recent studies have revealed that fluid flow is more widespread as previously expected and easily overlooked. This emphasises the need for more efficient and less laborious techniques for determination of essentially "invisible" fluid flow. Such techniques include efficient screening methods to decide whether fluid flow occurs within a study area, the localisation of discharge sites, and measurements of discharge rates and chemical composition of fluids. Improved seismic and, at least for emanation of freshwater or hypersaline fluids, geo-electrical tools might provide capabilities for time efficient investigations on a larger scale in the near future. Such attempts might include recognition of "weakening" of strata due to porewater overpressure by post processing of seismic records or geo-electric studies by large scale electrode arrays. In addition, future seagoing techniques which measure naturally occurring radio-nuclides and studies on short-lived Ra and Rn isotopes might allow less laborious tracer studies.

Studies on temporal variation of fluid discharge might require deployment of long-term benthic chamber systems or installation of submarine wells. Related to studies on fluid flow from accretionary prisms Westbrook et al. (1994) were able to install a CORK System (Circulation Observation Retrofit Kit) at the OPD drilling site Leg 146 (off Oregon). This system seals the bore hole and includes e.g. a thermistor string, recording the temperature within the entire borehole, and a pressure gauge. Sampling by submersible allowed studies on temporal variation within the system, hydrologeological pumping tests, direct measurement of fluid flow from the CORK, and sampling of fluids for geochemical studies (Davis et al. 1995; Screaton et al. 1995; Schlüter et al. 1998). The temporal variation of submarine groundwater discharge might be studied with a novel approach, using a well-system constructed by perforated screening, submersible pump etc. and specific fittings which is deployed by a vibro-corer system (Sauter et al. in press).

Besides improved techniques for detection, localisation, and sampling of fluids emanating from the seafloor enhanced strategies for data analysis seem essential. These include calculations of spatial budgets and assessments of the impact of fluid flow on biogeochemical cycles and physical properties of the sediment. A geostatistical approach combining field data and classification schemes considering different features like geological setting, sediment type, slope of the seafloor, or hydrography seems to be promising to characterise continental margins concerning potential importance of fluid flow.

Furthermore, the need for an improved understanding of the driving forces for fluid flow and its consequences for biogeochemical cycles and physical properties of sediments requires numerical, non-steady state 2D and 3D modelling. For example, microbial generation of methane or transport of thermogenic gases from depth, formation and release of gas bubbles, and precipitation of gas hydrates can not appropriately be considered by 1D models, which would require unrealistic simplifications regarding the spatial structure of the flow field or distribution of source/sink terms in the sediment. In contrast, transport-reaction modelling of the 2D or 3D structure of subsurface flow, along fractures or dispersed flow, allow more detailed information and prediction of the effect of fluid flow on the environment.

Upcoming needs for a better understanding of land-ocean interactions, as groundwater discharge and potential processes causing destabilisation of gas hydrates (e.g. in drowned permafrost regions or at accretionary prisms), requires intensive investigations of fluid flow through sediments. Combinations of easy-to-use techniques for localisation and quantification of fluid flow and improved 2D and 3D modelling might allow the required assessment of impacts of submarine groundwater discharge on nutrient and trace element inventories along continental margins characterised by low river run-off and the potential recovery of this groundwater resource for landside use.

Acknowledgements

For the invitation to the Workshop I am grateful to the Hanse Institute of Advanced Studies (Delmenhorst, Germany). Thanks to I. Mehser and D. Hebbeln for co-ordination of the Workshop. Discussions with the members of the ELOISE project Sub-GATE (Submarine Groundwater-fluxes And Transport-processes from methane rich coastal sedimentary Environments; ENV4-CT97-0631), E. Suess, B. Burnett and members of the Workshop helped significantly in shaping the ideas presented here.

References

Barry JP, Greene HG, Orange DL, Baxter CH, Robison BH, Kochevar RE, Nybakken JW, Reed DL, McHugh CM (1996) Biologic and geologic characteristics of cold seeps in Monterey Bay, California. Deep-Sea Res 43:1739-1762

Bohrmann G, Greinert J, Suess E, Torres M (1998) Authigenic carbonates from Cascadia subduction zone and their relation to gas hydrate stability. Geol 26:647–650

Bugna GC, Chanton JP, Cable JE, Burnett WC, Cable PH (1996) The importance of groundwater discharge to the methane budgets of nearshore and continental shelf waters of the northeastern Gulf of Mexico. Geochim Cosmochim Acta 60: 4735-4746

Burnett WC (1998) SCOR and LOICZ examine submarine groundwater discharge. IGPB Newsletter 36:13

Bussmann I, Suess E (1998) Groundwater seepage in Eckernförde Bay (Western Baltic Sea): Effect on methane and salinity distribution of the water column. Cont Shelf Res 18: 1795-1806

Cable JE, Burnett WC, Chanton JP, Weatherly GL (1996) Estimating groundwater discharge into the north-eastern Gulf of Mexico using radon-222. Earth Planet Sci Lett 144:591-604

Carson B, Seke E, Paskevich V, Wolmes ML (1994) Fluid expulsion sites on the Cascadia accretionary prism: Mapping direct diagenetic deposits with processed GLORIA. J Geophys Res 99: 959-969

Corbett DR, Chanton J, Burnett W, Dillon K, Rutkowski C, Fourqurean J (1999) Patterns of groundwater discharge into Florida Bay. Limnol Oceanogr 44: 1045-1055

Dando PR, Hughes JA, Leahy Y, Taylor LJ, Zivanovic S (1995) Earthquakes increase hydrothermal venting and nutrient inputs into the Aegean. Cont Shelf Res

15: 655-662

Davis EE, Becker K, Wang K, Carson B (1995) Long-term observations of pressure and temperature in Hole 892B, Cascadia Accretionary Prism. Proc ODP Sci Results 146:299-311

Henry P, Foucher JP, Le Pichon X, Sibuet M, Kobayashi K, Traits P, Chamot-Rooke N, Furuta T, Schultheiss P (1992) Interpretation of temperature measurements from the Kaiko-Nankai cruise: Modelling of fluid flow in clam colonies. Earth Planet Sci Lett 109:355-371

Hovland, M, Judd AG, 1988. Seabed Pockmarks and Seepages. Graham & Trotman, London

Hüttel M, Ziebis W, Forster S (1996) Flow-induced uptake of particulate matter in permeable sediments. Limnol Oceanogr 41:309-322

Jahnke RA, Nelson JR, Marinelli RL, Eckman, JE (2000) Benthic flux of biogenic elements on the south-eastern US continental shelf: Influence of pore water advective transport and benthic microalgae. Cont Shelf Res 20:109-127

Johannes RE (1980) The ecological significance of the submarine discharge of groundwater. Mar Ecol Prog Ser 3:365-373

Kohout F (1966) Submarine springs: A neglected phenomenon of coastal hydrology. Hydrol 26:391-413

Kvenvolden KA, Ginsburg GD, Soloviev VA (1993) Worldwide distribution of subaquatic gas hydrates. Geo-Mar Lett 13:32-40

Laroche J, Nuzzi R, Waters R, Wyman K, Falkowski P, Wallace D (1997) Brown tide blooms in Long Island's coastal waters linked to interannual variability in groundwater flow. Glob Change Biol 3:397-410

Lee DR (1977) A device for measuring seepage flux in lakes and estuaries. Limnol Oceanogr 22:140-147

Le Pichon X, Foucher JP, Boulegue J, Henry P, Lallemant S, Benedetti M, Avedik F (1990) Mud volcano field seaward of the Barbados accretionary complex: A submersible survey. J Geophys Res 95:8931-8943

Linke P Suess E, Torres M, Martens V, Rugh WD, Ziebis W, Kulm LD (1994) In situ measurement of fluid flow from cold seeps at active continental margins. Deep-Sea Res 41:721-739

MacDonald IR, Guinasso NL, Sassen R, Brooks JM, Lee L, Scott KT (1994) Gas hydrate that breaches the seafoor on the continental slope of the Gulf of Mexico. Geol 22:699-702

Mienert J, Posewang J, Lukas D (2001) Changes in the hydrate stability zone on the Norwegian margin and their consequence for methane and carbon releases into the oceanosphere. In: Schäfer P, Ritzrau W, Schlüter M, Thiede J (eds) The Northern North At-

lantic: A Changing Environment. Springer, Berlin pp 259-280

Moore WS (1996) Large groundwater inputs to coastal waters revealed by [226]Ra enrichments. Nature 380: 612-614

Moore WS (1999) The subterranean estuary: A reaction zone of ground water and sea water. Mar Chem 65: 111-125

Moore WS, Shaw TJ (1998) Chemical signals from submarine fluid advection onto the continental shelf. J Geophys Res 103:21543-21552

Moore JC, Vrolijk P (1992) Fluids in accretionary prisms. Rev Geophys 30:113-135

Olu K, Lance S, Sibuet M, Henry P, Fiala-Médioni A, Dinet A (1997) Cold seep communities as indicators of fluid expulsion patterns through mud volcanoes seaward of the Barbados accretionary prism. Deep-Sea Res 44:811-841

Sauter E, Laier T, Andersen C, Dahlgaard H, Schlüter M (in press) Sampling of sub-seafloor aquifers by a temporary well for CFC age dating and natural tracer investigations. J Sea Res

Sayles FL, Dickinson WH (1991) The Seep Meter: a benthic chamber for the sampling and analysis of low velocity hydrothermal vents. Deep-Sea Res 38: 129-141

Schlüter M, Linke P, Suess E (1998) Geochemistry at a sealed deep-sea borehole of the Cascadia Margin. Mar Geol 148:9-20

Screaton EJ, Carson B, Lennon GP (1995) Hydrogeologic properties of a thrust fault within the Oregon accretionary prism. J Geophys Res 100:20025-20035

Shaw TJ, Moore WS, Kloepfer J, Sochaski MA (1998) The flux of barium to the coastal waters of the southeastern USA: The importance of submarine groundwater discharge. Geochim Cosmochim Acta 62:3047-3054

Suess E, Carson B, Ritger SD, Moore JC, Kulm LD, Cochrane GR (1985) Biological communities at vent sites along the subduction zone off Oregon. In: Jones ML (ed) The Hydrothermal Vents of the Eastern Pacific: An Overview. Vol. 6, Bull Biol Soc Washington, pp 475-484

Suess E, Bohrmann G, von Huene R, Linke P, Wallmann K, Lammers S, Sahling H, Winckler G, Lutz RA, Orange D (1998) Fluid venting in the eastern Aleutian subduction zone. J Geophys Res 103:2597-2614

Suess E, Torres ME, Bohrmann G, Collier RW, Greinert J, Linke P, Rehder G, Trehu A, Wallmann K, Winckler G, Zuleger E (1999) Gas hydrate destabilization: Enhanced dewatering, benthic material turnover and large methane plumes at the Cascadia convergent margin. Earth Planet Sci Lett 170:1–15

Valiela I, D'Elia C (1990) Groundwater inputs to coastal waters. Biogeochem 10: 328 p

Vanderborght JP, Wollast R, Billen G (1977) Kinetic models of diagenesis in disturbed sediments. Part I. Mass transfer properties and silica diagenesis. Limnol Oceanogr 22:787-793

Vogt PR, Gardner J, Crane K (1999) The Norwegian-Barents-Svalbard (NBS) continental margin: Introducing a natural laboratory of mass wasting, hydrates, and ascent of sediment, pore water, and methane. Geo-Mar Lett 19:2-21

von Huene R, Scholl DV (1991) Observations at convergent margins concerning sediment subduction, erosion, and the growth of continental crust. Rev Geophys 29:279-316

Westbrook GK, Carson B, Musgrave RT (1994) Proc. Ocean Drill. Program, Initial Rep 146:609p

Wever TF, Abegg F, Fiedler HM, Fechner G, Stender IH (1998) Shallow gas in the muddy sediments of Eckernförde Bay, Germany. Cont Shelf Res 18: 1715-173

Macrobenthic Activity and its Effects on Biogeochemical Reactions and Fluxes

R.R. Haese

Institute of Earth Sciences, University of Utrecht, P.O. Box 80021, 3508 TA Utrecht, The Netherlands
corresponding author (e-mail): rhaese@geo.uu.nl

Abtract: The impact of macrobenthic activity on the geochemistry of surface sediments is reviewed to provide conceptual insights on animal-sediment relations for benthic ecologists, paleoceanographers applying paleo-redox proxies and geochemists interested in the broad area of early diagenesis. It is pointed out that conceptual models for the geochemical implications of macrobenthic activity are relatively well understood but that quantitative approaches are largely lacking. Consequently, particular attention is directed to *in situ* and *ex situ* methods to derive rates of macrobenthic activity. From this literature study it becomes clear that benthic fauna studies and geochemical studies have rarely been integrated. However, this is essential to fully understand the impact of the temporal and spatial variable benthic assemblages on important issues such as organic matter mineralization and metal mobilization in ocean margin sediments. The effects of macrobenthic activity are highly diverse and concern dissolved and solid phase distributions. With respect to nutrient cycling and organic matter mineralization the most important effects arise from bioirrigation. Burrows and tubes are flushed with oxic bottom water which increases the total surface area for aerobic respiration, nitrification and denitrification. In addition, active pumping increases the efflux of dissolved species and creates radial diffusion which is not accounted for when fluxes are quantified from (vertical) pore water profiles by means of molecular diffusion. Since metal diagenesis is ultimately related to solid phase redistributions, e.g. across redox boundaries, bioturbation plays an important role. The depth distribution of bioturbatory activity depends on the feeding strategy of the prevailing fauna which varies significantly.

Introduction

Coastal and ocean margin sediments are of particular biogeochemical interest due to the high accumulation of organic carbon, C_{org}, which is a prerequisite for abundant microbial and macrobenthic life. As a consequence of biological activity a complex interplay of transport and reactions between dissolved and solid phase is occurring. Physically and biologically induced transport drives chemical fluxes across redox-boundaries which separate different benthic compartments such as bottom water / nepheloid layer, oxic, suboxic and sulfidic sediment. Each compartment is characterized by its specific chemical conditions and the respective microbial community. Schematic two-dimensional representations of the distribution of such compartments are shown in Figure 1. If macrofauna is absent one can envision the sedi- ment as a chemically layered structure (Fig. 1a) where each layer is a benthic compartment. Macrofauna cause a dissection of such layers by forming (vertical) housing structures (Fig. 1b) and excreting fecal pellets (Fig. 1c). The latter are particularly nutritious for microorganisms which is the reason why sulfate reduction has been found in the oxic layer (Jørgensen 1977).

Animal – sediment interactions are manifold and cause a wide range of effects, which are of significance for many disciplines of marine sciences. The following principal processes are geochemically important: The search for food and the construction of burrows (mucus lined walls) and tubes (consolidated walls) cause a mixing of solid and dissolved phase, so called bioturbation, which is particularly relevant for solid phase fluxes. Bio-

From WEFER G, BILLETT D, HEBBELN D, JØRGENSEN BB, SCHLÜTER M, VAN WEERING T (eds), 2002,
Ocean Margin Systems. Springer-Verlag Berlin Heidelberg, pp 219-234

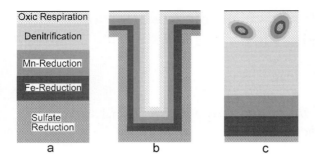

Fig. 1. Schematic representation of the distribution of benthic biogeochemical compartments in **a)** non-bioturbated **b)** bioirrigated and **c)** fecal pellet enriched surface sediment (redrawn after Aller 1982).

irrigation is the most relevant transport mode for the dissolved phase apart from molecular diffusion. The former describes the active pumping of bottom water to deeper parts of the sediment and vice versa within discrete burrow and tube structures to sustain oxic living conditions and to feed on suspended particles. Particularly in recent years the significance of biologically induced particle exchange between the sediment and the bottom water has been studied (Graf 1999). Filter feeders significantly enhance the sediment accumulation rate, which is referred to as biodeposition. The ejection of particulate and dissolved phases into the bottom water induces bioresuspension, which significantly increases and alters the composition of the suspension load.

It will be shown that the above mentioned processes are of environmental relevance because they significantly influence the sedimentary uptake of O_2, the C_{org} mineralization and thus the CO_2 production rate, the efflux of nutrients (NO_3^-, PO_4^{3-}, Si) into the bottom water, as well as the mobility of (trace) metals. Macrobenthic biomass is linearly correlated with the primary productivity in estuaries (Herman et al. 1999) which implies particular relevance of macrobenthic activity in organic rich environments such as coastal and ocean margin sediments. Such systems are typically strongly exposed to regional and temporal variations of C_{org} input, bottom water temperature, salinity and oxygen concentrations, as well as to physical events such as resuspension by current or wave action. Consequently, the interpretation of single obser-

vations deserves special considerations concerning the question of steady versus non-steady state conditions.

This overview on effects of macrobenthic fauna on geochemical reactions and fluxes attempts to summarize the most important concepts and observations, which includes biological, geochemical and modeling results. It is the aim of this manuscript to explain major methods of determinations and to deduce process-orientated cause-and-effects relationships. Quantifications by means of abundances of e.g. macrofauna and rates of e.g. mineralization reported in the literature vary by orders of magnitude depending on the specific conditions and therefore they will be scarcely discussed in this study.

Principles of biologically induced particle and solute transport

The strategy of food search and burrow construction - in other words the mode of bioturbation - differs considerably with respect to the induced displacement of matter. Organisms feeding at the sediment surface and facultative on suspended matter are called 'surface deposit feeders'. Typically, these are small tube-dwelling polychaetes and tubicolous amphipods, such as *Capitella capitata* and *Streblospio benedicti*, which can reach up to 10.000 individuals per m^2 (Rhoads et al. 1978). Rachor and Bartel (1981) report on the ecology of the surface deposit feeding spoonworm *Echiurus echiurus* in the SE North Sea (German Bight) where the abundance of individuals was found to increase by a factor of 100 after particularly cold and stormy winters. This agrees well with the frequently cited concept of successional stages of the benthic fauna assemblage as described by Rhoads et al. (1978). As shown in Figure 2 densely populated surface deposit feeders comprise the pioneering stage after a major physical, biological or chemical perturbation, i.e. a resuspension event caused by currents or waves. At this stage, bioturbation is negligible but bioirrigation is intense to maintain low concentrations of reduced species such as NH_4^+ and HS^- within the tubes. An equilibrium stage in temperate climate is characterized by 'deep deposit feeders'

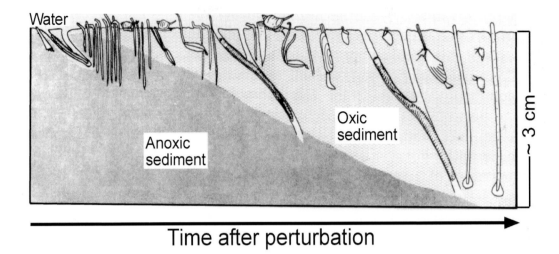

Fig. 2. Successional stages of benthic fauna assemblages after sediment perturbation (redrawn after Rhoads et al. 1978). After a biological, chemical or physical perturbation the pioneering assemblage of macrofauna predominantly consists of bioirrigating species whereas the equilibrium stage is more diverse and comprises particularly bioturbating species.

dominated by polychaetes such as *Nereis sp.* and *Nephtys incisia* and bivalves such as *Nucula annulata* and *Yoldia limatula*. In this case bioturbation becomes important and can comprise random mixing of the surface sediment as well as 'conveyor-belt mixing'. The latter is carried out by polychaetes such as *Clymenella torquata* and *Owenia fusiformis* (Rhoads et al. 1978, Matisoff 1995). They build discrete vertical burrows, ingest in deeper parts of the sediment and egest at the sediment surface, which creates fecal mounds. Due to the preferential vertical transport within discrete channels their induced mixing of matter is referred to as 'conveyor-belt transport'. Reversed conveyor-belt mixing is carried out by the ophiuroid *Amphiura filiformis*. It must be stressed that the proposed concept by Rhoads et al. (1978) is meant to be idealistic and that in reality parts of the faunal assemblage are at different levels of succession. As a result, the observed assemblage resembles a mosaic of different stages. Alternatively to the concept of (time) successive assemblages Pearson (in press) presents a review on faunal groups and their bioturbatory behaviour as a function of variable environmental conditions (e.g. salinity, temperature, current flow, rate and quality of detrital food) to provide new conceptual

insights in benthic community structures and distributions.

Indirectly induced solute transport and biogeochemical reaction zones are particularly important in highly permeable sediments which is referred to as bioroughness (Huettel and Gust 1992). In front of the upstream and behind the down-stream side of a mound oxygen-rich water is forced into the sediment due to increased pressure. Underpressure is found at the sediment/water interface of the upper downstream side which induces an upward advection of solute carrying reduced species such as ammonia, dissolved iron and manganese (Fig. 3) (Huettel et al. 1998). A high resolution oxygen depth profile transect across a mound formed by *Calliannassa truncata* in sandy sediment revealed a maximum oxygen penetration of 3 centimeters in contrast to an oxygen penetration depth of about 0.5 centimeter at a position unaffected by the mound (Ziebis et al. 1996).

Methods to determine transport rates

In order to evaluate the impact of macrobenthic activity on geochemical reactions and fluxes biologically induced transport rates need to be known. Methods for such determinations must be appli-

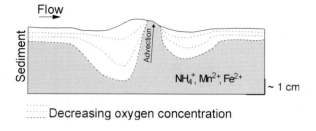

Fig. 3. Schematic illustration of bottom water flow induced distribution of dissolved constituents across a mound in sandy sediment. Increased pressure in front and behind the mound forces oxygen-rich water into the sediment and underpressure at the downstream side gives rise to upward flow of pore water containing reduced species.

cable under *in situ* conditions as well as under laboratory conditions where the activity of single organisms can be studied. Radioisotopes, dyed particles, and chlorophyll-*a*, Chl-*a*, will be discussed here with respect to bioturbation. Conservative, dissolved tracers will be introduced as an approach to quantify bioirrigation. The mathematics for the interpretation of the tracer measurements will be discussed along.

Rate determinations of bioturbation

Radioisotopes such as ^{234}Th ($t_{1/2}$ = 24.1 days) and ^{210}Pb ($t_{1/2}$ = 22.3 years) are short-lived on a geologic time scale and are typically used for the *in situ* determination of bioturbation. ^{234}Th and ^{210}Pb are daughter products of the ^{238}U decay series. Once these isotopes are formed in the water column they are rapidly scavenged by any particulate matter. When interpreting sedimentary profiles of excess activity in terms of bioturbation one relies on a constant flux of matter to the sediment for the time span represented by the radiotracer. The latter can be estimated by multiplying the half-life time by a factor of ~ 4-5. Profiles of ^{234}Th and ^{210}Pb contain information on processes, which affected the sediment during the last 50-100 days and 50-100 years, respectively. In slowly accumulating sediments with moderate to low bioturbation rates ^{210}Pb is useful whereas ^{234}Th becomes depleted in the uppermost millimeters.

In contrast, in coastal and shelf sediments where rates of accumulation and bioturbation are high, ^{234}Th exhibits a gradient of activity in the surface sediment, which is suitable for the quantification of mixing. Apart from bioturbation physical processes such as resuspension, winnowing and focusing may affect the depth distribution of a tracer. The latter two processes refer to temporal and/or spatial changes in sedimentation resulting in a hiatus or increased accumulation, respectively.

Radiotracer distributions are most frequently interpreted in terms of diffusive redistribution of solid particles. Clearly, particles do not diffuse and the distribution of organisms and their activity is not random. Yet, the effect of large numbers of organisms on transport can be described mathematically by the diffusion analogy. The general one-dimensional advection-diffusion equation for the distribution of a particle-associated radionuclide is given as

$$\frac{\partial A}{\partial t} = \frac{\partial}{\partial z}\left[D_b\left(\frac{\partial A}{\partial z}\right)\right] - \omega\left(\frac{\partial A}{\partial z}\right) - \lambda A \qquad (1)$$

with
A = activity of the radionuclide [dpm cm^{-3} of bulk sediment]
t = time [yr]
D_b = particle diffusion coefficient [cm^2 yr^{-1}]
z = depth in the sediment, positive downward [cm]
ω = vertical (advective) transport rate [cm yr^{-1}]
λ = decay constant [yr^{-1}]

ω, the rate of advective transport, has been used for different processes. If bioturbation can be excluded it can be used to calculate the sedimentation rate (Nittrouer et al. 1983). Alternatively, ω has been used to quantify vertically directed transport by conveyor-belt feeders (Gerino et al. 1998). Conveyor-belt feeders typically feed at depth and defecate at the sediment surface. Vice versa, reversed conveyor-belt feeders harvest at the sediment/water interface and defecate at depth. Both transport modes are referred to as 'nonlocal mixing' (Boudreau 1986). The first term on the right side of Eq. 1 describes the activity change at a given depth, z, due to non-directional bioturbation, the second term represents the change of activity at a given depth due to vertically-directed advective transport, and the third term accounts for the radioactive decay.

Assuming steady state conditions ($\partial A/\partial t = 0$), one can solve Eq. (1) given the appropriate boundary conditions ($A=A_0$, z=0; A$\rightarrow$0, z$\rightarrow \infty$; e.g. Gerino et al. 1998):

$$A_z = A_0 \cdot \exp\left[\frac{\omega - \sqrt{\omega^2 + 4\lambda D_b}}{2D_b} \cdot z\right] \qquad (2)$$

Nittrouer et al. (1983) solved Eq. (1) for D_b under the assumption of $\omega^2 \ll \lambda \cdot D_b$ and obtained

$$D_b = \lambda \left(\frac{z}{\ln\dfrac{A_0}{A_z}}\right)^2 \qquad (3)$$

with A_0 and A_z annotating the activities at an upper and a lower level of the activity gradient.

According to Gerino et al. (1998) it is impossible to differentiate D_b and ω with ^{234}Th as a steady state tracer and therefore they set $\omega = 0$ and estimated D_b for the depth interval 0 to z by using a least-squares fit of an exponential function to the measured radiotracer profile. In contrast, Soetaert et al. (1996) applied a suit of more complex models to ^{210}Pb profiles from shelf, slope and deep sea sediments and could successfully differentiate diffusive from nonlocal transport. Their FORTRAN-based programs are available at http://www.nioo.knaw.nl/homepages/soetaert/soetaert.htm. Soetaert et al. (1996) could demonstrate that the proportion of directly injected matter from the sediment/water interface into a discrete sediment layer (downward directed advective transport) versus the flux to the sediment surface decreased with water depth analogous with D_b.

The depth distribution of radioisotopes is by far the most commonly used method to derive D_b in the field. However, a steady-state flux and benthic activity must be assumed which is particularly doubtful when applying a long-living tracer relative to biological processes, e.g. ^{210}Pb, in environments exposed to frequent changes in external forces such as temperature and organic matter fluxes.

A good overview of laboratory experiments using artificial radioisotopes is given by Matisoff (1995). In this report Matisoff also describes a set-up where multiple layers of ^{137}Cs enriched sediment are used to derive depth-dependent D_b values for the marine bivalve *Yoldia limatula*. His results give evidence for decreasing D_b values in the top six centimeters. Such results have not been achieved elsewhere and therefore the depth-dependent distribution of D_b remains an open question.

Dyed particles, so called luminophores, have been successfully applied in field and in laboratory experiments to derive bioturbation coefficients (Mahaut and Graf 1987; Gerino et al. 1998). In contrast to the discussed steady-state input of radio tracers luminophores of different size categories are instantaneously deposited on top of the sediment. After an incubation time of a few weeks cores are sectioned into depth intervals and the depth distribution of luminophores can determined by counting under a microscope with UV light. Equation 1 was originally solved for such non-steady state conditions by Officer and Lynch (1982):

$$C(z,t) = \frac{1}{\sqrt{\pi \cdot D_b}} \exp\left[\frac{-(z - \omega \cdot t)^2}{4D_b \cdot t} - k \cdot t\right] -$$

$$\frac{\omega}{2D_b} \exp\left[\frac{\omega \cdot z}{D_b} - k \cdot t\right] erfc\left[\frac{z + \omega \cdot t}{\sqrt{4D_b \cdot t}}\right] \qquad (4)$$

with
C = Concentration, e.g. [particles cm^{-3}]
t = time [d]
k = first-order reaction constant [d^{-1}] = 0

In this case, Gerino et al. (1998) succeeded in discriminating biodiffusive mixing, D_b, and bioadvection by conveyor-belt feeders. The spreading of the initial tracer maximum was used to derive D_b, whereas the migration of the peak maximum downwards was taken to calculate advective transport by conveyor-belt feeders. Clearly, luminophores are well applicable in the field and in the laboratory to trace biodiffusive and bioadvective transport, as well as particle selective feeding behaviour. However, this approach is very time consuming.

Chl-a can be used as a tracer of organic matter input and bioturbation if its degradation constant in the sediment is known (Sun et al. 1991). According to Sun et al. (1993) Chl-a degradation is represented by a first-order reaction rate which is primarily temperature dependent. Strictly speaking, it is the temperature dependent activity of micro- and meiofauna under oxic conditions, which primarily drives Chl-a degradation. Consequently, incubations as described by Sun et al. (1993) are necessary to first derive the degradation constant which then can be used to calculate biodiffusion coefficients and organic matter fluxes from Chl-a profiles. Analogous to Nittrouer et al. (1983) it was shown by Sun et al. (1991) that the advective term of Eq. 1 is negligible in the sediments they studied and that Eq. 3 could be used to calculate D_b. In this case, k_d, the experimentally derived degradation rate constant, replaces λ and the concentration, C, replaces the activity, A. Notice that C is then also given in mass per volume of total sediment. The flux of Chl-a to the sedi-ment surface can be calculated according to:

$$J = \left(C_0 - C_\infty\right)\sqrt{\frac{k_d}{D_b}} \qquad (5)$$

with C_0 and C_∞ representing the total concentration at the sediment surface and at depth where no change of concentration is apparent.

The inherent advantage of Chl-a is its direct association with organic matter whose reactivity is a central issue in the field of early diagenesis. Its disadvantage is the absence of labile Chl-a during winter months of mid-latitude regions, thus a monitoring of benthic activity throughout the year fails (Gerino et al. 1998).

Rate determination of bioirrigation

Conservative (non-reactive) dissolved constituents are typically used to trace the advective transport within the irrigated surface sediment layer. Here I describe the most common approach by the addition of Br to the bottom water in closed sediment/water incubation experiments. During the incubation Br is progressively transported into the sediment by molecular diffusion and bioirrigation. This is mathematically described by Aller (1983) as:

$$\frac{\partial C}{\partial t} = \frac{\partial}{\partial z}\left[D_s \frac{\partial C}{\partial z}\right] - \alpha \cdot (C - C_{BW}) \qquad (6)$$

with
D_s = the molecular diffusion coefficient in the sediment [cm^2 yr^{-1}]
α = irrigation exchange coefficient [yr^{-1}]
C, C_{BW} = concentration at depth and in the bottom water [moles cm^{-3} (of total sediment)]

In this case boundary conditions are defined as:

$C(x=0, t>0) = C_{BW}$

$\dfrac{\partial C}{\partial x}(x = x_{max}, t > 0) = 0$

$C(x, t = 0) = C_0$

Apparently, Eq. 6 is similar to Eq. 1 due to the diffusion term. The advective term of Eq. 6 is expressed by an irrigation (advective) exchange coefficient multiplied by a concentration difference. The irrigation coefficient integrates over the depth interval, z, and comprises different factors concerning properties and flushing of burrows (Martin and Banta 1992). Martin and Banta (1992) refined α to achieve a depth-dependent exchange coefficient by assuming an exponential decrease in the irrigation activity. However, this approach deserves further proof as assemblages dominated by tube dwelling species most likely produce a subsurface, gaussian-like maximum of irrigation instead of an exponentially decreasing irrigation profile. Most recently, an inverse model for the computation of statistically optimal depth-dependent bioirrigation coefficients was developed (Meile et al. 2001). It can be run with profiles of reactive or conservative tracers. For the four investigated sites a decrease of α with depth was generally observed and at two sites discrete subsurface maxima were revealed (Fig. 4).

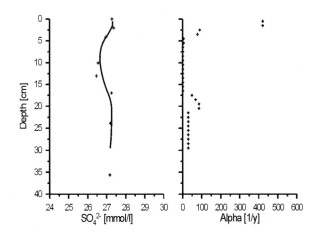

Fig. 4. Depth distribution of measured (crosses) and modeled (line) sulfate in sediments of the Washington shelf. The inverse modeling approach allowed to compute depth-dependent bioirrigation coefficients, α (diamonds), revealing a decrease in irrigation activity with depth in the upper 5 centimeters and a discrete zone of increased irrigation in the subsurface (redrawn from Meile et al. 2001).

Macrobenthic impact on sediment geochemistry

In the absence of macrobenthic fauna and physical reworking geochemical reactions are purely driven by microbial metabolism and subsequently induced chemical reactions. Under such conditions transport is restricted to molecular diffusion. In order to decipher experimentally the impact of macrobenthic activity on the sediment geochemistry reaction rates and fluxes are either compared between situations with and without macrobenthic fauna or determined in sediment plug incubations simulating different path lengths of diffusion analogous to different burrow densities (Aller and Aller

1998). Suppressing macrobenthic activity can be achieved in three ways:

• One may remove macrobenthic fauna by sieving (250-1000 μm mesh depending on definition) which destroys the original sediment properties. In this case one has to wait for a considerable time until compaction is negligible and a new chemical steady-state is established (McCall et al. 1986; Matisoff 1995).

• Freezing (< -20 °C) kills macrobenthic fauna whereas microorganisms survive and reestablish their metabolic activity to a given ambient temperature within days. However, freezing results in the release of NH_4^+ due to lysis (Aller et al. 1998) and other effects on the state of organic matter. Sun et al. (1991) could demonstrate that Chl-*a* is released upon freezing-thawing from a bound, hardly-degradable state to a rapidly degradable fraction. In other words this procedure increases the degradability of Chl-*a*. Additionally, the dead fauna serves as superior microbial nutrition as can be seen by the rapid formation of black spots and patches around dead bodies due to the formation of ironsulfides.

• Anoxic incubation of slices of sediment from discrete depth intervals interrupts fluxes between benthic compartments, and one can remove macrofauna by hand. This technique is most widely used to determine rates and pathways of microbial mineralization at different depth levels (Canfield et al. 1993).

The macrobenthic significance for O_2, nutrients, and C_{org} mineralization

The importance of macrofauna for oxygen fluxes across the water/sediment interface has been studied in coastal and deep-sea sediments as well as in laboratory incubation experiments enriched with organisms of one species. The diffusive flux can be calculated from the pore water profile measured by microelectrodes and compared to total fluxes across the water/sediment interface measured with benthic chambers. As illustrated in Figure 5, the measurement for the determination of the diffusive flux is carried out in between burrows and tubes by means of a microelectrode.

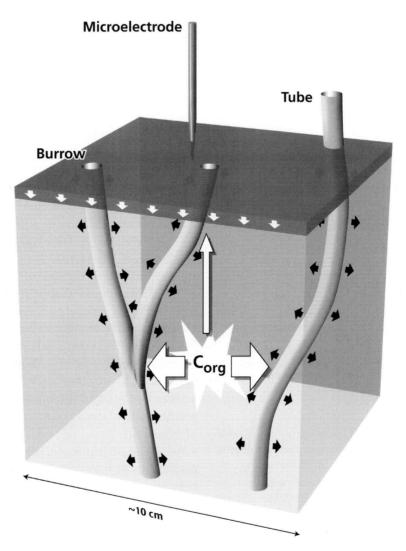

Fig. 5. Sketch of sediment dissected by vertical burrows and tubes. Three-dimensional diffusion of mineralization products (e.g. NH_4^+, Mn^{2+}, Fe^{2+}, HS^-) results in much higher horizontal fluxes (thick arrows) than in vertical direction (thin arrows) in case of densely spaced burrows and tubes. The arrows indicate oxygen fluxes across the water/sediment and burrow/sediment interface.

Burrows and tubes are usually sustained oxic by active pumping of macrofauna. Particularly in deeper parts of the sediment where the mineralization of organic matter produces reduced species such as NH_4^+, Mn^{2+}, Fe^{2+} and HS^- the latter diffuse laterally towards the biogenic housings (Fig. 5) where the oxic interface additionally serves for the (partial) oxidation of such species as discussed below.

In systems where the supply of labile C_{org} limits oxic respiration (e.g. deep-sea sediments) Glud et al. (1994) found a linear correlation between the diffusive oxygen uptake and the C_{org} content (0.02-0.18 g cm^{-3}). They also observed a rough correlation between the oxygen uptake induced by macrofauna with the dry weight of macrofauna. Archer and Devol (1992) studied the impact of bottom water oxygen concentration on sediments of the Washington shelf and slope where labile C_{org} (> 0.25 g cm^{-3}) is not limiting aerobic respiration. For shelf sediments with high bottom water oxygen concentrations they found that the biological flux was generally higher than the diffusive flux. In contrast, in slope sediments exposed to O_2 bottom water concentrations of typically < 100 µM they found that the diffusive flux was equal or higher than the biologically induced flux. Apart from this shift in the ratio of the two flux modes the diffusive flux decreased with decreasing O_2 bottom water concentration. This can be attributed to reduced microbial aerobic respiration since anaerobic mineralization and the reoxidation of its products by O_2 should not be influenced by changes in the oxygen bottom water concentration. Reduced microbial aerobic respiration was also found by Forster et al. (1995) under hypoxic conditions. In contrast to Archer and Devol's (1992) results the laboratory results by Forster and colleagues also revealed a macrobenthic compensation for the decreased oxygen concentrations by an increased irrigation rate. However, this reaction to O_2-stress may be a transitional consequence and on a mid to long time scale macrobenthic fauna and thus irrigation activity is expected to decrease.

In environments with high O_2 bottom water concentrations and high C_{org} contents the uptake of O_2 is primarily determined by the available surface area and the given temperature. The presence of burrows in sediments increases the total oxic surface area (Fig. 1b and 3) which serves oxic metabolism. Hargrave and Phillips (1977) could demonstrate experimentally that a linear correlation between microbial oxygen consumption and the substrate surface area exists. At a temperature of 10 °C they found an O_2 consumption of 135 +/– 25 µmol O_2 cm^{-2} yr^{-1} which is surprisingly similar to the mean diffusive flux of 180 µmol O_2 cm^{-2} yr^{-1} as determined by Archer and Devol (1992) in Washington shelf sediments with a bottom water temperature of 10 °C and O_2 bottom water concentrations ≥ 100 µM. Similarly, Forster et al. (1995) report an O_2 diffusive flux of 157 µM cm^{-2} yr^{-1} for sediments of the Skagerrak with a bottom water temperature of 6.5 °C and O_2 bottom water concentrations of 250 µM. For a temperature range of 1 – 20 °C it has been shown that aerobic metabolism increases as a log-function of temperature under O_2-saturated conditions, however direct deduction of aerobic metabolism as a function of surface area and temperature failed due to changes in the microbial composition and biomass (Hargrave and Phillips 1977). Additionally,

in sediments oxygen is not only consumed by aerobic metabolism but also by the oxidation of reduced species such as NH_4^+, HS^-, Fe^{2+} and Mn^{2+}.

As a consequence of aerobic metabolism at oxic burrow and tube walls it has been shown that nitrification and denitrification are enhanced in and adjacent to macrofaunal burrow and tube walls. In burrows and tubes of single species as well as in sediment in between the biogenic structures, Mayer et al. (1995) determined the capacity of a unit of sediment to oxidize NH_4^+ when NH_4^+ and O_2 are not limiting, the so called nitrification potential. They found enhanced nitrification capacity for biogenic housings relative to the surrounding sediment by a factor of 1.5 to 61. They related the high variability of nitrification potential within different biogenic structures to differences in sediment NH_4^+ concentrations as well as to irrigation behaviour, which is characterized by irrigation rate and duration. Apart from the increase in NO_3^--production at burrow and tube walls irrigation is a rapid and effective pathway e.g. for NO_3^- to be transported to the bottom water and vice versa into the sediment. As a result, increased fluxes of NO_3^- are observed in the presence of irrigating fauna (e.g. Matisoff et al. 1985). However, the produced NO_3^- also diffuses into ambient anoxic sediment where denitrification becomes enhanced. This bacterial activity stimulation adjacent to the biogenic housings is believed to cause the observed increases in NO_3^- and NH_4^+ production (McCall et al. 1986, Aller 1988). Aller (1988) estimates an increase of bulk sediment NH_4^+ production (= macrobenthic excretion plus microbial stimulation through mucus and fecal pellet excretion) by macrofauna of 20-30 %. Additionally, the denitrification - nitrification ratio increases by a factor of 2 in most natural organic-rich sediments due to irrigated burrows which implies less availability of N for the plankton community and changes in the global N cycle over geologic times (Aller 1988). Aller and Aller (1998) studied the increase of mineralization due to the presence of oxygenated burrows and found significant increases at densities > 800 burrows per m^2. Fluxes of the N cycle between bottom water, oxic and anoxic sediment including an irrigated burrow are represented in Figure 6 according to Kristensen (1988). Apart from burrows and their ambient sediment macrofauna itself has been shown to be enriched in N utilizing bacteria. As early as in 1980, Chatarpaul et al. isolated nitrifying and denitrifying bacteria from body walls and guts of oligochaetes.

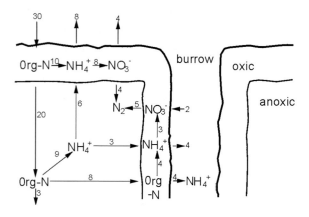

Fig. 6. Fluxes of N-species in [gN m^{-2} yr^{-1}] calculated for a Danish coastal sediment containing *Nereis virens* with an individual density of 700 ind. m^{-2} (redrawn from Kristensen 1988).

Similar to the above discussed species O_2, NO_3^- and NH_4^+, Calvero et al. (1991) showed that total versus diffusive flux ratios of PO_4^{3-} in coastal sediments vary between 1.5 and 7 which gives evidence for the importance of irrigated burrows for the reflux PO_4^{3-} to the bottom water. The impact of macrofauna and its activity on reaction rates of P is difficult to investigate due to numerous interacting sources and sinks such as organic P, P_{org}, P adsorbed to and co-precipitated with iron(oxy)-hydroxides, and P bound to fluorapatite and phosphorites. However, Aller and Aller (1998) could demonstrate by plug incubations that an increase in P_{org} formation is found in anoxic parts of irrigated sediment due to diffusive openness. Consequently, an increase in retention of P_{org} versus total-P by a factor 2-3 due to the presence of macrofauna can be explained which was originally observed by Ingall et al. (1993). Analogous to the N cycle, this is of relevance for the global P cycle and its variations during geologic times.

Silicate is an essential nutrient for the planktonic community to form skeletons. It is dissolved in the sediment due to undersaturation, the respective rate, R_{opal}, is related to the degree of undersaturation according to Rabouille et al. (1997):

$$R_{opal} = -(1-\phi)K_B(x)B_{Si}\frac{Si_{eq} - Si}{Si_{eq}} \quad (7)$$

with

ϕ = porosity
$K_B(x)$ = depth-dependent dissolution kinetic constant
B_{Si} = particulate biogenic silica concentration
Si = dissolved Si concentration
Si_{eq} = Si concentration at saturation with respect to opal

High radial diffusive fluxes towards irrigated burrows (Fig. 5) and subsequent advection to the bottom water results in high biologically induced transport versus diffusive transport. As a result of this increased removal from pore water the actual pore water concentration remains lower than in non-irrigated sediments which increases the dissolution rate of opal (Matisoff et al. 1985).

Clearly, evidence for stimulated bacterial growth and activity coupled to irrigated burrows and tubes is indirectly given by increased rates of aerobic respiration, nitrification and denitrification as well as by increased P_{org} formation. Independent estimates of enhancement derived from increased NH_4^+ (Aller 1988) and CO_2 (Herman et al. 1999) production in organic rich sediments agrees surprisingly well. It is concluded that in the presence of macrofauna the amount of C_{org} mineralization is increased by up to 20-30 %. Direct evidence of increased microbial abundance was shown by Aller and Aller (1986). The bacterial standing stock in the vicinity of burrows was found to be almost doubled relative to ambient sediment. This bacterial enrichment has consequences for the distribution of higher fauna since similar enrichments of meiofauna such as nematodes and foraminifera were found to be associated with higher abundances of microorganisms. Reasons for increased bacterial abundances and mineralization rates associated with macrofaunal activity are compiled in Aller and Aller (1998):
• Biogenic housings increase the surface area for the rapid supply of electron acceptors.

• Metabolites inhibiting microbial activity are rapidly removed through irrigated burrows.

• Macrofauna graze on microbes which stimulates their reproduction.

• Excretion of mucus by means of fecal pellets and burrow linings forms a nutritious substrate.

The macrobenthic significance for Fe, Mn, trace metals and S

The influence of macrofauna on constituents such as O_2, NO_3^-, NH_4^+, and Si is predominantly coupled to bioirrigation since they are transported in dissolved form. In contrast, Fe, Mn, trace metals and S are redistributed in surface sediments by bioirrigation and bioturbation, as they may be present in the solid and dissolved phases. Strictly speaking, bioturbation contributes to the transport of dissolved and solid phases. Yet, molecular diffusion coefficients are typically two orders of magnitude higher than bioturbation coefficients and therefore, the redistribution of the dissolved phase by bioturbation can be neglected.

In Figure 7 the influence of bioturbation on the typical development of the Fe and Mn solid phase profile is shown. At a certain depth Fe- and Mn-(oxy)hydroxides are reduced and Fe^{2+} and Mn^{2+} are released into the pore water. Due to the oxidation by NO_3^- or O_2 these reduced species become reoxidized at some depth above the depth of reduction. Between the depth of reduction and reoxidation an upward directed diffusive flux is induced. Depending on the relation of particulate phase flux by bioturbation and dissolved flux by molecular diffusion the depth profile of solid phase Fe or Mn may vary significantly. The higher the proportion of dissolved flux the higher will be the Fe or Mn solid phase enrichment at a discrete depth. The more intense bioturbation the broader and less distinct will be the enrichment in the solid phase. This representation of the sedimentary Fe and Mn cycle is one-dimensional and as such it can be quantified by means of fluxes. However, as shown in Figure 5 molecular diffusion is three-dimensional and directed to sites of reduced concentration, e.g. oxygenated burrows and tubes serving as sinks for reduced species such as Fe^{2+}, Mn^{2+}, NH_4^+, HS^- or dissolved trace metals. The significance of this horizontal versus vertical transport is related to the spacing of burrows (and tubes) and the depth of reduced species release. Consequently, dissimilatory reduction rates of Fe(III) and Mn(IV) quantified by vertical depth profiles inherently underestimate the true values in bioirrigated sediments. Deducing SO_4^{2-} reduction rates in surface sediments by using pore water gradients fails in bioirrigated sediments because SO_4^{2-} is highly enriched in the bottom water (28 mM) which rapidly replenishes relatively small amounts of consumed SO_4^{2-} through burrow or tube transport (Furukawa et al. 2000). Additionally, produced HS^- has been shown to be reoxidized by O_2 or Mn(IV) whose availability at depths of SO_4^{2-} reduction depends on bioirrigation in the case of O_2 and on bioturbation and concurrent dissimilatory Mn reduction in case of Mn(IV). As a result, it has been found that up to 95% of produced S(-II) becomes reoxidized in ocean margin sediments such as in the Kattegat (Jørgensen and Bak 1991). In this sense, S(-II) is represented by acid (1 N HCl over 30 minutes) volatile sulfide (AVS) which comprises FeS and $\Sigma H_2 S_{aq}$. Consequently, as a result of macrobenthic activity only a small fraction of produced S(-II) is buried in the form of FeS_2.

Dissimilatory Fe- and Mn-reduction are important pathways of C_{org} degradation and it has been shown that their contribution to total C_{org} degradation varies significantly in different environments. Apart from the bioavailable fraction (for review see Haese 2000) it is bioturbation which controls the dissimilatory Fe- and Mn-reduction rate due to the replenishment of Fe- and Mn-oxides at depths where this reaction becomes favorable. Consequently, it has been shown that in sediments of high Mn-oxide accumulation and moderate bioturbation such as in the Panama Basin (Aller 1990) and in the deepest part of the Skagerrak (Canfield et al. 1993) dissimilatory Mn-reduction is the primary process of C_{org} degradation. In such cases O_2 is primarily used for the reoxidation of Mn(II).

An enhancement of Mn and possibly Fe fluxes from the sediment to the bottom water due to bioirrigation can be suspected based on results by Sundby et al. (1986). They found effluxes of the

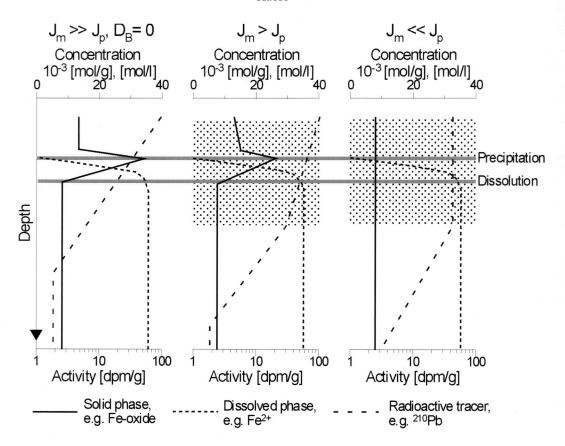

Fig. 7. Schematic representation of the contribution of particulate transport (J_p) by bioturbation and dissolved transport by diffusion (J_m) for the development of the Fe- and Mn- solid phase profiles (solid lines). The shaded area represents the homogeneously bioturbated zone. In all three cases is the pore water profile (dotted lines) kept constant, and thus, the rate of release into pore water and the flux into the horizon of precipitation also remain constant. The profile of any radioactive tracer is given along (stippled line) (redrawn from Haese 2000).

redox sensitive elements Mn, Fe and Co across the sediment/water interface only when oxygen fluxes were reduced. This is particularly given in sediments with burrows and tubes which are flushed intermittently so that no thin diffusive boundary layer prevails within the biogenic housings and oxygen concentrations decrease between episodes of flushing.

Trace metals (TM) such as Cd, Cu, Zn, and Pb are closely coupled to the Fe and Mn cycle due to the high adsorption capacity of TM to Fe- and Mn-(oxy)hydroxides. At the same time they form insoluble complexes with $S(-II)_{aq.}$ at ΣH_2S concentrations < 1 μM which is typically below the detection limit. Roots of vascular plants such as *Spartina maritima* can be considered as analogs

to burrows or tubes since they introduce O_2 into ambient anoxic sediments. Sundby et al. (1998) could demonstrate that Fe, Mn, and TM enriched concretions can form around the roots within a few weeks and are enriched relative to ambient sediment by a factor of 2 for Fe and Mn and by a factor of 5-10 for TM. This clearly illustrates the importance of lateral versus vertical molecular diffusion in the presence of oxygen injection through biological structures for Fe, Mn, and TM.

The significance of macro- and megabenthic activity in cold seep environments

Cold seep environments have been studied with respect to fluid flow, the production of high concentrations of sulfide (> 10 mM) through anaerobic methane oxidation and the occurrence of macro- and megafauna and their symbionts for more than 15 years (e.g. Kulm et al. 1986). From these early observations it was clear that the cold seep fauna which typically consists of big mussels and tube worms (Sibuet and Olu this volume) must carry out intensive bioirrigation for two major reasons: Mainly endosymbiontic sulfur oxidizing bacteria live associated with the megafauna and serve as primary food source for them. As such bacteria need sulfide and oxygen to sustain their metabolism the interface between the oxic burrow and the ambient sulfidic sediment represents an ideal living realm. The sulfide flux from the ambient sediment to the housing of the fauna is very high and needs to be counterbalanced to prevent toxic effects on the fauna. Wang and Chapman (1999) argue that fauna which is permanently exposed to sulfide has achieved different metabolic adaptations and reacts with high irrigation rates to cope with the sulfide. However, until recently such behavioural response to the extreme environment has not been quantified and evaluated in the light of fluxes in surface sediments. Wallmann et al. (1997) were the first ones known to the author who determined nonlocal mixing coefficients, α, in cold seep environments. They conclude that depth-integrated α varies between 20 and 30 yr^{-1} in the Aleutian subduction zone which is sufficient to dominate the fluid flow across the sediment/water interface even at advective flow rates of 3.4 +/- 0.5 m yr^{-1}. Such α-values are comparable to moderately bioirrigated, organic rich coastal sediments and significantly influences the sedimentary oxygen uptake, the reoxidation of ammonia and sulfide as well as the depth of carbonate precipitation in case of cold seep environments.

Summary and conclusions

Most studies on the degradation of organic matter in ocean margin and deep-sea sediments are still based on data sets excluding rates of macrobenthically induced particulate and dissolved phase redistribution. However, this is particularly important in organic rich sediments where high macrofauna activity and high rates of mineraliza-

tion occur contemporaneously. In order to stimulate measurements of macrobenthic activity an overview of fundamental experimental and model approaches are provided herewith.

Effects of macrobenthic activity are manifold and specific for the observed species. One of the most prominent effects is the introduction of oxygen through bioirrigated burrows and tubes into ambient anoxic sediments. This induces aerobic respiration, nitrification and denitrification at and adjacent to the biogenic structure. As a result, the bacterial and meiofauna abundance and consequently the total C_{org} mineralization rate at burrow walls are increased relative to the ambient sediment. In addition to burrow walls, fecal pellets serve as sites of enhanced C_{org} mineralization. Independent studies have estimated an increase of up to 20-30% of C_{org} mineralization due to the activity of macrofauna. For reduced species such as NH_4^+, HS^-, Fe^{2+}, and Mn^{2+} the redox interface of burrow and tube walls represents a site of oxidation and therefore an important link for their sedimentary cycle. The rate of biogenic silicate dissolution and the efflux of Si to the bottom water are increased by bioirrigation due to enhanced removal from the interstitial water. Bioturbation is most important for the rapid downward transport of labile organic matter, as well as bioavailable Fe- and Mn-oxy(hydr)oxides. This enhances anaerobic pathways of organic matter degradation, and O_2 becomes increasingly important as an oxidant for reduced species instead of its role as the electron acceptor for aerobic respiration. Dissimilatory Mn-reduction has been found to be the most important pathway of C_{org} mineralization in regions characterized by high Mn-oxide accumulation and moderate bioturbation. The fate of the sedimentary S cycle is inherently coupled to bioturbation and bioirrigation due the reoxidation of HS^- and FeS in anoxic sediments by O_2 and Mn(IV). The replenishment of the latter two constituents at sites of S(-II) oxidation is strongly enhanced by macrofaunal activity. As a result of the enhanced S(-II) oxidation only a small fraction of the produced S(-II) is buried.

Although we have a good mechanistic understanding of the effects of macrofaunal activity on biogeochemical reactions and fluxes we fail to

quantify them in a spatial and temporal changing environment. Almost all model approaches for the quantification of reactions and fluxes are one-dimensional, and yet, three-dimensional diffusion and laterally dependent reaction rates are evident in sediments with infauna particularly at burrow densities > 800 per m^2 (Fig. 5). Temporal changes of macrobenthic activity have been identified at different time scales: Cycles of burrow flushing and resting occur within one hour, intensified feeding after a deposition event of organic rich material occurs within days and changes in the macrobenthic assemblage and abundances of individual species take place within months. At the moment it remains unclear
• what are the consequences of (regularly reoc-curring) non-steady state conditions at the different time scales,
• how quickly does the pore water geochemistry readjust to given changes and
• is it relevant to study effects of non-steady state i.e. with respect to the mineralization of organic matter?
Such questions can only be answered by coupled 2D/3D transport and reaction modeling under well-defined non-steady state conditions. The spatial distribution of fauna and their activity should be addressed with statistical approaches.

Animal – sediment relations have been a scientific issue for more than 20 years, yet for many important infauna the manner and rate of particulate and dissolved phase redistribution remains unclear. Typically, laboratory incubation experiments with single species are carried out to answer such questions; however, such studies will never cover all important species in ocean margin sediments and the transfer of laboratory based data to the field situation may be ambiguous. Instead, as a first approximation it is suggested to search for a general relation between the average body weight of specific types of 'activists' (e.g. random mixer, conveyor-belt mixer, surface deposit feeder), and their induced transport rate as function of temperature and supplied labile organic matter. The mathematical description of non-random redistribution of particulate and dissolved phase (nonlocal mixing) has only been started in the last few years and deserves further investi-

gations to empirically derive their significance. The lateral displacement has not been approached at all until now.

In order to make quantitative predictions on macrobenthic activity and its impact on C_{org} mineralization in ocean margin sediments we need to know more about the development of the macrobenthic assemblage as a function of surrounding conditions. In other words ecological information needs to be integrated into predictive geochemical models. Such models must clearly include extreme environments such as cold seeps for which bio-irrigation has been identified to significantly contribute to the fluid flow and sedimentary oxygen utilization.

Acknowledgements

I want to thank I. Mehser for the excellent management and hospitality during the Hanse Workshop. Critical reading and comments by S. Kasten, C. Hensen and P. Van Cappellen are highly appreciated. This study is part of the SMILE project (NWO, 750.297.01c) and represents NSG publication number 20010201.

References

Aller RC (1982) The effects of macrobenthos on chemical properties ofmarine sediment and overlying water, In: McCall PL, Tevesz MJS (eds) Animal-Sediment Relations: The Alteration of Sediments. Plenum Press Inc, New York, pp 53-102

Aller RC (1983) The importance of the diffusive permeability of animal burrow linings in determinig marine sediment chemistry. J Mar Res 41:299-322

Aller RC (1988) Benthic fauna and biogeochemical processes in marine sediments: The role of burrow structures. In: Blackburn TH, Sørensen J (eds) Nitrogen Cycling in Coastal Marine Environments (SCOPE), Wiley Press Inc, Chichester, pp 301-338

Aller RC (1990) Bioturbation and manganese cycling in hemipelagic sediments. Phil Trans R Soc Lond 331: 51-68

Aller JY, Aller RC (1986) Evidence for localized enhancement of biological activity associated with tube and burrow structures in deep-sea sediments at the HEBBLE site, western North Altantic. Deep-Sea Res 33:755-790

Aller RC, Aller JY (1998) The effect of biogenic

irrigation intensity and solute exchange on diagenetic reaction rates in marine sediments. J Mar Res 56:905-936

Aller RC, Hall POJ, Rude PD, Aller JY (1998) Biogeochemical heterogeneity and subsurface diagenesis in hemipelagic sediments of the Panama Basin. Deep-Sea Res (I) 45:133-165

Archer D, Devol A (1992) Benthic oxygen fluxes on the Washington shelf and slope: A comparison of *in situ* microelectrode and chamber flux measurements. Limnol Oceanogr 37:614-629

Boudreau BP (1986) Mathematics of tracer mixing in sediments: II. Nonlocal mixing and biological conveyer-belt phenomena. Am J Sci 286:199-238

Canfield DE, Jørgensen BB, Fossing H, Glud R, Gundersen J, Ramsing NB, Thamdrup B, Hansen JW, Nielsen LP, Hall POJ (1993) Pathways of organic carbon oxidation in three continental margin sediments. Mar Geol 113:27-40

Chartarpaul L., Robinson JB, Kaushik NK (1980) Effects of tubificid worms on denitrification and nitrification in stream sediments. Can J Fish Aquat Sci 37:656-663

Clavero V, Niell FX, Fernandez JA (1991) Effects of *Nereis diversicolor* O.F. Muller abundance on the dissolved phosphate exchange between sediment and overlying water in Palmones River estuary. Est Coast Shelf Sci 33:193-202

Forster S, Graf G, Kitlar J, Powilleit M (1995) Effects of bioturbation in oxic and hypoxic conditions: A microcosm experiment with North Sea sediment community. Mar Ecol Prog Ser 116:153-161

Furukawa Y, Bentley SJ, Shiller SJ, Lavoie DL, Van Cappellen P (2000) The role of biologically-enhanced pore water transport in early diagenesis: An example from carbonate sediments in the vicinity of North Key Harbor, Dry Tortugas National Park, Florida. J Mar Res 58:493-522

Gerino M, Aller RC, Lee C, Cochran JK, Aller JY, Green MA, Hirschberg D (1998) Comparison of different tracers and methods used to quantify bioturbation during a spring bloom: 234-thorium, luminophores and chlorophyll-a. Est Coast Shelf Sci 46:531-547

Glud RN, Gundersen JK, Jørgensen BB, Revsbech NP, Schulz HD (1994) Diffusive and total oxygen uptake of deep-sea sediments in the eastern South Atlantic Ocean: *in situ* and laboratory measurements. Deep-Sea Res (I) 41:1767-1788

Graf G (1999) Do benthic animals control the particulate exchange between bioturbated sediments and benthic turbidity zones? In: Gray JS, Ambrosa W Jr, Szaniawska A (eds) Biogeochemical Cycling and Sediment Ecology. Kluwer BV, Deventer, pp 153-159

Haese RR (2000) The reactivity of iron, In: Schulz HD, Zabel M (eds) Marine Geochemistry. Springer, Berlin pp 233-261

Hargrave BT, Phillips GA (1977) Oxygen uptake of micro-bial communities on solid surfaces. In: Cairus J (ed) Aquatic Microbial Communities. Garland Press Inc, New York, pp 545-587

Herman PMJ, Middelburg JJ, Van de Koppel J, Heip CHR (1999) Ecology of estuarine macrobenthos. Adv Ecol Res 29:195-204

Huettel M, Gust G (1992) Impact of bioroughness on interfacial solute exchange in permeable sediments. Mar Ecol Progr Ser 89:253-267

Huettel M, Ziebis W, Forster S, Luther GW III (1998) Adective transport affecting metal and nutrient distributions and interfacial fluxes in permeable sediments. Geochim Cosmochim Acta 62:613-631

Ingall ED, Bustin RM, Van Cappellen P (1993) Influence of water column anoxia on the burial and preservation of carbon and phosphorous in marine shales. Geochim Cosmochim Acta 57:303-316

Jørgensen BB (1977) Bacterial sulfate reduction within reduced microniches of oxidized marine sediments. Mar Biol 41:7-17

Jørgensen BB, Bak F (1991) Pathways and microbiology of thiosulfate transformations and sulfate reduction in a marine sediment (Kattegat, Denmark). Appl Environ Microbiol 57:847-856

Kristensen E (1988) Benthic fauna and biogeochemical processes in marine sediments: Microbial activities and fluxes, In: Blackburn TH, Sørensen J (eds) Nitrogen Cycling in Coastal Marine Environments. (SCOPE), pp 275-299

Kulm LD, Suess E, Moore JC, Carson B, Lewis BT, Ritger SD, Kadko DC, Thornburg TM, Embley RW, Rugh WD, Massoth GJ, Langseth MG, Cochrane GR, Scamman RL (1986) Oregon subduction zone: Venting, fauna, and carbonates. Science 231:561-566

Mahaut M-L, Graf G (1987) A luminophore tracer technique for bioturbation studies. Oceanol Acta 10: 323-328

Martin WR, Banta GT (1992) The measurement of sediment irrigation rates: A comparison of Br- tracer and 222Rn/226Ra disequilibrium techniques. J Mar Res 50:125-154

Matisoff G (1995) Effects of bioturbation on solute and particle transport in sediments. In: Allen HE (ed) Metal Contaminated Aquatic Systems. Ann Arbor Press Inc, Ann Arbor, pp 202-272

Matisoff G, Fisher JB, Matis S (1985) Effects of benthic macroinvertebrates on the exchange of solutes between sediments and freshwater. Hydrobiol 122:19-33

Mayer MS, Schaffner L, Kemp WM (1995) Nitrification potentials of benthic macrofaunal tubes and burrow walls: effects of sediment NH_4^+ and animal irrigation behavior. Mar Ecol Prog Ser 121:157-169

McCall PL, Matisoff G, Tevesz MJS (1986) The effects of a unionid bivalve on the physical, chemical, and microbial properties of cohesive sediments from Lake Erie. Am J Sci 286:127-159

Meile C, Koretsky CM, Van Cappellen P (2001) Quantifying bioirrigation in aquatic sediments: An inverse modeling approach. Limnol Oceanogr 46:164-177

Nittrouer CA, DeMaster DJ, McKee BA, Cutshall NH, Larsen IL (1983) The effect of sediment mixing on Pb-210 accumultion rates for the Washington continentel shelf. Mar Geol 54:201-221

Officer CB, Lynch DR (1982) Interpretation procedures for the determination of sediment parameters from time-dependent flux inputs. Earth Planet Sci Lett 61:55-62

Pearson TH (in press) Functional group ecology in soft sediment marine benthos: the role of bioturbation. Oceanography and Marine Biology: An annual review

Rabouille C, Gaillard J-F, Tréguer P, Vincendeau M-A (1997) Biogenic silica recycling in surficial sediments across the Polar Front of the Southern Ocean (Indian Sector). Deep-Sea Res (II) 44: 1151-1176

Rachor E, Bartel S (1981) Occurrence and ecological significance of the spoon-worm *Echiurus echiurus* in the German Bight. Veröff Inst Meeresforsch Bremerh 19:71-88

Rhoads DC, McCall PL, Yingst JY (1978) Disturbance and production on the estuarine seafloor. Am Sci 66:577-586

Sibuet M, Olu-Le Roy K (2002) Cold seep communities on continental margins: Structure and quantitative distribution relative to geological and fluid venting patterns. In: Wefer et al. (eds) Ocean Margin Systems. Springer, Berlin pp 235-251

Soetaert K, Herman PMJ, Middelburg JJ, Heip C, de Stigter HS, van Weering TCE, Epping E, Helder W (1996) Modeling 210Pb-derived mixing activity in ocean margin sediments: Diffusive versus nonlocal mixing. J Mar Res 54:1207-1227

Sun M, Aller RC, Lee C (1991) Early diagenesis of chlorophyll-a in Long Island Sound sediments: A measure of carbon flux and particle reworking. J Mar Res 49:379-401

Sun M-Y, Lee C, Aller RC (1993) Laboratory studies of oxic and anoxic degradation of chlorophyll-*a* in Long Island Sound sediments. Geochim Cosmochim Acta 57:147-157

Sundby B, Anderson LG, Hall POJ, Iverfeldt Å, Rutgers van der Loeff MM, Westerlund SFG (1986) The effect of oxygen on release and uptake of cobalt, manganese, iron and phosphate at the sediment-water interface. Geochim Cosmochim Acta 50:1281-1288

Sundby B, Vale C, Caçador I, Catarino F, Madureira MJ, Catarino M (1998) Metal-rich concretions on the roots of salt marsh plants: mechanism and rate of formation. Limnol Oceanogr 43:245-252

Wallmann K, Linke P, Suess E, Bohrmann G, Sahling H, Schlüter M, Dählmann A, Lammers S, Greinert J, von Mirbach N (1997) Quantifying fluid flow, solute mixing, and biogeochemical turnover at cold vents of the eastern Aleutian subduction zone. Geochim Cosmochim Acta 61:5209-5219

Wang F, Chapman PM (1999) Biological implications of sulfide in sediment – A review focusing on sediment toxicity. Env Toxic Chem 18:1526-2532

Ziebis W, Forster S, Huettel M, Jørgensen BB (1996) Complex burrows of the mud shrimp *Callianassa truncata* and their geochemical impact in the sea bed. Nature 382:619-622

Cold Seep Communities on Continental Margins: Structure and Quantitative Distribution Relative to Geological and Fluid Venting Patterns

M. Sibuet[*] and K. Olu-Le Roy

DRO/Département Environnement Profond, Ifremer, Centre de Brest, BP 70, 29280 Plouzané Cedex, France
* corresponding author (e-mail): msibuet@ifremer.fr

Abstract: Cold seep ecosystems occur on active and passive continental margins. Chemosynthesis-based communities depend on autochtonous and local chemical energy and produce organic carbon in large quantities through microbial chemosynthesis. The high organic carbon production leads to the large size of the fauna and the high biomass of the communities. The remarkable abundance of giant tubeworms (vestimentiferans) and large bivalves (i.e. *Vesicomyidae, Mytilidae* and others) is one of the most striking features of such communities and one of the best indicators or tracers of fluid emissions at the seafloor. Cold seep communities are known since about 15 years and have shown that the chemoautotrophy and many symbiont containing organisms are not unique to hydrothermal vents. Ecosystem characteristics and functioning in continental margin habitats are incompletely understood and we do not know how detritus and chemosynthesis-based ecosystems interact. There is a clear need of more field investigations. But with progress in deep-sea submersible technology, our understanding continues to grow. Following a recent review that focused on biogeographical trends and comparisons with hydrothermal vent communities (Sibuet and Olu 1998), we review here the ecology of chemosynthesis-based communities from several cold seep areas. Our synthesis addresses biodiversity and abundance fluctuations and distribution patterns linked to geological and fluid venting features. The diversity of the "symbiotic" fauna expressed as species richness decreases with ocean depth. Species composition is an indicator of the biotope variability. The spatial extension of active seep areas is highly variable from hundreds of square meters to several hectares. Three distinct categories of cold seep sites are recognised. The shape, density and biomass of aggregations reflect the intensity of fluid flow, and characterise fluid circulation and different expulsion pathways through geological structures.

Introduction

Complex ecosystems occur on active and passive continental margins. Their complexity is linked to the environment which favours either detritus based communities, usually with a high biodiversity of mainly small sized species, or chemosynthesis based communities. Detritus-based communities are fed indirectly by photosynthetic production in the upper layers of the ocean and may be coupled to seasonal variations of phyto-zooplankton cycles. In contrast, chemosynthesis based communities depend on autochtonous, local chemical energy which is the basis for organic carbon production in large quantities (Fig. 1). This high organic carbon production leads to the large size of the organisms and the high biomass of the communities. As an example, the remarkable abundance of giant tubeworms (vestimentiferans) and large bivalves (i.e. *Vesicomyidae, Mytilidae*) is one of the most striking features of such communities and one of the best indicators or tracers of the existence of an original style of life linked to fluid emission at the seafloor. Cold-seep communities are known since about 15 years and have shown that the chemoautotrophy and many symbiont containing organisms are not unique to hydrothermal vents. Justification for work on continental margin habitats still lies in the need to answer fundamental questions: 1. How can biological systems adapt to extreme conditions and

From WEFER G, BILLETT D, HEBBELN D, JØRGENSEN BB, SCHLÜTER M, VAN WEERING T (eds), 2002, *Ocean Margin Systems*. Springer-Verlag Berlin Heidelberg, pp 235-251

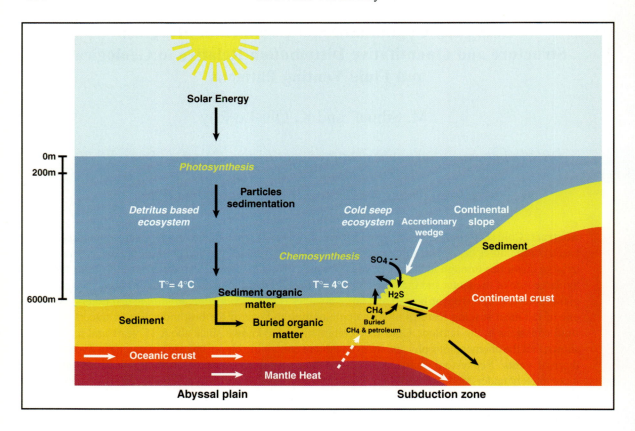

Fig. 1. Diagram showing geological context and geochemical fluxes which control margin ecosystems.

fragmented habitats? 2. How variable are the biological responses to the physical and chemical characteristics of the fluid? 3. How does the biological community respond (from the bacterial level to the megafaunal level) to temporal and episodic variations of cold seeps ? 4. How are biomass and biological production linked to fluid flow rates and how does the biological and chemical processes modify the organic carbon cycle in the deep ocean?

Ecosystem characteristics and functioning in continental margin habitats are incompletely understood and we do not know how detritus and chemosynthesis-based ecosystems interact. There is a clear need of more field investigations. But with progress in 1) deep-sea submersible technology, 2) technologies to measure physico-chemical properties and 3) technologies to assess spatial and temporal variability of communities and the environmental conditions, our understanding continues to grow. The next approaches need to be interdisciplinary, with *in situ* observations, sampling and

measurements of multiple processes. Following a recent review that focused in biogeographical trends and comparisons with hydrothermal vent communities (Sibuet and Olu 1998), we review here the ecology of chemosynthesis-based communities from several cold seep areas. Our synthesis addresses biodiversity and abundance fluctuations and shows distribution patterns linked to geological and fluid venting features.

Cold seeps and their geological settings

Seep communities have been identified in about twenty five deep-sea areas located in the Atlantic Ocean, the Eastern and Western Pacific Ocean and the Mediterranean Sea on both passive and active margins at depths ranging between 400 and 7000 m depth. Only 14 sites have been sufficiently explored to provide data for ecological analysis (Table 1 and Fig. 2).

Abbreviated name of the zones (and sites)	Location of the zones (and sites)	Depth (m)	References used for biological data
Atlantic Ocean			
Bar. –N	Barbados prism (13°49'N) (Atalante, Manon, Cyclope)	4700-5000	(Olu et al. 1997)
Bar. –S	Barbados prism (10-11°N) (Orenoque, El Pilar)	1000-2000	(Jollivet et al. 1990); (Olu et al. 1996b)
Lou-l	Gulf of Mexico, Louisiana lower continental slope (Alaminos canyon)	2200	(Carney 1994)
Lou-u	Gulf of Mexico, Louisiana upper continental slope	400-1000	(MacDonald et al. 1989); (MacDonald et al. 1990a); (MacDonald et al. 1990b); (MacDonald et al. 1990c); (Carney 1994); (Rosman et al. 1987)
Flo	Gulf of Mexico, Florida escarpment	3500	(Hecker 1985); (Paull et al. 1984)
Mediterranean Sea			
Med	Eastern Mediterranean	1700-2000	(Corselli and Basso 1996); (Aloisi et al. 2000, Salas and Gofas (pers.com.)
Eastern Pacific Ocean			
Ale	Aleutian trench	3200-5900	(Suess et al. 1998)
Ore	Oregon prism	2000-2400	(Suess et al. 1985)
Mon -b	Monterey Bay	600-1000	(Barry et al. 1996),
Mon-v	Monterey Fan Valley	3000-3600	(Barry et al. 1996).
Per (MAF and MSS)	Peruvian margin –Paita Zone (Main Active Field and Middle slope Scarp),	2300-5100	(Olu et al. 1996a)
Western Pacific Ocean			
Jap.	Japan subduction zones (Japan and Kurile trenches)	3800-6000	(Laubier et al. 1986); (Juniper and Sibuet 1987); (Ohta and Laubier 1987); (Sibuet et al. 1988)
Nan- t	Eastern Nankai trough, (Tenryu canyon)	3600- 3850	(Juniper and Sibuet 1987); (Sibuet et al. 1988)
Nan -p	Nankai prism	2000	(Sibuet et al. 1990); (Lallemand et al. 1992)
Sag	Sagami Bay	900-1200	(Hashimoto et al. 1989); (Fujikura et al. 1995); (Fujikura et al. 1999)

Table 1. Well explored cold seeps.

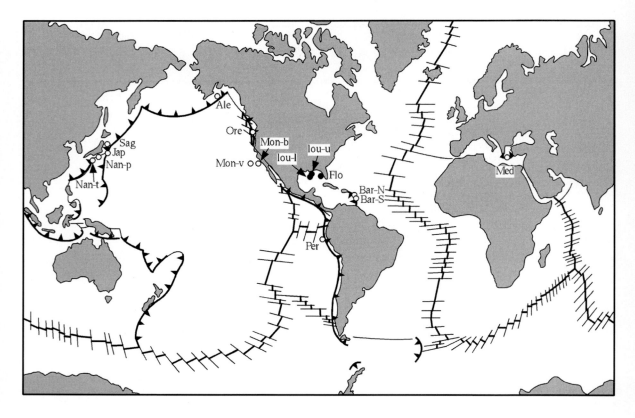

Fig. 2. Location of cold seeps on active (lines with triangles) and passive margins considered in this review. Open dots indicate sites on subduction zones and black dots indicate sites in the Gulf of Mexico passive margin.

The exploration of sites on passive margins is still very limited. The only well explored region is the northern Gulf of Mexico, where detailed biological descriptions have been made. These studies highlighted the spatial distribution of populations of bivalves (*Mytilidae*, *Vesicomyidae*), of vestimentiferan tube worms, and of the unusual habitat of a polychaete *Hesionidae* worm living at the surface of gas hydrates (Desbruyères and Toulmond 1998). This margin is a rich oil and gas province. In the context of passive margins, cold seeps and associated benthic life and carbonate mounds seem to mainly occur in areas where in the deep large amounts of organic matter are present, i.e mostly on continental margins with a potential oil and gas content. Future exploration linked with areas of deep offshore hydrocarbon interest should contribute to the discovery of other communities based on hydrocarbon seeps. It is likely that our knowledge on the spatial and geographical distribution of seeps will greatly expand.

Many more locations of cold seep communities have been identified on active than on passive margins. This is caused by a stronger interest of the geoscience community for subduction processes in relation to seismic risks for coastal human populations (for example in Japan). Studies on active fault systems within accretionary prisms, mud volcanoes and decollement surfaces within the sediment will allow a better understanding of fluid circulation and release in the sea water. The presence of active reverse faults in accretionary wedges combined with the compaction of sediment during the construction of accretionary wedges is responsible for relatively continuous fluid expelling and maintenance of cold seep benthic communities. The presence of cold seep communities associated with active geological features has been useful to locate and identify different types of geological features on the sea bottom (Sibuet et al. 1988; Le Pichon et al. 1990; Lewis and Cochrane 1990; Olu et al. 1996a, 1997) and to quantify fluid flow through time (Henry et al. 1992, 1996).

Diversity of symbiont containing fauna in the different zones and variation with depth

Chemosynthesis is the primary carbon production process for these large faunal communities. At the seafloor, both methane (transported in the fluids) and hydrogen sulphide (produced in the sediment using sea water sulphate) provide energy for microbial chemosynthesis. Organic matter synthesis occurs in symbiotic associations between bacteria and invertebrate hosts. Known symbioses include: 1. Pogonophoran tube worms which keep their sulphur-oxidising symbionts in a specialised organ, the trophosome; 2. Species of some bivalve families which host their symbionts (methanotroph and/or sulphur oxidiser) in the gill tissue; 3. some sponge

species which host methanotroph symbionts extra- and intra-cellularly. Whereas direct evidence of chemoautotrophy has been published for only some species, several species from known genera can be considered to be "assumed symbiont-containing species". The listed species (Table 2) is an excerpt of that published by Sibuet and Olu (1998) as only the families of assumed symbiont containing species directly linked to fluid emissions are indicated and the surrounding fauna is not. This list is updated with recent species identifications.

The total number of 61 species is certainly underestimated as more taxonomic studies are needed, but already shows the high diversity of the chemosynthesis-based fauna in cold seep environments compared to the known species of the same phyla at hydrothermal vents. Considering the 36 named species, the distribution of the species shows a great endemism at species level but not

Species list	Atlantic Ocean					Med	Eastern Pacific					Western Pacific			
	Bar-N	Bar-S	Lou-l	Lou-u	Flo		Ale	Ore	Mon-b	Mon-v	Per	Jap	Nan-t	Nan-p	Sag
Phylum Porifera															
Class Demospongiae															
Family Cladorhizidae															
Cladorhiza sp.	+														
Family Hymedesmiidae															
Hymedesmia sp.				+											
Family Suberitidae															
* *Rhizaxinella pyrifera*						+									
Phylum Pogonophora/Obturata															
(Vestimentifera)															
Family Lamellibrachiidae															
Lamellibrachia barhami			+					+	+						
* *Lamellibrachia* sp.		+		+		+									+
Family Escarpiidae															
Escarpia laminata			+		+										
Escarpia cf. *laminata*		+													
Escarpia cf. *spicata*														+	

Table 2. Known or assumed symbiont-containing species in the major cold seep zones. See Table 1 for abbreviated names of cold seep zones. See references of biological and taxonomic studies in Sibuet and Olu (1998). Species with * are recent identifications coming from Medinaut cruise results on Mediterranean mud volcanoes, Pogonophoran specimens identification by Southward (personal communication), bivalve specimens identification by Gofas and Salas (personal communication), from Gulf of Mexico for *Mytilidae* by Gustafson et al. 1998, from Barbados mud volcanoe for *Mytilidae* by Cosel and Olu 1998 and *Vesicomyidae* by Peeks et al. 1997), and from Japan for *Thyasiridae* by Fujikura et al. 1999.

Species list	Atlantic Ocean						Eastern Pacific					Western Pacific			
	Bar-N	Bar-S	Lou-l	Lou-u	Flo	Med	Ale	Ore	Mon-b	Mon-v	Per	Jap	Nan-t	Nan-p	Sag
Escarpia sp.		+		+											
Family uncertain															
gen. sp.															+
Phylum Pogonophora/Perviata															
Family Polybrachiidae															
Galathealinum sp.			+												
Polybrachia sp.							+			+					
Family Siboglinidae															
Siboglinum sp.						+									
Family Oligobrachiidae															
Family Spirobrachiidae															
Spirobrachia sp.							+								
Family uncertain															
gen. sp.		+									+				
Phylum Pogonophora/Monilifera															
Family Sclerolinidae															
Sclerolinum sp.	+						+								
Sub Phylum uncertain															
gen.sp.							+								
Class Bivalvia															
Family Solemyidae															
Acharax johnsoni															+
Acharax caribea			+												
Acharax sp.										+	+				
Solemya sp.								+	+						
gen. sp.	+	+					+								
Family Mytilidae															
Bathymodiolus aduloides															+
Bathymodiolus japonicus															+
Bathymodiolus platifrons															+
* *Bathymodiolus boomerang*		+													
Bathymodiolus sp. B.		+													
* *Bathymodiolus childressi*			+	+											
* *Bathymodiolus brooksi*			+		+										
* *Bathymodiolus heckerae*					+										
* *Tamu fisheri*				+											
* *Idas macdonaldi*				+											
Idas modiolaeformis						+									

Table 2. cont.

Species list	Atlantic Ocean					Med	Eastern Pacific					Western Pacific			
	Bar-N	Bar-S	Lou-l	Lou-u	Flo	Med	Ale	Ore	Mon-b	Mon-v	Per	Jap	Nan-t	Nan-p	Sag
Family Lucinidae															
Lucinoma atlantis				+											
Lucinoma yoshidai															+
*Lucinoma sp.				+		+									
Myrtea sp.						+									+
Family Thyasiridae															
* Thyasira striata						+									
Conchocoele disjuncta															+
Conchocoele sp.		+												+	
Thyasira olephila				+											
* Parethyasira kaireiae												+			
*Maorithyas hadalis												+			
Family Vesicomyidae															
Calyptogena kaikoi													+	+	
Calyptogena laubieri													+	+	
Calyptogena nautilei												+			
Calyptogena phaseoliformis												+			
Calyptogena extenta										+					
Calyptogena soyoae															+
Calyptogena packardana									+						
Calyptogena ponderosa				+											
Calyptogena sp. 1											+				
Calyptogena sp. 2											+				
Calyptogena sp.	+				+								+	+	
Vesicomya stearnsii									+						
Vesicomya cordata				+											
Vesicomya sp.		+													
* C. pacifica/V. lepta complex								+	+		+				
*C. kilmeri/V. gigas complex								+	+						
*Isorropodon perplexum						+									
gen. sp.														+	

	Bar-N	Bar-S	Lou-l	Lou-u	Flo	Med	Ale	Ore	Mon-b	Mon-v	Per	Jap	Nan-t	Nan-p	Sag
Number of symbiont-containing species Total number of species: 61 (including 36 identified and named species) Tozal number of genera: 25	4	9	4	13	4	8	5	3	6	3	5	3	4	6	10

Table 2. cont.

at genus level. While the *Mytilidae* bivalves have a more restricted bathymetric range and are mainly found between 400 and 2000 m water depth (with the exception of one species at 3270 m at the Florida escarpment), the *Vesicomyidae* bivalves species are observed down to 5800 m, such as in the Japan Trenches (Juniper and Sibuet 1987; Sibuet et al. 1988, Ogawa et al. 1996) and Peru Trench (Olu et al. 1996a). Only thyasirids have been observed in the deepest known site along the Japan trenches at 7326 m depth (Fujikura et al. 1999). Diversity can be high in some shallow locations such as on the Louisiana Slope and in Sagami Bay. Figure 3 shows the number of species found at each site

and at different waters depths and indicates a decrease with depth, since at the deepest sites only one species is present. The presence of only one species can be caused by bathymetric constraint or by a homogeneous habitat, whereas the coexistence of several species characterizes heterogenous biotopes. Indeed, differences in sulfide and methane concentration can influence the species composition due to differences in species biology and physiology such as symbiotic associations (methanotrophic and/or sulphur-oxidising bacteria) and physiological characteristics such as sulfide binding affinity. This has been suggested for *Vesicomyidae* species (Barry et al. 1997).

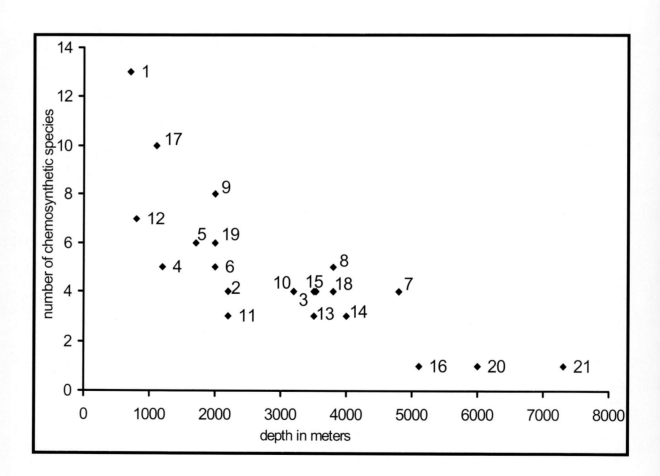

Fig. 3. Species richness versus depth. 1:Louisiana upper slope, 2: Louisiana lower slope, 3: Florida escarpment, 4 to 7: Barbados prism El Pilar (4), Orénoque A (5), Orénoque (6), 13°40' N area (7), 8:Laurentian fan, 9: Mediterranean ridge, 10: Aleutian trenches, 11: Oregon subduction zone, 12 and 13: Monterey Bay (12) and fan valley (13), 14: Mid- American trenches, 15 and 16: North Peruvian margin, middle slope scarp (15) and lower scarp (16), 17 to 21: Japan subduction zones, Sagami Bay (17), Nankai trough (18), Nankai prism (19), Japan trench (20, 21).

Different fluid flow and habitat (soft mud or concretions) may influence the type of communities composed by distinct species of *Vesicomyidae* bivalves (Sibuet et al. 1990), *Mytilidae* bivalves or both (Olu et al. 1996b). Therefore several species can live together owing to their different physiological adaptations and different environmental preferences (nature of the substrate, intensity of fluid expulsion, methane or sulphide affinities, chemical concentration). Biological features such as growth rates (Barry et al. 1997) and maximum adult size (Cosel and Olu 1998) are highly variable and may be a response to the variety of environmental conditions. Biodiversity of cold seep communities reflects the biotope variability in space and probably also in time.

Spatial distribution and variation of biological activity in terms of abundance

The exploration of some cold seep areas by submersible and the detailed analyses of video records of near bottom surveys allow the spatial distribution of biological communities (bivalve communities and pogonophoran aggregations) to be determined at different scales. Based on several parameters of distribution, a classification is proposed. Most of the active sites are recognised from the existence

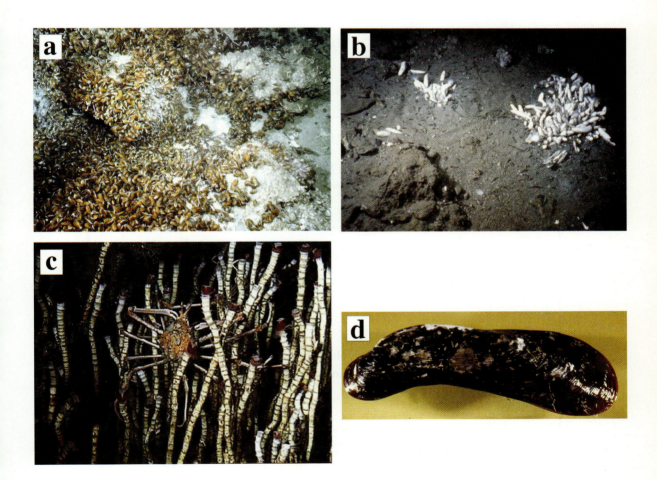

Fig. 4. Cold seep communities: **a)** mussel field dominated by *Bathymodiolus* sp. at El Pilar dome (Barbados accretionary prism); **b)** *Calyptogena phaseoliformis* aggregates from the Japan trench; **c)** vestimentiferan bushes in the Gulf of Mexico at the Louisiana upper continental slope (C. Fisher, personal communication); **d)** example of a very large bivalve (35 cm length), *Bathymodiolus boomerang*, from Barbados Orenoque ridge.

of aggregates or clusters of fauna (Fig. 5). These clusters are of various sizes and shapes. In a few areas only dispersed living organisms of large size are observed.

The community density varies within clusters and also between active sites. A site is generally representative of a geological feature, for example mounds, volcanoes, ridges, scarps, and normal or reverse faults. Estimations of the spatial extension of these enriched habitats are largely dependent on the number of submersible surveys, and the length of transects at each site. In some cases the active area or field have been approximately estimated.

Based on the values of quantitative parameters and distribution features, three distinct categories of cold seep sites have been established (Table 3):

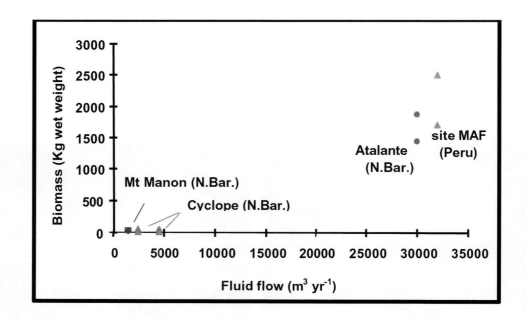

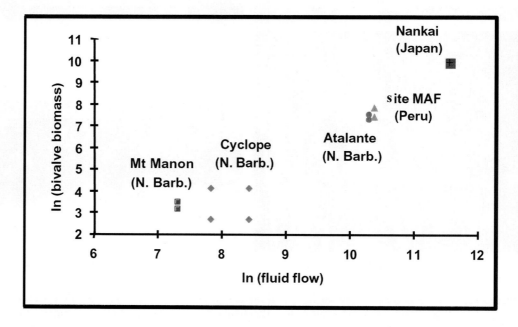

Fig. 5. Relationship between biomass and fluid flow.

a) sites of very high biological activity with dense clusters on well-defined fields

Zone: site	Cluster size (m²)	Density in clusters (ind./m²)	Active area observed (m²) (rough estimation)	Mean density in active area (ind./m²)
Jap (Nankai prism)	1-10	10-1000	12 300	
Jap.: Tenryu canyon	1-20	10->100	1600	
Per.: MAF	1-50	10-1000	1000	40
Barb.-S: El Pilar (dome 2)	2-50	710 to >1000	780 (minimum)	>370
Barb.-S: Orenoque (A) ridge	0.2-3	50-1670	12 600	0.21
Barb.-S: Orenoque (B) domes dome 9 thanatocoenose	0.2-1 20	45-290	1800	0.44
Sag.	1 to 70	200 to 1000	6000	
Barb.-N: Manon at the top	1-12	50-160	3 800	0.24
Mon.-B clam field	1 to 7		300	

b) Sites of high biological activity with dispersed clusters over large areas

Zone: site	Cluster size (m²)	Density in clusters (ind./m²)	Active area observed (m²) (rough estimation)	Mean density in active area (ind./m²)
Jap.: trench	0.5-1	5-50	8000	
Barb. N: Atalante	0.2-7	10-150	470 000	0.1
Barb.-N: Volcano A	5	30-50	170 000	0.031
Mon.- B Crushmore	0.25-3		10^6	
Lou-u: Bush hill	1-20		10^6	
Lou-u: Mussel beach	50		10^5	
Lou-u: Pogonophora field	1-7 300			
Per. MSS	1-10	1-10	200 000	

c) Sites of low biological activity with dispersed organisms over large areas

Zones: site	Cluster size (m²)	Density in clusters (ind./m²)	Active area observed (m²) (rough estimation)	Mean density in active area (ind./m²)
Cyclope	10-20	0.1-10	320 000	0.01
Lou.-u (vesicomyids)		2.3-2.7		

Table 3. Classification of cold seeps with respect to density and spatial distribution of chemosynthesis based communities.

• Sites with very high biological activity, dense clusters of high densities forming large and well defined fields;

• Sites of high biological activity with dispersed clusters of high densities over very large areas;

• Sites of relatively low biological activity with dispersed organisms over large areas.

The density is very high in some aggregates with more than 1600 individuals/m² (various size mussels) on the Barbados mud volcano (Olu et al. 1996b) and 1000 individuals/m² in the Sagami Bay (Hashimoto et al. 1989), the Nankai trough (Sibuet et al. 1990; Henry et al. 1992) and the Peruvian large clam field (Olu et al. 1996a). Spatial distribution of chemosynthetis based communities on various mud volcanoes have helped to characterise fluid circulation and different expulsion pathways for distinct types of mud mounds. For example, on diatreme type (e.g. Atalante mud volcano on the Barbados accretionary prism) a central conduit devoid of fauna is observed and only the peripheral part of the mud volcano is colonised, whereas on diapirs (e.g. Manon mound) the seepage and biological activity are limited to the summit with no seepages on the flanks (Olu et al. 1997). Localised clam fields and dense patches may indicate that fluids are expelled through conduits, whereas more dispersed clusters and very large fields are associated with diffuse flow through the sediment coverage (Sibuet et al. 1988). The type of distribution may reflect the nature and volume of expelled fluids. This relationship between faunal distribution and fluid expulsion intensity has to be further analysed, but these categories of spatial distribution on various scales may help to conduct the appropriate field measurements in the future. There is a clear need of temporal observation in order to analyse the duration of seeps and its consequences for benthic ecosystems. Different evolution stages of seeps have been already suggested according to temperature measurements, faunal distribution patterns and demographic features, such as the presence of thanatocoenoses (dead clam field for example) and the proportion of dead and living clams in a single community.

Biomass estimation in relation to fluid flow

Considering the difficulty to obtain quantitative data from sea bottom observation, we only compare sites for which quantitative data have been obtained by similar methods. Data obtained from Nautile (submersible) cruises in the Pacific and Atlantic oceans are presented (Table 4, Fig. 5). Biomass estimates were deduced from the relationship between shell length and wet weight of specimens. Measurements on photos and videos of shell lengths are used when the number of collected shells was not sufficient to allow the size distribution of a bivalve population to be determined. Mean densities were estimated on video and photo records by using geometric relationship knowing the altitude or dimensions of tools within the field of view.

On a local scale, fluid flow velocity in seeps (in m/yr) below clam aggregations (sometimes called clam colonies) can be determined from measurements of temperature at shallow depth (0-50 cm) within the sediment. Temperature measurements have been made by instrumented corers operated from the submersible. The T-Naut temperature probe, developed at IFREMER (Henry et al. 1992), allowed the determination of the temperature gradient in the superficial sediment down to 50 cm below the seafloor. In the Nankai trough, four types of biological aggregations have been proposed in order to classify the observed biological activity. Each of which appear to have a characteristic thermal signature. T-Naut temperature values in different types of clam beds defined by their specific composition, density, diameter, and bivalve shell lengths (Sibuet et al. 1990; Henry et al. 1992) have shown that the maximum rates were estimated to be 100 m/yr in "type A colonies", i.e. a total fluid flow of 103.000 m³/yr for "type A colonies" (Henry et al. 1992). Biomass was estimated to be 20 kg/m² for "type A colonies", the total biomass for the 1030 m² of "type A colonies" is therefore estimated to be 20.600kg. The velocity of upward fluid flow (in m/yr) and the flow pattern appear to be correlated with the intensity and type of biological activity. On a local scale, rapid flow of diluted fluid is characteristic of the most dense and highest biomass aggregations of *Calyptogena*, while slow

Zone	Site or Structure	Surface of active area (m²)	Total bivalve biomass (kg wet weight)	Expelled volume of fluid (m³/yr)
Barbados North	Atalante	470*10³	1440-1866	30000
	Cyclope	200*10³	15-61	2500-4500
	Mound Manon (summit)	4*10³	24-32	1500
	Volcano A	170*10³	48-181	-
Barbados South	Ride	17*10³	119-185	-
	Dome 13	1.5*10³	31-61	-
	Dome 2	760	2000-3200 (minimum)	-
North Peruvian margin	site MAF	1000	1700-2500	32000
North Peruvian margin	MSS bend (scarp)	200*10³	215-317	-
Jap. Nankai prism			20600	103000

Table 4. Estimation of total bivalve biomass at different sites.

seepages of methane (or sulfide) rich fluid favour the development of dispersed clams populations and bacterial mats (Sibuet et al. 1990; Henry et al. 1992).

At a larger scale, the total volume of fluid expelled through a geological structure or a site (i.e. a mud volcano or an active field along a scarp of hundreds to thousands of m²) has been estimated over the whole area covered by bivalve communities of known biomass. Fluid flows were estimated using a model of heat transfer and temperature measurements obtained *in situ* from instrumented cores operated from the ship or with the Nautile T-Naut probes (Henry et al. 1992,

1996). In the North Barbados prism (Atalante and Cyclope mud volcanoes) heat flow values lead to fluid flow estimates of about 30 000 m³/yr in the area colonised by the vesicomyids for the Atalante mud volcano, and 2500 to 4500 m³/yr for the Cyclope mud volcano (Henry et al. 1996). At the top of Manon Mound, no heat flow measurements were available and fluid flow volume was estimated to be 1500 m³/yr at maximum, based on a flow of 100-1000 l/day through each "vent", i.e. each clam bed (Henry et al. 1996). In the main active field along the Peruvian margin (MAF site), the T-Naut temperature gradient measurements allow to estimate a velocity of upward fluid flow of 1.1 m/day or 400 m/yr (Olu et al. 1996a). Based on the method of Henry et al. (1992) for the Nankai prism, the area of fluid release was considered to be the same as the area covered by clam beds (i.e. about 60 m²) and the total fluid volume that nourished clam beds at this site is thus about 24 000 m³/yr. This flux has been assumed to be 32 000 m³/yr when including the fluid flow in the dispersed fields of bivalves.

In Figure 5, the relationship between fluid flow and bivalve biomass indicates that high fluid flow sustains a biomass of the same order of magnitude in different geological contexts: a mud volcano (Barbados prism), an anticlinal ridge (Nankai prism) and a slide scar scarp (Peru slope), whereas weak fluid flow sustains a low biomass, such as the ones on the Manon Mound and Cyclope volcano.

When comparing the different study sites, there is an exponential relationship (linear in log-log presentation) between fluid flow and biomass, which indicates that high flows sustain a bivalve biomass proportionally higher than that found in weak flows. The duration of the fluid emission has probably an influence on the development of the chemosynthesis based communities. As an example, the Cyclope and Atalante mud volcanoes are the same type of structure (diatrem) and are at two different stages of activity (Henry et al.1996; Lance et al. 1998). The chemosynthetis based communities obviously had developed more recently on the Cyclope volcano because the renewal of the surface could have occurred only a few tens of years ago for the Cyclope volcano and a few thousands of years ago for the Atalante volcano (Olu et al. 1997). It is suggested that the Cylope community had not reached the same "maturity" as the Atalante community. This could explain the difference in biomass produced for the same quantity of expelled fluid. Other factors may influence the biomass production as well as the fluid/biomass relationship, such as methane concentration in the fluid, chemical and bacterial processes in the sediment that consume and produce methane and sulfide, and the occurrence of conduits or diffusion processes. However, our data interpretation is a first approach which shows that fluid flow can be related to biomass in different geological contexts.

Conclusion

This synthesis summarises present ecological knowledge on cold seep ecosystems and shows patterns in community distribution and organisation obtained from the comparison of some well studied cold seep environments. Considering that the ecology of cold seep environments is a relatively new research field, it should be stressed that these initial data are obtained through detailed examination of video and photo documentation of submersible cruises that can be used for scientific purpose when adequately analysed. Images can be an excellent complement to sampling. Whereas traditional sampling techniques are inefficient means of data gathering in such localised and fragmented habitat, near bottom submersible surveys make it possible to locate and selectively sample these. New approaches and new types of investigations were necessary to use video recordings in a quantitative way in addition to species identifications. Microcartography reconstructions and image mosaic of faunal assemblages can be used as tools to reveal spatial community structure and distribution patterns.

A total of 14 relatively well explored cold seep sites have been considered. This comparative study shows that faunal composition, abundance, biomass and spatial distribution are significant indicators of seepage. In particular our present understanding indicates that:

• Large invertebrates of a few typical families are directly linked through symbiosis to chemosynthetic production and are the major visible indicators of fluid expulsion.

• The diversity of the "symbiotic" fauna expressed as species richness shows a decreasing trend with depth. Only one species occurs at the very deep cold seep sites in the Japan trench, where *Calyptogena phaseoliformis* was found at one seep site at 6000 m depth and one *Thyasiridae* species was found at another seep site at 7000 m depth. Up to 13 species can live together in a same area (Louisiana upper slope).

• Species composition is an indicator of the biotope variability. Indeed, faunistic differences within a site can be explained by various habitat conditions, variety of substrates, differences in trophic requirements and in symbiotic bacteria, either linked to methane rich fluid and/or to a secondary product, sulphide.

• The shape and density of aggregations, generally on the scale between one to several meters, reflect the intensity of fluid flow and the type of small-scale geological structure which supports the fluid expulsion. Localised and dense faunal aggregations may indicate that fluids are expelled through conduits, whereas dispersed distributions over large areas are more probably linked to diffuse flow through a sediment coverage.

• A comparison between the biomass of different types of aggregations and the fluid discharge rate showed a positive relationship which could be modelled quantitatively.

• Spatial distributions of biological activity on a large scale (several km^2) characterises fluid circulation and different expulsion pathways through geological structures. This is particularly the case for mud volcanoes for which distinct types of distribution occur, as e.g. diatremes and diapirs (with or without a central depression and large conduit).

• Total biomass and total fluid expelled through a large seep site (mud volcano, ridge, scarp, fault), seem to be related. However, representative video surveys and *in situ* measurements have been done in only a few sites and there is a clear lack of temporal observations.

Future work

Whereas our knowledge on cold seep ecosystems is increasing, there are still many unanswered questions to be answered which need further investigation, new strategies of *in situ* work and further interdisciplinary research:

• As symbiosis and chemoautotrophic bacteria are the major cause of high biomass (several kg/m^2) it is important to understand: 1. How environmental factors influence the establishment of the different symbiont-containing invertebrates? 2. How the diversity of symbiotic and non-symbiotic species is controlled? 3. How the species colonize fragmented habitats? 4. What are the relationships with the fauna present in other chemosynthesis based ecosystems, such as hydrothermal vents?

• How does the fauna respond to the chemical composition of the fluid, to the fluid flow intensity and to its temporal variation? No data exist on the duration of fluid expulsion in cold seeps and on the consequences of temporal variability.

• What are the animal–substrate interactions, and their consequences for substrate modifications and chemical fluxes?

• High carbon production is evident, but until now only biomass data exist; estimates of rate of production and investigations on fundamental biological features, physiology and demographic characteristics are necessary. The dynamics and function of cold seep ecosystems, and the role of species diversity, and the different trophic strategies, in driving the communities are not known.

• What are the spatial extensions of cold seep ecosystems and the consequences of the high biological production on the surrounding detritus based ecosystem which dominates the deep-sea environment? Such processes need to be understood, as all these systems have a role in the organic carbon cycle in the ocean.

References

Aloisi G, Asjes S, Bakker K, Charlou JL, De Lange G, Donval JP, Fiala-Médioni A, Foucher JP, Haanstra R, Haese R, Heijs S, Henry P, Huguen C, Jelsma B, de Lint S, van der Maarel M, Mascle J, Muzet S, Nobbe G, Pancost R, Pelle H, Pierre C, Polman W, de Senerpont Domis L, Sibuet M, van Wijk T, Woodside J, Zitter T (2000) Linking Mediterranean brine pools and mud volcanism. Eos 81:625-632

Barry JP, Greene HG, Orange DL, Baxter CH, Robison BH, Kochevar RE, Nybakken JW, Reed DL, McHugh CM (1996) Biologic and geologic characteristics of cold seeps in Monterey Bay, California. Deep-Sea Res 43: 1739-1762

Barry JP, Kochevar RE, Baxter CH (1997) The influence of pore-water chemistry and physiology in the distribution of vesicomyid clams at cold seeps in Monterey Bay: Implications for patterns of chemosynthetic community organization. Limnol Oceanogr 42:318-328

Carney RS (1994) Consideration of the oasis analogy for chemosynthetic communities at Gulf of Mexico hydrocarbon vents. Geo-Mar Lett 14:149-159

Corselli C, Basso D (1996) First evidence of benthic

communities based on chemosynthesis on the Mediterranean Ridge (Eastern Mediterranean). Mar Geol 132:227-239

Cosel von R, Olu K (1998) Gigantism in *Mytilidae*. A new Bathymodiolus from cold seep areas on the Barbados accretionary prism. C R Acad Sci Paris 321:655-663

Desbruyères D, Toulmond A (1998) A new species of hesionid worm, *Hesiocaeca methanicola* sp. nov. (Polychaeta: *Hesionidae*), living in ice-like methane hydrates in the deep Gulf of Mexico. Cah Biol Mar 39:93-98

Fujikura K, Hashimoto J, Fujiwara Y, Okutani T (1995) Community ecology of the chemosynthetic community at Hatsushima site off Sagami Bay, Japan. Jamstec J Deep-Sea Res 11:227-241

Fujikura K, Kojima S, Tamaki K, Maki Y, Hunt J, Okutani T (1999) The deepest chemosynthesis-based community yet discovered from the hadal zone, 7326 m deep, in the Japan Trench. Mar Ecol Prog Ser 190:17-26

Gustafson RG, Turner RD, Lutz RA, Vrijenhoek RC (1998) A new genus and five new species of mussels (Bivalvia, *Mytilidae*) from deep-sea sulfide/hydrocarbon seeps in the Gulf of Mexico. Malacologia 40:63-112

Hashimoto J, Ohta S, Tanaka T, Hotta H, Matsuzawa S, Sakai H (1989) Deep-sea communities dominated by the giant clam, *Calyptogena soyoae*, along the slope foot of Hatsushima Island, Sagami Bay, central Japan. Palaeogeogr Palaeoclimatol Palaeoecol 71:179-192

Hecker B (1985) Fauna from a cold sulfur-seep in the Gulf of Mexico: Comparison with hydrothermal vent communities and evolutionary implications. Bull Biol Soc Wash 6:465-473

Henry P, Foucher JP, Le Pichon X, Sibuet M, Kobayashi K, Tarits P, Chamot-Rooke N, Furuta T, Schultheiss P (1992) Interpretation of temperature measurements from the Kaiko-Nankai cruise: Modeling of fluid flow in clam colonies. Earth Planet Sci Lett 109:355-371

Henry P, Le Pichon X, Lallemant S, Lance S, Martin JB, Foucher JP, Fiala-Médioni A, Rostek F, Guilhaumou N, Pranal V, Castrec M (1996) Fluid flow in and around a mud volcano field seaward of the Barbados accretionary wedge: Results from Manon cruise. J Geophys Res 101:297-323

Jollivet D, Faugères JC, Griboulard R, Desbruyères D, Blanc G (1990) Composition and spatial organization of a cold seep community on the south Barbados accretionary prism: Tectonic, geochemical and sedimentary context. Prog Oceanogr 24:25-45

Juniper SK, Sibuet M (1987) Cold seep benthic communities in Japan subduction zones: Spatial organization, trophic strategies and evidence for temporal evolution. Mar Ecol Prog Ser 40:115-126

Lallemand SE, Glaçon G, Lauriat-Rage A, Fiala-Médioni A, Cadet JP, Beck C, Sibuet M (1992) Sea-floor manifestations of fluid venting and related tectonics at the top of a 2000 metres deep ridge in the eastern Nankai accretionary wedge. Earth Planet Sci Lett 109:333-346

Lance S, Henry P, Le Pichon X, Lallemant S, Chamley H, Rostek F, Faugères JC, Gonthier E, Olu K (1998) Submersible study of mud volcanoes seaward of the Barbados accretionary wedge: Sedimentology, structure and rheology. Mar Geol 145:255-292

Laubier L, Ohta S, Sibuet M (1986) Découverte de communautés animales profondes durant la campagne franco-japonaise KAIKO de plongées dans les fosses de subduction autour du Japon. C R Acad Sci Paris Sér III 303:25-29

Le Pichon X, Foucher JP, Boulègue J, Henry P, Lallemant S, Benedetti M, Avedik F, Mariotti A (1990) Mud volcano field seaward of the Barbados accretionary complex: A submersible survey. J Geophys Res 95:8931-8943

Lewis BTR, Cochrane GC (1990) Relationship between the location of chemosynthetic benthic communities and geologic structure on the Cascadia subduction zone. J Geophys Res B 95:8783-8793

MacDonald IR, Boland GS, Baker JS, Brooks JM, Kennicutt II MC, Bidigare RR (1989) Gulf of Mexico hydrocarbon seep communities II. Spatial distribution of seep organisms and hydrocarbons at Bush Hill. Mar Biol 101:235-247

MacDonald IR, Callender WR, Burke Jr RA, McDonald SJ, Carney RS (1990a) Fine-scale distribution of methanotrophic mussels at a Louisiana cold seep. Prog Oceanogr 24:15-24

MacDonald IR, Reilly II JF, Guinasso Jr NL, Brooks JM, Carney RS, Bryant WA, Bright TJ (1990b) Chemosynthetic mussels at a brine-filled pockmark in the northern Gulf of Mexico. Science 248:1096-1099

MacDonald IR, Guinasso Jr NL, Reilly JF, Brooks JM, Callender WR, Gabrielle SG (1990c) Gulf of Mexico hydrocarbon seep communities: VI. Patterns in community structure and habitat. Geo-Mar Lett 10:244-252

Ogawa Y, Fujioka K, Fujikura K, Iwabuchi Y (1996) En echelon patterns of *Calyptogena* colonies in the Japan Trench. Geol 24:807-810

Ohta S, Laubier L (1987) Deep biological communities in the subduction zone of Japan from bottom photographs taken during "Nautile" dives in the Kaiko project. Earth Planet Sci Lett 83:329-342

Olu K, Duperret A, Sibuet M, Foucher JP, Fiala-Médioni A (1996a) Structure and distribution of cold seep communities along the Peruvian active margin: Relationship to geological and fluid patterns. Mar Ecol Prog Ser 132:109-125

Olu K, Sibuet M, Harmegnies F, Foucher JP, Fiala-Medioni A (1996b) Spatial distribution of diverse cold seep communities living on various diapiric structures of the southern Barbados prism. Prog Oceanogr 38:347-376

Olu K, Lance S, Sibuet M, Henry P, Fiala-Medioni A, Dinet A (1997) Cold seep communities as indicators of fluid expulsion patterns through mud volcanoes seaward of the Barbados accretionary prism. Deep-Sea Res 44:811-841

Paull CK, Hecker B, Commeau R, Freeman-Lynde RP, Neumann C, Corso WP, Golubic S, Hook JE, Sikes E, Curray J (1984) Biological communities at the Florida escarpment resemble hydrothermal vent taxa. Science 226:965-967

Peeks AS, Gustafson RG, Lutz RA (1997) Evolutionary relationships of deep-sea hydrothermal vent and cold-water seep clams (Bivalvia: *Vesicomyidae*): results from the mitochondrial cytochrome oxidase subunit I. Mar Biol 130:151-161

Rosman I, Boland GS, Baker JS (1987) Epifaunal aggregations of *Vesicomyidae* on the continental slope off Louisiana. Deep-Sea Res 34:1811-1820

Sibuet M, Olu K (1998) Biogeography, biodiversity and fluid dependence of deep-sea cold seep communities at active and passive margins. Deep-Sea Res 45:517-567

Sibuet M, Juniper SK, Pautot G (1988) Cold-seep benthic communities in the Japan subduction zones: geological control of community development. J Mar Res 46:333-348

Sibuet M, Fiala-Médioni A, Foucher JP, Ohta S (1990) Spatial distribution of clams colonies at the toe of the Nankai accretionary prism near 138°E. International conference on fluids in subduction zones and related processes, Paris, Nov. 5-6, 1990

Suess E, Carson B, Ritger SD, Moore JC, Jones ML, Kulm LD, Cochrane GR (1985) Biological communities at vent sites along the subduction zone off Oregon. Bull Biol Soc Wash 6:475-484

Suess E, Bohrmann G, Von Huene R, Linke P, Wallmann K, Lammers S, Sahling H (1998) Fluid venting in the eastern Aleutian subduction zone. J Geophys Res 103:2597-2614

The Importance of Mineralization Processes in Surface Sediments at Continental Margins

M. Zabel*, C. Hensen

Universität Bremen, Fachbereich Geowissenschaften, 28359 Bremen, Germany
corresponding author (e-mail): mzabel@uni-bremen.de

Abstract: This paper is intended to give a brief overview covering the main aspects of mineralization and preservation of organic carbon in continental margin sediments. It is not meant to be a comprehensive overview of the whole subject. Instead, we will summarise the relevant subjects, present data from a number of well studied sites from different areas of the world ocean and focus on the aspects of lateral sediment advection, the role of oxygen minimum zones and the preservation/ dissolution of calcium carbonate. We also summarise data compiled in different studies and compare it to global estimates to be able to better evaluate the role of mineralization in ocean margin sediments for the world oceans.

Introduction

Organic carbon in continental margin systems - a global perspective

Ocean margins or more specific continental shelves and slopes are the main link between the two major biogeochemical systems of the continents and the open oceans. It is increasingly realised that they are important from a global view since they are sites of increased biological activity with high turnover rates of nutrients and carbon. Furthermore, they might be affected by human activities in highly populated areas or in regions with a high resource potential. Generally, they are characterised by elevated primary productivity compared to average oceanic surface waters (cf. Antoine et al. 1996; Behrenfeld and Falkowski 1997) inducing a high export flux to the sediment surface (Fig. 1). Together with large inputs of land-derived detrital material this leads to sedimentation rates of up to 60 cm/kyr making them the most important regions of organic carbon and calcium carbonate burial. About 50% of the global carbonate and more than 80% of the global organic carbon accumulation takes place in the coastal zone of the continents (Milliman 1993; Schneider et al. 2000 and references therein; Ver et al. 1999 and references therein). The importance of continental margin sediments regarding their burial and mineralization potential has been outlined since more than two decades. Berner (1982) and Jørgensen (1983) estimated that coastal and continental slope sediments are responsible for about 96% of organic carbon burial and about 98% of carbon mineralization in global marine sediments covering about 40% of the total surface area (Middelburg et al.1993 and references therein). Quite a number of more recent studies confirm the estimates given above so that there are no general doubts on the magnitude of the processes occurring in ocean margin sediments, but as a further result a number of questions have arisen from these studies which have already been addressed and must be continued in future work. Among others, the following aspects are major tasks concerning the study of ocean margin sediments:

The role of terrestrial organic material

The largest amount of terrestrial organic carbon (TeOC) is supplied to the oceans by rivers. The amount of dissolved organic carbon (DOC) by river input is sufficient to maintain the DOC turnover in global oceans and the particulate organic carbon (POC) delivered is as much as the total amount of

From WEFER G, BILLETT D, HEBBELN D, JØRGENSEN BB, SCHLÜTER M, VAN WEERING T (eds), 2002, *Ocean Margin Systems.* Springer-Verlag Berlin Heidelberg, pp 253-267

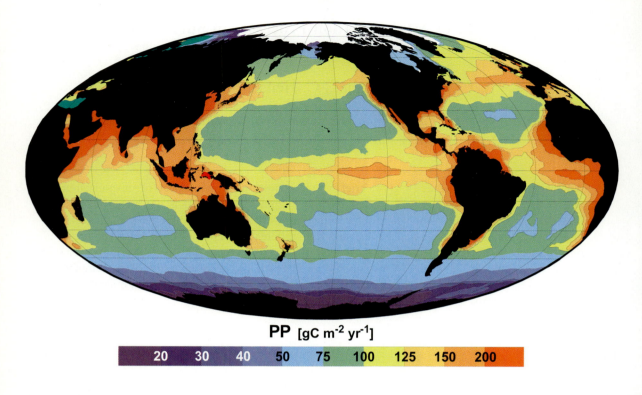

Fig. 1. Map representing the mean annual primary production in g C m^{-2}yr^{-1} (after Antoine et al. 1996).

organic carbon being buried in marine sediments (Hedges et al. 1997 and references therein). A recent re-estimate by Schlünz and Schneider (2000) gives a total mass of 430*10^{12} g TeOC delivered by rivers into the shallow ocean, of which about 50% are contributed by the tropical rivers of Oceania, South America, and Africa. About the same amount is additionally made up by particulate and dissolved inorganic carbon (Liu et al. 2000). A general believe is that terrestrial organic matter is refractory and therefore to a large extent "inert" to remineralization by marine heterotrophs. The results compiled by Schlünz and Schneider (2000), however, indicate that the continental margin sediments accumulate a 50% mixture of marine and terrestrial organic matter summing up to a total amount of about 100*10^{12} g. This in turn requires that an amount of about 380*10^{12} g TOC is remineralised or exported to the open ocean where it might accumulate in dispersed form over vast areas. Indeed, there is much evidence for TeOC being extensively mineralised in the ocean and in marine sediments, but it is widely unknown where it occurs or difficult to identify the proc-

esses underneath (Hedges et al. 1997). In this regard, results of Ransom et al. (1998) from the Californian continental margin indicate no preferential preservation of terrestrial organic matter over that of marine origin. To resolve this problem the investigation of the fate of river derived carbon in the ocean must be a major topic for ocean margin research.

Mineralization or preservation? Which are the driving forces and pathways of organic matter degradation?

Whether organic carbon is buried for some millions of years in marine sediments or is cycled between the sediments and the ocean has major impact on the levels of CO_2 and O_2 over geological time scales (e.g. Ransom et al. 1998; Hartnett et al. 1998). The most relevant regions to study the processes are the continental margin areas since open ocean areas are generally limited in organic carbon and preservation efficiency is low. The relevance of this subject has attracted many researchers aiming to quantify the rates of organic carbon oxidation and the pathways of degradation or to find out the controlling parameters that govern the degree of preservation. As a result, a huge number of studies has been published until today without finally giving the clue to the problem, but providing us with a number of empirical relationships relating organic burial or preservation efficiency to (1) primary productivity or export productivity, (2) sedimentation rate, (3) surface area of the sediment particles, (4) carbon degradation rates and pathways, (5) differences in quality of the organic matter, (6) oxygen level of the bottom water or (7) exposure time to oxygen (see below).

How is the link between continent and open ocean working in the whole system? - Regionalisation of benthic fluxes

On a global scale the coastal zone has been neglected despite its importance in biogeochemical cycles. Most GCMs (General Circulation Models) do not consider the coastal zone or continental margins as a specific reservoir. Recent studies within LOICZ and JGOFS indicate, however on a sparse data base, that coastal areas might act as a nitrate sink – due to intense denitrification - and seem to be neutral or a weak sink concerning the CO_2 budget (Liu et al. 2000). On the other hand, there is a substantial input of alkalinity from the shallow ocean to the deep-sea (better defined as shelf export). According to Milliman and Droxler (1996) about 17% of the observed oceanic alkalinity flux can be attributed to as "margin derived". At present state the role of continental margins is not very well known and results are not very much constrained. This is mainly because continental margins are highly complex systems with high spatial and temporal variability. The compilation by Liu et al. (submitted) points out that, for example, coarsely gridded ΔpCO_2 data (maps) between ocean and atmosphere on a global scale miss most of the continental margins and can therefore not reflect processes and fluxes related to them. This variability and small scale heterogeneity is also conveyed to the sedimentary environment. Apart from small scale regional studies, global or ocean wide estimates for benthic fluxes of oxygen and nutrients (Jahnke 1996; Hensen et al. 1998) exclude

the continental margins above 1000 m water depth until today due to the above reasons. More detailed studies in this regard are therefore highly demanded.

Effects of anthropogenic activities

Since the beginning of this century human impacts on the environment increased exponentially, exemplary due to burning of fossil fuel. Most studies, however, focus on the importance of CO_2- release for global warming and resulting natural hazards as well as on the handling of contaminated materials on the continents. The study of the marginal seas in this regard has intensified over the last decade coming out with results that increasing pressure might also be expected for adjacent deep-sea areas when dumping of unwanted or poisonous materials is shifted from the coastal to the proximal open ocean due to administrative controls (Kennish 1997). In a recent model study Ver et al. (1999) investigated the effect of C-N-S-emissions from fossil fuel burning, changes of land-use activities leading to increased organic matter export from the land, sewage emissions containing reactive C-N-P-compounds, the application of fertilisers, and the rise in global temperature during the past three centuries on coastal margin systems and came out with the first three factors being the most important ones in this regard.

Mineralization pathways in surface sediments

It is generally accepted that over the whole seafloor oxic respiration is the most important organic carbon oxidation pathway in marine sediments (Canfield 1993). On a global scale sulfate reduction (SR) is considered to be the next important process followed by denitrification. Corresponding to Table 1, which gives an overview of global estimates of the capacities of the single oxidants for bacterial organic carbon mineralization, the other processes only contribute a very minor portion to the total amount.

However, this global "ranking" of importance can drastically change and show a high variability, when examining continental margin sediments

Oxidant	Carbon Mineralization [10^{14} g C yr^{-1}]	Percentage
O_2	23 [a]	~57
	17 [b]	(total minus anoxic)
	11 [c]	
Nitrate	0.93 [a]	~11
	0.96 [d]	
	2.4 - 3.0 [e]	
SO_4	1.3 - 13.0 [b]	~26
	4.4 [f]	
CH_4	0.44 - 0.9 [f]	3.8
Mn-oxides	0.23 - 0.46 [f]	1.4
Fe-oxides	0.07 - 0.23 [f]	0.6
Total	18 - 31 [e]	100
	26 [g]	

References: [a] Jørgensen (1983), [b] Henrichs and Reeburgh (1987), [c] Jahnke (1996), [d] Christensen *et al.* (1987), [e] Middelburg *et al.* (1997), [f] Canfield (1993), [g] Smith and Hollibaugh (1993)

Table 1. Estimations of the global capacity of the different terminal electron acceptors for organic carbon oxidation.

separately. Which process regionally and temporally dominates, depends mainly on the supply of organic matter and on the geochemical environment at the sediment-water interface. For example, when bottom water is depleted in dissolved oxygen, which is especially the case where the oxygen-minimum zones (OMZs) impinge on a continental slope and/or shelf, most organic matter degradation occurs via sulfate reduction. When sediments are permanently overlaid with anoxic water (euxinic sediments) organic matter is almost exclusively mineralised by SR. But also under less extreme conditions, oxygen is not necessarily

the dominant oxidant. Thamdrup and Canfield (1996) calculate that SR account for 56-79% of the total carbon mineralization in surface sediments (0-10 cm) below the OMZ at the central Chilean continental slope where oxygen concentrations are 105-220 µmol O_2/l. Only 15% of the carbon oxidation is coupled to the combined reduction of oxygen and nitrate (Fig. 2). However, both, the single turnover rates and relationships among them, are governed by seasonal and/or interannual variations in the controlling factors mentioned above. As an example, Ferdelman et al. (1997) resampled the same locations as Thamdrup and Canfield (1996), but determined up to ~66% and 50% lower SR rates within the OMZ and below. In this region, differences may be also caused by the episodic occurrence of the El-Niño phenomena.

In a recent review article, Thamdrup (2000) sums up the present knowledge of bacterial manganese and iron reduction in margin sediments. Based on all case studies conducted during the last decade, contributions of Fe reduction to organic carbon oxidation range from below detection (for example in the OMZ where Fe reduction is mainly coupled to oxidation of reduced sulfur) to >50%, with an average of 17% suggested for a wide range of continental margin sediments (cf. Fig. 2). In contrast to iron, Mn-oxides are much less important. Instead, Mn reduction is mostly coupled with the reoxidation of reduced Fe and sulfur compounds. Only when sediments are extremely enriched in manganese, microbial reduction of Mn-oxides can gain importance during early diagenetic degradation of organic matter (Thamdrup 2000). Nevertheless, at present no exact global estimation of the importance of the single terminal electron acceptors involved in benthic organic carbon degradation at continental margins exists. First of all, this is caused by the great heterogeneity of the available data base. Numerous geochemical studies were performed during the past decades, but investigations examining all relevant processes and components which are necessary to account for all different pathways of organic matter mineralization - including associated processes - are extremely rare. On the one hand most studies

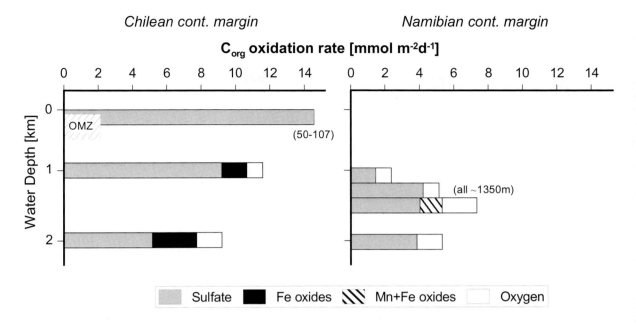

Fig. 2. Total mineralization rates and contributions of individual electron acceptors in continental margin sediments off Chile and Namibia (after Thamdrup 2000 and references therein).

either focus on special aspects like the behaviour of a single element during early diagenesis and neglect essential parameters which are indispensable for a complete description of the benthic system (first of all abiotic reoxidation of reduced species). On the other hand, the large influence of macrobenthic organisms on flux rates of dissolved constituents by active irrigation as well as on the sediment composition by particle mixing is difficult to quantify (cf. Haese this volume). This is the reason why these processes are often neglected in calculations of benthic turnover rates, although especially the efficiency of Fe and Mn reduction highly depends on infauna activities (cf. Thamdrup 2000). Additionally, *in situ* measurements of pore water concentration profiles and total consumption and release rates, respectively, are limited to a few selected areas of the ocean so far. Against this background, the great variety of highly complex continental margin systems prevents more exact global quantifications of the contribution of the single electron acceptors to organic carbon oxidation in these sedimentation regimes although our knowledge about the interrelations between biogeochemical processes has been considerably improved. Without considering biogeochemical interactions between single compounds, the im-

portance of continental margin sediments for ocean budgets is also expressed by the distribution of total flux rates of dissolved compounds across the sediment-water interface. Compared with the accumulation of particulate matter on the sea floor, they offer the only opportunity to determine whether a sediment serves as a specific sink or source. Caused by the great heterogeneity in the spatial distribution of data sets, maps showing global patterns of benthic flux rates without using empirical correlations for extrapolation do not yet exist. However, the few oceanwide examinations which only based on flux data show the expected connection (e.g. Zabel et al. 1998; Hensen et al. 1998) (Fig. 3). Although re-stricted to water depths >1000m, they clearly underline the great importance of continental margin sediments for the ocean carbon and nutrient cycles.

Influence of lateral advection on benthic turnover rates

The intensity of benthic remineralization processes is mainly influenced by 1) the quality (reactivity) and the quantity of the available organic substrate which accumulates on the sea floor, 2) the supply

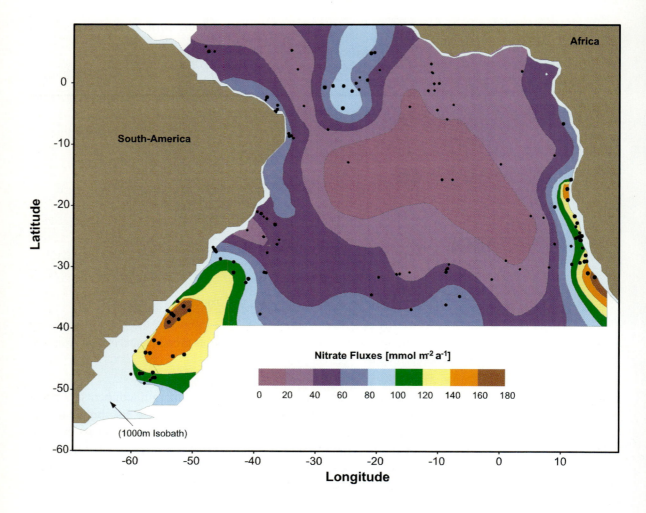

Fig. 3. Regional distribution of benthic nitrate release rates in the S-Atlantic below 1000m water depth (modified from Hensen et al. 1998). Fluxes were calculated from pore water profiles. Highest release rates of nitrate occur at the continental margins of Argentina/Uruguay and Namibia. In both areas distinct depocenters of organic carbon exist. Whereas fine-grained organic matter is focussed on the lower continental slope off Argentina and Uruguay (1800-3500) by lateral transport of particles from the shelf areas, the distribution pattern of benthic flux rates off Namibia are mainly attributed to high primary production due to intense wind-driven upwelling.

of oxidants (especially of dissolved oxygen), 3) the prevailing community of organisms, 4) the temperature, and 5) the sediment fabric. However, there is no doubt that the input of organic matter fundamentally depends on the dimension of the primary productivity in the euphotic surface water and the residence time of the settling particles, or rather efficiency of their decomposition in the water column. That is why benthic turnover rates are in general more intense in coastal regions than in the deep open ocean, where both, the predominantly oligotrophic conditions and the long particle transport to the sea floor result in comparatively lower accumulation rates of organic matter at the sea bed. Although the simple relation between the amount of organic carbon fixation by phototrophic algae in the surface water and benthic recycling and burial rates may be correct on a global scale in principle, numerous geochemical and oceanographic studies give evidence that this link is much more complex in detail. Especially at continental margins with intense upwelling of water masses and/or with strong prevailing contour currents lateral transport processes can cause a serious decoupling between the pattern of primary or rather export production and the distribution of benthic remineralization rates. Therefore, one has to be very cautious when transferring e.g. surface water productivity quantitatively to underlying marine sediments by only accounting for decomposition of organic matter expressed as a function of water depth. Lateral drift during the process of particle settling, as well as re-suspension and re-deposition of sediments result in the development of widespread "depocenters" at continental slopes or rises (e.g. Walsh 1991). Such depocenters of organic carbon at continental margins which were attributed to a predominantly cross-shelf export of fine-grained organic material beyond the shelf-slope break are described for many coastal regions all around the ocean (e.g. California continental margin – e.g. Reimers et al. 1992; Oman Margin – e.g. Pedersen et al. 1992; NW-Atlantic Margin – e.g. Anderson et al. 1994; SW-Atlantic margin – Hensen et al. 2000). The present knowledge of the mechanisms of lateral transport processes at continental margins and methods of their evaluation are extensively described and discussed in more detail by Thomsen (this volume). We want to focus on the effects of lateral advection on the benthic biogeochemical processes. Among others, investigations of microbial activity and benthic recycling rates in the mentioned areas reveal significant differences in the relationship between the bulk organic carbon content and the intensity of benthic remineralization processes. In general, the results indicate that, apart from the geochemical environment in the overlying bottom water (e.g. OMZ, see below), the range of particulate organic carbon supply and the maturity of the organic matter are essential controlling factors for benthic turnover rates. As an example, Figure 4 shows relationships between water depth, the total organic carbon (TOC) content of surface sediments and the benthic oxygen uptake at two continental margins within the North Atlantic where bottom water oxygen concentrations are near saturation. Although the integrated annual primary production on both adjacent shelf regions are quite similar (~470 mg C $m^{-2}yr^{-1}$ on the N-Carolina shelf and ~330 mg C $m^{-2}yr^{-1}$ on the

NW-European shelf/Goban Spur; Behrenfeld and Falkowski 1997) and depth related distributions of TOC evidence the lateral input of C_{org} at both locations (increasing TOC content with depth), a distinct depocenter zone only exists at the N-Carolina continental margin at depth of ~1000m.

Nevertheless, apart from this fundamental difference which may be caused by the respective current conditions, benthic oxygen consumption rates reveal an additional distinctive feature. Whereas oxygen respiration and TOC content are positive correlated at the western N-Atlantic margin, the opposite was observed at Goban Spur. Lohse et al. (1998) therefore suggest that the deposition of organic matter on Goban Spur is dominated by refractory, less degradable compounds. This assumption is supported by the lack of a distinct response of benthic flux rates to pulses of organic matter input, which were detected by accompanying sediment trap studies. Despite the results of Anderson et al. (1994) who reported that as much as half of the organic carbon that accumulates on the upper slope of the US-continental margin is refractory, the intense oxygen consumption rates in the depocenter zone indicate that the amount of the labile fraction is significantly higher than at Goban Spur. Despite of differences in methods of estimating carbon oxidation rates, this is reflected by the flux rates of 0.97 - 3.9 mol C $m^{-2}yr^{-1}$ for the N-Carolina depocenter (Jahnke and Jahnke 2000) and 0.58 – 0.65 mol C $m^{-2}yr^{-1}$ for the depth of maximum TOC at Goban Spur (Lohse et al. 1998). This brief comparison should point out the great variety of different types of continental margins. Lateral particle drift and re-deposition predominantly along the coast line (e.g. at the SE-African continental margin – Giraudeau et al. 2000) characterises an additional type. The given example demonstrates that especially at continental margins relationships between benthic turnover rates and their control parameters are highly complex and specific for a distinct region. Therefore any empirical function describing these relationships basing on a regional data set (c.f. Cai and Reimers 1995) (Fig. 5) are definitively not universally applicable to other areas where sedimentary and oceanographic conditions can be completely different. A suitable

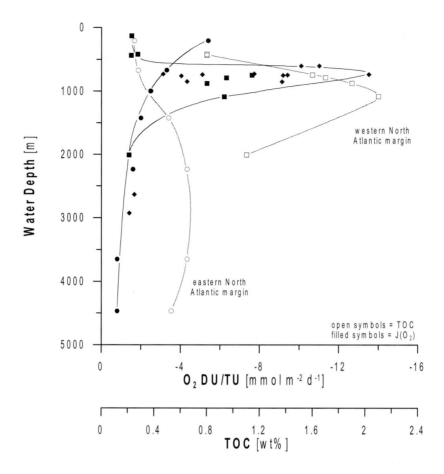

Fig. 4. Relationships between benthic oxygen uptake, water depth and the total organic carbon content at two different continental margins within the North Atlantic. Benthic chamber-based data for the western North Atlantic margin were given by Anderson et al. (1994) (■□) and Jahnke and Jahnke (2000) (◆). Results for the NW-European continental margin were taken from Lohse et al. (2000) (●○). Flux rates here were calculated from *in situ* measurements with microelectrodes. The high (small scale) lateral variability at depocenters is documented by the scatter of data for oxygen respiration (◆).

approach to solve this problem in regard to global estimations of ocean cycles which also include the coastal ocean may be the characterization of benthic (biogeochemical) provinces in analogy to the division of the surface ocean by Longhurst et al. (1995). For this purpose, however, the available data base seems to be still full of gaps.

What controls organic matter preservation in marine sediments?

Only a very minor fraction of the total primary production is finally buried in marine sediments.

Estimates range between 0.1-0.3% (de Baar and Suess 1993; Hedges and Keil 1995). This fraction usually increases as one approaches the continents due to increased primary production rates, input of terrestrial organic matter and shallower water depths (meaning a decrease in mineralization in the water column). Quite a number of studies exist suggesting a quantitative correlation between the burial of organic carbon and the input of organic carbon to the sediment surface or the sediment accumulation rate. Figure 6 shows data from a compilation by Henrichs (1992) where the input of organic carbon is related to both the mineralization and the organic carbon burial rate. Although

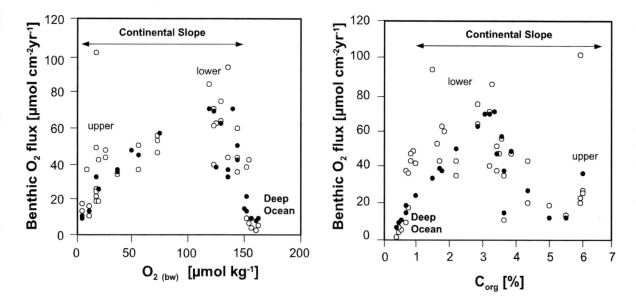

Fig. 5. Distribution of benthic oxygen fluxes across the continental slope in the Northeast Pacific related to **(a)** oxygen bottom water concentration and **(b)** organic carbon content in the surface sediments. Highest oxygen respiration occurs at the lower continental slope. Open circles are measured fluxes (JO_2), solid circles were calculated by applying the following equation : $JO_2 = \delta \cdot C_{org} \cdot [O_2]_{BW} / (126 + [O_2]_{BW})$, with $\delta = 44.4$ (redrawn after Cai and Reimers 1995).

a clear correlation is shown in this example meaning that high input of organic material means both, high mineralization and burial, regional variability is large and a global correlation to a single key variable does obviously not exist – (similar as for benthic flux rates as stated above).

The problem might be summarised best by the final statement of Canfield (1993) that overall, the processes controlling the preservation of organic carbon are not yet well understood. This situation has not changed overall because of the lack of detailed investigations and the heterogeneity of continental margin sediments although much progress has been made analysing possible control mechanisms. One of the most extensively discussed subjects in this regard is whether there exists a general difference between the burial potential under oxic and anoxic conditions. The most important anoxic pathway of organic matter degradation in marine environments is via sulphate reduction. Under normal marine conditions (oxic bottom water) in high accumulation areas oxic respiration and sulphate reduction are about equal in their importance Canfield (1993 and references

therein; Jørgensen 1982; see above). A vital discussion, however, still exists concerning the question whether oxygen deficient conditions (e.g. within OMZs) enhance carbon preservation. Generally, this idea is favoured (e.g. Demaison and Moore 1980; Canfield 1989, 1993; Van der Weijden et al. 1999) although there is evidence from field observations which are contradictory (Pedersen et al. 1992; Ransom et al. 1998). One possible mechanism to explain a better preservation under anoxic conditions could be that certain organic molecules might only be decomposed under oxic conditions although experimental results exist showing similar degradation rates for fresh organic matter exist under oxic as under anoxic conditions (Canfield 1993 and references therein). As stated above, many other mechanisms are possibly responsible for the enhancement or the decrease of organic matter preservation. As suggested by Ransom et al. (1998), surface area characteristics of the grains and the mineralogy of clays are supposed to be the overall driving force for the control of preservation. Hartnett et al. (1998) investigated different transects across the continental margin in

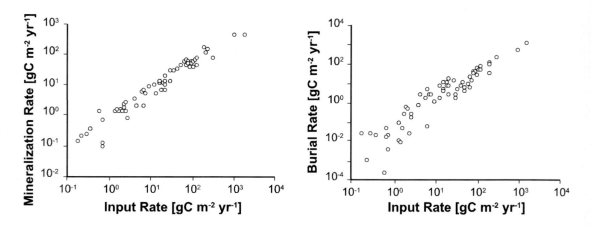

Fig. 6. Input rates of organic carbon plotted against carbon mineralization and carbon burial from different data sources (redrawn after Henrichs 1992).

the northern East Pacific and found a strong correlation between organic carbon burial efficiency and oxygen exposure time (Fig. 7).

Overall, we do not find the ongoing discussion on "the key parameter" which is finally controlling organic carbon preservation and especially the role of OMZs very helpful for the overall task. Apart from studies elucidating the mechanisms controlling the degradation or preservation as suggested for example the surface area and mineralogy of the sediments (Hedges and Keil 1995; Ransom et al. 1998) future efforts should be made mainly in order to gain more complete data sets on benthic remineralization, sedimentation rates, and net fluxes to the sediment surface related to primary production and lateral input. From our point of view it is not possible – but has been done in many studies – to calculate e.g. preservation efficiencies from the available data sets which are mostly lacking exact data on sedimentary carbon fluxes.

Effect of organic matter mineralization on calcite dissolution

One important fact related to organic matter decay has been left out up to this point: the dissolution of calcium carbonate. The dissolution of calcium carbonate in marine sediments is controlled by two major factors, the degree of undersaturation of the bottom waters with respect to calcite and

aragonite, and the reaction with respiratory carbon dioxide. Continental margin sediments are to a large extent located above the lysocline so that bottom waters are usually oversaturated at least with respect to calcite. Calcium carbonate dissolution due to CO_2 release from organic carbon respiration can be expressed as follows:

$$(CH_2O)_{106}(NH_3)_{16}H_3PO_4 + 138O_2 + 124CaCO_3 \rightarrow$$
$$230HCO_3^- + 16NO_3^{--} + HPO_4^- + 124Ca^{2+} + 16H_2O$$

where the equation refers to organic matter oxidation with Redfield C:N:P ratio and complete reaction of respiratory CO_2 with sedimentary calcium carbonate. It was long time neglected that carbonate dissolution by CO_2 derived from metabolic processes in marine sediments could play an important role. Indeed, Emerson and Bender (1981) found that degradation of organic matter at the sediment-water interface can have significant impact on the calcite dissolution, and hence, on the preservation of calcium carbonate in marine sediments. A number of recently published studies has identified this problem and generally focused on the differentiation between calcium carbonate dissolution by undersaturation of bottom waters and organic matter remineralization (cf. Schneider et al. 2000 for an overview). Data from ocean margins, strictly speaking from locations above the lysocline and with high input of

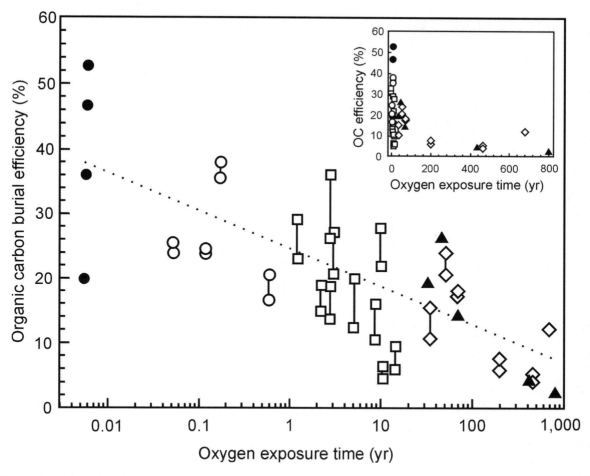

Fig. 7. Organic carbon burial efficiency plotted as a function of oxygen exposure time. Oxygen exposure time was calculated from oxygen penetration depth divided by the sedimentation rate. Data are from different transects across the continental margin in the northern East Pacific (Mexican shelf and slope, California slope, Washington shelf and slope). Filled circles represent data from sediments accumulating beyond a permanent OMZ. Insert is plotted with linear scales. Reprinted with permission from Nature (Harnett et al. 1995, Nature 391:572-574). Copyright (1998) Macmillan Magazines Limited.

organic matter are very sparse. However, Hales and Emerson (1997) report a 20-60% dissolution of the calcium carbonate flux to the sediment surface due to metabolically released CO_2 in sediments of the oligotrophic Ceará Rise (Equatorial Atlantic). Wenzhöfer et al. (in press) and Pfeifer et al. (2000) found that up to 90% of the calcium carbonate rain is dissolved at the sediment surface by this process in equatorial upwelling area of the Eastern Atlantic. Figure 8 shows measured data and modelling results of Pfeifer et al. (2000) showing clearly the relation between oxygen consumption and calcium carbonate dissolution within the first few centimetres of the sediment.

Owing to the low number of available investigations, estimates of total carbon release from marine sediments imply a great inaccuracy. This is the reason why only a few compilations exist until today. Table 2 summarises some more recent data compilations that should help to constrain the carbon flux from deep-sea sediments (>1000 m water depth). Three flux categories are given: The CO_2 produced due to oxic respiration, the alkalinity as a sum parameter for calcium carbonate dissolution and CO_2 from oxic respiration, and the dissolved organic carbon (DOC).

In spite of large uncertainties rough estimations of the global benthic CO_2 release based on global benthic oxygen fluxes (Jahnke 1996) and benthic

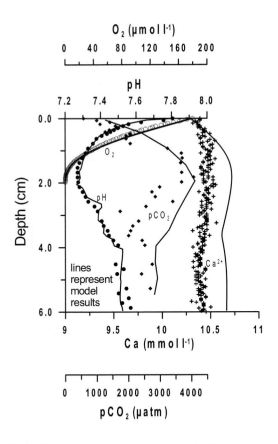

Fig. 8. *In situ* measurements and modelling results from station GeoB 4906 (Eastern Equatorial Atlantic) modified after Pfeifer et al. (2000). Oxygen consumption causes major increase in Ca^{2+} concentration and decrease of pH and pCO_2 due to calcite dissolution.

nitrate release (from oxic respiration) in the South Atlantic are available (Schneider et al. 2000). Berelson et al. (1994) estimate the benthic alkalinity input to the deep ocean for the Pacific and the Indo-Pacific. They suggest that most of the carbonate dissolution in the deep ocean occurs within the sediments (85%). However, it has to be admitted that the proportions between $CaCO_3$ dissolution in the water column and in the sediment are still under debate and that estimates for dissolution in the sediment range between 0-90% (Archer 1996). A global projection for benthic alkalinity flux based on the calculations of Berelson et al. (1994) using the surface areas for the global deep ocean (Table 1) seems not appropriate. The extension of their results from the Pacific and the Indian Ocean to the Atlantic Ocean which

resuls in an ocean wide dissolved carbon flux of $120*10^{12}$ mol yr^{-1} from deep-sea sediments, may be critical because of the completely different deep-water conditions in the Indo-Pacific and the Atlantic. Despite this problem the global estimate extending the Berelson et al. (1994) approach is in agreement to that of Mackenzie et al. (1993). In contrast, that given by Morse and Mackenzie (1990) (and also by Milliman 1993) seems to be drastically underestimated. Comparing the estimates of total carbon release based on benthic alkalinity fluxes with those from organic matter decay one can assume that metabolic CO_2 may contribute about 50% ($60*10^{12}$ mol yr^{-1}) to the total carbon release of $12*10^{12}$ mol yr^{-1} from deep-sea sediments, meaning that the other 50% come from $CaCO_3$ dissolution. This is consistent with model results of Archer (1996) who proposed an upper limit of global calcite dissolution flux from deep-sea sediments of up to $54*10^{12}$ mol yr^{-1}. Thus, the total carbon release from deep-sea sediments of about $120*10^{12}$ mol yr^1 seems to be a relatively good approximation regarding all sources of uncertainty. However, the estimates for the pelagic realm are not very well constrained on a global base. As indicated above, this situation is even worse for the continental margin areas, because of the – already stressed – stronger heterogeneity of the depositional environment. Interesting aspects deserving more attention in future studies are the enhanced preservation of calcium carbonate in sediments within OMZs due to missing or strongly decreased oxic respiration or dissolution within platform carbonates triggered by sulphide reoxidation (Ku et al. 1999).

Arising questions and future tasks

Based on the discussion we have formulated some relevant questions and suggestions that should be addressed in future research on ocean margin sediments:

• How much calcium carbonate is dissolved on continental shelves and slopes? Is a global re-estimate necessary due to metabolically induced calcite dissolution?

• Systematic study – and data compilation – of existing data sets on the role of OMZs for the

Parameter	Area Production	Flux	Source
Respiratory CO_2[1]	South Atlantic	6 - (11)	after Hensen et al. (1998)
	Global	60 - (110)	after Hensen et al. (1998)
		40	after Jahnke (1996)
Calcite Dissolution	Global	27 - 54	Archer (1996)
Alkalinity / TCO_2	Pacific	55	Berelson et al. (1994)
	Indo-Pacific	91	Berelson et al. (1994)[2]
	Global	100	after Berelson et al. (1994)
		120	after Berelson et al. (1994)[2]
		120	Mackenzie et al. (1993)
		5	Morse and Mackenzie (1990)
DOC	Atlantic[3]	4	Otto (1996)
	Global[3]	18	Otto (1996)

[1] Estimated as oxic respiration of organic matter.

[2] Using data of Broecker and Peng (1987).

[3] Including water depths above 1000m.

Table 2. Estimated carbon fluxes from deep-sea sediments (below 1000 m) in 10^{12} mol yr^{-1} (modified from Schneider et al. 2000).

preservation of calcium carbonate and organic carbon.

• Increasing use of anthropogenically derived substances as tracers to characterise mechanisms or time scales of processes relevant in marine surface sediments.

• Time series analysis of benthic fluxes in key areas to overcome seasonal variations for the predictions of annual estimates and global budgets.

• Investigations of carbon burial efficiency generally miss exact data on carbon fluxes arriving at the sea floor and thus an adequate determination of sedimentation rates. This must be overcome by coupled investigations within the water column, the benthic boundary layer and the sediments.

• In this context the importance of lateral advection, re-suspension and re-deposition on the quantity of particle accumulation and burial effi-

ciency as well as on particle quality/reactivity is almost unknown.

References:

Anderson RF, Rowe GT, Kemp PF, Trumbores S, Biscaye PE (1994) Carbon budget for the mid-slope depocenter of the Middle Atlantic Bight. Deep-Sea Res II 41:669-703

Antoine D, André JM, Morel A (1996) Oceanic primary production; 2. Estimation at global scale from satellite (coastal zone color scanner) chlorophyll. Glob Biogeochem Cycl 10(1):57-69

Archer DE (1996) A data-driven model of the global calcite lysocline. Glob Biogeochem Cycl 10:511-526

Behrenfeld MJ, Falkowski PG (1997) Photosynthetic rates derived from satellite-based chlorophyll concentration. Limnol Oceanogr 42(1):1-20

Berelson WM, Hammond DE, McManus J, Kilgore TE

(1994) Dissolution kinetics of calcium carbonate in equatorial Pacific sediments. Glob Biogeochem Cycl 8:219-235

Berner RA (1982) Burial of organic carbon and pyrite sulfur in the modern ocean: Its geochemical and environmental significance. Am J Sci 282:451-473

Broecker WS, Peng T-H (1987) The role of CaCO3 compensation in the glacial to interglacial atmospheric CO2 change. Glob Biogeochem Cycl 1:15-29.

Canfield DE (1989) Sulfate reduction and oxic respiration in marine sediments: implications for organic carbon preservation in euxinic environments. Deep-Sea Res 36:121-138

Canfield DE (1993) Organic matter oxidation in marine sediments. In: Wollast R, Mackenzie FT, Chou L (eds) Interactions of C, N, P, and S in Biogeochemical Cycles and Global Change, NATO ASI Series 14, Springer, Berlin pp 333-363

Cai WJ, Reimers CE (1995) Benthic oxygen flux, bottom water oxygen concentration and core top organic carbon content in the deep northeast Pacific Ocean. Deep-Sea Res 42:1681-1699

Christensen JP, Murray JW, Devol AH, Codispoti LA (1987) Denitrification in continental shelf sediments has major impact on the oceanic nitrogen budget. Glob Biogeochem Cycl 1:97-116

De Baar HJW, Suess E (1993) Ocean carbon cycle and climate change - An introduction to the interdisciplinary Union Symposium. Glob Planet Change 8: VII-XI

Demaison GJ, Moore GT (1980) Anoxic environments and oil source bed genesis. Org Geochem 2:9-31

Emerson S, Bender M (1981) Carbon fluxes at the sediment-water interface of the deep-sea; calcium carbonate preservation. J Mar Res 39:139-162

Ferdelman TG, Lee C, Pantojy S, Harder J, Bebout BM, Fossing H (1997) Sulfate reduction and methanogenesis in a *Thioploca*-dominated sediment off the coast of Chile. Geochim Cosmochim Acta 61:3065-3079

Giraudeau J, Bailey GW, Pujol C (2000) A high-resolution time-series analyses of particle fluxes in the Northern Benguela coastal upwelling system: Carbonate record of changes in biogenic production and particle transfer processes. Deep-Sea Res II 47:1999-2028

Haese RR (2002) Macrobenthic activity and its effects on biogeochemical reactions and fluxes. In: Wefer et al. (eds) Ocean Margin Systems. Springer, Berlin pp 219-234

Hales B, Emerson S (1997) Calcite dissolution in sediments of the Ceara Rise: *In situ* measurements of porewater O_2, pH, and $CO_{2(aq)}$. Geochim Cosmo-

chim Acta 61(3):501-514

Hartnett HE, Keil RG, Hedges JI, Devol AH (1998) Influence of oxygen exposure time on organic carbon preservation in continental margin sediments. Nature 391:572-574

Hedges JI, Keil RG (1995) Sedimentary organic matter preservation: an assessment and speculative synthesis. Mar Chem 49:81-115

Hedges JI, Keil RG, Benner R (1997) What happens to terrestrial organic matter in the ocean. Org Geochem 27(5/6):195-212

Henrichs SM (1992) Early diagenesis of organic matter in marine sediments: progress and perplexity. Mar Chem 39:119-149

Henrichs SM, Reeburgh WS (1987) Anaerobic mineralization of marine sediment organic matter: Rates and the role of anaerobic processes in the oceanic carbon economy. Geomicrobiol J 5(3/4):191-237

Hensen C, Landenberger H, Zabel M, Schulz HD (1998) Quantification of diffusive benthic fluxes of nitrate, phosphate and silicate in the Southern Atlantic Ocean. Glob Biogeochem Cycl 12(1):193-210

Hensen C, Zabel M, Schulz HD (2000) A comparison of benthic nutrient fluxes from deep-sea sediments off Namibia and Argentina. Deep-Sea Res II 47:2029-2050

Jahnke RA (1996) The global ocean flux of particulate organic carbon: Areal distribution and magnitude. Glob Biogeochem Cycl 10:71 - 88

Jahnke RA, Jahnke DB (2000) Rates of C, N, P and Si recycling and denitrification at the US Mid-Atlantic continental slope depotcenter. Deep-Sea Res 47: 1405-1428

Jørgensen BB (1982) Mineralization of organic matter in the sea bed - the role of sulphate reduction. Nature 296:643-645

Jørgensen BB (1983) Processes at the sediment-water interface. In: Bolin B, Cook RB (eds) The Major Biogeochemical Cycles and their Interactions. SCOPE, Wiley, New York, pp 477-515

Kennish MJ (1997) Practical handbook of estuarine and marine pollution. CRC Press, Boca Raton, Florida, 524 p

Ku TCW, Walter LM, Coleman ML, Blake RE, Martin AM (1999) Coupling between sulphur recycling and syndepositional carbonate dissolution: Evidence from oxygen and sulphur isotope composition of pore water sulphate, South Florida Platform, USA. Geochim Cosmochim Acta 63(17):2529-2546

Liu KK, Atkinson L, Chen CTA, Gao S, Hall J, Macdonald RW, Talaue McManus L, Quiòones R (2000) Are continental margin carbon fluxes significant to the global ocean carbon budget? EOS 81(52):641-644

Lohse L, Helder W, Epping EHG, Balzer W (1998) Recycling of organic matter along a shelf-slope transect across the NW European Continental Margin (Goban Spur). Prog Oceanogr 42:77-110

Longhurst A, Sathyendranath S, Platt T, Caverhill C (1995) An estimate of global primary production in the ocean from satellite radiometer data. J Plankton Res 17(6):1245-1271

Mackenzie FT, Ver LM, Sabine C, Lane M, Lerman A (1993) C, N, P, S global biogeochemical cycles and modelling of global change. In: Wollast R, Mackenzie FT, Chou L (eds) Interactions of C, N, P, and S in Biogeochemical Cycles and Global Change. NATO ASI Series, 14, Springer, Berlin pp 1-61

Middelburg JJ, Vlug T, Jaco F, Van der Nat WA (1993) Organic matter mineralization in marine systems. Glob Planet Changes 8:47-58

Middelburg JJ, Soetaert K, Herman PMJ (1997) Empirical relationships for use in global diagenetic models. Deep-Sea Res 44(2):327-344.

Milliman JD (1993) Production and accumulation of calcium carbonate in the ocean: Budget of a nonsteady state. Glob Biogeochem Cycl 7(4):927-957

Milliman JD, Droxler AW (1996) Neritic and pelagic carbonate sedimentation in the marine environment: Ignorance is not bliss. Geol Rdsch 85:496-504

Morse JW, Mackenzie FT (1990) Geochemistry of Sedimentary Carbonates. Elsevier, Amsterdam, 707 p

Pedersen TF, Shimmield GB, Price NB (1992) Lack of enhanced preservation of organic matter in sediments under the oxygen minimum on the Oman Margin. Geochim Cosmochim Acta 56:545-551

Pfeifer K, Hensen C, Adler M, Wenzhöfer M, Strotmann B, Schulz HD (2000) Modeling of subsurface calcite dissolution regarding respiration and reoxidation processes in the equatorial upwelling off Gabon. Abstract, Goldschmidt Conference, Oxford, September 3rd-8th 2000

Otto S (1996) Die Bedeutung von gelöstem organischen Kohlenstoff (DOC) für den Kohlenstofffluß im Ozean. Ph.D. Thesis, Berichte 87, Fachbereich Geowissenschaften, University of Bremen, 150 p

Ransom B, Kim D, Kastner M, Wainwright S (1998) Organic matter preservation on continental slopes: Importance of mineralogy and surface area. Geochim Cosmochim Acta 62(8):1329-1345

Reimers CE, Jahnke RA, McCorkle DC (1992) Carbon fluxes and burial rates over the continental slope and rise of central California with implications for the global carbon cycle. Glob Biogeochem Cycl 6:199-224

Schlünz B, Schneider RR (2000) Transport of terrestrial organic carbon to the oceans by rivers: Reestimating flux- and burial rates. Int J Earth Sci 88:599-606

Schneider RR, Schulz HD, Hensen C (2000) Marine carbonates: Their formation and destruction. In: Schulz HD, Zabel M (eds) Marine Geochemistry. Springer, Berlin pp 283-307

Smith SV, Hollibaugh JT (1993) Coastal metabolism and the oceanic organic carbon balance. Rev Geophys 31:75-89

Thamdrup B, Canfield DE (1996) Pathways of carbon oxidation in continental margin sediments off central Chile. Limnol Oceanogr 41:1629-1650

Thamdrup B (2000) Bacterial manganese and iron reduction in aquatic sediments. Adv Microbial Ecol 16:41-84

Thomsen L (2002) The benthic boundary layer. In: Wefer et al. (eds) Ocean Margin Systems. Springer, Berlin pp 143-156

Van der Weijden CH, Reichart GJ, Visser HJ (1999) Enhanced preservation of organic matter in sediments deposited within the oxygen minimum zone in the northeastern Arabian Sea. Deep-Sea Res I 46:807-830

Ver LM, Mackenzie FT, Lerman A (1999) Carbon cycle in the coastal zone: effects of global perturbations and change in the past three centuries. Chem Geol 159:283-304

Walsh JJ (1991) Importance of continental margins in the marine biogeochemical cycling of carbon and nitrogen. Nature 350:53-55

Wenzhöfer F, Adler M, Kohls O, Hensen C, Strotmann B, Boehme S, Schulz HD (in press) Calcite dissolution driven by benthic mineralisation in the deep-sea: In situ measurements of Ca^{2+}, pH, pCO_2, O_2. Geochim Cosmochim Acta

Zabel M, Dahmke A, Schulz HD (1998) Regional distribution of diffusive phosphate and silicate fluxes through the sediment-water interface: The eastern South Atlantic. Deep-Sea Res 45:277-300

Numerical Modelling of Transport Processes in the Subsurface

R. Hinkelmann[*], R. Helmig

Universität Stuttgart, Institut für Wasserbau,
Lehrstuhl für Hydromechanik und Hydrosystemmodellierung, Pfaffenwaldring 61,
70550 Stuttgart, Germany
[*]corresponding author (e-mail): reinhard.hinkelmann@iws.uni-stuttgart.de

Abstract: In this paper an overview is given about numerical modelling of transport processes in the subsurface. Numerical models enable a better process understanding, can determine local or global budgets and are able to make predictions for changing conditions. The physical and mathematical model concepts are introduced for subsurface transport processes in a single phase, water, and, briefly in two phases, water and gas. The last model concept is required to simulate flow and transport processes when e.g. free gas occurs. In many cases, flow and transport processes must be considered in two or three dimensions. Finite-Difference, Finite-Element and Finite-Volume-Methods are discussed together with stabilization techniques which are required because of the advection / dispersion character of the basic equations. The authors give recommendations to the different methods. In the future the models can be further developed to simulate non-isothermal multiphase / multicomponent processes, which occur e.g. around gas hydrates.

Introduction

Importance of numerical simulation in the subsurface

The numerical simulation of flow and transport processes in subsurface systems plays an important role in a number of environmental and economical fields. This is exemplarily illustrated for groundwater and ocean margin systems. Groundwater is the main source of drinking water in many countries all over the world. For example, about 70 % of drinking water in Germany comes from groundwater. Therefore, the occurence and spreading of substances such as nutrients or contaminants must be observed and predicted to guarantee a high water quality. In ocean margin systems salt water intrusion processes have to be taken into account as well as submarine discharge of freshwater and release of free gas (e.g. methane) into the coastal zone.

Remarks on the term model

In the following, the term model will be briefly explained. A model describes processes in a system. The conceptual model reduces the occuring physical, chemical and microbiological processes to those which are of interest and importance for the problems and questions to be investigated. A mathematical model transfers the conceptual model to a mathematical formulation using balance equations for mass, momentum, and energy as well as equations of state. This procedure leads to partial differential equations for the unknown variables.

The derivations are based on a continuum approach corresponding to the definition of a REV (representative elementary volume) or on a particle approach regarding the 'fate' of each particle in space and time. Mathematical models can be deterministic refering to the formulations of motion or stochastic taking into account the statistical behaviour of state variables or parameters. Mathematical models can only be solved analytically for a limited number of simple cases. For a more flexible, general approach, numerical models are required, because they enable the treatment of general geometries as well as initial and boundary conditions. The result is a simulation program which can be used for the computation of the specified processes.

From WEFER G, BILLETT D, HEBBELN D, JØRGENSEN BB, SCHLÜTER M, VAN WEERING T (eds), 2002, *Ocean Margin Systems.* Springer-Verlag Berlin Heidelberg, pp 269-294

Application fields of numerical models in the subsurface

Models can be applied for different problems and questions. Models can enable or simplify the process understanding, e.g. at sites of submarine groundwater discharge or of fluid venting. Dominant parameters and processes can be detected or estimated, e.g. by numerically seperating processes in (strongly) coupled systems. As measurements are only able to show the present state, models can make predictions for changing situations and conditions. An important prerequisite is the previous calibration and validation of the model. With their help, a quantification of mass fluxes of water, nutrients or gas is possible for local, regional or large scale considerations. Models can be used to determine global budgets by integrating fluxes over domains.

Classifications and definitions

The movement and spreading of substances in the subsurface is depending on the aquifer characteristics, the flow conditions as well as the properties of the substances etc. The most important topics are explained in the following:

Fractured and pore water aquifers are distinguished. In fractured systems, high flow velocities and fast transport occur in the fractures, while the matrix shows a high storage capacity and slow transport processes - and vice versa. In pore water aquifers the flow and transport processes do not differ that much, compared to fractured aquifers.

In soils, the part between the water table in the subsurface and the surface of the earth or, more generally spoken, the space of the subsurface which is mainly filled with air is called the unsaturated zone. We speak about the saturated zone, if we regard the part below the water table or the space which is fully filled with water, respectively. In the unsaturated zone, the transport of substances is mainly vertically directed, while in the saturated zone the spreading is more horizontally or three dimensional.

If substances are soluable in water, we have multicomponent transport in the water phase (one phase). If the substances are in a first approach not soluble in water, as in case of oil, PCB or methane gas, they build a separate phase, and we talk about multiphase flow. If additionally it is taken into account that such phases are slightly soluble in each other, multiphase / multicomponent processes must be considered.

Ideal tracer have no influence on the flow field, and they show no adsorption, reaction or decay. If substances have an influence on the fluid properties, e.g. on the density or viscosity, then the transport processes are coupled with the flow. In coastal aquifers salt water intrusion causes density driven flow.

Generally, flow and transport processes in the subsurface are three-dimensional. But in many cases the considerations can be reduced to two and sometimes even to one dimension depending on the behaviour of the dominant processes and parameters. Different aspects must be taken into account, if simulations are carried out on a small / local or a large / regional scale.

Restrictions and contents of this paper

In this paper we will deal with transport processes which are dominant in coastal aquifer systems or in ocean margin systems, respectively. Consequently, we focus on pore water aquifers and the saturated zone. A large part is concerned with multicomponent transport in the water phase, but two-phase / three-component processes are also addressed (see Fig. 1). Reaction, adsorption and decay processes are not covered here. But substances which cause density driven flow are considered. The major part deals with vertically two-dimensional systems, taking into account effects on different scales. An extension to three dimensions is comparatively easy in most cases.

In chapter two the physics of multicomponent transport processes in single phase and two phase subsurface systems are explained. Finite-Difference, Finite-Element and Finite-Volume as well as other methods are discussed in chapter three. Three examples of the different model concepts and numerical schemes are shown in chapter four. Finally, conclusions are drawn.

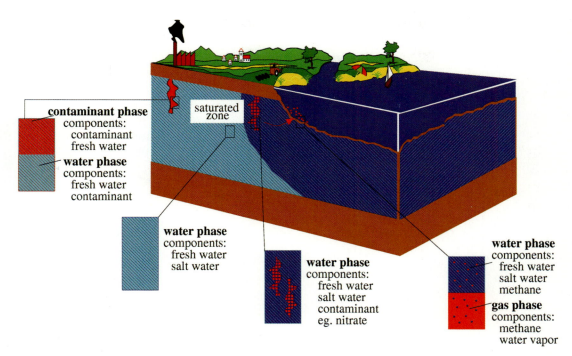

Fig. 1. Flow and transport processes in coastal aquifer.

Physical and mathematical model concept

Representative elementary volume

The physical and mathematical model concepts are based on the assumption of representative elementary volumes (REVs). If we want to simulate flow and transport processes in the subsurface on the macroscale (length scale L in Fig. 2), it is not possible to describe these processes on the microscale (length scale d), i.e. between single pores. An upscaling is required where the processes are averaged over certain volumes, the REVs (length scale l). Thus the porous subsurface structure can be treated by common means of continuum mechanics. On the one hand, these volumes must be large enough to avoid inadmissible fluctuations of the regarded properties (see top in Fig. 2). On the other hand, they must be small enough to detect spatial variations of the averaged properties (see bottom Fig. 2). The REV idea is the most common approach for porous media and subsurface structures, respectively.

Multicomponent transport processes in the water phase

The flow field is one of the important input parameters for the transport simulation. Its determination is briefly described here. Further information is given e.g. in Bear (1972). The piezometric head h [m] is defined as:

$$h = \frac{p}{\rho g} + z \qquad (2.1)$$

where p [N/m^2] stands for the pressure, ρ [kg/m^3] for the density, g [m/s^2] for the gravity constant, and z [m] for the reference geodetic head. For slow laminar flow in a porous medium Darcy's law is valid:

$$\underline{v} = -\underline{\underline{K}}_f \ grad \ h$$

$$\begin{bmatrix} v_x \\ v_z \end{bmatrix} = -\begin{bmatrix} K_{f,xx} & K_{f,xz} \\ K_{f,zx} & K_{f,zz} \end{bmatrix} \begin{bmatrix} \partial h/\partial x \\ \partial h/\partial z \end{bmatrix} \qquad (2.2)$$

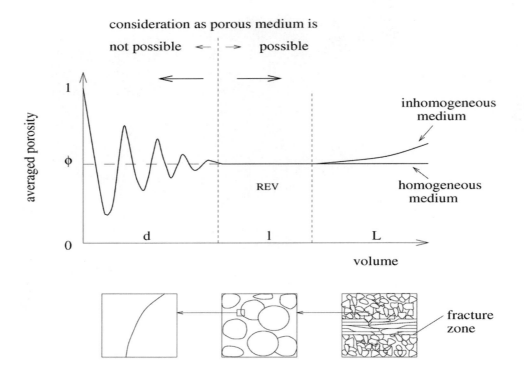

consideration as porous medium is

not possible ← → possible

Fig. 2. Definition of an REV (modified from Bear 1972 and Helmig 1997).

In this equation, K_f [m/s] is the tensor for the hydraulic conductivity, which is symmetric. For large scale simulations, the groundwater systems are often inhomogeneous and anisotropic. This is caused by different layers, geological deposition or by the shape of the material. A material consisting of long flat grains has a higher hydraulic conductivity in the length direction of the grains compared to the perpendicular direction. If the subsurface material is isotrop, this tensor simplifies:

$$k_{f,xx} = k_{f,zz} = k_f \quad , \quad k_{f,zx} = k_{f,xz} = 0$$

(2.3)

If the aquifer is homogeneous, $\underline{\underline{K}}$ is independent of the spatial location. The hydraulic conductivity tensor can be determined by experiments in the laboratory or in the field. If the main flow directions do not coincide with the global coordinate directions, the experiments are carried out in the main flow directions, and then the parameters can be transformed into the global coordinate system.

The continuity equation for mass leads to:

$$\frac{\partial(\phi\rho)}{\partial t} + \frac{\partial(\rho\, v_x)}{\partial x} + \frac{\partial(\rho\, v_z)}{\partial z} = q$$

$$\frac{\partial(\phi\rho)}{\partial t} + div(\rho\, \underline{v}) \quad = \quad q$$

(2.4)

where ϕ [-] denotes the porosity of the porous medium, which is the ratio of the void space volume which can be filled with water to the whole volume, and q [kg/m³s] represents a source term. The density and porosity are slightly depending on the pressure. The first term in equation 2.4 generally is replaced by:

$$\frac{\partial(\phi\rho)}{\partial t} = \frac{\rho\, S_0}{\rho_0\, g}\frac{\partial p}{\partial t}$$

(2.5)

S_0 [1/m] stands for the specific storage coefficient and and ρ_0 [kg/m³] for a reference density. If the density is a function of a tracer, an equation of state must be given. Oldenbourg and Pruess (1995) determine the density of saline water ρ by the salinity s [-], the fresh-water density ρ_f [kg/m³] and the density of concentrated brine ρ_s [kg/m³]:

$$\frac{1}{\rho} = \frac{1-s}{\rho_f} + \frac{s}{\rho_s}$$ (2.6)

If the density is constant, equation 2.4 reads with the help of equation 2.1 and 2.5 as follows:

$$S_0 \frac{\partial h}{\partial t} + \frac{\partial v_x}{\partial x} + \frac{\partial v_z}{\partial z} = q$$

$$S_0 \frac{\partial h}{\partial t} + div\ \underline{v} = q$$ (2.7)

If in addition we have stationary conditions, equation 2.7 is simplified:

$$\frac{\partial v_x}{\partial x} + \frac{\partial v_z}{\partial z} = q$$

$$div\ \underline{v} = q$$ (2.8)

Using Darcy's law (eq. 2.2), equation 2.8 or 2.7 leads to one equation for the unknown piezometric head which is further treated and solved with the techniques described later. In an additional step, the Darcy or filter velocity is determined applying a multiplication of the hydraulic conductivity tensor with the gradient of the piezometric head. This procedure causes that the accuracy of the flow vector is one order of magnitude worse when compared to the piezometric head. The average flow velocity in the void space u [m/s] is higher than the Darcy velocity, and it is determined by:

$$\underline{u} = \underline{v} / \phi$$ (2.9)

In the following transport processes are introduced. The main transport mechanisms in porous media are advection / convection, diffusion and dispersion. Other processes such as reaction, adsorption and decay are not considered here. Advection or convection describes the movement of a tracer with the flow field in horizontal or vertical direction without changing the shape of the

concentration isoareas (see Fig. 3). Diffusion processes are a result of Brown's molecular movement which leads to a compensation of concentration differences, and thus causes transport processes in the direction of lower concentrations. Diffusion is a pure physical processes - in contrast to dispersion. Therefore, a diffusive spreading of a substance is independent of the direction (see Fig. 3).

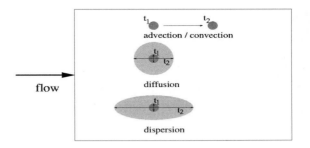

Fig. 3. Advection / convection, diffusion and dispersion processes.

Dispersion represents all transport effects which are caused by inhomogeneities of the flow field below the REV scale.

On the micro scale, dispersion is a result of the velocity profile within a pore, the variable pore sizes as well as the flow processes around the grains (see Fig. 4, left). For bigger REVs, fluctuating velocities due to inhomogeneities of the aquifer lead to macro dispersion (see Fig. 4, middle and right). It is obvious that dispersion is scale dependent.

Advective or convective flux of a tracer with a concentration c [-,mg/l,...] in a porous medium is given by:

$$\underline{F}_{adv} = \phi_e\ c\ \underline{u}$$ (2.10)

In this equation ϕ_e stands for the effective porosity which only takes into account mobile water. When compared to the porosity, the effective porosity is always smaller since it does not consider water in dead-end-pores or bound water at the grain surfaces. For small scale considerations, there are some empirical relationships for the effective porosity, for larger scales this parameter is unknown a priori, and must be estimated (see Kinzelbach 1992).

Diffusive flux of a tracer in a porous medium is determined by the first Fickian law with molecular diffusion coefficient D_{mol} [m²/s]:

$$\underline{F}_{dif} = -\phi_e \, D_{mol} \, grad \, c = -\phi_e \, D_{mol} \begin{bmatrix} \partial c/\partial x \\ \partial c/\partial z \end{bmatrix}$$
$$(2.11)$$

The hampering of the spreading by the grains is taken into account by the multiplication with the effecitve porosity. Further approaches are given in Bear and Bachmat (1990).

For the description of the dispersive transport, the velocity vector and the concentration are splitted in space in an REV average (expressed by an overbar) and a fluctuation (expressed by apostrophe):

$$\underline{u} = \underline{\bar{u}} + \underline{u}' \quad , \quad c = \bar{c} + c' \qquad (2.12)$$

Now we return to the advective and diffusive transport and integrate the splitting of the state variables:

$$\underline{F} = \phi_e \left[c\underline{u} - D_{mol} \, grad \, c \right] =$$
$$\phi_e \left[(\bar{c} + c')(\underline{\bar{u}} + \underline{u}') - D_{mol} \, grad \, (\bar{c} + c') \right]$$
$$= \phi_e \begin{bmatrix} \bar{c}\,\underline{\bar{u}} + \bar{c}\,\underline{u}' + c'\,\underline{\bar{u}} + c'\,\underline{u}' - \\ D_{mol} \, grad \, \bar{c} - D_{mol} \, grad \, c' \end{bmatrix}$$
$$(2.13)$$

If a volume averaging is performed, the product terms, which contain a fluctuation once, disappear, and the last equation simplifies to:

$$\underline{F} = \phi_e \left[\bar{c}\,\underline{\bar{u}} - D_{mol} \, grad \, \bar{c} + \overline{c'\underline{u}'} \right] \qquad (2.14)$$

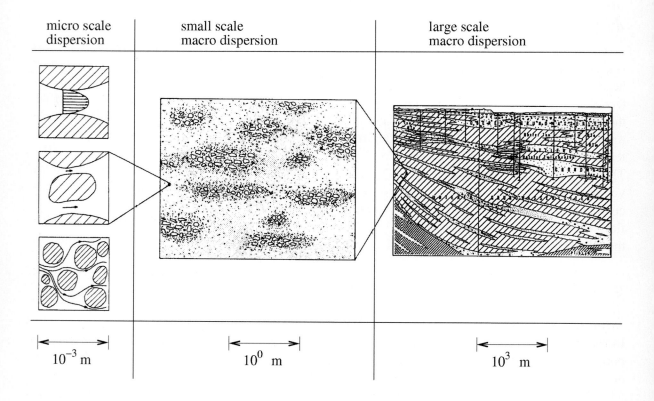

| micro scale dispersion | small scale macro dispersion | large scale macro dispersion |

10^{-3} m 10^0 m 10^3 m

Fig. 4. Reasons for the variability of the flow field on different spatial scales (modified from Kinzelbach 1992).

The product of the velocity fluctuation and the tracer fluctuation does not disappear. This is the dispersion flux term. Similar to the molecular diffusion, the dispersion is approached by the first Fickian law with the dispersion tensor $\underline{\underline{D}}_{mech}$ [m²/s]:

$$\overline{c'\underline{u}'} = -\underline{\underline{D}}_{mech}\ grad\ \overline{c} \quad \Leftrightarrow \quad \begin{bmatrix} \overline{c'u'_x} \\ \overline{c'u'_z} \end{bmatrix} =$$

$$-\begin{bmatrix} D_{xx} & D_{xz} \\ D_{zx} & D_{zz} \end{bmatrix}\begin{bmatrix} \partial\overline{c}/\partial x \\ \partial\overline{c}/\partial z \end{bmatrix}$$

$$(2.15)$$

It is obvious that the dispersion tensor is not a physical property of the fluid or the aquifer material. Therefore, it is called mechanical dispersion. In the following, the overbars are skipped. In a certain sense dispersion resembles turbulence with the difference that dispersion is the result of a spatial averaging, while a very common approach for turbulence is determined by temporal averaging. If the molecular diffusion coefficient is multiplied with the unit tensor, it is transformed to the molecular diffusion tensor $(D_{mol} \rightarrow \underline{\underline{D}}_{mol})$

The sum of the molecular diffusion and mechanical dispersion tensor is also called hydrodynamic dispersion tensor:

$$\underline{\underline{D}}_{hyd} = \underline{\underline{D}}_{mol} + \underline{\underline{D}}_{mech} \qquad (2.16)$$

If advection, diffusion and dispersion are added up, the complete flux vector is determined by:

$$\underline{F} = \phi_e\left[c\underline{u} - \left(\underline{\underline{D}}_{mol} + \underline{\underline{D}}_{mech}\right) grad\ c \right]$$

$$(2.17)$$

It should be mentioned that the (REV) scales for the flow and transport computations must match as well as horizontal and vertical averaging length scales, and that the way which is passed by a tracer plume must be large compared to the REV length scale.

The continuity equation for a tracer mass in a porous medium, applied to an Eulerian control volume, states that the sum of tracer mass inflow and outflow and sink or source terms is in equilibrium with the temporal change of the tracer mass in the regarded control volume.

$$-\frac{\partial F_x}{\partial x}\ dx\ dz - \frac{\partial F_z}{\partial z}\ dz\ dx + \tilde{r}\ dx\ dz =$$

$$\frac{\partial(c\phi_e)}{\partial t}\ dx\ dz \Leftrightarrow \frac{\partial c}{\partial t} + \underline{u}\ grad\ c -$$

$$div\left[\left(\underline{\underline{D}}_{mol} + \underline{\underline{D}}_{mech}\right) grad\ c\right] = r$$

$$(2.18)$$

In this equation $\tilde{r}/\phi_e = r$ [m³/s/m³, 1/s] stands for sink or source terms. If reaction, adsorption and decay processes are considered, they are modelled with this term.

Taking into account equation 2.16, the last equation is transformed to:

$$\frac{\partial c}{\partial t} + \begin{bmatrix} u_x \\ u_z \end{bmatrix}\begin{bmatrix} \partial c/\partial x \\ \partial c/\partial z \end{bmatrix} - \begin{bmatrix} \partial/\partial x \\ \partial/\partial z \end{bmatrix}$$

$$\begin{bmatrix} D_{xx} & D_{xz} \\ D_{zx} & D_{zz} \end{bmatrix}\begin{bmatrix} \partial c/\partial x \\ \partial c/\partial z \end{bmatrix} = r$$

$$(2.19)$$

The dispersion tensor given in equation 2.15 is anisotropic, even in an isotrop medium. It has diagonal shape, if one coordinate direction coincides with the flow direction.

$$\underline{\underline{D}}_{mech} = \begin{bmatrix} D_L & 0 \\ 0 & D_T \end{bmatrix} \qquad (2.20)$$

Generally, the dispersion in flow direction expressed by the longitudinal dispersion coefficient D_L [m²/s] is about one order of magnitude larger compared to the dispersion in the orthogonal direction expressed by the transversal dispersion coefficient D_T [m²/s].

The mechanical dispersion coefficients contain information about the aquifer and the flow field:

$$D_L = \alpha_L\left|\underline{u}\right| \quad , \quad D_T = \alpha_T\left|\underline{u}\right| \qquad (2.21)$$

In this equation α_L [m] and α_T [m] stand for the longitudinal and transversal dispersion length. The most common description for the dispersion tensor is given by Scheidegger:

$$D_{xx} = \alpha_L \frac{u_x^2}{|\underline{u}|} + \alpha_T \frac{u_z^2}{|\underline{u}|} + D_{mol}$$

$$D_{xz} = D_{zx} = (\alpha_L - \alpha_T) \frac{u_x u_z}{|\underline{u}|}$$

$$D_{zz} = \alpha_T \frac{u_x^2}{|\underline{u}|} + \alpha_L \frac{u_z^2}{|\underline{u}|} + D_{mol}$$

(2.22)

As already mentioned before, dispersion is scale dependent. In the laboratory, $0.0001 < \alpha_L < 0.01$m was measured for homogeneous sands and $0.07 < \alpha_L < 0.7$m was found for natural subsurface material, e.g. gravel. On the field scale, the dispersivities are about 4 up to 5 scales larger. This is a consequence of the macro disperison which results from the inhomogeneities of the aquifer and does not occur in the laboratory. Macro dispersion is increasing with the length of the transport way (see Fig. 5). This is caused by the influence of bigger inhomogeneities, but this effect is limited, as soon as the inhomogeneities are represented by the computation of the flow field. On the field scale, the molecular diffusion is often negligible compared to the mechanical disperion, i.e. $D_{mol} << D_L$ and $D_{mol} << D_T$. For practical applications on scales larger than in the laboratory, the dispersion coefficients are calibration parameters, and thus contain all other unknowns, i.e. lacking information about the geological structures or effects not included in the model concept. Further information is given in Kinzelbach (1992).

Multicomponent transport processes in the water and gas phase

In this section two-phase / multicomponent flow and transport processes in subsurface systems are briefly introduced. The scope of this paper is limited here to the two phases water and gas (subscripts w, g), and three components fresh-water, dissolved salt and methane (superscripts f, s, m). The continuity equation for the masses of fresh-water, dissolved salt and methane are given for each component a by:

$$\frac{\partial(\phi \rho_\alpha S_\alpha X_\alpha^\kappa)}{\partial t} + \nabla \left\{ \rho_\alpha \underline{v}_\alpha X_\alpha^\kappa - \phi D_\alpha^\kappa \nabla(\rho_\alpha X_\alpha^\kappa) \right\}$$
$$= q_\alpha^\kappa, \quad \alpha = w, g$$

(2.23)

In this equation ϕ denotes porosity [-], ρ density [kg/m³], S [-] saturation, X [-] mole fraction (corresponding to the tracer concentration c), t time, $\underline{v}$ [m/s] Darcy velocity vector, $\underline{\underline{D}}$ [m²/s] hydrodynamic dispersion tensor (see above) and q [kg/(m³s)] source or sink term. The velocity vector of each phase a is determined by the generalized Darcy law:

$$\underline{v}_a = -\frac{k_{r\alpha}}{\mu_\alpha} \underline{\underline{K}} (\nabla p_\alpha - \rho_\alpha g)$$

(2.24)

In the last equation k_r [-] stands for the relative permeability, μ [kg/(ms)] for the dynamic viscosity, $\underline{\underline{K}}$ [m²] for the absolute permeability tensor, p [Pa] for the pressure and g [m/s²] for the gravity constant. The relation between the permeability and hydraulic conductivity (see eq. 2.2, 2.3) is given by:

$$\underline{\underline{K}}_f = \frac{k_r}{\mu} \underline{\underline{K}}$$

(2.25)

The mass fractions add up to one in each phase:

$$X_\alpha^f + X_\alpha^s + X_\alpha^m = 1; \quad \alpha = w, g$$ (2.26)

In the porous medium the void space is completely filled with the two phases:

$$S_w + S_g = 1$$ (2.27)

The pressure difference of the two phases is a function of the capillary pressure p_c:

$$p_g - p_w = p_c$$ (2.28)

The mole fraction of water vapor in the gas phase is obtained by:

$$X_g^f = \frac{P_g^f}{P_g}$$

(2.29)

P_g^f is the partial pressure of the water vapor in the gas phase, and it equals to the saturation pressure P_{sat}^f.

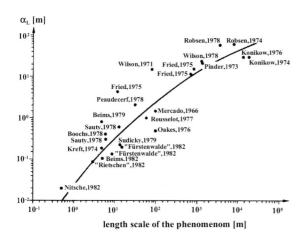

Fig. 5. Scale dependency of the dispersion (modified from Kinzelbach 1992).

The mole fraction of methane in the water phase is determined by Henry's law:

$$X_w^m = \frac{P_g^m}{H_w^m}$$ (2.30)

P_g^m is the partial pressure of methane in the gas phase and H_w^m the Henry coefficient. For both components in the gas phase the validity of the ideal gas law is assumed.

Dissolved salt is restricted to the water phase. Dalton's laws gives a relation between the total pressure and the partial pressures:

$$P_g = P_g^m + P_g^f$$ (2.31)

In addition, equations of state for the densities, e.g. the density of water as a function of dissolved salt, the viscosities and other variables must be prescribed. There is a choice between different constitutive relationships for the relative permeability / saturation and the capillary pressure / saturation. For a detailed description, the authors refer to Helmig (1997) and Class (2000).

Numerical methods for solving the transport equation

First, it should be mentioned that there do exist several analytical solutions for specified system geometries, flow conditions, physical parameters and boundary conditions. If a problem fits these constraints, such analytical solutions can or should

be applied. In a number of practical engineering cases, only rough estimations for the transport of a contaminant are looked for. Since the average movement of a contaminant is determined by the advective transport processes, diffusion and dispersion processes can be neglected in such cases. Thus the solution of the transport equation simplifies. Analytical and dispersion free solutions for the transport equation are discussed in Kinzelbach (1992).

A more general approach for solving the transport equation is given by numerical methods, also called discretization techniques, which are not limited to the constraints mentioned before. In the mathematical sense, the mixed advective / dispersive character of the transport equation leads to a mixed hyperbolic / parabolic type of partial differential equation.

In many cases the advective or hyperbolic part of the transport equation causes numerical difficulties and requires special stabilization techniques. This is illustrated for a sharp front problem in figure 6. If advection is dominant, the numerical solutions are not monotonous and have the tendency to oscillations, over and under shooting or to numerical (or artificial) diffusion (or dispersion). In the following the most common discretization techniques for solving transport equations together with several stabilization techniques are introduced.

For detailed information about numerical modelling of transport processes in the subsurface the authors refer to Bear and Bachmat (1990), Kinzelbach (1992), Finlayson (1992), Helmig (1997) and Malcherek (2000).

Finite-Difference-Methods (FDM)

The Finite-Difference-Method (FDM) is one of the oldest methods to solve partial differential equations. The computational domain is discretized by rectangular or quadrilateral cells (see Fig. 7). Often, the cell lengths dx and dz are constant or even $dx = dz$, but local refinement is also possible. Nodes, where the unknown variables are defined, are placed in the centers of the cells or in the intersection points of cell boundaries (see Fig. 7). From the geometrical point of view it is obvious that

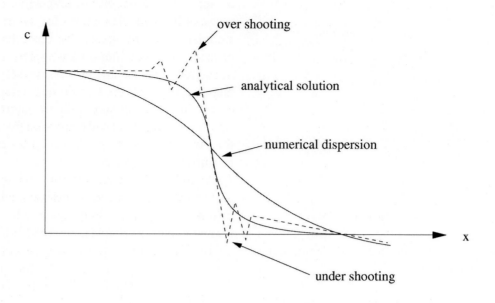

Fig. 6. Principal behaviour of oscillations and numerical dispersion.

complex boundaries or complex inner structures can only be reproduced in a very simplified way by step functions.

The basic idea of the FDM is to substitute the differential quotients by difference quotients. With the help of a Taylor series the first derivative of a function, e.g. the concentration c, after a variable, e.g. the x-direction, can be determined by three different difference quotients which are called forward, backward and central difference quotient (FD, BD, CD).

$$FD: \quad \frac{\partial c}{\partial x} = \frac{c_{i+1} - c_i}{\Delta x}$$

$$CD: \quad \frac{\partial c}{\partial x} = \frac{c_{i+1} - c_{i-1}}{2\Delta x}$$

$$(3.1)$$

$$BD: \quad \frac{\partial c}{\partial x} = \frac{c_i - c_{i-1}}{\Delta x}$$

The accuracy is of first order for the forward and backward differences quotient and of second oder for the central one (see Malcherek 2000).

The second derivative of the concentration after the x-direction is given by:

$$\frac{\partial^2 c}{\partial x^2} = \frac{c_{i+1} - 2c_i + c_{i-1}}{(\Delta x)^2} \qquad (3.2)$$

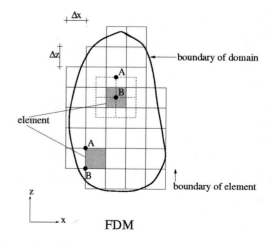

Fig. 7. Space discretization for the FDM.

For reasons of simplicity we consider only one-dimensional problems without sink or source terms, an extension to 2D is easy. We recall the transport equation (see eq. 2.18) and reduce it to 1D:

$$\frac{\partial c}{\partial t} + u \frac{\partial c}{\partial x} - D_{hyd} \frac{\partial^2 c}{\partial x^2} = r \qquad (3.3)$$

In the following we focus on two special cases of the transport equation and give some additional remarks and recommendations later. Detailed information is given in Malcherek (2000).

FDM - Stationary transport equation

If we apply e.g. the forward difference quotient in time and the difference quotient for the second derivative in space, we obtain in the stationary case:

$$u \frac{c_{i+1} - c_i}{\Delta x} - D_{hyd} \frac{c_{i+1} - 2c_i + c_{i-1}}{(\Delta x)^2} = 0 \qquad (3.4)$$

The three different difference quotients for the first derivative can be generalized by defining an upwind parameter α:

$$u \left[\alpha \frac{c_i - c_{i-1}}{\Delta x} + (1-\alpha) \frac{c_{i+1} - c_i}{\Delta x} \right] = 0 \qquad (3.5)$$

For $\alpha = 0$, 0.5, 1 we get the forward, central, backward difference quotient.

We introduce the Peclet number which is the ratio of advection multiplied with the discretization length to diffusion / dispersion:

$$Pe = \frac{|u| \Delta x}{D_{hyd}} \qquad (3.6)$$

The Peclet number is high, if advection dominates. Therefore, it is an important number to indicate the tendency to instabilities.
Using equation 3.6 equation 3.5 can be rewritten:

$$u \frac{c_{i+1} - c_{i-1}}{\Delta x} - D_{hyd} \left[1 + Pe \left(\alpha - \frac{1}{2} \right) \right]$$
$$\frac{c_{i+1} - 2c_i + c_{i-1}}{(\Delta x)^2} = 0 \qquad (3.7)$$

An artifical dispersion term Pe occurs for $\alpha \neq 0.5$ which leads to a smearing and damping of the front. If the forward or backward difference quotient ($\alpha = 0$ or 1) is applied, the artifical dispersion does not vanish. For $Pe > 2$ the artificial dispersion is even higher than the physical one. If the central difference quotient is applied ($\alpha = 0.5$), the artificial dispersion term vanishes independent of the Peclet number. In this case, it can be shown that the solution shows oscillations for $Pe > 2$, but it is smooth for $Pe \leq 2$. Overall, this simple example shows the dilemma of numerical simulation of advection / diffusion problems.

FDM - Advective transport equation

In this subsection we introduce explicit and implicit methods. With an explicit method, the solution for the new time step $n+1$ is completely determined by concentrations of the actual time step n with a 'simple formula' (see Fig. 8, left). Therefore, the computational effort for one time step is very low, but generally, there is a limitation of the time step size. When an implicit method is applied, the solution for the new time step is determined from concentrations of the actual and the new time step. As the concentrations on the new time level are depending on each other, a system of equation occurs (see Fig. 8, right). Therefore, the computational effort for one time step is high, but theoretically, there is no time step limitation.

The upstream method uses a forward difference quotient in time and the upstream node in space:

$$\frac{c_i^{n+1} - c_i^n}{\Delta t} + u \frac{c_i^n - c_{i-1}^n}{\Delta x} = 0 \quad for\ u \geq 0$$
$$(3.8)$$

$$\frac{c_i^{n+1} - c_i^n}{\Delta t} + u \frac{c_{i+1}^n - c_i^n}{\Delta x} = 0 \quad for\ u < 0$$
$$(3.9)$$

The accuracy in space and time is only of first order. Therefore, this method is characterized by high numerical diffusion.

In the diffusive (Lax) method the accuracy in space is increased to second order by using a central difference quotient. For reasons of stability, the time derivative at a node i must be determined by an average of $i-1$ and $i+1$. Thus we get:

$$\frac{2c_i^{n+1} - c_{i-1}^n - c_{i+1}^n}{2\Delta t} + u\frac{c_{i+1}^n - c_{i-1}^n}{2\Delta x} = 0$$

(3.10)

As already indicated by its name, this method shows much numerical diffusion.

When the Leap-Frog method is applied, the accuracy in space and time is of second order by using central differences in space and time. It must be mentioned, that the time levels $n-1$ and n are required for the solution on the time level $n+1$:

$$\frac{c_i^{n+1} - c_i^{n-1}}{2\Delta t} + u\frac{c_{i+1}^n - c_{i-1}^n}{2\Delta x} = 0 \qquad (3.11)$$

This method shows the tendency to numerical oscillations.

Finally, the Lax-Wendroff method which also has a second order accuracy in space and time is explained. It is a combination of the diffusive method with the Leap-Frog method.

First, half a time step is carried out with the diffusive method to the intermediate points (i-0.5,n+0.5) and (i+0.5,n+0.5). Second, the Leap-Frog method starts from the intermediate points to determine the solution in (i,n).

The Lax-Wendroff method sometimes has some slight oscillations. Nevertheless, it is the best of the introduced methods. It must be mentioned that the computational effort is about the double when compared to the Leap-Frog method. This method

also belongs to the so-called predictor-corrector methods which determine the solution in two steps. In a first predictor step a stable initial solution is computed. In a second corrector step the order of accuracy of the initial solution is improved

By a stability analysis it can be shown that explicit methods must obey the Courant number which is the ratio of the advective velocity to the 'mesh velocity':

$$C_r = \frac{|u|}{\Delta x / \Delta t} \leq 1 \qquad (3.12)$$

In the following implict methods are introduced. The Crank-Nicholson method uses a forward difference quotient in time and central differences in space which can be varied between the time levels n and $n+1$ by the Crank-Nicholson factor Q:

$$\frac{c_i^{n+1} - c_i^n}{\Delta t} + \Theta u\frac{c_{i+1}^{n+1} - c_{i-1}^{n+1}}{2\Delta x} +$$

$$(1-\Theta) u\frac{c_{i+1}^n - c_{i-1}^n}{2\Delta x} = 0$$

(3.13)

This scheme is stable for $\Theta \geq 0.5$. For $\Theta = 0.5$ the accuracy in space and time is of second order, otherwise of first order in time and second order in space. If $\Theta = 1$, the scheme is fully implicit.

The Preissmann scheme applies forward differences for the time derivative by averaging the values of two neighbor nodes, and forward differences for the spatial derivative by weighting these terms between the time levels $n+1$ and n with the Crank-Nicholson factor.

$$\frac{c_{i+1}^{n+1} - c_{i+1}^n + c_i^{n+1} - c_i^n}{2\Delta t} + \Theta u\frac{c_{i+1}^{n+1} - c_{i-1}^{n+1}}{\Delta x} +$$

$$(1-\Theta) u\frac{c_{i+1}^n - c_{i-1}^n}{\Delta x} = 0$$

(3.14)

The accuracy of this method is of second order in space and time.

FDM - Remarks and recommendations

All methods described up to now require initial and boundary conditions for the solution.

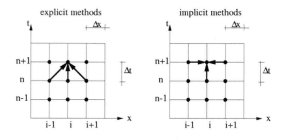

Fig. 8. Space and time discretization for the FDM.

Initial conditions must be known or estimated reasonably. Boundary conditions of Dirichlet type are for the advective flux and prescribe a concentration. Boundary conditions of Neumann type are for the dispersive flux and prescribe a concentration gradient. At closed boundaries, advective and dispersive fluxes are set to zero. At open inflow boundaries a concentration must be given. At open outflow boundaries it is possible to set the dispersive flux to zero. For further information the authors refer to Kinzelbach (1992).

The developer or user of a numerical method should check the convergence of the algorithm. For decreasing Δx and Δt, the solution must converge.

If we consider the instationary transport equation with advection and dispersion, another stability condition, the Neumann criterion, must be fullfilled:

$$\Delta t \leq \frac{\Delta x^2}{2D_{hyd}} \qquad (3.15)$$

In two-dimensional simulations, the Courant as well as the Neumann condition must be satisfied. Further details concerning stability are given in Kinzelbach (1992). For the second spatial derivatives higher order approximations (5 or 9 point stars) can be chosen (see Helmig 1997). In 2D computations numerical diffusion also occurs orthogonal to the flow direction which is called cross diffusion.

The FDM does not necessarily guarantee the mass conservation which is one of the big disadvantages of this method. The differential quotient in the basic equations are approximated by difference quotients, but the method does not imply conservativity.

Finally, some recommendations: The cell length should be chosen in a way that $P_e \leq 2$ is fullfilled in both spatial directions. The Courant number should be around 1 for both directions, even for implicit methods. For explicit methods the Neumann condition must be taken into account. The Lax-Wendroff or the Preissmann method should be applied. Two major disadvantages must be kept in mind: Mass conservation is not necessarily guaranteed. Complex boundaries and inner structures can only be discretized very roughly.

Finite-Element-Method (FEM)

The Finite-Element-Method (FEM) has been widely spread in engineering science for several decades. The computational domain is subdivided in many little finite elements, in 2D triangles or quadrilaterals, with nodes at the element edges. The unknown variables, here the concentrations, are defined at the nodes. The course of the unknown over the elements are determined by interpolation functions. For each element, the differential equation, here the transport equation, is solved by minimising an integral formulation. Then the single solutions are put together, in most cases to a system of equations. Finally, the solution for the whole system is determined taking into account initial and / or boundary conditions. Due to the unstructured meshes, it is very suitable for complex boundaries and complex inner structures (see Fig. 9). We recall the transport equation (see eq. 2.18):

$$\frac{\partial c}{\partial t} + \underline{u}\ grad\ c - div\left(\underline{\underline{D}}_{hyd}\ grad\ c\right) = r \qquad (3.16)$$

In the following we focus on the dispersive part of the transport equation. Then stabilization techniques are discussed for the full equation. Finally, we give some additional remarks and recommendations. There is a lot of literature to the FEM for advection / diffusion problems. The authors refer e.g. to Johnson (1992).

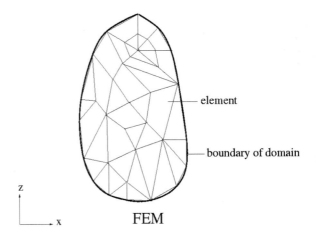

Fig. 9. Space discretization for the FEM.

FEM - Dispersive transport equation

The FEM is quite suitable for parabolic problems, e.g. the dispersive transport equation:

$$\frac{\partial c}{\partial t} - div \left(\underline{\underline{D}}_{hyd} \; grad \; c\right) = 0 \qquad (3.17)$$

First, we introduce the semidiscrete method which says that the space is discretized with FEM and the time with the FDM. This is the most common way for time dependent problems that are treated with FEM. For the time derivative a forward difference quotient is applied:

$$\frac{\partial c}{\partial t} = \frac{c^{n+1} - c^{n}}{\Delta t} \qquad (3.18)$$

In space the isoparametric concept is used. This means that the geometry and the unknown variables are desribed by the same interpolation functions. Here we focus on linear functions which lead to second order accuracy in space. The unknown variables (c) are approximated ($\tilde{c}$) by summing up the products of the interpolation functions N and the nodal values c, e.g. for the unknown c^{n}:

$$c^{n} \approx \tilde{c}^{n} = \sum_{i=1}^{4} N_i \; \hat{c}_i^{n} = N_1 \; \hat{c}_1^{n} + N_2 \; \hat{c}_2^{n} + N_3 \; \hat{c}_3^{n} + N_4 \; \hat{c}_4^{n} \qquad (3.19)$$

The interpolation functions are shown for a quadrilateral. The interpolation or shape function N_i equals 1 at node i and declines to 0 at all other nodes of the element (see Fig. 10).

If we insert this approximation (3.19) in the dispersive equation 3.17 no longer equals exactly 0, but an error or residual occurs.

$$\frac{\partial \tilde{c}}{\partial t} - div \left(\underline{\underline{D}}_{hyd} \; grad \; \tilde{c}\right) = \varepsilon \qquad (3.20)$$

The method of weighted residuals multiplies this residual with a weighting function W, integrates this over the computational domain Ω and forces the result to 0:

$$\int_{\Omega} W_j \; \varepsilon \; d\Omega = 0 \qquad (3.21)$$

Thus, the error vanishes in the average over the whole computational domain. When the Standard-Galerkin method is applied, the weighting function equals the interpolation function:

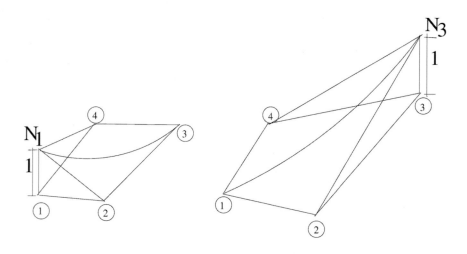

Fig. 10. Linear shape functions for quadrilaterals.

$$W_j = N_j \qquad (3.22)$$

We insert equations 3.20 and 3.22 into 3.21:

$$\int_\Omega N_j \left[\frac{\partial \widetilde{c}}{\partial t} - div \left(\underline{\underline{D}}_{hyd} \ grad \ \widetilde{c} \right) \right] d\Omega = 0$$

$$(3.23)$$

The Green-Gauß theorem is applied:

$$\int_\Omega N_j \frac{\partial \widetilde{c}}{\partial t} \ d\Omega - \int_\Gamma N_j \ \underline{\underline{D}}_{hyd} \ grad \ \widetilde{c} \ \underline{n} \ d\Gamma +$$

$$\int_\Omega grad \ N_j \ \underline{\underline{D}}_{hyd} \ grad \ \widetilde{c} \ d\Omega = 0$$

$$(3.24)$$

Γ denotes the boundary and $\underline{n}$ the normal vector. The integral along the boundaries Γ is set to zero.

The concentration is determined by the Crank-Nicholson factor Θ between the time levels n and $n+1$. For $\Theta = 0.5$ the accuracy in time is of second order. Otherwise it is of first order, e.g. for the explicit method ($\Theta = 0$) or the fully implicit method ($\Theta = 1$).

We include equation 3.18 and 3.19 and obtain:

$$\frac{\hat{c}_i^{n+1} - \hat{c}_i^n}{\Delta t} \ \underbrace{\int_\Omega N_i N_j d\Omega}_{} \ + \left[\Theta \hat{c}_i^{n+1} + (1-\Theta)\hat{c}_i^n \right]$$

mass matrix M_{ij}

$$\underbrace{\int_\Omega grad \ N_i \underline{\underline{D}}_{hyd \ i,j} \ grad \ N_j \ d\Omega}_{} = 0$$

dispersion matrix D_{ij} $\qquad (3.25)$

The derivatives of the interpolation functions after the cartesian coordinates must be computed. As they are functions of the natural coordinates, a transformation relation is required (see Fig.11) which is given by the Jacobi matrix:

$$\begin{bmatrix} \partial c/\partial r \\ \partial c/\partial s \end{bmatrix} = \underbrace{\begin{bmatrix} \partial x/\partial r & \partial z/\partial r \\ \partial x/\partial s & \partial z/\partial s \end{bmatrix}}_{Jacobi \ matrix} \begin{bmatrix} \partial c/\partial x \\ \partial c/\partial z \end{bmatrix} \Leftrightarrow$$

$$\begin{bmatrix} \partial c/\partial x \\ \partial c/\partial z \end{bmatrix} = \underline{\underline{J}}^{-1} \begin{bmatrix} \partial c/\partial r \\ \partial c/\partial s \end{bmatrix}$$

$$x = \sum_{i=1}^4 N_i \ x_i \quad , \quad z = \sum_{i=1}^4 N_i \ z_i \quad ,$$

$$d\Omega = dx \ dz = det \ \underline{\underline{J}} \ dr \ ds \qquad (3.26)$$

We insert the Jacobi matrix in equation 3.25 and get:

$$\int_\Omega N_i N_j \ det \ \underline{\underline{J}} \ dr \ ds \ \frac{\hat{c}_i^{n+1} - \hat{c}_i^n}{\Delta t}$$

$$+ \int \frac{grad \ N_i \ \underline{\underline{J}}^{-1} \ D_{hyd \ i,j} \ \underline{\underline{J}}^{-1} \ grad \ N_j \ det}{\underline{\underline{J}} \ dr \ ds} \left[\Theta \ \hat{c}_i^{n+1} + (1-\Theta) \ \hat{c}_i^n \right] = 0$$

$$(3.27)$$

As the integrands of the mass and diffusion matrices are no rational functions for any quadrilaterals, the integration is numerically carried out by Gauß-Point integration (see Johnson 1992). The procedure to compute the element matrices may seem to be very time consuming. But one has to take into account that many parts of the computation must be carried out only once for all quadrilateral elements. Only the parts which are affected by the Jacobi matrix must be determined for each element, and of course the numerical integration.

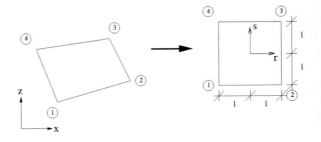

Fig. 11. Transformation from cartesian to natural coordinates.

The procedure described for one element must be carried out for every element. According to the neighborhood relationships the element matrices are put together to the global system matrix and the global right hand side terms.

$$\left[M_{ij} + \Theta D_{ij} \right] \hat{c}_i^{n+1} = \left[M_{ij} + (\Theta - 1)D_{ij} \right] \hat{c}_i^n \qquad (3.28)$$

FEM - Advective / dispersive transport equation

Principally, the advective part can be modelled in the same way as shown before with the Standard-Galerkin method, i.e. the weighting function equals the interpolation function (see eq. 3.22):

$$W_j = N_j$$

The solutions are oscillatory for high Peclet or Courant numbers. Therefore, the practical use is very limited.

An improvement is given by Petrov-Galerkin methods. The idea is to take more information from the upstream flow direction by chosing modified weighting functions:

$$W_j = N_j + \alpha T_j \qquad (3.29)$$

In the classical Petrov-Galerkin method a polynom T_j of higher degree is added, while the Streamline-Upwind-Petrov-Galerkin method (SUPG method) adds a polynom T_j of lower degree (see Fig. 12):

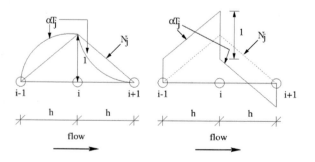

Fig. 12. Classical and Streamline-Upwind-Petrov-Galerkin method.

The modified weighting functions behave like artifical diffusion in the flow direction (not orthog-onal to the flow direction, thus no cross diffusion). Therefore, the oscillations disappear, but the front is damped, and a more or less slight over- or under shooting occurs. The upwind parameter α is free a priori. It must be chosen by the user or can be estimated by:

$$\alpha \approx \sqrt{\frac{Pe^2}{36 + Pe^2}} \qquad (3.30)$$

Generally, the SUPG method is superior to the classical one.

Another alternative are Taylor-Galerkin methods which have a third order accuracy in time by solving a defect corrected differential equation shown here for 1D:

$$\frac{\partial \tilde{c}}{\partial t} + u \frac{\partial \tilde{c}}{\partial x} - \left(D_{hyd} - \frac{\Delta x^2}{12} \frac{u^2}{D_{hyd}} \right) \frac{\partial^2 \tilde{c}}{\partial x^2} = 0 \qquad (3.31)$$

The defect term (right term in the bracket) contains the biggest error and is determined by a Taylor series. It behaves like artifical diffusion in the flow direction. Up to Pe=20 very good results are obtained. A big disadvantage results of the circumstance that the third order accuracy in time is only obtained on othogonal grids.

The most promising stabilization techniques are those which maintain the monotonicity of the solution, such as Slope limiter, Flux limiter or Flux-corrected transport methods (see Finlayson 1992).

FEM - Remarks and recommendations

The remarks concerning boundary conditions, convergence and stability as described for the FDM are also valid in context of the FEM.

The FEM guarantees global mass conservation which is a consequence of the formulation, because it states that the residual should vanish in the whole computational domain. In this respect the FEM is superior to the FDM. The authors recommend that the element length should be chosen in such a way that the Peclet number exceeds 2, but is not much greater than 10. The Courant number should be around 1, even for implicit methods. For explicit methods the Neumann condition must be taken into account. Upwind or monotonicity preserving methods should be applied for stabilisation. When compared to the FDM, the FEM has two important advantages: Mass conservation is guaranteed globally, and the unstructured meshes enable an excellent approximation of complex boundaries and inner structures.

Finite-Volume-Method (FVM)

The Finite-Volume-Method (FVM) has been established in engineering sciences for a few decades. Traditionally, it has its origin in the field of Navier-Stokes-Equations and technical flows and it is often applied to structured grids. If it is taken for rectangular grids, this method is also called Integral-Finite-Difference-Method (IFDM). Lately, the FVM has been used for unstructured meshes, and in this case it is comparable to the FEM (see Fig. 9). For every node a control volume is defined, and the differential equations are integrated over these control volumes. It is very common to define the control volume by the polygon built of the verticals on the mid points of the element edges (see Fig.13). Taking into account initial and / or boundary conditions, a solution is obtained. The unstructured meshes enable the FVM to approximate complex boundaries and complex inner structures very well.

In the following we discuss the treatment of the different terms together with some stabilization techniques for the advection. We explain the method for quadratic elements for reason of a better

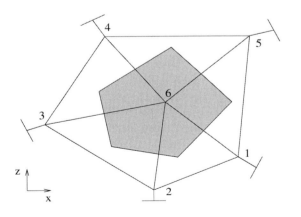

Fig. 13. Control volume for the FVM.

understanding, and we add some remarks concerning unstructured meshes. This subsection gives some general remarks and recommendations, too. For further reading the authors refer to Kinzelbach (1992) and Malcherek (2000).

FVM - Advective / dispersive transport equation

Again, we recall the transport equation (see eq. 2.18):

$$\frac{\partial c}{\partial t} + \underline{u}\ grad\ c - div\ (\underline{\underline{D}}_{hyd}\ grad\ c) = r$$

The integration of the differential equation over the control volume yields that the temporal change of the concentration added to the advective and dispersive fluxes is in equilibrium with sink / source terms:

$$\int_\Omega \frac{\partial c}{\partial t}\ d\Omega + \int_\Omega div\ \underline{F}\ /\phi_e d\Omega = \int_\Omega r\ d\Omega \tag{3.32}$$

$$with\ \underline{F}\ /\phi_e = \left(c\ \underline{u} - \underline{\underline{D}}_{hyd}\ grad\ c\right)$$

The Green-Gauß theorem (see eq. 3.24) is applied to the flux terms:

$$\int_\Omega \frac{\partial c}{\partial t}\ d\Omega + \int_\Gamma \underline{F}\ /\phi_e \cdot \underline{n}\ d\Gamma = \int_\Omega r\ d\Omega \tag{3.33}$$

In the same way as for the FEM, we introduce the semidiscrete method which says that the space is discretized with FVM and the time with the FDM. This is the most common way for time dependent problems that are treated with FVM. For the time derivative the same forward difference quotient as used for the FEM (see eq. 3.18) is applied:

$$\int_{\Omega_i} \frac{\partial c_i}{\partial t}\ d\Omega = \frac{c_i^{n+1} - c_i^n}{\Delta t}\ \Delta x\ \Delta z \tag{3.34}$$

The sink / source term is treated as follows:

$$\int_{\Omega_i} r_i\ d\Omega = r_i\ \Delta x\ \Delta z \tag{3.35}$$

For the dispersive terms we have to determine the gradients of c at the boundaries of the control volume.

$$\underline{F}_{dis}/\phi_e = -\underline{\underline{D}}_{hyd}\ grad\ c =$$

$$-\begin{bmatrix} D_{xx} & D_{xz} \\ D_{zx} & D_{zz} \end{bmatrix}\begin{bmatrix} \partial c/\partial x \\ \partial c/\partial z \end{bmatrix} =$$

$$-\begin{bmatrix} D_{xx}\dfrac{\partial c}{\partial x} + D_{xz}\dfrac{\partial c}{\partial z} \\ D_{zx}\dfrac{\partial c}{\partial x} + D_{zz}\dfrac{\partial c}{\partial z} \end{bmatrix}$$

$$(3.36)$$

This is illustrated for the x-direction in figure 14. The gradients in x-direction are replaced by central differences between i-$1,j$; i,j and i,j; i+$1,j$. The gradients in z-direction are determined by central differences between j+1 and j-1, at i-0.5 as an average of i-1, and i and at i+0.5 as an average of i and i+1. Thus we obtain:

$$\int_\Gamma F_{dis}^x/\phi_e\cdot n_x d\Gamma =$$

$$\int_\Gamma (F_{dis,i-0.5,j}^x + F_{dis,i+0.5,j}^x)/\phi_e\cdot n_x d\Gamma$$

$$-D_{xx}\frac{c_{i,j}-c_{i-1,j}}{\Delta x}\Delta z -$$

$$D_{xz}\left(\frac{c_{i,j+1}-c_{i,j-1}}{2\Delta z}+\frac{c_{i-1,j+1}-c_{i-1,j-1}}{2\Delta z}\right)\frac{\Delta z}{2}+$$

$$D_{xx}\frac{c_{i+1,j}-c_{i,j}}{\Delta x}\Delta z +$$

$$D_{xz}\left(\frac{c_{i+1,j+1}-c_{i+1,j-1}}{2\Delta z}+\frac{c_{i,j+1}-c_{i,j-1}}{2\Delta z}\right)\frac{\Delta z}{2}$$

$$(3.37)$$

The dispersivities must be determined at the boundaries of the control volume, e.g. for the first term in eqation 3.37 D_{xx} at i-$0.5,j$ and the last term in equation 3.37 at D_{xz} at i+$0.5,j$. In the same way the fluxes for the z-direction are obtained (see Kinzelbach 1992).

For the advective terms, the advective fluxes must be determined at the boundaries of the control volume. The most simple way to do this, is to use the central method which averages the values of the neighboring nodes.

$$\int_\Gamma F_{adv}^x/\phi_e\cdot n_x d\Gamma =$$

$$-\frac{1}{2}\left(u_{i-1,j}\,c_{i-1,j} + u_{i,j} + c_{i,j}\right)$$

$$\Delta z + \frac{1}{2}\left(u_{i,j}\,c_{i,j}+u_{i+1,j}+c_{i+1,j}\right)\Delta z$$

$$(3.38)$$

If the equations 3.34 - 3.38 are summed up, the complete IFDM / FVM using central weighting for the advection is obtained.

Up to now we have not mentioned at which time level the terms in the equations 3.35 - 3.38 are determined. This can be explicit, fully implicit or with a Crank-Nicholson factor Θ between the time levels n and n+1. The accuracy in space is of second order, while the accuracy in time is of second order for Θ=0.5 and of first order otherwise.

For unstructured meshes the dispersive and advective fluxes perpendicular to the boundaries of the control volume must be determined.

As the central method oscillates for medium and high advection, improved stabilization techniques are introduced in the next subsection.

FVM - Stabilization techniques

When the upstream method is applied, the fluxes are determined at the upstream nodes, shown here for the x-direction at i-$0.5,j$:

$$F_{adv}^x/\phi_e = \begin{cases} u_{i,j}\,c_{i,j} & for\quad u\geq 0 \\ u_{i-1,j}\,c_{i-1,j} & for\quad u<0 \end{cases}\quad(3.39)$$

This method is stable, but shows high numerical dispersion and is only of first order accuracy in space.

It is possible to switch between the upstream and the central method with an upwind parameter α ($0\leq\alpha\leq 1$), shown here for the x-direction at i-$0.5,j$:

$$F_{adv}^x/\phi_e = \alpha\,u_{i-1,j}\,c_{i-1,j} + (1-\alpha)u_{i,j}\,c_{i,j}$$

$$(3.40)$$

α can be estimated in the same way as shown for the FEM (see eq. 3.30):

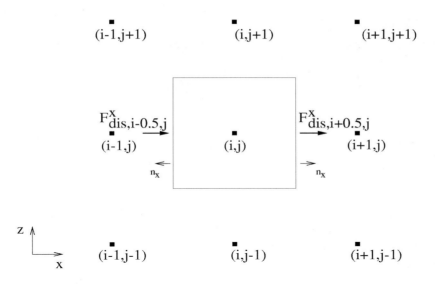

Fig. 14. Dispersive fluxes at the boundaries of the control volume in x-direction.

$$\alpha \approx \sqrt{\frac{Pe^2}{36 + Pe^2}}$$

Principally, it is also possible to chose the methods discussed in the context with the FDM, e.g. the Lax-Wendroff method.

The Quick method, which stands for quadratic upstream interpolation for convective kinetics, uses a quadratic function for the upstream flux (see Fig.15). For an equidistant mesh, the flux in x-direction is determined at *i-0.5,j* by:

$$F_{adv}^x / \phi_e = -\frac{1}{8} u_{i-2,j} \, c_{i-2,j} +$$

$$\frac{6}{8} u_{i-1,j} \, c_{i-1,j} + \frac{3}{8} u_{i,j} \, c_{i,j} \qquad (3.41)$$

As already mentioned for the FEM, a very promising group of stabilization techniques are those which maintain the monotonicity of the solution, i.e. Slope limiter, Flux limiter or Flux-Corrected-Transport methods (see Finlayson 1992).

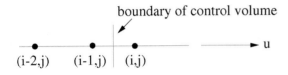

Fig. 15. Quick method.

FVM - Remarks and recommendations

The remarks concerning boundary conditions, convergence and stability given for the FDM are also valid in context of the FVM. The FVM guarantees local mass conservation which is a consequence of the formulation, because the differential equations are fulfilled in each control volume. In this case the FVM is superior to the FEM and FDM. The authors recommend, like for the FEM, that the element length should be chosen in such a way that the Peclet number exceeds 2, but is not much greater than 10. The Courant number should be around 1, even for implicit methods. For explicit methods the Neumann condition must be taken into account. Upwind or monotonicity preserving methods should be applied for stabilisation. When compared to the FEM and FDM, the FVM or IFDM has one important advantage: Mass conservation is guaranteed locally. When compared to the FDM, the unstructured meshes enable an excellent approximation of complex boundaries and inner structures for the FVM.

Some other methods and techniques

Another important method to solve advection / dispersion problems is given by Operator-Splitting methods, also called fractional step methods. The terms in the differential equations are splitted up into hyperbolic and parabolic ones. Then for each part a solution method specially suitable for the type of problem class is applied. The hyperbolic terms can be treated very well with the method of characteristics, the parabolic terms with the FEM (see e.g. Hinkelmann and Zielke 2000). In this context, Eulerian-Lagrangian-Localized-Adjoint methods or ELLAM methods are mentioned, as they show good results for advection / diffusion problems (see Celia et al. 1990). They are based on the method of characteristics and use special space-time weighting functions.

The methods described in the sections up to now are continuum methods and based on an Eulerian point of view. Particle methods which are based on a Lagrangian point of view can also be chosen for transport modelling. Here, the 'fate' of each tracer particle is followed in space and time in the computational domain. When the Random-Walk method is used, the advective movement is driven by the flow field and the dispersive movement is determined by a statistical approach. Particle methods have no problems with sharp fronts. However, in order to get reasonable results, a large number of particles must be simulated which causes a longer calculating time. Further information is given in Kinzelbach (1992).

Modelling in the subsurface is always concerned with uncertainties. Three different possibilities to deal with uncertainties in order to make reliable predications are addressed here: One can use conservative assumptions which have a high measure of security, but this may be not economical. Uncertainties in parameters and their influences on the simulation results can be estimated by sensitivity analyses. Finally, stochastical modelling can be applied which leads to bandwidths of results with mean values and standard deviations. As the flow field is the most important input parameter for the transport equation, it is common to examine the influence of different permeability distributions assuming the same mean value and the same standard deviation of the permeability and different correlation lenghts. Simulations can be carried out applying Monte-Carlo methods or Turning-Band methods. Stochastical methods for subsurface modelling are an

important research field which is gaining on importance. For further reading see Kinzelbach (1992) or Bárdossy (1992).

Adaptive methods aim at numerical solutions with a high accuracy and an optimum effort of CPU-time and storage requirement by automatically adjusting the mesh to temporally and spatially variable solution functions. They can be reasonably applied only to unstructured meshes, i.e. to the FEM or FVM. Among different techniques the h-adaptive methods were established for advection / diffusion methods (see Verfürth 1996). In the h-variant, the number of elements is increased by refining marked elements. Marking the elements is carried out by error estimators or indicators. For advection / diffusion problems, most of the error indicators are of heuristic nature, e.g. gradients of the concentrations. It is also possible to coarsen a mesh, e.g. if a refinement is no longer required at a certain location.

The implicit formulations lead to systems of equations with a sparse structure, which are generally non-symmetric and which may have a large number of unknowns. The authors recommend to apply the conjugate gradient solvers BiCGSTAB or GMRES for medium size problems and multigrid methods for large size problems (see e.g. Barret et al. 1994). For huge problem sizes parallel computers should be used (see Bastian 1996; Hinkelmann and Zielke 2000).

Applications

In this section, three examples for modelling single- and multicomponent subsurface transport processes in the water phase as well as in the water and gas phase are given based on the FDM, FEM or FVM. The examples demonstrate the use of different numerical modelling techniques for subsurface transport processes as well as possibilities and limitations.

Modelling of single-component transport processes in the water phase with the Finite-Difference-Method

In this advection / dispersion example which is taken from Kinzelbach (1992) an explicit FDM which uses central differences in space and forward differences in time ($O(\Delta x^2, \Delta t)$) is applied. The sytem set up is shown in figure 16. The computational domain of 2000x2000m is discretized with quadrilateral cells of 50x50m. The aquifer has a thickness of 25m, a hydraulic conductivity of K_f =300m/d (d=day) and a porosity ϕ=0.17. A diagonal flow field with u=1m/d and a source term with q=60m³/d are imposed. The source term is so small that it nearly does not influence the constant flow field. The concentration is set to zero at the left and upper boundary. At the right and lower boundary the concentration gradient is set to zero. The longitudinal dispersion length is set to α_L=100m, the transversal dispersion length to α_T=1m. The source has a high inflow concentration c=2000mg/l (milligram per liter). The Courant number is around Cr=1 and the Peclet number around Pe=0.5.

In figure 17 the concentration isolines 1, 20, 40, 60, 80 and 100 mg/l are shown after 1000d simulation time for the analytical solution on the left and the numerical solution with FDM on the right. A

qualitative agreement is observed. The numerically computed transversal spreading of the plume is higher, and due to mass conservation the spreading in flow direction is smaller compared to the analytical solution. The major reason for this is the cross dispersion as a consequence of the flow direction diagonal to the coordinate axis. An improvement is obtained when the cell length is reduced.

Modelling of single-component transport processes in the water phase with the Finite-Element-Method

In this advection / diffusion example, taken from Barlag et al. (1998), the semidiscrete FEM using the SUPG method for stabilization is applied. H-adaptive methods with a gradient error indicator were chosen to optimize the solution accuracy and the computational effort. The time step is adapted to a Courant number Cr=1 in the element which encloses the sharpest gradient. In the closed

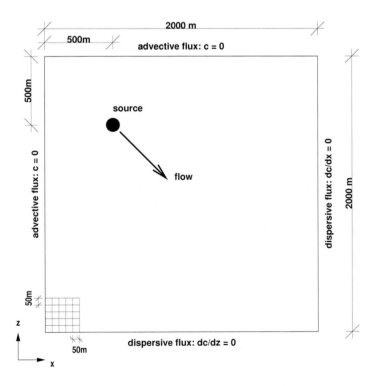

Fig. 16. System set up for the FDM example (modified from Kinzelbach 1992).

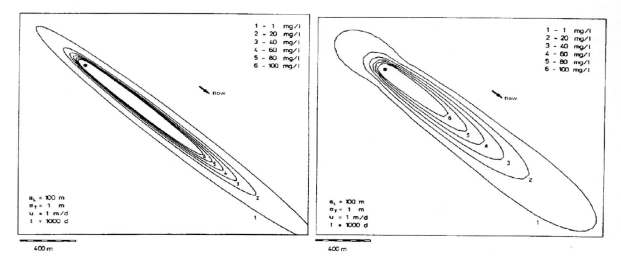

Fig. 17. Concentration isolines; left: analytic solution, right: FDM solution (modified from Kinzelbach 1992).

system shown in figure 18, a fracture is located in a matrix block between a source and a sink. The hydraulic conductivities and diffusion coefficients are K_f=0.1m/s, D_f=2·10⁻⁹m²/s for the fracture and K_m=10⁻¹¹m/s, D_m=10⁻¹²m²/s for the matrix. Dispersion is not considered. A piezometric head h=10m and an inflow concentration c=1.0% are given for the source, while the piezometric head for the sink is set to zero h=0m. High velocities occur near the source and in the fracture, while the velocities are very small in the right part of the system.

In figure 18 the adaptive meshes and concentration isoareas are shown. The SUPG method together with the adaptive method enabled a solution without oscillations. The number of elements (maximum 5000) adapted to the dynamics of the solution, which was therefore very efficient. At the time step T=10⁹s ≈ 31.8a (a=years) a high mesh resolution is shown in the area of the plotted isolines, i.e. the sharp front. At the time step T=3·10¹²s ≈ 95129.4a the mesh refinement has moved with the concentration front, and can now be found in a considerable distance to the source and near the fracture. In the closer surroundings of the source, the mesh has been coarsened. A comparable solution accuracy with a grid not using adaptive refinement requires a uniform, highly resoluted mesh with several times more elements.

Modelling of three-component transport processes in the water and gas phase with the Finite-Volume-Method

In this two-phase / three-component example taken from Hinkelmann et al. (2000) the semidiscrete FVM using the upstream method for stabilization and a fully implicit method in time is applied. The system shown in figure 19 was chosen for simulating flow and transport processes around a submarine groundwater discharge site (vent) in the Eckernförder Bucht, Baltic Sea, Germany.

The sediment is highly porous ϕ=0.9, and the permeability was estimated to K=10⁻⁸m². The system is closed at the left and right boundaries and at the lower boundary with the exception of the inflow opening where a fresh water (ρ_w=1000kg/m³, μ_w=10⁻³ Pas) discharge was determined as the boundary condition. According to vent sampler measurements the fresh-water discharge was q_w^f=10.8 l/(m²h). On the upper boundary, a twenty meter sea water column with a salinity of 7‰ is imposed and the methane concentration is set to zero. It is assumed that the salinity is mainly determined by chloride. For a given mole mass of water 0.018 kg/mol, 7‰ are equivalent to 0.007/0.018=0.39 mol/l=390 mmol/l (see Fig. 20). For the gas phase, the density was ρ_g=0.68kg/m³ and the dynamic viscosity μ_g=1.24·10⁻⁵ Pas. Disper-

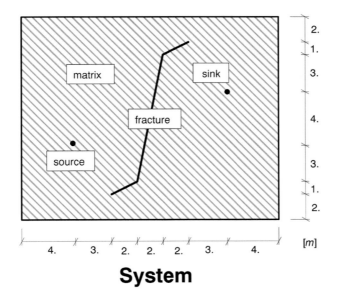

System

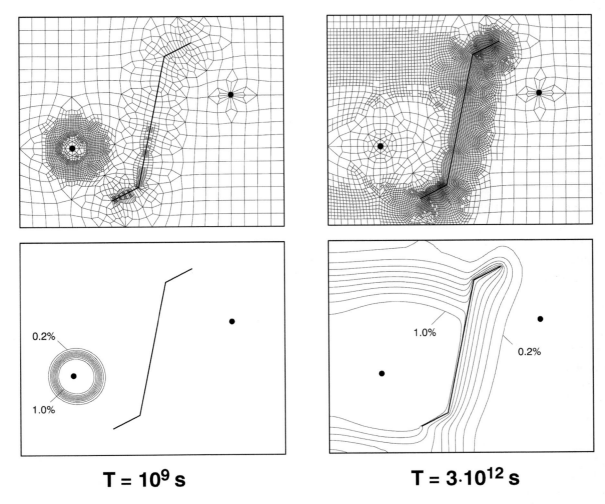

T = 10⁹ s   **T = 3·10¹² s**

Fig. 18. System set up, adaptive meshes and isoareas of concentration at $t=10^9$s ≈ 31.8a and $t=3 \cdot 10^{12}$s ≈ 95129.4a (modified from Barlag et al. 1998).

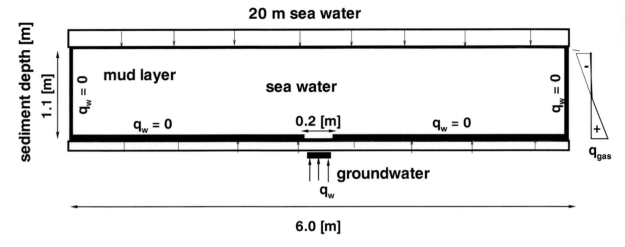

Fig. 19. System set up, initial and boundary conditions.

sion was neglected, only molecular diffusion was taken into account with $D_w^s = 6.6 \cdot 10^{-6}$ m/s, $D_w^m = 9.0 \cdot 10^{-10}$ m²/s and $D_g^f = 10^{-6}$ m²/s. Linear distributions of sink and source terms for methane in the water phase, which is caused by biological and chemical activities, are prescribed with a source of $q_w^m = 0.015$ mol/(m²h) at the bottom and a sink of $q_w^m = 0.05$ mol/(m²h) at the top. As initial condition, the system was filled with sea water (without methane). It must be mentioned that some model parameters and boundary conditions had to be estimated, as information was lacking.

The system was discretized with 60x13=780 elements. In figure 20 the steady state results are shown. In the left part of the figure, high methane concentrations in the water phase are observed in most lower parts of the mud layer. These concentrations are partially so high, that the maximum solubility of methane in the water is exceeded. Consequently, a gas phase (methane bubbles) occurs here. Only in the region around the inflow opening the advection is very high which leads to much lower methane concentrations. Therefore, no gas phase can occur. In the right part of figure 20, a comparison of computations and measurements in the discharge site (cross section A-A) is presented. The vertical methane distribution is nearly constant and the vertical salinity distribution decreases from sea water concentration at the top to low values at the bottom.

Overall, a reasonable agreement between computations and measurement was obtained.

Conclusions and recommendations

In this paper an overview is given about numerical modelling of transport processes in the subsurface. Numerical flow and transport models can be applied for different questions and problems. They enable a better process understanding, e.g. for processes occuring at sites of submarine groundwater discharge. With the help of numerical models, fluxes of water, nutrients or gas as well as local or global budgets can be estimated. As measurements are only capable to describe the present state, numerical models are able to make predictions for changing conditions and situations.

The physical model concept, which is based on representative elementary volumes (REV) together with the mathematical model concept, is discussed focussing on multicomponent flow and transport processes in a single phase, water, and briefly in two phases, water and gas. The last model concept is required, if the occurence and spreading of free gas is of importance. Generally, flow and transport processes in the subsurface are two or three dimensional.

The major part of this paper is concerned with numerical methods for solving transport equations. First, several variants of Finite-Difference-a)

b)

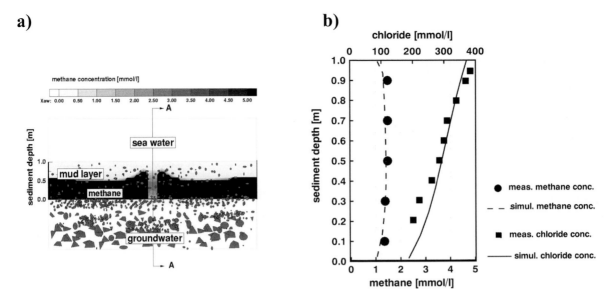

Fig. 20. Methane concentration in the water phase **a)** and concentration profiles in section A-A **b)**.

Methods, explicit and implicit ones with different orders of accuracy in space and time, are presented. Here, the authors recommend to chose the rectangular cell length in a way that a Peclet number smaller than 2 should be fulfilled in all spatial directions. The Courant number should be around 1 for all directions, even for implicit methods. For explicit methods the Neumann condition must be taken into account. The Lax-Wendroff or the Preissmann method should be applied. It must be kept in mind that mass conservation is not necessarily guaranteed, and complex boundaries and inner structures can only be discretized very roughly.

Second, the Finite-Element-Method is explained together with stabilization techniques for advection / dispersion problems. Here the authors share the opinion that the element length can be chosen in such a way that the Peclet number is greater than 2, but should not exceed 10. The Courant number should be around 1, even for implicit methods. For explicit methods the Neumann condition must be taken into account. The SUPG or monotonicity preserving methods, e.g. slope or flux limiter, should be applied. Mass conservation is guaranteed globally, and the unstructured meshes enable an excellent approximation of complex boundaries and inner structures.

Third, the Finite-Volume-Method is introduced also focussing on stabilization techniques for advection / dispersion problems. The Finite-Volume-Method can be applied to rectangular grids, than it is called Integral-Finite-Difference-Method, or to unstructured grids. The authors recommend the same things concerning Peclet, Courant and Neumann number as for the FEM. For stabilization, the Quick or monotonicity preserving methods should be applied. The advantages and disadvantages of rectangular or unstructured grids were also mentiond. The major advantage of this method is the fact that mass conservation is guaranteed locally.

Three examples demonstrate the use of different modelling techniques for flow and transport processes in the subsurface as well as possibilities and limitations.

Finally, the authors recommend to use Finite-Volume-Methods, because of the local mass conservation and the unstructured meshes. Stochastical modelling will gain on importance in order to be able to give confidence intervals of the simulation results.

In the future the models can be further developed to simulate non-isothermal multiphase / multicomponent processes, which occur e.g. around gas hydrates.

Acknowledgements

The authors are grateful to Dipl.-Ing. Daniel Sietoo, Dipl.-Ing. Björn Witte and cand.-Ing. Christian Gabriel for transfering the whole manuscript to the word processor and proofing the contents. Without their help, this paper could not be in the form presented now. The major part of the work was funded by the EU within the SubGATE project (ENV4CT970631).

References

Barlag C, Hinkelmann R, Helmig R, Zielke W (1998) Adaptive Methods for Modelling Transport Processes in Fractured Subsurface Systems. In Proceedings of 3rd International Conference on Hydroscience and Engineering, Cottbus

Bastian P (1996) Parallele adaptive Mehrgitterverfahren. Teubner Skripte zur Numerik. Teubner-Verlag, Stuttgart

Bárdossy A (1992) Geostatistical Methods: Recent Developments and Applications in Surface and Subsurface Hydrology. UNESCO, Paris

Barret R et al. (1994) Templates for the Solution of Large Linear Systems: Building Blocks for Iterative Methods. SIAM, Philadelphia

Bear J (1972) Dynamics of Fluids in Porous Media. American Elsevier, New York

Bear J, Bachmat Y (1990) Introduction to Modelling of Transport Penomena in Porous Media. Kluwer Academic Publishers, The Netherlands

Celia MA et al. (1990) Eulerian-Lagrangian Localized Adjoint Method for the Advection-Diffusion Equation. Advances in Water Resources, 13 (4):187-206

Class H (2000) Theorie und numerische Modellierung nichtisothermer Mehrphasenprozesse in NAPL - kontaminierten porösen Medien, Dissertation, Mitteilungen des Instituts für Wasserbau, Heft 105, Universität Stuttgart

Finlayson BA (1992) Numerical Methods for Problems with Moving Fronts. Ravenna Park Publishing, Seattle, Washington

Helmig R (1997) Multiphase Flow and Transport Processes in the Subsurface - A Contribution to the Modelling of Hydrosystems. Springer, Berlin

Hinkelmann R, Zielke W (2000) Parallelization of a Lagrange-Euler-Model for 3D free surface flow and transport processes. Computers & Fluids, 29 (3):301-325

Hinkelmann R, Sheta H, Helmig R, Sauter EH, Schlüter M (2000) Numerical simulation of water-gas flow and transport processes in coastal aquifers. International Symposium 2000 on Groundwater IAHR: New Science and Technology for Sustainable Groundwater Environment, Sonic City, Japan. In: Sato K, Iwasa Y (eds) Groundwater Updates. Springer, Berlin

Johnson C (1992) Numerical Solution of Partial Differential Equations. Cambridge University Press, New York

Kinzelbach W (1992) Numerische Methoden zur Modellierung des Transports von Schadstoffen im Grundwasser. 2. Auflage, R Oldenbourg Verlag, München, Wien

Oldenburg CM, Pruess K (1995) Dispersive transport dynamics in a strongly coupled groundwater-brine Flow System. Water Resources Research 31(2): 289-302

Malcherek A (2000) Numerische Methoden in der Hydrodynamik, Technical Documentation, Bundesanstalt für Wasserbau, Außenstelle Küste, Hamburg

Verfürth R (1996) A Review of A Posteriori Error Estimation and Adaptive Mesh Refinement Techniques. J Wiley & Sons and B G Teubner, New York, Stuttgart

Subsurface Fluid Flow and Material Transport

N. Kukowski[1*], M. Schlüter[2], R.R. Haese[3], C. Hensen[4], R. Hinkelmann[5], M. Sibuet[6], M. Zabel[4]

[1] GeoForschungsZentrum Potsdam, 14473 Potsdam, Germany
[2] Alfred-Wegener-Institut für Polar- und Meeresforschung, 27570 Bremerhaven, Germany
[3] University of Utrecht, Institute of Earth Sciences, 3508 TA Utrecht, Netherlands
[4] Universität Bremen, Fachbereich Geowissenschaften, 28359 Bremen, Germany
[5] Universität Stuttgart, Institut für Wasserbau, 70550 Stuttgart, Germany
[6] Ifremer, DRO/Département Environnement Profond, 29280 Plouzane Cedex, France
* corresponding author (e-mail): nina@gfz-potsdam.de

Abstract: Submarine subsurface fluid flow is ubiquituous with rates varying enormously with space and time. Therefore, fluid flow most probably is of paramount importance for the transport of matter and heat as well as to control fluxes between the subsurface and the ocean. However, rates to date still are only very poorly understood. In this study we therefore first identify major fluid flow systems including fault transport at active and passive margins, gas hydrate affected flow systems, or submarine groundwater discharge. Then, we describe geophysical and geochemical methods which are capable to better and quantitatively understand interacting submarine fluid flow systems and lastly, propose strategies for future research.

The importance of fluid flow in ocean margin systems

The submarine subsurface including shallow and deep sediments as well as the igneous crust is saturated with aqueous fluids which possess energy, e.g. through their hydraulic potential, in the form of heat, and chemical energy. The hydraulic potential varies with slope or the amount of over-pressuring, the subsurface temperature field is quite heterogeneous, and the aqueous fluids content varying amounts of dissolved compounds. Consequently, the properties of the fluids and amounts of their energy vary both, spatially and temporally. These variations force the subsurface fluids to move, and, therefore, may provide means to efficiently transport mass and energy. From the physical point of view, two modes of fluid flow are distinguished: i) diffuse flow, i.e. the percolation of aqueous fluids through the interconnected pore space at low rates as described by Darcy's law, and ii) focused flow, i.e. fluid motion along narrow zones, e.g. fault zones, of high transmissivity, often

described with a cubic flow law. Consequently, fluid flow occurs at very different rates and scales and is supposed to be most important in driving and controlling fluxes between storage systems like the sub-seafloor sediment reservoir and the ocean reservoir.

However, to date quantitative understanding of fluid flow systems still is very limited despite growing evidence for the widespread occurrence and importance of subsurface fluid flow at ocean margins. These issues point straight forward to i) the release of methane and trace gases to the ocean, ii) the formation of gas hydrates, iii) the development of unique macro- and microbiological communities at discharge sites, iv) enhanced sedimentary oxygen uptake and microbial carbon mineralisation in surface sediments due to macrofaunal activity, and v) submarine groundwater discharge. Regions of fluid flow are found in various settings and water depths along the slopes of active and passive continental margins. Typical indicators for fluid flow outlets are e.g. massive formation of authigenic carbonates, mud volca-

From WEFER G, BILLETT D, HEBBELN D, JØRGENSEN BB, SCHLÜTER M, VAN WEERING T (eds), 2002, *Ocean Margin Systems.* Springer-Verlag Berlin Heidelberg, pp 295-306

noes, pockmarks, and active (gas) venting at the seafloor, or anomalies in the concentration of bottom water constituents, fauna associated with chemosynthetic bacteria as well as submarine freshwater discharge in restricted near-shore areas.

The potential importance of subsurface fluid flow systems has been recognised e.g. with respect to the transfer of trace elements and trace gases from the deep subsurface to the marine hydrosphere, which may also affect slope stability (Mienert et al. this volume) and hydrocarbon formation (see below). However, only very few studies have been carried out to localise and survey discharge sites, determine flow rates and chemical compositions of fluids, or to apply numerical simulation to derive a detailed understanding of coupled reaction and transport processes considering fluid, heat, and chemical transport. Consequently, there is an enormous lack of information and at the same time strong need for extensive research concerning our quantitative understanding of subsurface fluid flow and its role for global (bio)geochemical budgets, regional effects on primary productivity, unique fauna assemblages, and the formation of gas hydrates with its consequences for potential rapid climate change and slope instability.

These issues involving fluid flow actually attract increasing attention, both, from a pure scientific and a socio-economic standpoint including future natural resources. The rising number of reports describing fluid flow from the seafloor emphasises the need for time and cost efficient techniques to localise discharge sites and to quantify the amount and composition of expelled fluids. For this, we foresee the need for innovative techniques (Table 1) such as automated underwater sampling

Techniques	Identification of possible structures (Site survey)	Identification unambiguous	Rate determination possible	Temporal resolution possible	Spatial resolution / distribution possible
High resolution seismics (2D + 3D)	**Yes** (suitable for site survey)	No	No	*Yes** (to a limited extent)	**Yes**
Bathymetry and side scan sonar survey	**Yes** (suitable for site survey)	No	No	*Yes** (to a limited extent)	**Yes**
Acoustic tomography (4D – seismic by hydrophones)	**Yes** (suitable for site survey)	No	No	**Yes**	**Yes**
Acoustic and laser observation of vents	*Yes* (to a limited extent)	**Yes**	No	**Yes**	No
Heat flow measurements	*Yes**	**Yes**	*Yes* (to a limited extent)	**Yes**	*Yes* (to a limited extent)
Direct flow determination (e.g. by flow meters, heat pulse technique)	*Yes**	**Yes**	**Yes**	**Yes**	No
Geoelectric (e.g. conductivity)	*Yes**	**Yes**	*Yes* (to a limited extent)	**Yes**	No
Seafloor monitoring (by mobile platforms, e.g. video survey by submersibles or remotely operating vehicles (ROVs))	*Yes**	**Yes**	**Yes** (with some techniques)	**Yes**	**Yes**
Determination of chemical constituents in the water column	*Yes**	**Yes**	*Yes* (to a limited extent)	**Yes**	*Yes* (to a limited extent)
Determination of chemical constituents in the pore water	No	**Yes**	**Yes**	*Yes* (to a limited extent)	No
Sedimentary composition (e.g. carbonate and baryte precipitations)	No	**Yes**	No	*Yes* (only on larger time scales ;>100-1000yrs)	*Yes* (to a limited extent)
Isotopic composition (of dissolved components, gases, minerals and shells)	No	**Yes**	No	*Yes* (on medium to large time scales)	*Yes* (to a limited extent)
Determination of radionuclides (e.g. ^{226}Ra, ^{222}Rn)	No	**Yes**	**Yes**	**Yes**	No
Observation of biota adapted to specific environment (e.g. species as indicators of seeps)	No	**Yes**	No	*Yes* (only on medium time scales, >several yrs.)	**Yes**
Remote sensing (e.g. sea surface roughness, temperature, chlorophyll)	**Yes** (suitable for site survey)	No	No	*Yes* (to a limited extent)	*Yes* (to a limited extent)

(*Yes** - only sensible when information from site survey methods exist)

Table 1. Techniques to find locations of fluid flow and to quantify flux rates

and measuring platforms, long-term installations of monitoring devices, and 2-3D transient simulation approaches to quantify fluid flow.

Settings of fluid flow systems at ocean margins

At ocean margins, fluid flow processes affect the sub-seafloor and the overlying water column in various specific environments and on varying regional scales. These settings are found along slopes of both, active and passive continental margins including accretionary wedges, where local features such as mud volcanoes, brines, or specific biological communities give evidence for fluid flow (Fig.1). Freshwater has been detected in the deep marine subsurface which results from processes such as gas hydrate dissociation and/or clay mineral dehydration. An additional environmental setting obviously affected by fluid flow is the near coastal area, where groundwater discharge via submarine springs or as dispersed flow from sediments occurs. Due to significant flow rates and large areas affected by fluid discharge, such fluid flow systems are potentially major contributors to the global fluid budget.

Despite the differences in such systems with regard to geological settings, driving forces, and scale, they share fluid flow as the main carrier of chemical species and heat. Additionally, hydrologic properties can be described by similar numerical approaches. Consequently, we herein consider fluid flow in active and passive margins and submarine groundwater discharge in near coastal areas together.

Sub-seafloor flow systems in the sediment reservoir

Fluid flow processes at convergent margins, as revealed from studies e.g. at the Barbados, Peruvian, or Cascadia margins, include diffuse fluid transport driven by tectonic dewatering resulting from sediment accretion and related compaction leading to pore water overpressure and seepage, and focused flow along thrust faults leading to fluid venting at the ocean bottom. Such vent sites are often characterised by unique biota and massive mineral precipi-

tations. Flow rates at vent sites are about 3 to 4 orders of magnitudes larger than diffuse flow rates, however, pathways of focused flow seem to be very restricted in number and size (Suess et al. 1998; Mann and Kukowski 1999). Therefore, the importance of diffuse flow for the regional or global budget may be at least as important than that of focused flow. However, until now, rates have only been measured in a very few studies and therefore driving forces still are not well understood. Due to the significantly different flow rates, reactive transport will proceed quite differently in both types of fluid flow systems.

Whereas at passive margins tectonic processes as driving forces are only of local importance, e.g. in connection with deep reaching fault zones, these margins nevertheless share many features with active margins. These include the roles of diffuse and focused flow as carriers for material and heat, the occurrence of gas hydrates, e.g. at Blake Ridge and the Norwegian continental slope, the occurrence of evaporites, and the transport of reactive chemical components like methane.

Mud volcanoes are good examples for the occurrence in different settings. So far, they have been mainly identified and studied at convergent margins, e.g. the Mediterranean Ridge or Barbados accretionary prisms, however, they also have been found on passive margins, e.g. the Haakon Mossby mud volcano field at the Norwegian continental slope.

In the following we describe key-features of specific fluid flow systems at ocean margins and develop concepts for their interdisciplinary investigation as is necessary in the next decades together with the development of new techniques to measure and quantify these processes.

Fluid flow derived from the deep subsurface

In the past two decades, it has been realised that several sites at active and passive continental margins are affected by fluid flow driven by different forces: (1) Due to the collision of plates lateral compression leads to the reduction of pore volume with causing an upward directed advection. (2) In the deep subsurface at temperatures >80°C clay minerals dehydrate and

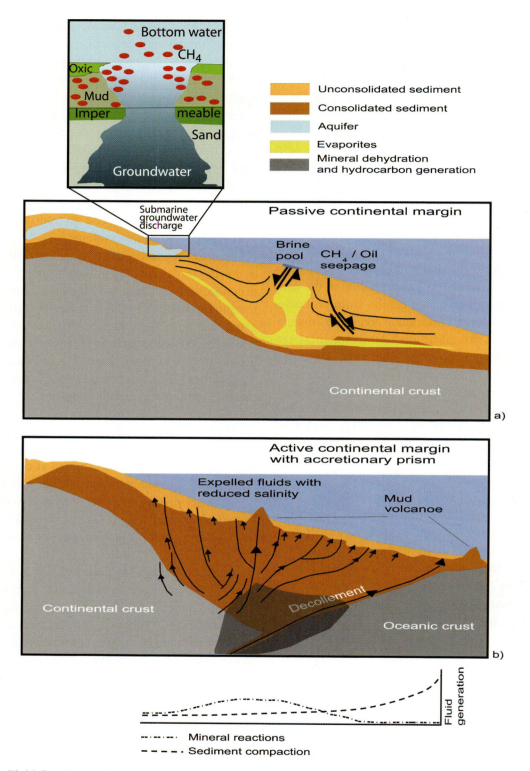

Fig. 1. Fluid flow in passive **(a)** and active **(b)** continental margin settings. Submarine groundwater discharge (SDG) is apparent in near-coastal areas. Fluid chemistry depends on the source and pathway of fluid migration. Brine seepage is common if evaporites occur in the subsurface, while expelled fluids in active margins often have reduced salinities relative to sea water due to dissociation of gas hydrates and/or dehydration of minerals in the deep subsurface. For the active continental margin, fluid generation is shown semiquantitatively along a transect (modified from Moore and Vrolijk 1992).

smectite is transformed to illite which creates a source of fresh water and induces density-driven flow. (3) Sediments of large basins particularly associated with salt tectonics are exposed to rapid compaction and consequently water extraction (Bolton and Maltmann 1998). (4) Dissociation of gas hydrates under increasing temperature and pressure conditions results in the release of fresh water. A network of thrusts and fractures with higher permeabilities than the surrounding sediment serve as preferred pathways of fluids (Fig.1). Related to these different scenarios a distinction between focused and diffuse flow regimes is required. Fluid flow rates differ by about four orders of magnitude between these two flow regimes. In analogy, sites of fluid escape from the sediment are referred to as vents and seeps, respectively.

The chemistry of the fluid expelled at the sediment surface is determined by its source location, pathways and transfer rates, as well as by chemical and microbiologically induced reactions during fluid transport. Prominent differences in concentration are found for conservative elements such as chloride due to a deep fresh water source (see above) or contact to evaporite (salt) deposits. At high temperature conditions (>100 °C), several fluid-rock interactions occur as boron release from siliciclastic rocks, ^{18}O enrichment in the fluid (Dia et al. 1999), and formation of thermogenic methane. Biogenic methane formed by microbial activity is widespread at continental margins and documents one important aspect of deep microbial activity. Both processes may lead to the re-entering of buried organic carbon to the global carbon cycle (Fig.2).

Formation and dissociation of gas hydrates in sediment bodies influenced by fluid flow is inferred indirectly by seismic observations of BSRs (bottom simulating reflector), and by sediment and pore water sampling showing depleted chloride concentrations. However, the latter indicator is not unequivocal, as the depletion may occur both, through *in situ* dissociation or by gas hydrate dissolution during core recovery. So far, only a few recoveries of gas hydrates from sediments were possible, e.g. by drilling Blake Ridge or during seafloor sampling at Cascadia. Thermodynamic and kinetic considerations are restricted to laboratory studies or theoretical approaches. A transfer of this knowledge to in-situ sediment conditions has not been achieved so far and is one aim of future studies.

In near surface sediments, the ascending methane-rich fluid comes in contact with sulfate-bearing sea water which causes anaerobic methane oxidation. The latter has far-reaching consequences for chemical fluxes across the sediment water interface and the stimulation of life: Part of the methane is oxidised and methane-derived carbon precipitates as calcium/magnesium-carbonate which forms crusts and mounds.

Biological indicators of fluid emission at the seafloor

Sub-seafloor fluid flow is essentially 'invisible' as only rarely plumes and high rates of discharge like at hydrothermal vents occur. However, most of the recent work on cold seeps and cold vents (opposite to hot vent as they are characterised by only small temperature anomalies relative to bottom water) has considered the unique vent communities and bacterial mats as the most visible and obvious indicators of seeps (Sibuet and Olu-Le Roy 2001). Like at hydrothermal vents remarkably large invertebrate species belonging to similar classes, families, and even genus use, through bacterial-invertebrate symbiosis, organic matter produced by chemosynthesis based on methane or on sulfide produced by anaerobic methane oxidation. Microbial chemosynthesis is the primary organic carbon production process for the very dense and large clusters of typical bivalves and tube worms (Fig.2). The best known endosymbiosis are between vestimentiferan tube worms and sulfur-oxidising bacteria, between bivalves (clams, mussels) and methanotrophs and/or sulfur-oxidising bacteria. The faunal density and biomass seem to be related to the fluid discharge rate at the local level. The diversity of the symbiotic fauna can be an indicator of the variability of fluid composition considering the differences between species trophic requirements, i.e. symbiotic bacteria can be either linked to methane-rich or to sulfide-rich fluids.

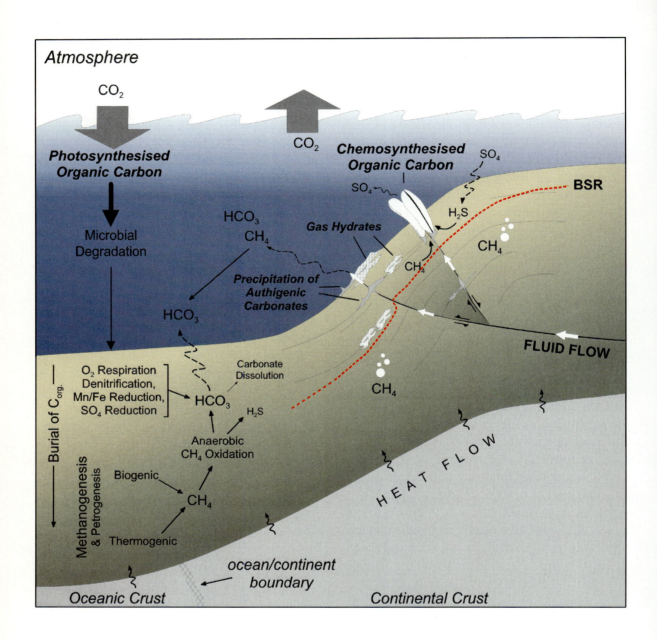

Fig. 2. Simplistic overview of carbon fluxes in the ocean and along a continental margin. On the right side generalised diagenetic pathways are illustrated. The occurrence or rather the importance of single processes depends on particular environmental conditions (e.g. on the flux of reactive organic carbon to the sea floor and the temperature gradient in the sediment column). Reaction rates of these processes are considerably influenced, especially where active fluid flow dominates the transport of dissolved and gaseous constituents. Such physically induced transport occurs as well on passive continental margins for example above salt domes, and on active continental margins (e.g. along fault zones of accretionary prisms). Here biotic communities interacting with cold seep fluids play an essential role for the cycling of components. Shells in the diagram represent symbiotic fauna which depends on chemosynthesis. The Bottom Simulating Reflector (BSR) is interpreted as the transition between sediments where gas hydrates are to be expected (above) and sediments containing free gas (below).

On a large scale (several km²), biological activity and the spatial distribution of biological clusters help to characterise fluid circulation and different expulsion pathways. By consuming methane, these vent and cold seep communities extract gases and contribute to limit methane discharge to the water column. However, the quantification and role of the biological activity remain to be understood in space and time.

Submarine Groundwater Discharge (SGD)

Present research on SGD in near coastal areas is concentrated on the localisation and determination of discharge sites and rates. Related to the applied tracers and techniques SGD has been investigated at different spatial scales like extended areas along continental margins (e.g., the Mid Atlantic Bight off the US) down to small scale features like submarine springs or pockmark locations. However, there is still a considerable lack of information about driving forces, mode of flow, discharge of freshwater compared to re-circulated seawater, chemical composition of fluids, and the impact of discharge on the benthic biota, primary production, and formation of larval stages of fish or crustaceans.

Related to the salinity of fluids two types of groundwater discharge from continental margin sediments have to be distinguished: hypersaline-saline fluid emanations and discharge of freshwater. For both types, flows might occur predominantly as focused flow (along fractures of karst systems) or as diffuse flow through porous media as sandy or muddy sediments. Discharge of hypersaline fluids were reported, for example, for the Gulf of Mexico, where high salinities are often associated to occurrences of salt diapirs in the subsurface.

Submarine discharge of modified (due to admixture of groundwater and ambient seawater) or pure freshwater has been observed in several regions of the world ocean including Tuckey Bay (Australia), Chile, small islands, large estuaries as the Ganges, the Baltic Sea, and the Mediterranean Sea. Submarine freshwater springs are mainly associated with karst flow. Emanations of more than 10,000 L m^{-2} d^{-1} where reported for the Mellisi coast, the coastal region of Lebanon, off the Italian coast or Cyprus. For regions predominately characterised by diffuse flow, as reported for regions in the North Sea, the Baltic Sea, and off Spain, discharge volumes of 200-400 L m^{-2} d^{-1} were observed by direct measurements. Since large areas are affected by diffuse flow, this mode of SGD may be of considerable relevance for the environment (Schlüter this volume).

In the context of fluid flow, freshwater discharge is of prior importance from a scientific perspective, e.g. for the assessment of land-ocean interaction, or the transfer of trace gases (e.g., CH_4, CO_2) from sediments into the water column and atmosphere. From a socio-economic perspective, SGD is relevant in terms of loss of freshwater to the sea, water quality in coastal areas, nutrient and pollutants influx to the marine environment, as a potential resource for aquacultures, and with respect to the potential recovery and on-land use for irrigation purposes.

Methods to identify and quantify fluid transport

Techniques to identify and measure fluid flow quantities

For the detection of sites affected by active fluid flow a number of techniques are known and have already been proven to be efficient tools to investigate fluid flow (Table 1). Nevertheless, until now, no single method seems to exist which allows the detection and quantification of flow rates for different temporal and spatial scales, ranging from site specific studies at vent sites to budgets integrated over large areas. Since it is of prime interest to quantify fluid flow and to estimate its importance for large scales or even ocean wide budgets, it is generally necessary to apply a whole suite of different techniques. For example, on one hand, exact measurements of flux rates of specific constituents over time at single locations are required, on the other hand, information on the total number of locations and their regional distribution needs to be obtained for a quantitative approach.

Usually, general information on possible locations of sites of fluid flow and their spatial distribution is needed before more detailed investigations can be started. This prerequisite information can be obtained by application of hydroacoustic or remote sensing methods (Table 1). The knowledge of prominent bathymetric features (e.g. pockmarks) or seismic structures (e.g. fractures, faults, or transparent zones) might then be followed by more diagnostic methods as a survey of the distribution of relevant constituents within the water column resulting in a positive or negative proof for active fluid flow. If measurements indicate active fluid flow, more specific methods might then be applied that are suitable for the determination of transport rates. One important new technique in this regard might be the application of autonomous mobile platforms allowing time and cost efficient monitoring programs to record the spatial and temporal development of flow rates.

Numerical methods for simulating flow, transport and reaction processes in the subsurface

Subsurface fluid flow is a highly dynamic process affecting the spatial and temporal distribution of components, e.g. nutrients, trace elements, or dissolved gas, as well as gaseous phases. Whereas direct measurements are only capable to describe the present state, transient numerical models – if properly calibrated – can be predictive also in case of changing conditions and situations. This allows a quantification of mass fluxes of water, nutrients and contaminants etc. for local, regional and large scales and therefore also to estimate global budgets by integrating fluxes over domains. Most important, numerical simulation enables a better understanding of coupled processes, e.g. fluid and heat transport or fluid and reactive transport, by determining the main controlling parameters and selecting dominant processes. A necessary prerequisite for numerical simulation are field data in order to run, calibrate and validate the model.

The model concept extracts the governing physical, chemical and microbial processes, which are of interest and importance for the problems and

questions that should be investigated. Generally, processes of flow, the interaction with the deforming porous structure (tectonics, landslides), diffuse and focused mass transport, conductive and advective heat transport as well as chemical and biological reaction processes need to be considered in ocean margin systems. Additionally, transport and reaction processes in the water phase should be also taken into account. Consequently, single phase flow and multicomponent transport and reactions are distinguished and changing fluid properties, e.g. density as a function of e.g. temperature and salinity, should be incorporated into algorithms as they drive fluid flow. Furthermore, different fluid phases, e.g. water and gas, may occur, if the phases are not or only slightly soluble in each other resulting in the need of multiphase models. Mass transfer of components needs to be taken into account, e.g. by assuming a thermodynamic equilibrium. I.e. a dissolved gas in the water phase can switch into the gas phase (e.g. in methane rich sediments) or methane in the solid phase can be transferred into the gas phase (e.g. in gas hydrates). In this context, the terms two-phase or multiphase flow and multicomponent transport and reactions are used. The model concept is transferred to a mathematical formulation by applying the balance equations for mass, momentum and energy, which results in a set of partial differential equations which are numerically solved by discretisation techniques, which make the treatment of general initial and boundary conditions as well as complex geometries possible (Hinkelmann and Helmig; Soetaert et al. both this volume). Numerous numerical simulation tools already exist today, however, the state of the art, the application range and the limitations differ a lot for the different model concepts.

To run a model, a number of data concerning the structures in the computational domain and around the boundaries are required. For flow simulations, the locations of layers, fault zones and fractures, i.e. the spatial distribution of hydraulic conductivity, porosity and storage term, as well as prescribed hydraulic heads or discharges at open boundaries are needed. If a coupling with the deformation is investigated, Young's modulus and

Poisson's ratio as well as the pressure distribution at open boundaries need to be specified. If two or more phases occur, the relationships between capillary pressure and saturation as well as relative permeability and saturation additionally need to be known. Simulating transport requires information about diffusivities and dispersivities as well as concentrations or fluxes at open boundaries. For simulation of reaction processes, reaction constants need to be known. It should be mentioned that flow, transport and reaction processes can be weakly or strongly coupled, and possibly need to be considered on different temporal and spatial scales. During the calibration of a model, additional data are required at control points in the computational domain, e.g. hydraulic head or concentration, in order to fit certain parameters which are a priori unknown and which are assumed not to change in the predictions.

(Single-phase) Water flow and transport processes are comparatively well understood, and a number of suitable 2D and 3D numerical models are used for small and large scale simulations. With some restrictions, this is also the case for density driven flow. Models which describe the interaction of fluid flow and the deforming porous medium are in an early stage and have hardly been applied to ocean margin systems. In recent years a number of 1D geochemical and biological reaction models have established. However, it has become obvious that 1D is a too strong simplification for many processes and problems, which are more realistically described by two or even three spatial dimensions, e.g. advection (see Haese this volume). Multiphase flow processes as well as non-isothermal multiphase / multicomponent flow and transport processes are only partly understood so far. Due to the very complex processes, only a few numerical simulators have been developed, and, generally, application is limited to smaller scales.

Finally, it should be mentioned that numerical models can not do magic and, overall, the accuracy and reliability of their predictions cannot be better than the amount and quality of laboratory and field data they are based on. Nevertheless, we would like to emphasise the need for numerical models which are able to describe and simulate different environmental settings affected by fluid flow for assessment of potential changes due to e.g. temperature increase of bottom waters, changing of hydraulic heads, or anthropogenically caused disturbances of boundary conditions of fluid flow systems.

Consequences and recommendations

Based on the present stage of knowledge, a quantitative investigation of the importance of fluid flow for global scale transport in the marine subsurface is not possible to date. Little is known about driving forces and consequences of early diagenetic pathways at the great variety of continental margin systems (Zabel and Hensen this volume). The regional effect of input by fluid flow on selected ocean margin areas, however, might be obtained by a coupled approach applying a consortium of techniques to estimate fluid flow rates and transport rates of relevant constituents supplied to the ocean as CH_4, N, P, S, and Si. This has to include investigations of "background" flux rates of these constituents supplied by river run-off and re-mineralisation processes as well as to determine flux rates of these elements in the water column and within the sediments. In this regard, future research initiatives will only be successful when interdisciplinary activities, including geological, geochemical, geophysical, modelling, oceanographic, sedimentological and biogeochemical methods in certain key regions are combined (cf. recommendation and background papers of working groups 1 "Microbial systems in sedimentary environments" and 3 "Margin building – regulating processes" this volume).

Due to different regional settings and the importance of fluid flow for the exchange of matter between sediments and the ocean and the variety of specific related problems, for the key-subjects defined earlier in this report, the following specific recommendations should be taken into consideration, as they are fundamental for a better characterisation of discharge sites and the assessment of the impact of fluid flow on ocean margin systems:

CH₄ and CO₂ fluxes from the deep geo-/biosphere to the oceanosphere

CH$_4$ and CO$_2$ fluxes from the deep geo-/biosphere to the oceanosphere

Pathways and transfer rates of greenhouse gases such as methane and carbon dioxide from deeper parts of the continental crust to the oceans are of great socio-economic interest since the driving processes control the formation and dissociation of gas hydrates, the accumulation of free gas in the subsurface and gas escape across the sediment/water interface. Reliable data of methane concentration in pore water are required to assess the earlier named processes which can only be obtained by pressurised sampling equipment. Due to a lack of such high quality data the study of gas hydrates has widely been limited to indirect observations in natural settings and experimental / theoretical approaches in home-based laboratories. With respect to fluxes of gases across the sediment/water interface and their impact on climate change a two-fold approach is proposed: (1) A process-orientated study should unravel the relative importance of carbon fixation by carbonate precipitation as a consequence of anaerobic methane oxidation and the formation of biomass versus the release of gases into the water column at seep and vent locations. (2) A regional and temporal orientated study is required to assess the impact of released gases on a large scale. For such an approach remote operated platforms, long-term moorings and 2D/3D coupled transport-and-reaction models are needed. The proposed studies on gas hydrate formation will also provide information required for slope stability assessment and the role of methane release for rapid climatic changes in the geologic past.

Fluid chemistry as a key to fluid flow pathways and origin

Fluid chemistry as a key to fluid flow pathways and origin

The chemical composition of fluids is the only direct evidence for the origin and pathways of fluid flow. However, primary chemical signatures are imprinted by chemical and microbiologically induced reactions and mixing of fluids from different sources. Increased knowledge on influences of temperature, pressure and mineral assemblages can be achieved from systematic categorisation of fluid composition and their respective conditions of origin. Such categorisation is lacking although extensive information are available from hydrocarbon, geothermal and research drillings. Additional information can be derived from thermodynamic calculations. The implications of this field of study are manifold: (1) Fluid origin in accretionary prisms will explain the relative significance of fresh water formation by dehydration of clay minerals, de-watering by consolidation and sea water circulation within the prism. (2) Improved knowledge on chemical conditions in the subsurface will help to localise high priority areas for research in the deep biosphere. (3) The potential of geothermal energy exploitation depends on the chemical composition of the used fluid apart from a high geothermal gradient. Precipitation of minerals due to pressure and temperature reduction during fluid ascend and during the re-injection of cold fluid into the subsurface must be excluded. Insight on the chemical composition of fluids as a result of the geologic setting will help to explore areas of geothermal energy exploitation.

Biota at seeps and vents at active and passive margins

Biota at seeps and vents at active and passive margins

The critical issues for future research on seep communities rely on the identification and quantification of the faunal (microbiota, benthic fauna) interactions with fluids on the sea floor and also in the subseafloor. It is important to consider the existence of deep bacteria (deep biosphere), the formation of methane through methanogenetic processes by anaerobic archaea and particularly the oxidation of methane under anoxic conditions. Further, the relative importance of biogenic formation of methane versus thermogenic origin needs to be evaluated.

Despite the evident role of symbiont bacteria their characterisation is hampered by the difficulty to culture and isolate them from host tissue. It is important to understand how the nature of the fluid influences the establishment of the different symbiotic associations, how they influence the diversity of the megafauna community directly linked to chemosynthetic production.

Whereas the signature of seeps at the sea floor is easily recognised through the presence of bacteria mats and areas of high density of biomass, the variations in terms of species composition as well as abundance already observed have to be analysed in detail with an interdisciplinary approach (biology and geochemistry) to understand the relationships between the community structure and the chemical concentration of various compounds in the fluids and the intensity of the fluid flow. In several known seep sites, living organisms have been considered as directly linked to chemosynthesis through bacterial symbionts and are indicators of active seep sites, whereas dead clam fields (cementeries) are evidence of cessation of fluid expulsion. The demographic structure of the population of the different species can be indicator of the more or less regularity of the fluid emissions is space and time. The duration of fluid flow and of the subsequent chemosynthetic based ecosystems are not known. No data exist neither on time variation and duration of fluid expulsion nor on consequences of time variability to the biological activity.

The temporal variation as well as the spatial variation of the biological activity has to be estimated through long term observation and regular visit (remotely operating vehicle (ROV) imagery survey very near the bottom) of marked locations. Ecological studies need to be undertaken at the species level, at the population level and at the community level to find qualitative and quantitative indicators for seep patterns. *In situ* experimental approaches using benthic chambers are also needed to enable estimates of the rate of organic matter production and transformation, the role of the original and rich ecosystems in the global carbon cycle as well as their role in the limitation of green house gases (methane) through the methanotrophy. Spatial and long term integrated studies including fluid flow measurements, chemical concentration investigations in the fluids and biological analyses, i.e. molecular taxonomy of microorganism and of representative invertebrates, bio-markers, larval dispersion and physiology, are suitable in selected areas representative of the various types of seeps and geological structures.

Submarine groundwater discharge

In near coastal zones along continental margins, the discharge of groundwater from the seafloor may be of major interest from the perspective of environmental concerns, and of the "loss" of drinking water to the sea. To assess the regional significance of SGD, time and cost efficient techniques are necessary which allow exploration of sites affected by SGD and to quantify discharge rates for a wide range of environmental settings including muddy and karst sediments and hydrological systems around small islands. Based on such surveys, the composition and temporal variation of emanating fluids has to be quantified by long-term benthic observations. In combination with numerical simulations, such studies might allow assessment of potential impacts on SGD systems due to nutrients, pollutants, and trace gases introduced from the onshore groundwater recharge area to the marine environment and possibly related to the potential use of emanating ground-water.

Numerical simulation

In recent years numerical models which describe flow, transport and reaction processes in ocean margin systems have merged to a powerful tool, as they are used e.g. to determine global budgets and they are the first priority tool which enables to make predictions, to simulate possible future evolution of fluid flow processes, and to characterise possible future scenarios. Additionally, they can help to understand complex coupled processes.

(Single-phase) Water flow and (multicomponent) transport models are already widely used in ocean margin systems. Models which are able to describe the interaction of fluid flow and the deforming porous structure and which are already used in other disciplines, e.g. mechanics or soil mechanics, should be applied and further developed in the field of ocean margin systems. Such models can be used for systems with tectonically driven deformations or for systems with landslide problems.

Geochemical and biological reaction models are well established in 1D. For systems where e.g. advection can not be neglected these models are not sufficient as they cannot account for lateral transport and therefore need to be further developed to 2D and 3D. This holds true for large and small scale spatial investigations, e.g. fluid flow in accretionary prisms and bioirrigation at fauna-rich sites. Multiphase flow processes and (non-)isothermal multiphase / multicomponent flow and transport processes are only partially understood. Due to the highly complex processes, the application range of such models is generally limited to smaller scales. They were already successfully applied to SGD where an interaction of fresh water, salt water and methane gas flows occurs. In the future, such models should be further developed to simulate flow and transport processes in gas hydrate systems. Finally, it needs to be mentioned that the quality of model predictions strongly depends on the quality and amount of laboratory and field data.

Acknowledgements

We would like to thank I. Mehser, who organised the workshop which gave the frame for this manuscript, and D. Hebbeln, who did the editorial work for this volume, as well as A. Hampel for carefully and critically reading a close-to-final version of this manuscript.

References

Bolton A, Maltman A (1998) Fluid-flow pathways in actively deforming sediments: The role of pore fluid pressures and volume change. Mar Petrol Geol 15: 281-297

Dia AN, Castrec-Rouelle M, Boulègue J, Comeau P (1999) Trinidad mud volcanoes: Where do the expelled fluids come from? Geochim Cosmochim Acta 63:1023-1038

Haese RR (2002) Macrobentic Activity and its Effects on Biogeochemical Reactions and Fluxes. In: Wefer et al. (eds) Ocean Margin Systems. Springer, Berlin pp 219-234

Hinkelmann R, Helmig R (2002) Numerical Modelling of Transport Processes in the Subsurface. In: Wefer et al. (eds) Ocean Margin Systems. Springer, Berlin pp 269-294

Mann D, Kukowski N (1999) Numerical modelling of focussed fluid transport in the Cascadia accretionary prism. J Geodynamics 27:359-372

Mienert J, Berndt C, Laberg JS, Vorren TO (2002) Slope Instability of Continental Margins. In: Wefer et al. (eds) Ocean Margin Systems. Springer, Berlin pp 179-193

Moore JC, Vrolijk P (1992) Fluids in accretionary prisms. Rev Geophys 30:113-135

Schlüter M (2002) Fluid Flow in Continental Margin Sediments. In: Wefer et al. (eds) Ocean Margin Systems. Springer, Berlin pp 205-217

Sibuet M, Olu-Le Roy K (2002) Biological Indicators for Seeps. In: Wefer et al. (eds) Ocean Margin Systems. Springer, Berlin pp 235-251

Soetaert K, Middelburg J, Wijsman J, Herman P, Heip C (2002) Ocean Margin Early Diagenetic Processes and Models. In: Wefer et al. (eds) Ocean Margin Systems. Springer, Berlin pp 157-177

Suess E, Bohrmann G, von Huene R, Linke P, Wallmann K, Lammers S, Sahlig H, Winckler G, Lutz RA, Orange D (1998) Fluid venting in the Aleutian subduction zone. J Geophys Res 103:2597-2614

Zabel M, Hensen C (2002) The Importance of Mineralisation Processes in Surface Sediments at Continental Margins. In: Wefer et al. (eds) Ocean Margin Systems. Springer, Berlin pp 253-267

Benthic Biodiversity Across and Along the Continental Margin: Patterns, Ecological and Historical Determinants, and Anthropogenic Threats

J.D. Gage

Scottish Association for Marine Science, Dunstaffnage Marine Laboratory,
Oban PA34 4AD UK
corresponding author (e-mail): jdg@dml.ac.uk

Abstract: Deep water environments along the continental margin show mainly depth-related, but also significant along-slope, patterns in the composition, biomass and diversity of the benthic biota. Superimposed on this broad-scale pattern, local-scale habitat variability, including nature of the sediment, hydrodynamics and topography, can also be detected. However, along-slope variability is poorly understood, because only a relatively small number of sites have as yet been investigated and because sampling effort is generally sparse and poorly replicated. We are unable to predict benthic biology with a high enough resolution to provide useful precision for specific areas. Furthermore, the benthic biology associated with highly localised conditions, such as areas of seafloor fluid seepage or canyons, remains largely unknown, while knowledge of patterns in smaller benthic size classes is less well developed than for larger size classes. Progress in knowledge of benthic biodiversity on the continental margin is severely constrained by present taxonomic deficiency, with an increasingly large imbalance between available taxonomic capability and the extraordinarily rich, yet significantly undescribed, biodiversity. There is an urgent need for the application of molecular genetic tools to problems of understanding vertical and horizontal ranges of organisms. Little is presently known, even though studies of morphological variability have been applied to some groups of animals. Understanding of relevant processes operating both at (1) ecological time scales in maintaining high local-scale species richness, and at (2) historical time scales influencing speciation, needs to be developed before meaningful assessments of long-term anthropogenic impacts can be made. Yet, such impacts, particularly those from presently unregulated deep-sea trawling for non-quota fish stocks along the European continental margin may have already changed the ecosystem from its previously pristine state. Deep-sea trawling causes a direct physical impact, which is known already to have damaged slow-growing cold-water corals in some areas. Deep-sea fishing also may have an effect on the background ecosystem, including the benthos, as a result of return to the sea of biomass discarded from the catch. Although care needs to be taken in interpreting the outcome of specific anthropogenic impacts, knowledge of (1) likely sensitivities of species and (2) overall ecosystem resilience, will be informed by studies of responses to natural disturbance.

Introduction

Traditional marine biology has long been conditioned by what is most accessible. Our knowledge has perhaps been overly influenced by what can be learned from intertidal and shallow subtidal systems that represent only a vanishingly small proportion of the total ocean bed and its biological populations. On the shore, easier access and greater visibility of biota, either attached to, or hiding under rocks on the shore or in the shallow subtidal accessible to scuba, has provided a rich fund of knowledge compared to what we know about life in deeper waters. Here, direct observations might only be possible from hugely expensive manned submersible and deep-diving ROVs. Yet most of what we know of the rich biodiversity at the continental margin and beyond is still very largely based

From WEFER G, BILLETT D, HEBBELN D, JØRGENSEN BB, SCHLÜTER M, VAN WEERING T (eds), 2002,
Ocean Margin Systems. Springer-Verlag Berlin Heidelberg, pp 307-321

on what can be retrieved in trawls and grabs lowered to the bottom on hundreds or thousands of metres of wire rope.

I shall restrict discussion here to the ecosystem on the continental margin that is covered with a drape of sediment, rather than exposed bedrock. The sediment covered areas in any case are probably the most important in terms of area, and are part of the most extensive, continuous habitat on the face of this planet, the deep-sea bed. This habitat used to be regarded as a biological desert because of its severely food-limited biomass.

Yet if we could land on the abyssal plain and fly up the bottom of the continental margins, we would notice that as depth shallows, biomass as indicated by epifauna and bioturbation increases together with environmental variety. In deep water we would see a sediment landscape dominated at the small-scale by biogenic features, such as pits and mounds, created by burrowing animals (Fig. 1d). These animals exploit the increasing fall-out of organic material from enhanced surface water production above. We would notice that in places, deep canyons dissect the slope to provide (1) impressive topographic relief and (2) the focus of relatively intense down-slope transports. Approaching the shelf edge we would start being buffeted by water motions that by moulding and transporting sediment create a different sort of landscape, often current-smoothed and sometimes dominated by fields of dynamic bed forms. Here organisms may be abundant, such as suspension feeders (Fig. 1b), some with bizarre morphologies e.g. glass sponges and xenophyophores. Xenophyophores, the largest protozoans on Earth (Fig. 2a, b), probably trap particles moving in the current by means of a pseudopodial web. Biota become less obvious as grain size increases and the landscape becomes dominated by bedforms created by currents, and by coarse, mixed deposits and large boulders where only encrusting and swimming animals are common (Fig. 1a). These conditions continue up to the shelf/slope break. This changing landscape, together with the impingement, in places, of surface-water mass structure on the slope, such as oxygen minimum zones, causes major changes in benthic processes and hence the composition and structure of populations. This complex environmental hetero-geneity, by creating geographic isolation and by forcing species to adapt to changing conditions, acts in generating biodiversity. On an historical time scale the continental margin may act as a source of biodiversity to adjacent areas in deeper water, and, perhaps, also for shallower water at high latitudes.

In this background paper I shall review these patterns and the processes thought to generate biodiversity patterns as inherent parts of ocean margin systems. The likely controlling processes span time scales from those associated with ecology – the relationship of organisms to their environment – to the longer term associated with evolutionary radiation. Shorter term processes affecting biodiversity will also include hydrodynamics and spatial and temporal patterns in organic flux, while longer term processes will include plate tectonics and processes responsible for re-distributing sediment and moulding relief (Table 1). Hence, the potential for enrichment of our present understanding by interdisciplinary discussions is substantial.

The suite of strong vertical gradients in physical, chemical and biological parameters from the shallow continental shelf to the abyssal deep-sea help create one of the greatest environmental gradients on the face of the earth, the gradient associated with increasing depth on the continental margins. In the section below I shall attempt to summarise what we know of the predominant ecological patterns associated with this environment, and how these vary both across and along the continental margin.

However, one caveat is worth repeating. Much of what we know depends critically on secure taxonomy. The large-scale *ASCAR* programme for deep-sea macrobenthos sampling undertaken in the 1980s on the continental slope off New England estimated that 58% of the 798 species sorted from a total of 90,677 individual macrofaunal organisms were undescribed (Grassle and Maciolek 1992). Yet this area is one of the best known in the world. From this study a conservative guess may be that well more than half of the macrobenthic species that might be encountered in new sampling on the continental margin may turn out to be undescribed. Clearly a healthily de-

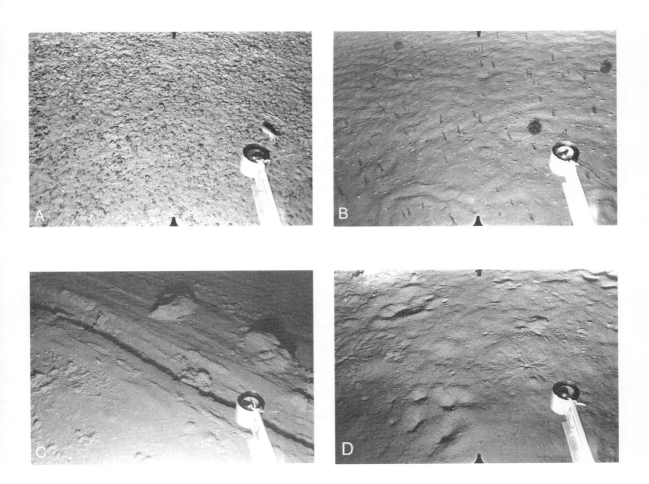

Fig. 1. Seabed photographs (oblique view covering about 1.2 m wide by 2.5 m from front to back) from the continental margin off the Western Isles of Scotland. **a)** 300 m depth, benthic habitat influenced by hydrodynamic activity, showing a fairly uniform field of pebble gravel and sand. Although there appears to be evidence of active hydrodynamic activity at the time of the photograph, other photographs taken nearby show ripple bed forms along with evidence of transport of sand-sized sediment particles. The sparse benthic fauna consists of motile sediment-surface dwelling organisms, such as the small octopus and brittle star visible just above the compass at middle right, animals encrusting rocks and virtually no burrowing fauna. **b)** 700 m, muddy sand smoothed and marked by extensive biogenic activity. The tyre-tread like tracks are made by the large surface-deposit feeding sea urchin *Spatangus raschi* (one is visible just above the compass). Dense population of the brittle star *Ophiocten gracilis*, and field of an unidentified cerianthid anemone, with tentacle crowns oriented facing upstream into the current are also visible. **c)** 885 m depth, showing a muddy bed off St Kilda strongly marked (the deep furrow running diagonally across the view) by demersal trawling. Clods of recently disturbed sediment that have fallen back to the seabed are also visible in the foreground, while the larger mounds visible further back are probably also disturbed in this way. The small, white brittle stars visible in the foreground are specimens of *Ophiocten gracilis*. **d)** 1,500 m depth, showing a soft muddy bottom with numerous biogenic surface traces indicating extensive re-working by burrowed deposit feeders. The radial pattern surrounding a central hole visible to the left of the top of the compass is an example of this activity made by a polychaete, or possibly echiuran, worm. There is no evidence of hydrodynamic activity from any photograph taken at this depth level at this site, so that the micro-landscape is entirely biogenic.

	Primary/secondary factors	Dependent modifying factors	Independent modifing factors
Factors influencing food and habitat availability	Depth, Latitude (insolation?) Water movement	Environmental variability – upwelling, oxygen minima, local turbulence, sediment transport, eddy kinetic energy generating benthic storms, episodic down slope transports (e.g. turbidity currents)	Stochastic, local-scale variability derived from biotic interactions e.g. competition and predation, Stochastic recruitment
Time scale	Long term (years x 10^6)	Medium (years x 10^4)	Short (years)
Structural scale of impact on community	Regional species richness; Mega-structure (distribution of major trophic and motility groupings in communities)	Macro-structure (partitioning of functional groups which define successional phenomena and life-history characteristics)	Local scale species richness; Micro-structure, Inter- and intra-species interactions creating patchiness

Modified from Pearson and Rosenberg (1987)

Table 1. Conceptual scheme showing how scales of habitat variability influence food and habitat availability of benthic biodiversity at the continental margin.

veloping taxonomy must accompany all future studies of the ecology and sensitivies of benthic organisms on the continental margin.

Depth related pattern in benthic biomass on the continental margin

Description of pattern

The best-documented depth-related pattern is that of the exponential decline in biomass of macro-benthos (metazoans larger than about 0.25-0.5 mm up to about 1 cm or more) with increasing depth. There is considerable overall scatter, but studies show a coherent, site-related pattern. It is not known whether these follow any regional- to latitudinal-scale pattern, but a relationship to productivity on the adjacent continental shelf as well as local conditions, such as upwelling (see below), more directly affecting productivity on the continental slope, would be expected. Although data are more sparse for larger and smaller size classes,

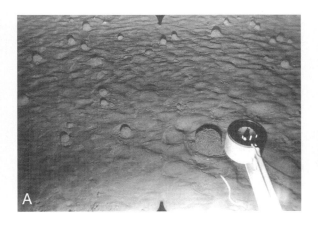
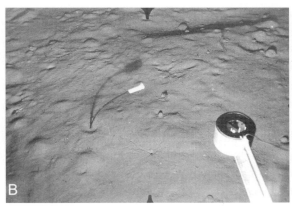

Fig. 2. Epifaunal organisms vulnerable to mechanical disturbance and smothering by resuspended sediment photographed on the continental margin off the island of Lewis. **a)** Rippled muddy sea bed at 1,108 m depth showing a field of xenophyophores (probably the species *Syringammina fragilissima*). **b)** Highly bioturbated seabed at 1,295 m showing cone-like mounds made by burrowing benthic animals. The large stalked structure is a hexactinellid, or glass sponge, *Hyalonema*. The glass-fibred stalk of the animal raises it off the seabed into better feeding conditions. Both photographs were taken in the course of an environmental assessment of Block 154/1 undertaken by SAMS for Enterprise Oil Ltd in 1998.

similar trends seem to apply to both smaller (meio- and microbenthos) and larger (megabenthos) populations, although the relationship to the better-known pattern with macrobenthos is unclear. However, some studies have shown that the reduction in biomass may not be monotonically decreasing, but may decline and then increase at upper slope depths to decline again below, a trend which can be related to enhanced, flow-related feeding conditions (e.g. Fig. 1b)

This pattern is for all intents and purposes stable through time. A small amount of seasonal variability in density, but not in biomass, of benthic organisms at the EU-funded *OMEX* study site in the NE Atlantic, was recorded from mass seasonal settlement of echinoids and brittle stars. One of these, the brittle star *Ophiocten gracilis*, may form dense populations on the upper continental margin in the NE Atlantic with settlement of larvae over a large depth range on the margin and the adjacent deep-sea basin (Gage and Tyler 1981; Lamont and Gage 1998). However, the dynamics of such species appears to be unusual, and no significant seasonal changes in abundance of fauna could be detected during the extensive *ASCAR* programme on the US Atlantic continental slope (see Grassle and Maciolek 1992).

Causes of pattern

The depth-related decline in benthic biomass is classically ascribed to mid-water consumers, which degrade the quality and quantity of sedimented particles from surface production. Regional-to-local scale variability in this pattern may arise from a multitude of causes. Coastal-zone productivity and land run-off may, in places, be important additional sources of input of organic material for benthic populations on the continental slope. Effects of land run-off are particularly noticeable in terms of enhanced terrigenous sedimentation near the mouths of major rivers, such as the Amazon, carrying high suspended particle loads. Upwelling on the continental margin may cause higher than expected benthic standing crops by "fuelling" from high surface production. Enhanced pelagic consumer utilisation, caused by elevated water temperature by reducing particle flux to the bottom, probably also causes the markedly reduced benthic biomass observed in the Red Sea and in the Mediterranean Sea. But in some areas, such as the Arabian Sea, upwelling-driven surface production causes an oxygen minimum that can affect the benthic biology on the continental margin. Although reduced consumer utilisation in mid-water may cause more organic matter to reach the bottom to stimulate production, hypoxia occurring where the oxygen minimum impinges on the bottom initially increases, but eventually reduces, benthic biomass (Levin and Gage 1998) following the classical model of Pearson and Rosenberg (1978). Enrichment also may occur through predation by benthopelagic fish of mesopelagic micronekton populations impinging on the continental slope, and through defaecation and mortality of seasonally migrating fish species tracking the continental margin, such as shoals of blue whiting moving along the continental margin in the NE Atlantic. Up- and down-slope movement of benthopelagic biomass may occur as a result of ontogenetic migration of demersal fish, such as synapho-branchid eels. Furthermore, the non-viable post-larval export of larvae from the upper slope populations of *Ophiocten gracilis* can be regarded as organic export, with significant quantities of larvae found in both sediment traps and bottom samples at mid-basin depths off the continental margin.

Where such features exist submarine canyons have long been suspected to serve as important conduits for organic material, and because of this, they support considerably higher biomass than adjacent slope areas (e.g. Vetter 1994). Variability may be associated with tidal currents and as such occur within the turnover time of individuals in populations. But mass wasting caused by slope instability may, by providing a stimulus to high benthic productivity, cause elevated biomass (e.g. Schaff et al. 1994). However, large events, such as those triggered by earthquakes, may result in mass extinction caused by smothering varying, but sometimes huge, areas with re-suspended sediment. These events are followed by re-colonisation, probably, at least initially, at enhanced levels of biomass because of the enhanced availability of organic carbon.

Hydrodynamics influencing particle aggregation and transport are pervasive sources of variability in benthic biomass on the continental margin. Areas experiencing strong near-bottom currents flowing against the continental margin, such as along-slope flowing slope currents, may support biomass levels well beyond that expected on

the basis of depth and the benthic standing crop found at less energetic sites nearby (e.g. Flach and Thomsen 1998). Highly localised upper-slope areas exposed to breaking internal waves are also thought to enhance particle flux and hence benthic biomass (e.g. Klitgaard et al. 1997).

Depth-related changes in benthic community composition on the continental margin

Description of pattern

In contrast to the almost imperceptible change in benthic community on the abyssal plain, species composition may alter rapidly with changing depth on the continental slope. Given the changing ecological and physical conditions this seems unsurprising. Describing this pattern of changing composition in biodiversity in objective terms has provided more of a challenge than for species richness or biomass, which can be summarised as single values for particular sites or depth levels.

Comparisons are frequently difficult because different workers have addressed differing size classes and have employed differing sampling methods. Moreover, the terminology applied to summarise trends has sometimes been based on incomplete knowledge and may reflect the author's subjective expectation as much as real pattern. While cosmopolitan deep-water species are commonly eurybathic, stenobathic species (restricted to a narrow bathymetric range) have a restricted geographic distribution. Furthermore, while species living deeper than 2,000 m might be either steno- or eurybathic, species occurring deeper than about 2,500-3,000 m ("true abyssal species") possess narrower bathymetric ranges, while the species occurring at less than 2,500-3,000 m ("bathyal species", have large vertical ranges. However, Pineda (1993) established from published data that maximum vertical range must be geometrically dependent on mean depth of occurrence when the upper and lower boundaries of theoretically possible range (the sea surface, and the deepest sea floor) are taken into account. These boundaries constrain vertical ranges to be narrow on the upper continental slope (200-1,500 m) depths, expanding with depth to a maximum at continental rise depths (1,500-3,500 m) and then narrowing again on the lower continental rise (3,500-4,500 m) and abyssal plain (4,500-6,000 m) depths.

This agrees with the concept of two major faunal provinces or depth zones below the shelf-break that determine faunal composition. On the one hand is the slope or bathyal zone incorporating elements from both shallow and very deep water; on the other is the true deep-sea fauna characteristic of the abyssal zone. The upper limit of the deeper zone was thought to lie roughly at 1,000 m depth and extending on to the abyssal plain. But with increasing depth-related resolution of sampling later workers have moved the upper limit of the abyssal much deeper, with an intervening, "semi-abyssal" zone, or a bathymetrically extended bathyal zone. This is the zone that later work showed to widen from low to high latitudes in response to increased seasonal mixing and deepening of the permanent thermocline. This may result in a fauna on the Polar shelf essentially similar to that of the slope in lower latitudes. This was developed from the earlier idea of equatorial submergence, based on the concept of narrow temperature tolerances with latitudinally converging bathymetric limits between the two from low to high latitudes. Sensitivity to even small temperature change was, until quite recently, thought to determine a "stenothermal" bathyal fauna characteristic of the relatively stable regime existing below the depth of the permanent thermocline. However, there were also recognised a limited suite of "eurythermal" species which although often numerous contribute little to faunistic characterisation of vertical zones. These are the species which although having their main centre of distribution on the slope may also be found in so-called "archibenthal enclaves" close inshore in deep fjords, or in deep submerged caves along the coast (e.g. Vacelet et al. 1994).

These largely notional zones based on subjective synthesis of records over wide geographic areas have mostly been confirmed in subsequent objective analysis using multivariate analysis. Using such techniques it is possible also to recognise the step-like changes that reflect zone boundaries (e.g. Gage 1986), and more subtle non-repeating, sequential changes in species replacement where peaks in population distribution may overlap over the depth gradient (see Carney et al. 1983). However, different major taxa have different patterns of zonation, with gastropods and cumacean crustaceans in particular having narrower depth distributions than either polychaetes or bivalve molluscs.

Pattern along the continental margin at the larger scale is much more difficult to recognise from the limited amount of comparable data available covering a large enough scale. Sadly incomplete, analysis of some faunal groups in the depth-related transects worked at various sites in the N and S Atlantic using standard methods by Sanders and his associates show a varying pattern depending on taxon. Cumacea, for example, show considerably less similarity at the same depth level between transects than protobranch molluscs (Jones and Sanders 1972; Sanders 1977).

Controls of depth-related distributions

Temperature sensitivities may be seen as a confirmation in the phenomenon of equatorial submergence. However, experimental transplants of fauna from deep water to shallow water, or from shallow to deep water, have indicated that survival may be determined more by physical or biological constraints in larval dispersal, rather than by tolerances of the adult. These observations, along with the presence of flourishing populations of hexactinellid (normally bathyal) sponges in submerged caves along the Mediterranean coast (Vacelet et al. 1994), suggest that hydrostatic pressure or the precise temperature sensitivity assumed in previous zoogeographic work cannot always be important. Indeed, experiments on thermal tolerances of larvae of bathyal echinoids with planktotrophic early development have shown a variable response depending on species.

However, it is known that hydrostatic pressure may, through effects on membrane structure and permeability and in the kinetics of enzyme systems, imply adaptational commitment to a prescribed depth range (reviewed by Somero 1992). This seems to be reflected in observed effects of hydrostatic pressure and temperature on the embryogenesis and early development of planktotrophic larvae of echinoids (Young et al. 1996).

Le Danois (1948) related vertically zoned faunal associations along the European margin to physiography and sediment types found at different depths. Studies of coarse sediments and vigorous hydrodynamic environments associated with the shelf edge and outer continental shelf using new and more effective sampling gear have made it clearer that the physiographic shelf break may provide a change only in properties associated with different water masses. Although many studies have confirmed the shelf break as a zoogeographic boundary, in the stormy N Atlantic, where the zone of winter mixing may extend as deep as 800 m, muddy sediment may not appear until 500-600 m depth, and current-moulded sediment bed forms until below 1,000 m (e.g see Fig. 2a). Even if quieter conditions may prevail to twice this depth, deep boundary currents at the base of the slope may again generate current-moulded bed forms, and in certain places propagated eddy energy may generate periodic benthic storms (e.g. Thistle et al. 1985). Sediment type, which may itself reflect other more directly acting agents such as flow conditions, remains a useful proxy for environmental factors likely more directly to govern the composition and distributions of benthic communities on the continental margin.

With improved knowledge of ecology, such as feeding type, it has been possible to link changes in distribution of individual species to environment, particularly hydrodynamic conditions involving resuspension and transport of organic particles. For example the, often dramatic, changes in relative proportions of suspension and deposit feeding macrobenthos seem to closely track prevailing flow conditions. Other changes may be more gradual and more difficult to detect. For example, biological interaction, such as competition, might control vertical range of species. Animals farther up the food chain are thought to be less sensitive to predatory reduction than prey populations, and hence are more subject to competitive crowding and variability in populations that might compress vertical range compared to the prey populations that would have wider vertical ranges (Rex 1977).

Along-slope changes in species composition

Perhaps the most spectacular features, submarine canyons, are a major source of along-slope discontinuity, yet remain poorly known in terms of resident biodiversity. Where their benthic biology has been explored they may show much modified benthic populations, with species common on adjacent sections of slope absent or at reduced density, whereas other species may be found only in or near canyons (e.g. Rowe 1972). Although not now active as major conduits for material eroded from the continents as they were at times of reduced sea level, canyons are still dynamic environments in terms of hydrodynamics and associated down slope processes. They may channel large quantities of sediment and organic detritus from the adjacent continental shelf. Local factors, such as seabed seepage, changes in sediment perhaps associated with currents, and declivity, have been suspected to cause some of the dramatic changes observed in seabed photographs.

Observations from towed camera sledges and manned submersibles have revealed faunal changes occurring on the continental margin at much smaller spatial scales than previously possible. In some locations high-density zones or fields of certain organisms, such as sponges and xenophyophores, have been detected (e.g. Fig. 2a) perhaps associated with hydrodynamic turbulence causing enhanced particle flux.

At the larger scale, Sanders (1979) proposed a relationship between vertical range, zoogeographic range, physiological tolerance and tendency to speciate. Species with largest vertical range would have the broadest zoogeographic range, greatest physiological tolerances and least tendency to speciate. Conversely, taxa with narrow vertical range have restricted zoogeographic range, least physiological tolerances and the greatest tendency to speciate. Testing this idea with data currently available is still difficult, but could be rewarding in terms of a better understanding of processes controlling faunal patterns on the continental margin. Here it is easy to envisage how the strong environmental gradients may be important in providing the selective pressures which drive speciation in bottom-living animals. Wilson and

Hessler (1987) suggested that the proximity to variability associated with surface layers of the ocean, variable topography and down-slope processes and energetic hydrodynamics will provide enhanced opportunities for speciation. Phenotypic variation in gastropod molluscs seems to decrease sharply down-slope from the shelf edge, and this is supported by strong across-slope genetic differentiation from DNA sequencing (Etter et al. 1997). In some populations there appears to be high levels of gene flow (e.g. Hensley et al. 1995). The brittle star species, *Ophiomusium lymani*, which has fairly wide bathymetric limits, is an example of this, and is thought to have a cosmopolitan distribution along the continental margins of the world's oceans. The little genetic heterogeneity detected from different sites along the continental margin seems unrelated to depth. The variability observed may have come about through fragmentation of suitable habitat along the continental margin as a result of environmental and topographic discontinuities. Creasey and Rogers (1999) provide an overview of what little is known on the population genetics of deep-sea animals.

Studies on the thermal and pressure tolerances of planktonic larvae of shallow- and deep-water sea urchins suggest potential for colonisation into deeper water (Tyler and Young 1998). There is a long-held idea that the Antarctic shelf is a major source of evolutionary radiation into the deep-sea via the continental margin. Tyler et al. (2000) interpret a potential for deep-sea invasion from the deep Antarctic continental shelf from the low-temperature dependent pressure tolerance of embryos of the shallow-water Antarctic species *Sterechinus neumayeri*. Hence, as discussed earlier, present distributions may more reflect constraints on dispersal or emigration of populations than by a restricted range of environmental tolerances (Gage and Tyler 1991).

Diversity gradients across and along the continental margin

Description of patterns

Intuition may suggest that only a limited suite of highly specialised species might be able to tolerate the extreme physical conditions and impoverished nutritional input of the deep-sea. This was countered by work in the 1960s along a transect from the continental shelf over the continental margin into deep water in the NW Atlantic off New England which demonstrated precisely the opposite (Sanders and Hessler 1969). However, the pattern of, contrastingly low, diversity on the shelf does not apply everywhere. High macrobenthic species diversity is also shown in muddy sediments in deep, inshore fjord environments in the NE Pacific (e.g. Sanders 1969), and on the outer continental shelf of the North Sea in the NE Atlantic (e.g. Gray 1994).

In later work focussing on a limited number of taxa, Rex (1983) argued from plots of species occurrences against depth in the NW Atlantic that numbers of species increase from shallow water to peak at mid-slope, but decline again on the continental rise and onto the abyssal plain. His fitted parabola suggested that peaks occurred at differing depth with different faunal groups. In the NE Atlantic the position is less clear. Even if sampling has not extended to the continental shelf, there seems to be a mid-slope peak in diversity of polychaete worms and other macrobenthic fauna in the Rockall Trough (Gage et al. 2000). Slightly further north, in the Faroer-Shetland Channel, downslope peaks in biodiversity seem to be associated with boundaries in water mass structure. But further south in the NE Atlantic, diversity increased monotonically with increasing depth from the outer continental shelf, over the shelf edge and right down to the continental rise at the *OMEX* site on the Goban Spur (Flach and de Bruin 1999).

There are fewer data available for benthic fauna smaller than macrobenthos, but samples from different depths indicate a non-linear relationship between diversity and depth (Boucher and Lambshead 1994), with bathyal and abyssal environments showing higher diversity than in shallow-water habitats. The megafauna, although in general better known than the macrofauna, shows a somewhat variable pattern at different sites, but like demersal (bottom-living) fish, seem to show greatest diversity at slope depths (reviewed in Gage and Tyler 1991).

Causes of high species richness in the deep-sea

The causes operating at the ecological time scale allowing such high numbers of species to co-exist at the local scale are still unresolved, but are unlikely to be simple. As results of more studies relating species richness to differing environmental conditions on the continental margin emerge, so do the number of potentially implicating factors. Historical and geographic factors, such as glacial extinctions (e.g. Dahl 1972) and area of ocean basins also must be of great importance (Rex et al. 1997), even if the role of regional processes in setting local species diversity is far from clear, even in terrestrial communities.

Assemblages of species as are typically observed in a sample, and the expression of pattern between them along an environmental gradient such as depth, may be seen, as on land, to be the result of interplay between ecological processes and environmental variability at local and regional scales. At the local scale, observed pattern, such as macrobenthic biodiversity and aggregation among species, is thought to reflect a mosaic of variously interacting populations fluctuating through time and space. At the regional scale, pattern is the result of aggregation in local processes from an interacting set of community patches mediated by environmental change.

In the much better-known assemblages of rocky intertidal habitats, where food and recruitment is rarely limiting, community structure and dynamics are primarily driven by negative effects, such as competition, disturbance and predation. Assemblages of subtidal soft sediments are, by contrast, much less limited by space and recruitment, or by competitive interactions mitigated by large predators, than by sediment composition/hydrodynamics, and food supply (rate and spatial pattern). Limitations from spatial and temporal variability in detrital food supply may be particularly acute in deep water because food supply characteristically limits population growth. The prevailing paradigm of high local-scale species richness in the deep-sea emphasizes the importance of spatial patchiness. This may result from long-lasting alterations of bottom topography by larger benthos and fluctuations in bottom flow which re-distribute detrital material, and by more conspicuous patchiness caused by macroalgae and seagrass lying on the bottom. This patchiness is thought to permit the continued coexistence of a high diversity of species and higher taxa (Grassle 1989). The large surface area with few barriers to dispersal, but with very low densities of larvae, are also features which, through a highly stochastic recruitment of larvae to open, patchy habitat, are thought to contribute towards high species richness in the deep-sea. The open environment is well suited to passive transport of propagules, allowing widely separated sub-populations determined by a temporal mosaic of small-scale patches of organic enrichment and disturbance to contribute to local populations despite the large distances traversed. High species richness is hence maintained, against a background of low productivity, by spatio-temporal patchiness and by sporadic, discrete disturbance events occurring against a background of relative constancy in environment

However, supporting data from seabed sampling has proved difficult to obtain. For example, directed sediment sampling from a manned submersible at scales of 0.1 – 100 m of biogenic features such as pits and mounds on the continental margin, suggest that sediment heterogeneity exerts only minor influence on community structure. The benthic assemblages appeared homogeneous at these small scales (Schaff and Levin 1994).

Causes of down-slope patterns in faunal diversity

The underlying reasons driving such observed pattern has generated much discussion. Possible causes suggested by positive correlation include environmental gradients in nutritive flux, biological interactions of various sort and environmental heterogeneity, such as particle-size diversity. However, it is well to remember that such patterns in nature may not necessarily have an underlying causal explanation in terms of biology. For example, the total area of the habitat itself, as in other sorts of habitat, has been shown as a reasonable predictor of total species richness. Apparent patterns may even be spurious. Supposedly parabolic patterns in species diversity may result from random re-ordering of depth-range data within the geometric constraints of the two spatial end-members. The parabolic pattern can be annulled by removal of species with large vertical range simply because they overlap more with others at the mid-point of their range (Pineda and Caswell 1998).

There are few data addressing variability in benthic diversity along, rather than across, the continental margin. In the *ASCAR* study a 180-km long section of the continental slope off New Jersey and Delaware showed very little change in diversity in 160 samples from 9 stations at 2,100 m depth, but the samples continued to accumulate previously uncollected species right along the transect (Grassle and Maciolek 1992).

At the larger scale comparisons from site to site on the continental margin are extraordinarily difficult to make because of differences in approach and protocol by different workers. One of the few studies conducted using the same methodology by the same group of workers was that of Sanders and his associates in the Atlantic. One problem with this study is that samples are essentially non-quantitative so that differences may simply reflect differences in sampling efficiency related to local conditions such as sediment. Rex et al. (1993) have used these data in various comparisons, which they argue demonstrate a latitudinal gradient in species diversity on the continental slope in the Atlantic Ocean.

Whatever the validity of this, one thing is clear; the continental margin, mainly through higher faunal densities, seems to support a considerably greater density in species richness than the rest of the deep-sea. What factors of the continental slope may particularly contribute towards high species richness? Even if genetic variability is highest along the centre of a species' range (reflecting weak selective pressures) environmental complexity will, by causing breaks in suitable habitat, promote differentiation amongst peripherally isolated populations according to the Founder Principle of Ernst Mayr. This would cause "leapfrog" patterns in geographic variation of species along the continental margin. This may have occurred several times through time with differing subsets of the original genotype to give rise to several new species to enhance the regional species pool. Clearly, mass extinctions, such as occurred through glaciation events, will re-set this process at the regional level causing large-scale pattern in species richness that is unrelated to ecological-scale processes, and which may, or may not, be related to latitude.

Effects of anthropogenic intervention on biodiversity on the continental margin

Introduction

There is a growing scale of intervention by man in the deep waters of the continental margin which poses a threat of disturbance to what is perceived as a pristine ecosystem. Because of its rich biodiversity, perceived very low rates of population turnover, and the highly tuned partitioning of habitat, the deep-sea benthic community of the continental margin was considered to be part of an ecosystem particularly sensitive and vulnerable to perturbation. Research over the past two decades has changed this view by revealing a rapid response to environmental forcing such as nutritive flux and a probably largely stochastic community assembly at the local scale from an "open" environment. Furthermore, there is a range of rates in population processes broadly similar to that in shallow water. Nevertheless, imparting precision to prediction of likely impact has proved difficult because the remoteness and under-study of the deep-sea provides insufficient knowledge to provide the level of risk assessment currently available in the shallow-water environment. We shall consider from what has been discussed above what aspects of biodiversity on the continental margin may be most vulnerable to disturbance. Studies of parts of the deep-sea that are subject to relatively frequent, large-scale natural hydrodynamic disturbance, such as benthic storms, or are stressed by periodic or permanent hypoxia, are informative. These areas show a similar generic response of reduced species richness and increased dominance by the most common species or by early growth stages (e.g. Thistle et al. 1985; Levin and Gage, 1998). While care needs to be taken in interpreting specific impacts, we can be encouraged, from parallels with responses observed in shallow-water benthic communities, that study of natural disturbance can provide a useful starting point for prediction of anthropogenic impacts.

Nature of likely impacts

Let us disregard for the moment the unknown sensitivities to changes in their environment of large non-motile epifauna (particularly groups almost unique to the deep-sea, such as xenophyophores, glass sponges and cold-water corals). Let us also assume that sensitivities to environmental change of macrobenthos in deep-sea sediments are broadly similar to those of the similar taxa in muddy sediments, for example, on the continental shelf in the North Sea. It then seems reasonable to expect that indices of community structure now routinely applied in monitoring studies on the continental shelf will be equally informative in deep water. Because the most common species in the deep-sea show little variance between sample replicates or seasons, their relative abundance ought to be a relatively sensitive measure of man-derived perturbation. This should be expressed by reduced species richness, through the elimination of rare species, and increased community dominance by population expansion of a few relatively tolerant species. There would also be reduction in mean body size of dominant species as a result of a demographic shift to younger individuals characteristic of rapidly expanding populations. This characteristic has already been found in deep-sea areas subjected to periodic benthic storms (Thistle et al. 1985).

However, the prediction of impacts at the organism, rather than community or population, level is more difficult than in shallow water. This mainly arises from difficulties in collecting and maintaining deep-water organisms alive under controlled conditions. The need to reproduce the hydrostatic environment of deep-sea animals, even if presently possible, imposes a major constraint to experimentation of sub-lethal effects without large expense relative to budgets considered acceptable in shallow-water work. However, despite what might be inferred from their restricted bathymetric distribution, the successful live culture of carefully collected organisms shows that at least some deep-sea organisms may flourish at, or near, atmospheric pressure. The ongoing public exhibit of a range of living deep-sea organisms at the Monterey Bay Aquarium in California should also be mentioned here along with successful laboratory culture of the deep-water coral *Lophelia pertusa* in aquaria in Norway and at the Dunstaffnage Marine Laboratory in Scotland. However, it does remain possible that organisms that require high hydrostatic pressure to survive may include those that exist under a more narrow range of environment, and it is these species that could be more vulnerable to disturbance.

Environmental assessment and potential impacts from oil/gas prospecting at the continental margin

Clearly potential sensitivities of individual species should be considered against the most likely source of environmental perturbation. Disregarding for the moment physical impacts on the seabed caused by demersal trawling and concentrating on offshore oil/gas exploration activities, the greater isolation from the surface compared to the continental shelf such as the North Sea, should reduce impacts from platform discharges by greater dilution and dispersion. Although part of wider risk assessment it is assumed here there are adequate safeguards against hydrocarbon leakage from sea bed pipelines and well-head structures, and against catastrophic escape of hydrocarbons from deep reservoirs to the sea bed triggered by drilling. The chief remaining impact will be from particulate material discharged to the seabed. Of this, the vast bulk will derive from the disposal of drill cuttings. Present use of low-toxicity drilling muds, will localise effects of any dissolved chemical contaminants associated with discharge of drill cuttings to a smaller area of seabed than otherwise. However, inevitably, as material accumulates over time, environmental degradation will occur, as previously described for benthos near production platforms in the North Sea. Chemical contamination, such as oil derived from drill cuttings, will be difficult to predict at the deep-sea bed, even if dispersal into the water column achieves greater dilution of material than in shallow water. Physical impacts of cuttings discharge on the seabed may be slightly easier to predict, but its accumulation causing smothering of populations will depend both on the rate of discharge and on the hydrodynamic regime at the benthic boundary. Models of deposition need to take into account variability in flow, particularly occasional high-energy motions occurring over a longer time scale. Some of these are unique to the continental slope, caused by internal tidal mixing. Although, occurring over a longer time scale, such events will have a controlling effect on dispersion of fine-grained material, rather than those driven by the tidal cycles. Potential sensitivities of biota, even if thought predictable at species level, will need to be assessed against this. The natural occurrence of such events does suggest that benthic biota at the continental margin will already be adapted to cope with varying, and occasionally high, levels of resuspension.

We can at present only speculate on the rate of recovery of an area where there has been local extinctions of benthic species. It would appear logical that rates of re-colonisation by individual species will depend on currents as much as modes of early development. Those species having direct or brooded development of early stages might take longer to migrate back to a cleared area than species with water-borne propagules, whether in near-bottom currents or after migration to the surface. However, the presence of along-slope currents (such as the slope current off Ireland and Scotland), prevailing direction, and flow strength will be important in rate of recovery. There are very few data from analysis of recovery of experimentally disturbed areas in deep water. Results showing a broadly similar response to that in shallow coastal communities come from the *DISCOL* (*Dis*turbance and Re*COL*onisation) study in a manganese nodule province in the abyssal SE Pacific. This long-term study is particularly interesting because the nature of the impact (experimentally ploughing the seabed causing sediment re-suspension and deposition) is relevant to smothering effects caused by drill cuttings, and the long time-span over which recovery was monitored. Large megafauna directly impacted were eliminated, but after 7 years even sessile fauna

were present again in the impacted area. With macrofauna, there was rapid recovery of total abundance to 82% of undisturbed abundance, but loss in species richness was still detectable after 3 years (Borowski and Thiel 1998).

In environmental assessment, because it is important in the first instance to provide a baseline for monitoring possible impacts, it is important to apply protocols tuned to the characteristics of the deep-sea environment. These protocols should in principle minimize potential bias and maximise comparability. This follows agreed practice in shallow seas where adoption of generally accepted limits for sieves, for example, has long been a guiding principle in maximising comparability of data. However, a divergence in sieve-size protocols for washing samples has arisen among shallow and deep-sea benthic biologists as a result of a perception of smaller body size among deep-sea benthic organisms. This now affects comparability across the two environments. A recent study has shown that if a similar sieve size, 0.5 mm, is applied, the proportion of total possible retention of biomass and numbers of individuals is rather similar to that shown in previous studies of inshore and continental shelf benthos. But whereas all or nearly all species of inshore macrobenthos are retained by the 0.5-mm sieve, in deep-sea samples 14% of those retained by a 0.25 mm sieve had been missed by the 0.5 mm sieve (Gage et al. in press).

Impact of deep-sea fishing

It is ironic that, without any kind of environmental impact assessment, we have seen a greatly increased effort in deep-sea trawling for a variety of non-quota species, the stocks of many of which have already become depleted (Koslow et al. 2000). Evidence for the intensity of this activity frequently may be seen in linear marks now commonly seen in acoustic side scan records from all along the continental margin off Scotland. These are caused by the passage of the heavy trawls towed over the seabed (Fig. 1c). Trawl marks probably are gouged by the trawl doors and typically run parallel to depth contours. Their abundance seems to relate positively to known trawling effort. The impact on the small-bodied infaunal community is difficult to predict. But the effect of trawling, either through direct physical crushing or by smothering by disturbed sediment, on fields of larger epifauna such as xenophyophores (Fig. 2a) and delicate stalked animals such as glass sponges (Fig. 2b) and cold-water corals, clearly would be severe. Xenophyophores are characteristically so physically delicate that they are extremely difficult to collect intact. There is sadly already evidence of serious damage to delicate forms of life, such as *Lophelia* coral and other epifauna off the coast of Norway and the west coast of Sweden (Dr J.H. Fosså, Institute for Marine Research, Bergen, Norway; Dr Tomas Lundalv, Tjaerno Marine Laboratory, Sweden, personal communications). Similar effects can be expected to be accumulating in deeper water at other locations on the continental margin.

Furthermore, the ecosystem effects of the return of discarded dead biomass (discards may at least equal the total catch retained) is at present conjectural. At present we are in the dangerous position of being on the sidelines of a large-scale natural experiment. Present knowledge is unable to give an answer to the question how much of this submerged natural wilderness, evidently so rich in benthic biodiversity, has already been modified, damaged, or even permanently lost, as a result of deep-sea trawling on the continental margin.

References

Borowski C, Thiel H (1998) Deep-sea macrofaunal impacts of a large-scale physical disturbance experiment in the Southeast Pacific. Deep-Sea Res II 45:55-81

Boucher G, Lambshead PJD (1994) Ecological biodiversity of marine nematodes in samples from temperate, tropical, and deep-sea regions. Conserv Biol 9:1594-1604

Carney RS, Haedrich RL, Rowe GT (1983) Zonation of fauna in the deep-sea. In: Rowe GT (ed) The Sea, Vol. 8. J Wiley & sons, New York pp 371-398

Creasey SS, Rogers AD (1999) Population genetics of bathyal and abyssal organisms. Adv Mar Biol 35:1-151

Dahl E (1972) The Norwegian Sea deep water fauna and its derivation. Ambio Spec Rep 2:19-24

Etter RJ, Chase MR, Rex MA, Quattro J (1997) Evolution

in the deep-sea: A molecular genetic approach. In: Eighth Deep-Sea Biology Symposium, Monterey, California 1997, 31 p. Monterey Bay Aquarium Res Inst, Monterey, USA (Abstract)

Flach E, Bruin W de (1999) Diversity patterns in macrobenthos across a continental slope in the NE Atlantic. J Sea Res 42:303-323

Flach E, Thomsen L (1998) Do physical and chemical factors structure the macrobenthic community at a continental slope in the NE Atlantic? Hydrobiologia 375/376:265-285

Gage JD (1986) The benthic fauna of the Rockall Trough: Regional distribution and bathymetric zonation. Proc R Soc Edinb Sect B (Biol Sci) 88:159-174

Gage JD, Tyler PA (1981) Non-viable seasonal settlement of larvae of the upper bathyal brittlestar *Ophiocten gracilis* in the Rockall Trough abyssal. Mar Biol 64:163-172

Gage JD, Tyler PA (1991) Deep-Sea Biology: A Natural History of Organisms at the Deep-Sea Floor. Cambridge University Press, Cambridge 504 p

Gage JD, Lamont PA, Kroeger K, Paterson GLJ, Gonzalez Vecino JL (2000) Patterns in deep-sea macrobenthos at the continental margin: Standing crop, diversity and faunal change on the continental slope off Scotland. Hydrobiologia 440:261-271

Gage JD, Hughes DJ, Lamont PA, Gonzalez Vecino JL (in press) Sieve size influence in estimating biomass, abundance and diversity in deep-sea macrobenthos. Mar Ecol Prog Ser

Grassle JF (1989) Species diversity in deep-sea communities. Trends Ecol Evol 4:12-15

Grassle JF, Maciolek N (1992) Deep-sea species richness: Regional and local diversity estimates from quantitative bottom samples. Am Nat 139:313-341

Gray JS (1994) Is the deep-sea really so diverse? Species diversity from the Norwegian continental shelf. Mar Ecol Prog Ser 112:205-209

Hensley RT, Beardmore JA, Tyler PA (1995) Genetic variance in *Opiomusium lymani* (Ophiuroidea: Echinodermata) from lower bathyal depths in the Rockall Trough (northeast Atlantic). Mar Biol 121: 469-475

Jones NS, Sanders HL (1972) Distribution of Cumacea in the deep Atlantic. Deep-Sea Res 25:297-319

Klitgaard AB, Tendal OS, Westerberg H (1997) Mass occurrences of large sponges (Porifera) in Faroe Island (NE Atlantic) shelf and slope area: Characteristics, distribution and possible causes. In: Hawkins et al. (eds) The Responses of Marine Organisms to their Environment. Southampton University Press, Southampton pp 129-142 (Proceedings of the 30th European Marine Biology Symposium)

Koslow JA, Boehlert GW, Gordon JDM, Haedrich RL, Lorance P, Parin N (2000) Continental slope and deep-sea fisheries: Implications for a fragile ecosystem. ICES J Mar Sci 57: 548-557

Lamont PA, Gage JD (1998) Dense brittle star population on the Scottish continental slope. In: Mooi R, Telford M (eds) Echinoderms. Balkema, Rotterdam pp 377-382

Le Danois E (1948) Les Profondeurs de la Mer. Payot, Paris

Levin LA, Gage JD (1998) Relationships between oxygen, organic matter and diversity of bathyal macrofauna. Deep-Sea Res II 45:129-163

Pearson TH, Rosenberg R (1978) Macrobenthic succession in relation to organic enrichment and pollution in the marine environment. Oceanogr Mar Biol Annu Rev 16:229-311

Pearson TH, Rosenberg R (1987) Feast and famine: Structuring factors in marine benthic communities. In: Gee JHR, Giller PS (eds) Organization of Communities. Blackwell, Oxford pp 373-395

Pineda J (1993) Boundary effects on the vertical ranges of deep-sea benthic species. Deep-Sea Res 40:2179-2192

Pineda J, Caswell H (1998) Bathymetric species-diversity patterns and boundary constraints on vertical range distributions. Deep-Sea Res II 45:83-101

Rex MA (1977) Zonation in deep-sea gastropods: The importance of biological interactions to rates of zonation. In: Keegan BF, Ceidigh PO, Boaden PJS (eds) Biology of Benthic Organisms. Pergamon Press Oxford pp 521-530

Rex MA (1983) Geographic patterns of species diversity in the deep-sea benthos. In: Rowe GT (ed) The Sea. Vol. 8, J Wiley & sons, New York pp 453-472

Rex MA, Stuart CT, Hessler RR, Allen JA, Sanders HL, Wilson GDF (1993) Global-scale latitudinal patterns of species diversity in the deep-sea benthos. Nature 365:636-639

Rex MA, Etter RJ, Stuart CT (1997) Large-scale patterns of species diversity in the deep-sea benthos. In: Ormond RFG, Gage JD, Angel MV (eds) Marine Biodiversity: Patterns and Processes. Cambridge University Press, Cambridge pp 94-121

Rowe GT (1972) The exploration of submarine canyons and their benthic faunal assemblages. Proc R Soc Edinb Sect B (Biol Sci) 73:159-169

Sanders HL (1969) Marine benthic diversity and the stability-time hypothesis. Brookhaven Sym Biol, 22:71-81

Sanders HL (1977) Evolutionary ecology and the deep-sea benthos. Acad Nat Sci Philadelphia Spec Publ 12:223-243

Sanders HL (1979) Evolutionary ecology and life-history patterns in the deep-sea. Sarsia 64:1-7

Sanders HL, Hessler RR (1969) Ecology of the deep-sea benthos. Science 163:1419-1424

Schaff TR, Levin LA (1994) Spatial heterogeneity of benthos associated with biogenic structures on the North Carolina slope. Deep-Sea Res II 41:901-918

Somero GN (1992) Adaptations to high hydrostatic pressure. Annu Rev Physiol 54:557-577

Thistle D, Yingst JY, Fauchald K (1985) A deep-sea benthic community exposed to strong bottom currents on the Scotian Rise (Western Atlantic). Mar Geol 66:91-112

Tyler PA, Young CM (1998) Temperature and pressure tolerances in dispersal stages of the genus *Echinus* (Echinodermata: Echinoidea): Prerequisites for deep-sea invasion. Deep-Sea Res II 45:253-277

Tyler PA, Young CM, Clarke A (2000) Temperature and pressure tolerances of embryos and larvae of the Antarctic sea urchin *Sterechinus neumeyeri* (Echinodermata: Echinoidea): Potential for deep-sea invasion from high latitudes. Mar Ecol Prog Ser 192:173-180

Vacelet J, Boury-Esnault N, Harmelin J-G (1994) Hexactinellid cave: A unique deep-sea habitat in the scuba zone. Deep-Sea Res 41:965-973

Vetter EW (1994) Hotspots of benthic production. Nature 372: 47

Young CM, Devin MG, Jaekle WB, Ekaratne SUK, George SB (1996) The potential for ontogenetic vertical migration by larvae of bathyal echinoderms. Oceanol Acta 19:263-271

Wilson DF, Hessler RR (1987) Speciation in the deep-sea. Annu Rev Ecol Syst 18:185-207

Molecular Ecology and Evolution of Slope Species

A.D. Rogers

School of Ocean and Earth Science, University of Southampton,
Southampton Oceanography Centre, European Way, Southampton, SO14 3ZH, UK
corresponding author (e-mail): a.d.rogers@soc.soton.ac.uk

Abstract: In the present paper, genetic studies on organisms that are associated with continental slopes and slope-like habitats were reviewed. This indicated that few slope-dwelling species exhibit homogenous spatial genetic structure over wide geographic scales. The species that do exhibit homogenous genetic population structures tend to inhabit isolated topographic features such as seamounts, plateaus and mid-ocean ridges. Such a genetic structure may reflect the success of species with a dispersive life history in colonising such fragmented habitats. Most studies on slope dwelling species indicate genetic differentiation between populations on oceanic, regional and more local scales. In such cases topographic and hydrographic factors have been considered as important in structuring populations. Sometimes strong temporal variance in allele frequencies may have been a factor in causing differences in allele frequencies between areas. Sibling or cryptic species are commonly found amongst deep-sea taxa especially when comparisons of samples from different depths are made. The reasons for this are unclear but may be associated with speciation driven by selection exerted by increasing pressure or other factors correlated with depth. Alternatively, it is likely that historical processes have been important in speciation processes on the continental slope. More detailed studies are required of both intraspecific and interspecific genetic variation of slope species. Such studies should be on the scale of species distributions and should include the collection of associated biological and environmental data that is of relevance when interpreting genetic structure. The increasing availability of genome sequence data and genomic technology may allow scientists to study significant genes that have played a role in speciation processes in the deep-sea. Extensive phylogenetic work across multiple taxa may be necessary to identify common historical factors that have influenced speciation in the deep-sea.

Introduction

One of the most surprising recent discoveries in marine ecology has been that deep-sea sediments harbour a startling diversity of species of small animals (e.g. Grassle and Maciolek 1992). Estimates, based on extrapolations from quantitative studies of benthic macrofauna and meiofauna, indicate that globally this diversity may reach tens of millions of species (e.g. Grassle and Maciolek 1992). Whilst the area of surveyed seabed is small, some general patterns of distribution of species in the deep-sea have been observed. Firstly, diversity within the benthic macrofauna is not evenly distributed amongst phyla. Polychaetes, crustaceans (pericarids) and bivalve molluscs are the most diverse groups (e.g. Grassle and Maciolek 1992). Species diversity generally appears to show a parabolic distribution with depth reaching a maximum in the bathyal zone, on the continental slope. However, the density of all size-classes of animals appears to gradually decrease with increasing distance from the continental slope (e.g. Gage and Tyler 1991; Gage 1996; for reviews of patterns of species diversity in the deep-sea).

Estimates of global species diversity, in the deep-sea, have been contentious (Gage 1996). One of the main arguments against extrapolations from quantitative studies is that though local species diversity is high, deep-sea animals have very wide or even global geographic distributions. This line of thought appears to arise from the old perception of the

From WEFER G, BILLETT D, HEBBELN D, JØRGENSEN BB, SCHLÜTER M, VAN WEERING T (eds), 2002, *Ocean Margin Systems.* Springer-Verlag Berlin Heidelberg, pp 323-337

deep-sea habitat as being homogenous over very wide areas (e.g. Marshall 1979). The only environmental gradients are those associated with increasing depth, causing a stratified distribution of species down-slope but not across-slope (see Gage and Tyler 1991 for review of species replacement with depth). Support for wide-scale species distributions also comes indirectly from theories that attempt to explain high species diversity in the deep-sea. The patch mosaic theory, for maintenance of high species diversity, postulates that random low-intensity recruitment from a diverse pool of larvae, colonise patches that become available after disturbance (Gage 1996). This may imply that many deep-sea species have relatively long-lived planktonic larvae that can travel long distances to colonise rare patches of available habitat.

Studies of the morphology of the continental margins have shown that they do not present a smooth shallow gradient into the abyssal zone, as previously suggested (e.g. Marshall 1979), but represent topography as complicated, and in some areas as extreme, as that found on land. The slope is carved into segments by canyons, slumps, seamounts, ridges and even microcontinents (e.g. Heezen and Hollister 1971). The currents and water masses that impinge on the continental margins are also extremely complex. For example, along the northern European continental margin, the relatively warm water of Atlantic origin overlies extremely cold water from the Arctic. Boundaries between such water masses are often correlated with changes in composition of benthic fauna (e.g. Bett 1999) and are also often sites of vigorous hydrodynamic activity associated with internal wave formation. Such hydrodynamic forces can lead to the occurrence of extreme resuspension events such as benthic storms (e.g. Hollister and McCave 1984).

The occurrence of spatial and temporal patterns of nutrient supply to the deep-sea is well known. In temperate regions, the spring phytoplankton bloom in surface waters reaches the deep-sea bed as a seasonal pulse of phytodetritus. Tidal currents may impinge on the continental margin at critical angles of slope and can also cause internal wave

formation that leads to increased vertical mixing and a narrow band of increased primary productivity in overlying waters (e.g. Pingree and Mardell 1981). Extreme oxygen minimum zones (OMZs) occur along the continental slopes of the margins of some of the world's oceans and these have a dramatic influence on the composition of the benthic fauna (see Rogers 2000 for review). Hydrocarbon seeps are local features that also occur on the slope in some parts of the world.

Many of these physical and biological factors, influence the distribution of species inhabiting the continental slope. It is therefore reasonable to suggest that such factors can act as barriers to dispersal between conspecific populations on the slope. If barriers to dispersal exist, then reproductive isolation is likely to occur between populations, a prerequisite of allopatric speciation. Furthermore, environmental factors, such as extremely low oxygen concentrations, may exert selection (Rogers 2000). Such a scenario is incongruent with a view of global species distributions for deep-sea animals.

The processes that maintain high species diversity in the bathyal zone are not necessarily identical to those that have caused speciation. Historical factors are increasingly recognised as critical in shaping contemporary patterns of species diversity and even intraspecific genetic variation. Throughout ancient history, global climate has fluctuated between warm and cold phases. During cooler periods, the oceans were ventilated by cold, dense, oxygen rich water sinking at the poles (as today). With the onset of a warmer climate, ventilation from the poles may have decreased or even ceased altogether. During these periods the entire deep-ocean may have been subjected to elevated temperatures and hypoxic conditions (reviewed by Rogers 2000). This may have had the effect of driving abyssal species into the bathyal zone. Patterns of surface productivity also changed dramatically during warm periods, with surface waters becoming very nutrient poor and clear. Fall in sea level, associated with glaciations may have restricted the distribution of shelf species. Falling temperatures and increasing turbidity levels are likely to have caused the mass extinction of warm-water (reef) faunas, causing them to retreat to refugia (Rogers 2000 and refs. therein). Increasing competition with bathyal species invading shelf waters because of lower temperatures may have enhanced this retreat. Through such climatic fluctuations it is likely that the bathyal zone accumulated species from both shallower and deeper waters and, furthermore, acted as a source for recolonisation of both the shelf and abyssal zones when conditions ameliorated (Rogers 2000).

Molecular genetic techniques provide an objective method of assessing diversity. For some taxa, especially microbial organisms, classification of "species" is based on DNA sequence similarity. For larger organisms, the widespread occurrence of cryptic or sibling species has been well documented, even for deep-sea species. This means that for many groups, genetic studies are the only accurate way to assess what represents intraspecific and interspecific morphological variation. Population genetic studies allow the estimation of reproductively effective migration between populations. Mitochondrial phylogeography and other DNA-based methods of studying deep phylogeny provide a window on evolutionary processes. It is through these techniques that we may be able to unravel some of the events that have led to high species diversity on the continental slope. In this paper genetic studies of slope species to date are reviewed and future areas for research are discussed.

Molecular genetic techniques

Genetic methods for studying deep-sea populations can be divided into biochemical and molecular methods. Biochemical methods are based on assessment of genetic variation through detecting variation in gene products, i.e. protein sequence. Molecular methods analyse variation in genomic DNA. For a thorough review of genetic methods used in studies of marine metazoa see Rogers (2001).

Allozyme electrophoresis

All enzymes are proteins with an amino acid sequence that is coded for by specific genes. Alter-

ations in the DNA sequence coding for a protein, through mutation, can result in a change in the amino acid sequence and an alteration in the charge/mass ratio of a protein. Alterations in protein structure of metabolically active proteins (enzymes) can be detected by separation of different forms of a protein (alleles) by electrophoresis followed by staining the protein using specific histochemical reactions. Details of the methods employed for allozyme electrophoresis are provided by a number of authors (e.g. Murphy et al. 1990). Specific enzymes are marked by applying specific histochemical stains to the electrophoretic medium following electrophoresis (e.g. Murphy et al. 1990).

Allozymes are usually coded for by unlinked genetic loci scattered throughout the genome. The loci have co-dominant alleles that are usually inherited in a Mendelian fashion. This means that allozyme data can be used for a variety of analyses at the intraspecific level including estimates of genetic variation, congruence of genotype frequencies to Hardy-Weinberg expectations, genetic distance, spatial genetic structure and effective population size. Allozymes can also be used to differentiate between closely related species and can provide systematic information up to the level of confamilial genera (see Thorpe 1982; Murphy et al. 1990).

Allozyme electrophoresis is an old technique but it has some advantages when compared to other molecular genetic markers. Several loci can be scored, for a large number of specimens, relatively cheaply. However, because the technique is decreasing in use it is often the case that electrophoresis equipment has to be specially built for this method, especially when using thick gels that are sliced so that a set of samples can be scored for multiple enzymes. Another major advantage is that there are a very large number of published allozyme studies. Because of this large body of data, measures of genetic identity and distance can often provide an accurate guide to the taxonomic relationships of sample populations (e.g. Thorpe 1982). However, there are exceptions to this in certain taxa, e.g. birds where there is less than expected genetic differentiation associated with speciation (e.g. Thorpe 1982). This may occur because the speciation process is rapid for some taxa.

One of the disadvantages of allozyme electrophoresis is that material must be fresh or frozen. This is a problem for studies of deep-sea taxa that are often collected from remote locations on research vessels. While low temperature freezers may be available on ships, transport of material from ports to genetic laboratories is often extremely difficult, especially given regulations on transport of liquid nitrogen or dry ice. Proteins are also rendered useless for allozyme studies by fixation and preservation of tissues in formalin or ethanol.

DNA extraction

All molecular genetic methods rely on analysis of variations in genomic DNA. Usually genomic DNA must be extracted from tissue prior to further stages in analysis (but not always – see Rogers 2001). Where high quality DNA is required this is usually obtained through a phenol-chloroform extraction. For invertebrates that produce a lot of mucous (e.g. molluscs) additional steps such as the use of CTAB may be required. Particularly difficult taxa, such as echinoderms may require the use of commercial kits to obtain sufficient quantities of high quality DNA. Where lower quality DNA is sufficient, short extraction protocols using resins (e.g. Chelex) or lysing solutions (e.g. Microlysis ® Microzone Ltd., Lewes, East Sussex, U.K.) may be utilised. This dramatically decreases time and cost for extractions if population studies, requiring large sample sizes, are undertaken.

DNA may also be extracted from tissues preserved in ethanol or buffer. The use of such preservatives decreases the logistical problems of transporting specimens from remote locations and also simplifies storage of material in the laboratory. Formalin fixation shears DNA into low molecular weight fragments. Such material is tractable to molecular genetic analysis but DNA extraction often requires stringent procedures and a high level of care to prevent contamination of samples. Quality of DNA from formalin fixed tissue is generally unpredictable. Despite the high cost and lengthy procedures involved in extracting DNA from formalin preserved material, it does open up

a vast quantity of museum material for deep-sea studies (e.g. France and Kocher 1996; Chase et al. 1998; Etter et al. 1999). As the collection of such material is very expensive, such an approach may be viable for addressing some questions relating to deep-sea taxa (e.g. phylogenetic studies).

Polymerase chain reaction (PCR)

PCR is used in the majority of molecular genetic techniques. It creates millions of copies (amplification) of a specific stretch of DNA sequence through enzyme-mediated replication of complementary strands. Each end of the sequence is delineated by primers, short oligonucleotides (usually 20-25 bp in length) complementary to the genomic DNA. The PCR reaction consists of three basic phases (Fig. 1). The first is denaturation of double stranded genomic DNA by heating to approximately 94-96°C. This is followed by an annealing step whereby primers anneal to complementary stretches of the single stranded DNA molecules (temperature 50-60°C). The final phase is extension, where complementary bases are added onto the end of the primer molecule in the presence of the enzyme Taq polymerase. This cycle is repeated usually up to 30 times, amplifying the desired segment of DNA. Some methods use random primers to generate multiple DNA fragments from throughout the genome (fingerprinting, see below).

Direct sequencing of DNA

Most molecular genetic methods indirectly assess variation in genomic DNA. The only method that reveals actual DNA variation is sequencing (Fig. 1). At present sequencing is usually achieved by PCR amplification of the region of interest. Waste components of the PCR reaction are removed from the amplified fragments usually by using a PCR purification kit. This is followed by a second PCR reaction using only a single primer and flourescently labelled nucleotides (cycle sequencing). This reaction produces fragments of the original amplified DNA sequence of all sizes up to the entire length of sequence. Each fragment is end labelled with a fluorescent nucleotide.

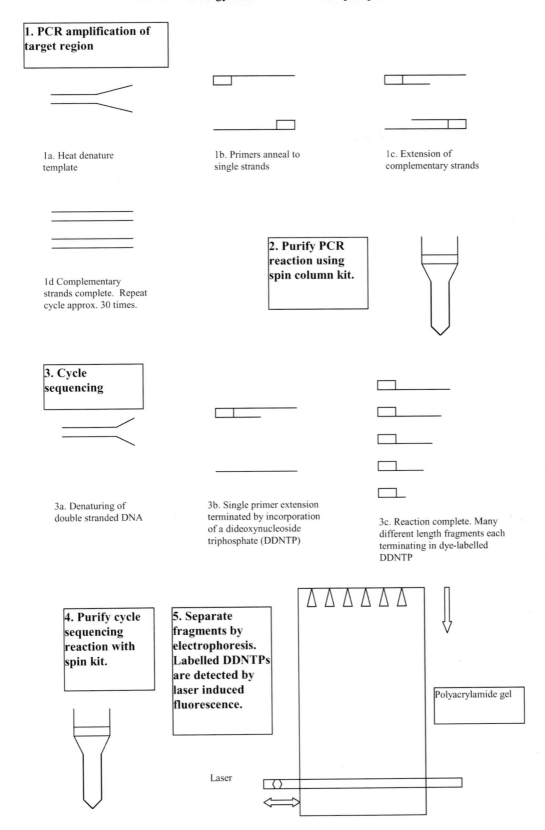

Fig. 1. Current method of DNA sequencing from the PCR stage. Note sequence detection is achieved with an automated sequencer (e.g. Applied Biosystems 377).

The products of the cycle sequencing reaction are purified and then separated by polyacrylamide electrophoresis in an automated sequencing machine. The smallest fragments migrate fastest. As fragments run along the gel they pass a laser that causes the end labelled nucleotide to fluoresce. There are four wavelengths of end label corresponding to four nucleotides (adenine, guanine, cytosine, thymine) that are detected and translated into the DNA sequence.

DNA sequencing maybe targeted at a variety of coding (exons) and non-coding (introns) DNA regions. Universal primers (amplify a target region for a wide variety of taxa) are available for many DNA regions. Coding sequences tend to evolve more slowly than non-coding regions because changes in exons may alter protein sequence and cause a deleterious mutation. This means that a range of DNA regions is available with different levels of resolution ranging from sequences that show intraspecific variation to those that are conserved across distantly related taxa.

Mitochondrial genes have been found to be particularly valuable in studies of population dynamics and evolutionary history. Mitochondrial genes are usually only inherited maternally and do not undergo recombination. Because of this clonal mode of inheritance, mitochondrial genes have an effective population size approximately one quarter that of nuclear genes. This means that random genetic drift has a greater effect on mitochondrial genes and differences tend to arise between reproductively isolated populations more readily than in nuclear genes. Mitochondrial genes also tend to have a higher mutation rate than other genes (e.g. Brown et al. 1979). Interpretation of mitochondrial sequence data may also be simpler than nuclear sequences. This is because heterozygous sites in nuclear sequences maybe difficult to score using automated sequencing methods and for some nuclear multigene families, often targeted for studies of intraspecific variation or phylogeny, intragenomic variation may occur. Studies that investigate the processes that govern the geographic distribution of mitochondrial lineages are known as mitochondrial phylogeography (e.g. Avise et al. 1987).

Indirect assessment of DNA sequence variation

Because sequencing is expensive and time-consuming it is not always practical for population studies in which large numbers of individuals are screened. Several alternative methods of assessing DNA sequence variation have been developed. The oldest of these is restriction fragment length polymorphism analysis (RFLP). This entails digesting DNA with restriction enzymes that have specific sequence recognition sites, usually of 4-8 bp in length. This produces fragments of DNA of different lengths depending on whether restriction sites are present or absent and on length variation of the sequence between restriction sites. Fragments are separated according to size by electrophoresis. Other methods are available to provide indirect assessment of DNA sequence variation. Such methods include Single-Strand Conformation Polymorphism (SSCP), Denaturing Gradient Gel Electrophoresis (DGGE) and Heteroduplex Analysis. These methods separate short stretches of DNA according to differences in sequence and have all been used in studies of intraspecific genetic variation. These methods are reviewed in Rogers (2001).

Repetitive sequences

There are several types of repetitive DNA sequence. The most widely used type of repetitive sequences to date, are microsatellites. These are tandemly repeated short DNA sequences between 1-6 bp long. The most commonly used microsatellites in genetic studies are dinucleotide repeats such as $(GT)_n$. As a result of a high mutation rate, microsatellites show polymorphism in the number of repeats. This polymorphism gives rise to the use of microsatellites as co-dominant, single-locus genetic markers. As microsatellites are generally non-coding, these markers are usually selectively neutral. Microsatellites are now widely employed as population-level markers and have been developed or used for some deep-sea species (e.g. Oke et al. 1999).

Minisatellites are tandem repeat sequences of up to 65 bp in length. They are not as widely employed as microsatellites in population studies but many multilocus DNA fingerprinting probes target these regions.

Fingerprinting

Fingerprinting includes several techniques that have different genetic characteristics, all of which produce highly polymorphic banding patterns after gel electrophoresis. These techniques are generally applied to studies of intraspecific variation, particularly in relation to reproductive behaviour and parentage. The oldest method is multilocus DNA fingerprinting (e.g. Jeffreys et al. 1985). Newer methods of fingerprinting are usually based on PCR with random primers that are entirely or partially degenerate. These methods include Random Amplified Polymorphic DNA (RAPDs), also known as Arbitrarily Primed PCR (AP-PCR), Amplified Fragment Length Polymorphism (AFLP) analysis, Direct Amplification Length Polymorphism (DALP) and Simple Sequence Repeat or Inter Simple Sequence Repeat analysis (SSR or ISSR). All these methods have different genetic bases and different characteristics and are re-viewed in Rogers (2001). RAPDs can suffer from problems with co-migration of bands and repeatability of amplifications because of sensitivity to template quality (see Rogers 2001). Some of the other methods show improved amplification quality and repeatability as primers are aimed at specific types of DNA sequence (i.e. tandemly repeated sequences).

Genetic studies on species from continental margins

Intraspecific variation

There have been relatively few studies of intraspecific genetic variation on species inhabiting continental slopes. Most studies relate to stock structure in commercially valuable species (Creasey and Rogers 1999). If slope species have very wide geographic ranges, large population sizes and dispersive life histories, then populations should only show limited spatial genetic structure

(limited or shallow phylogeographic structure – Avise et al. 1987). This should be especially the case for species of demersal and pelagic fish and cephalopods that occur along continental margins as they have a high mobility as adults and generally have a dispersive larval phase.

There are surprisingly few examples of species from continental margins that show widespread panmixia. The only examples where species do show widespread genetic homogeneity are from species, which inhabit seamounts and banks or ridges. These include sequencing, RFLP and SSCP studies on the pelagic armourhead, *Pseudopentaceros wheeleri*, from the Hawaiian Ridge (Martin et al. 1992), the alfonsino, *Beryx splendens*, from banks and seamounts around New Caledonia (Indian Ocean), the Chatham Rise off New Zealand (Pacific Ocean), and the Galicia Bank and Bay of Biscay (Hoarau and Borsa 2000), and the wreckfish, *Polyprion americanus*, from seamounts, plateaus and oceanic islands in the northern and southern Atlantic and southern Pacific Oceans (Sedberry et al. 1996). Whilst results should be treated with caution as in some cases sample sizes from sites are small (e.g. Hoarau and Borsa 2000 for Atlantic), they generally show genetic homogeneity on a regional basis with some genetic differentiation in populations from the northern and southern hemispheres (Sedberry et al. 1996) or from the Atlantic and Pacific Oceans (Hoarau and Borsa 2000).

All three of these fish species are associated with isolated topography. Whilst life histories are poorly understood it appears that they all exhibit extremely dispersive life histories that may represent adaptation to exploiting highly fragmented habitats separated by large tracts of ocean. As such they may not be treated as typical continental margin species but as species that may exploit continental slopes at some stage of their life history.

Studies of intraspecific genetic variation in the majority of other commercial species that inhabit continental slope habitats show some evidence of spatial genetic structure, either between regions or on a more local scale. In the North Atlantic, the roundnose or rock grenadier, *Coryphaenoides rupestris*, is found on the continental slope at depths of 400-2000 m (e.g. Whitehead et al. 1986). Like many deep-sea fish, *C. rupestris* is long-lived and highly vulnerable to modern fishing methods. Catches have been in decline since the 1970s and though some of this resulted from limited quotas set on halibut fisheries (of which *C. rupestris* is a by-catch) it is likely that populations have been overexploited (e.g. Koslow et al. 2000). Genetic studies using allozymes have reported significant genetic differentiation of populations located in different areas (e.g. Logvinenko et al. 1983; Duschenko 1988). On the basis of genetic data, Logvinenko et al. (1983) suggested that there were separate stocks of *C. rupestris* on the Northeast and Northwest Atlantic slope and the northern and southern mid-Atlantic Ridge. Duschenko (1988) not only found evidence of genetic differentiation between populations but also genetic heterogeneity among different size classes of fish from the same population. Such differences in genotype frequencies between different cohorts have been observed in other species of fish and have been connected with low effective population sizes (e.g. Ruzzante et al. 1996). The results of these studies must be treated with some caution as they relied on low numbers of enzyme loci (Creasey and Rogers 1999). However, it would seem that topography has an important influence on population structure in *C. rupestris*, with genetically distinct stocks inhabiting geographically separated regions of shelf / mid-ocean ridge. It has also been suggested that temperature affects maturation time in *C. rupestris*, possibly also contributing to reproductive isolation between high and low latitude populations (Creasey and Rogers 1999).

The veined squid, *Loligo forbesi*, is generally considered as a shelf, rather than a slope species but it does occur around offshore banks and oceanic islands in the North Atlantic Ocean. Recent studies on genetic variation of veined squid from the European continental shelf using allozymes, DGGE analysis of the mitochondrial COIII gene and microsatellites have all shown a homogenous population structure (e.g. Brierley et al. 1995; Shaw et al. 1999). However, microsatellite analysis has revealed significant genetic differentiation between populations on the European continental

shelf compared with offshore banks (Faroes and Rockall banks) and oceanic islands (Azores). This was explained through isolation by distance of these populations and hydrographic isolation of offshore topographic highs. The demersal habits of this species along with the requirement to lay eggs on the seabed are probably also partially responsible for reproductive isolation of offshore populations.

The Greenland halibut, *Reinhardtius hippoglossoides*, is fished in deep waters in both the North Atlantic and North Pacific and is also considered as overexploited (Koslow et al. 2000). In the North Atlantic both allozyme and sequencing data have indicated that sampled populations (Gulf of St Lawrence, Grand Banks, Newfoundland and Labrador Slope) are genetically homogenous apart from that in the Gulf of St Lawrence (Fairbairn 1981; Vis et al. 1997). This may have been related to topographic isolation of populations within the Gulf or it may have been a result of temporal genetic variation arising from small effective population size. Comparisons of populations from the North Atlantic and North Pacific have shown low but significant genetic differentiation (F_{ST} = 0.084 - 0.091; Fairbairn 1981). In the North Pacific, significant genetic heterogeneity has been found amongst populations from the Bering Sea and the Sea of Okhotsk (D'yakov 1991). It was suggested that both topographic and hydrographic barriers to dispersal were responsible for the observed spatial genetic structure. The halibut has a long-lived pelagic larva and populations were expected to show genetic homogeneity (Vis et al. 1997). Genetic studies of the walleye pollack, *Theragra chalcogramma*, Dover sole, *Microstomus pacificus*, and the rockfish, *Sebastolobus alascanus* and *S. altivelis* have also shown significant genetic heterogeneity between areas in the North Pacific (Mulligan et al. 1992; Bailey et al. 1997; Stepien 1999). Topographic and hydrographic barriers to dispersal as well as larval retention by ocean gyres have all been cited as being responsible for the observed genetic structure.

There have been several studies on commercial species of fish inhabiting the continental slope and offshore seamounts off the coasts of Australia and New Zealand. One of the most valuable commercial species and the one that has been most extensively studied is the orange roughy, *Hoplostethus atlanticus*. This species was originally fished from the continental slope around New Zealand, with exploitation starting in 1979. As catches on slopes decreased and fishing technology developed, populations located on banks, ridges and seamounts were targeted, with new fisheries developing around Australia and in the Atlantic (Clark 1999). Orange roughy are extremely long lived and take a long time to reach maturity. Furthermore, recruitment to populations tends to be sporadic (Koslow et al. 2000) and female fish do not spawn every year once they reach maturity. Fish are targeted over seamounts, while feeding or spawning, leading to a very intense fishing pressure on exploited populations. The result of this has been rapid depletion of almost all exploited populations of orange roughy (Clark 1999).

Biochemical and molecular genetic studies on orange roughy have shown contradictory results, even using the same genetic methods (see Creasey and Rogers 1999 for review). However, more recent allozyme and RFLP studies (of entire mitochondrial genome) around the coast of New Zealand have detected significant genetic differentiation between populations from the eastern vs. western and northern vs. southern areas (Smith et al. 1996; Smith and Benson 1997). In the allozyme study, genetic differences were detected between populations within the Chatham Rise area and between samples taken at different times (spawning and non-spawning period; Smith and Benson 1997). The RFLP study suggested that a barrier to gene flow between populations maybe related to the presence of the subtropical convergence lying between these sample sites. There was also evidence for more local genetic subdivision amongst populations (Smith et al. 1996). Interestingly, the ling, *Genypterus blacoides*, another commercially fished slope species, also shows a pattern of genetic differentiation between populations congruent with the subtropical convergence acting as a barrier to gene-flow between northern and southern populations (Smith and Francis 1982).

Contrasting results in genetic studies of orange roughy populations may arise for a number of reasons. One possibility is that some genetic markers are not sufficiently variable to detect genetic differences between populations. Analysis of molecular variance has also indicated that intrapopulation diversity is high in this species, probably reflecting large population sizes and periodic genetic exchange between populations. In such a case, studies are particularly prone to effects of low sample size. Allozyme studies have revealed significant temporal genetic variation within roughy populations (Smith and Benson 1997). Temporal variance in allele frequencies may indicate low effective population size resulting in a high level of random genetic drift. However, roughy are extremely long lived and unless a large proportion of sampled populations are replaced on an annual basis, it is difficult to account for temporal genetic variation in this way. Orange roughy show synchronous reproductive behaviour and it is possible that adults may show some degree of homing to natal breeding grounds. During periods of the year when spawning does not take place reproductively isolated units may be mixed on feeding grounds. Such a pattern of behaviour that influences genetic structure has been detected in mackerel in the North Atlantic (Nesbo et al. 2000). Such behavioural patterns may generate differences in spatial and temporal genetic variation of samples depending on when and where they are collected. Overall, cumulative data from genetic studies suggest that orange roughy populations do show genetic differences on a regional scale and are likely to be related to the association of populations with areas such as banks and seamounts. Higher resolution genetic markers, such as microsatellites may prove useful in furtherstudies of orange roughy in the future (see Oke et al. 1999).

The blue grenadier or hoki, *Macruronus novaezelandiae*, forms important commercial fisheries off Australia and New Zealand where it may be found on the continental slope between 200-700 m depth (see Creasey and Rogers 1999 for review of biology). However, unlike many deepwater fisheries, catches appear to be sustainable for hoki populations. This is thought to be because this species show many population characteristics of shallower water fish (Koslow et al. 2000). Hoki spawn above the seabed over canyons, with male and female fish often separating out into layers in the water column (e.g. Zeldis et al. 1998). The location of spawning concentrations in such defined geographic areas gives rise to the possibility of reproductive isolation and genetic differentiation between stocks. Allozyme and mtDNA studies have been carried out on populations around New Zealand and southern Australia (Smith et al. 1981; Milton and Shaklee 1987; Baker et al. 1995; Smith et al. 1996). On a regional scale, genetic differentiation between samples of hoki has not been detected. However, significant genetic differentiation has been detected in most comparisons of populations from Australia and New Zealand (Milton and Shaklee 1987; Baker et al. 1995). It would appear that the Tasman Sea is an effective barrier to gene flow in this species. This is not surprising as most stages of the life history are associated with the continental margin (e.g. Zeldis et al. 1998). A similar pattern of spatial genetic structure off New Zealand and Australia has been detected for the black oreo, *Allocyttus niger* (Ward et al. 1998). However, a homogenous genetic structure has been detected for the smooth oreo, *Pseudocyttus maculates*, whilst genetic differentiation between populations of the warty oreo, *A. verrucosus*, has been demonstrated not only between Australia and New Zealand but also within the Australian region (Ward et al. 1998).

The overall picture arising from studies of fish populations on continental slopes is that they do show significant spatial genetic structure within oceanic regions (e.g. North Pacific) and certainly show strong differentiation between oceans. The main exceptions to this appear to be some species that inhabit seamounts that have an extremely dispersive life history pattern. There are fewer studies of benthic invertebrate populations from continental slopes. As these are generally perceived as being less mobile than fish populations, strong spatial genetic structure maybe expected within regions.

The first studies on deep-sea invertebrate populations were carried out using allozyme electrophoresis (Creasey and Rogers 1999). Studies of a

single enzyme locus in *Ophiomusium lymani* from the Hatteras submarine canyon in the Northwest Atlantic Ocean reported genetic differentiation between bathymetrically segregated samples, but no differentiation between populations separated by 200 km (Doyle 1972). However, individuals from the deepest station in this study were collected in a different year to the others. Given our present knowledge of temporal genetic variation in populations of marine invertebrates (e.g. Hedgecock 1994) it is possible that this study did not reflect spatial genetic variation in bathymetrically separated populations (Creasey and Rogers 1999). Subsequent allozyme studies on *O. lymani* revealed only limited levels of genetic differentiation between populations (e.g. Hensley et al. 1995). Despite this species showing an abbreviated lecithotrophic development, populations appear to be panmictic over scales of hundreds of kilometres (Creasey and Rogers 1999).

Other studies on slope-dwelling invertebrates have shown marked genetic differentiation between populations even over small geographic scales. Two studies have been carried out on the Californian continental borderlands. In this area, the continental slope is characterised by basins, between which lie shallow sills. Five polymorphic enzyme loci were analysed for populations of the deep-sea gastropod, *Bathybembix bairdii* (Siebenaller 1978). Genetic differentiation between populations located at 910-1156 m depth was found to be fairly low (F_{ST} = 0.004-0.011). However, much stronger genetic differentiation was detected between these populations and one located on the upper slope (F_{ST} = 0.023-0.030).

Some studies have revealed significant genetic differentiation between populations located on the continental slope but not associated with bathymetry. The spider crab, *Encephaloides armstrongi*, inhabits the continental slope in the northern Indian Ocean from Oman to the Bay of Bengal (Creasey et al. 1997). This species is particularly associated with the extreme oxygen minimum zone that is found in this region and startling densities of *E. armstrongi* have been observed on benthic photographs (Creasey et al. 1997). Off the coast of Oman, significant genetic differentiation was detected between populations located between 350-650 m. Whilst this occurred between populations separated by only 10s km it was not strictly associated with depth. It was considered that such a genetic structure might have arisen through non-depth associated topographic variation on the slope or from behaviour of the crabs. Majid spider crabs are well known for sex-specific migratory behaviour patterns and during sampling relatively few female crabs were taken (Creasey et al. 1997). Interestingly, a juvenile population was sampled above the oxygen minimum zone at approximately 150 m depth. This population was genetically distinct from adult populations and raises the possibility of very strong temporal genetic variation in this species.

A similar study on two bathymetrically separated populations of the squat lobster, *Munidopsis scobina*, from the Oman slope, below the oxygen minimum zone, also revealed significant genetic differentiation (Creasey et al. 2000). Significant differences were also found in the size frequencies of the two populations as well as in the sizes of the eggs carried by females. Whilst generalisations are difficult with such a small number of study populations this study did raise the possibility that the proximity of the oxygen minimum zone to the two populations influenced allele frequencies directly or through selecting for size in adult squat lobsters (Creasey et al. 2000). Oxygen minimum zones certainly have a unique fauna and therefore must play a significant role in evolution of slope species in some geographic regions (Rogers 2000). Whilst conditions in such regions are extreme (low oxygen and presence of H_2S) food availability is high and this energy supply may allow some species to develop adaptations to dysaerobic conditions (Rogers 2000). Oxygen minimum zones may also act as a horizontal or vertical barrier to gene flow between populations located on the continental slope or in the surrounding water column (Rogers 2000).

Interspecific genetic variation

France (1994) reported on allozyme studies on the deep-sea amphipod *Abyssorchomene sp.*, from four shallow sill basins (920-1854 m depth) and one deep-sill basin (1834-1933 m depth). Genetic differentiation between populations in the shallow-sill basins was low ($F_{ST} = 0.002$-0.011; calculated in Creasey and Rogers 1999 from France 1994). However, extremely strong genetic differentiation was found between the shallow-sill and deep-sill basin populations. So marked were genetic and morphological differences between these populations that it was considered they probably represented separate species (France 1994).

Studies of bivalve and gastropod molluscs from the continental slope in the western Atlantic have also revealed extremely high levels of genetic differentiation between populations separated by very small horizontal distances but located at different depths (10s - 100s km; Chase et al. 1998; Etter et al. 1999). These studies were carried out by partial sequencing of the mitochondrial 16S rDNA region, sometimes utilising formalin fixed material up to 30 years old. Initial studies on the bivalve *Deminucula atacellana* showed extremely high levels of differentiation in populations above 2500 m depth, compared to those below 2500 m (separated by 134 km horizontal distance). Haplotype diversity was highest at depths ranging between 1600 m and 2500 m, coinciding with the zone of maximum species diversity on the continental slope (e.g. Gage 1996). This species is known to have a pelagic larva and 134 km is thought to be well within the range of dispersal. However, it was suggested that selection associated with the pressure gradient might have been the cause of the observed genetic heterogeneity.

A subsequent study on the gastropod *Frigidoalvania brychia* and the bivalves *Nuculoma similis*, *Deminucula atacellana* and *Malletia abyssorum* used similar genetic methods (Etter et al. 1999). Again, extremely high genetic differentiation was detected between populations at different depths, but only horizontally separated by 10s or 100s km. Levels of genetic differentiation were similar to those associated with interspecific comparisons in shallow water molluscs (Etter et al. 1999). It was considered likely that in this case the studied "species" in fact represented complexes of sibling species whose distribution is separated by depth. Such results indicate that high estimates of species diversity at mid-slope depths (e.g. Grassle and Maciolek 1992) may be underestimated (Etter et al. 1999).

The question arises as to what has driven speciation associated with depth. Evidence has indicated that the pressure gradient may be involved in natural selection and subsequent speciation. Work on deep-sea fish has shown that pressure may exert selection on protein structure as species evolve to live at increasing depths on the continental slope (e.g. Siebenaller and Somero 1989). Experiments on larval development have also demonstrated that the larvae of shelf and slope species are adapted for growth at certain pressures and temperatures (e.g. Tyler and Young 1998). Alternatively, species may have evolved in allopatry, with subsequent range expansion bringing species into the same geographical area but not necessarily into the same depth range.

An initial allozyme study on the deep-sea scavenging amphipod *Eurythenes gryllus* detected significant genetic differentiation between populations inhabiting the abyssal plain and those inhabiting the summit of a seamount in the Pacific Ocean (Bucklin et al. 1987). DNA sequencing of the mitochondrial 16S rDNA region, on fresh and archived specimens of *E. gryllus* revealed that this species comprised a complex (France and Kocher 1996). Populations of *E. gryllus* below 3200 m depth were identical all around the world, whilst above this depth a variety of species existed that were endemic to oceans and geographic regions. The most likely explanation for such a pattern of species distribution is that the abyssal species recently invaded the deep-sea and spread around the entire world, probably via cold water sinking from the poles. Such an invasion was probably a recent event. *E. gryllus* from slope depths, however, speciated on a regional basis, as at this depth significant barriers to geneflow exist (see above). In other words, the current distribution of the *E. gryllus* species complex reflects historical evolutionary processes. Such evolutionary processes are likely to have been strongly influenced by climate change, acting through oceanographic processes and by plate tectonics (see Rogers 2000).

Conclusions and directions for future work

Despite the range of molecular genetic methods that are available to biologists, relatively few studies have been carried out on deep-sea organisms, especially those from non-vent habitats. This means that detecting common patterns of genetic variation amongst taxa and analysing factors that influence this variation is very difficult. The lack of studies is partially a reflection of difficulties in obtaining samples from the deep-sea of sufficient number, quality and geographic and temporal coverage to give biologically meaningful results for species that have large populations with a potentially wide distribution. There is an urgent future requirement for more detailed studies of the genetic variation of slope species covering a wider area of potential species distribution and utilising multiple markers with differing levels of resolution. The present review, however, has shown that some general observations of the genetic structure of slope populations can be made.

Studies of spatial genetic variation reveal that even in mobile species such as fish and cephalopods, populations are often structured within geographical regions and are certainly structured on an oceanic scale. Some species of fish inhabiting remote, isolated habitats, such as seamounts, appear to show little genetic differentiation over extremely large geographic distances. The majority of less mobile benthic invertebrates show spatial genetic population structure even down to small geographic scales. Various mechanisms may prevent the occurrence of panmixia in continental slope populations. These include physical barriers to distribution and gene flow, such as the continents and large expanses of deep water. On a smaller scale, gene flow between populations can be disrupted by the hydrographic structure of the overlying water column, topography, such as seamounts and canyons and by oxygen minimum zones. The behaviour of deep-sea species, especially with regards to spawning may also act to separate and cause reproductive isolation between populations.

Studies of the genetic structure of populations, coupled with simultaneous observations of bottom topography, seabed composition and physical and chemical structure of the overlying water column will be necessary to identify factors that are important in affecting distribution and dispersal of species. This will require that sampling for genetic studies is carefully factored in to cruise plans as at present samples are taken on a more or less random basis for genetic studies. Once such basic data becomes available more detailed studies on specific adaptations at the genetic level to environmental variables will become possible. Current efforts in genome sequencing and new methods in genomic research are likely to provide a rich source of candidate genes that may play an important role in the adaptation of deep-sea species to their environment. Base-line data will also be critical in predicting the short-term consequences of climate change on populations of slope species.

Several studies on slope species have identified the occurrence of cryptic species and it is likely that genetic methods will further increase current estimates of deep-sea diversity. However, the potential use of molecular methods in the identification of particularly "difficult" groups of metazoans is severely underestimated and underutilised. Some groups of meiofauna are highly speciose, numerically important and are likely exert an important influence on the biology and biogeochemistry of slope sediments (e.g. Nematoda, Foraminifera and Platyhelminthes). The study of such groups is severely inhibited by our ability to describe and identify species. The development of molecular methods to identify species from environmental samples, coupled with the use of interactive databases has the potential to open these groups to detailed analysis of species diversity and distribution. If sampling for such studies are coupled with other environmental observations a better understanding of the role that these species (and species diversity) play(s) in slope processes can be reached.

Clues that past events may have had a profound influence on species diversity and intraspecific genetic variation indicate that exciting developments may be expected in this area in future years. Changes in climate, over geological timescales, have probably exerted a profound influence on speciation in the marine environment. Targeting of molecular phylogenetic studies on specific groups

of animals, inhabiting the deep-sea, provide one of the few ways to analyse long-term climatic effects on species diversity in the marine environment. The opening up of museum collections to molecular genetic methods means that there is a large body of material for such studies that is unlikely to be recollected because of financial constraints. Such studies are likely to be most powerful when coupled with a biogeographic approach.

Finally, there is an urgent requirement for funding agencies, biologists and regulating bodies to recognise molecular genetic approaches as an integrated part of resource management. As well as estimating the fundamental unit of biodiversity (genetic variation), molecular genetic techniques allow analyses of patterns of species distribution, dispersal, migration, recruitment and reproduction in marine species. For fisheries managers, genetic techniques provide an objective method of stock identification. Such data is often impossible to obtain for deep-water species in any other way.

References

Avise JC, Arnold J, Ball RM, Bermingham E, Lamb T, Neigel JE, Reeb CA, Saunders NC (1987) Intraspecific phylogeography: The mitochondrial DNA bridge between population genetics and systematics. Ann Rev Ecol Syst 18:489-522

Bailey KM, Stabeno PJ, Powers DA (1997) The role of larval retention and transport features in the mortality and potential gene flow of walleye pollack. J Fish Biol 51 (Suppl A):135-154

Baker CS, Perry A, Chambers GK, Smith PJ (1995) Population variation in the mitochondrial cytochrome b gene of the orange roughy *Hoplostethus atlanticus* and the hoki *Macruronus novaezealandiae*. Mar Biol 122:503-509

Bett BJ (1999) RRS Charles Darwin Cruise 112C, 19th May - 24th Jun, 1998 Atlantic Margin Environmental Survey: Seabed survey of the deep-water areas (17th Round Tranches) to the North and West of Scotland. NERC/University of Southampton, Southampton Oceanography Centre

Brierley AS, Thorpe J, Pierce GJ, Clarke MR, Boyle PR (1995) Genetic variation in the neritic squid *Loligo forbesi* (Myopsida: Loliginidae) in the northeast Atlantic Ocean. Mar Biol 122:79-86

Brown WM, George M, Wilson AC (1979) Rapid evolution of animal mitochondrial DNA. Proc nat Acad Sci USA 76:1967-1971

Bucklin A, Wilson RR Jr, Smith KL Jr (1987) Genetic differentiation of seamount and basin populations of the deep-sea amphipod *Eurythenes gryllus*. Deep-Sea Res 34:1795-1810

Chase MC, Etter RJ, Rex MA, Quattro JM (1998) Bathymetric patterns of genetic variation in a deep-sea protobranch bivalve, *Deminucula atacellana*. Mar Biol 131:301-308

Clark M (1999) Fisheries for orange roughy (*Hoplostethus atlanticus*) on seamounts in New Zealand. Oceanol Acta 22:593-602

Creasey S, Rogers AD (1999) Population genetics of bathyal and abyssal organisms. Adv Mar Biol 35:1-151

Creasey S, Rogers AD, Tyler PA, Young C, Gage J (1997) The population biology and genetics of the deep-sea spider crab, *Encephaloides armstrongi* Wood-Mason 1891 (Decapoda: Majidae). Phil Trans R Soc B 352:365-379

Creasey S, Rogers AD, Tyler PA, Gage J, Jollivet D (2000) Genetic and morphometric comparisons of squat lobster, *Munidopsis scobina* (Decapoda: Anomura: Galatheidae) populations, with notes on the phylogeny of the genus *Munidopsis*. Deep-Sea Res II 47:87-118

Doyle RW (1972) Genetic variation in *Ophiomusium lymani* (Echinodermata) populations in the deep-sea. Deep-Sea Res 19:661-664

Duschenko VV (1988). The formation of the commercial stock of the north Atlantic rock grenadier. Can Trans Fish Aquat Sci No. 5340, pp 21

D'yakov YP (1991) Population structure of Pacific black halibut, *Reinhardtius hippoglossoides*. J Ichthyol 31:16-28

Etter RJ, Rex MA, Chase MC, Quattro JM (1999) A genetic dimension to deep-sea biodiversity. Deep-Sea Res 46:1095-1099

Fairbairn DJ (1981). Biochemical genetic analysis of population differentiation in Greenland halibut (*Reinhardtius hippoglossoides*) from the Northwest Atlantic, Gulf of St. Lawrence, and Bering Sea. Can J Fish Aquat Sci 38:669-677

France SC (1994) Genetic population structure and gene flow among deep-sea amphipods, *Abyssorchomene spp.*, from six California continental borderland basins. Mar Biol 118:67-77

France SC, Kocher TD (1996) Geographic and bathymetric patterns of mitochondrial 16S rRNA sequence divergence among deep-sea amphipods, *Eurythenes gryllus*. Mar Biol 126:633-643

Gage JD (1996) Why are there so many species in deep-sea sediments? J Exp Mar Biol Ecol 200:257-286

Gage JD, Tyler PA (1991) Deep-Sea Biology: A Natural History of Organisms at the Deep-Sea Floor. Cambridge University Press, Cambridge, UK 504 p

Grassle JF, Maciolek NJ (1992) Deep-sea species richness: Regional and local diversity estimates from quantitative bottom samples. Amer Nat 139:313-341

Hedgecock D (1994) Does variance in reproductive success limit effective population sizes of marine organisms. In: Beaumont AR (ed) Genetic and Evolution of Aquatic Organisms. Chapman & Hall, London pp 122-134

Heezen BC, Hollister CD (1971) The Face of the Deep. Oxford University Press, New York 659p

Hensley RT, Beardmore JA, Tyler PA (1995) Genetic variance in *Ophiomusium lymani* (Ophiuroidea: Echinodermata) from lower bathyal depths in the Rockall Trough (northeast Atlantic). Mar Biol 121:469-475

Hoarau G, Borsa P (2000) Extensive gene flow within sibling species in the deep-sea fish *Beryx splendens*. C R Acad Sci III Vie 323:315-325

Hollister CD, McCave IN (1984) Sedimentation under deep-sea storms. Nature 309:220-225

Jeffreys AJ, Wilson V, Thein SL (1985) Individual specific "fingerprints" of human DNA. Nature 316:76-79

Koslow JA, Boehlert GW, Gordon JDM, Haedrich RL, Lorance P, Parin N (2000) Continental slope and deep-sea fisheries: implications for a fragile ecosystem. ICES J Mar Sci 57:548-557

Logvinenko BM, Nefedov GN, Massal'skaya LM, Polyanskaya IB (1983) A population analysis of rock grenadier based on the genetic polymorphism of non-specific esterases and myogenes. Can Trans Fish Aquat Sci No. 5406, pp 16

Marshall NB (1979) Developments in Deep-Sea Biology. Blandford Press, Poole, Dorset, 566 p

Martin AP, Humphreys R, Palumbi SR (1992) Population genetic structure of the armourhead, *Pseudopentaceros wheeleri*, in the North Pacific Ocean: Application of the polymerase chain reaction to fisheries populations. Can J Fish Aquat Sci 49:2368-2391

Milton DA, Shaklee JB (1987) Biochemical genetics and population structure of blue grenadier, *Macruronus novaezelandiae* (Hector) (Pisces: Merluccidae), from Australian waters. Austr J Mar Freshwater Res 38:727-742

Mulligan TJ, Chapman RW, Brown BL (1992) Mitochondrial DNA analysis of walleye pollock, *Theragra chalcogramma*, from the eastern Bering Sea and Shelikof Strait, Gulf of Alaska. Can J Fish Aquat Sci 49:319-326

Murphy RW, Sites JW, Buth DG, Haufler CH (1990) Proteins I: Isozyme electrophoresis. In: Hillis, DM, Moritz C (eds) Molecular Systematics. Sinauer Associates Inc, New York pp 45-146

Nesbo CL, Rueness EK, Iverson SA, Skagen DW, Jakobsen KS (2000) Phylogeographic and population history of Atlantic mackerel (*Scomber scombrus L.*): A genealogical approach reveals genetic structuring among eastern Atlantic stocks. Phil Trans R Soc B 267:281-292

Oke CS, Crozier YC, Crozier RH, Ward RD (1999) Microsatellites from a teleost, orange roughy (*Hoplostethus atlanticus*), and their potential for determining population structure. Mol Ecol 8:2145-2147

Pingree RD, Mardell GT (1981) Slope turbulence, internal waves and phytoplankton growth at the Celtic Sea shelf break. Phil Trans R Soc Lond 302 (A):663-682

Rogers AD (2000) The role of the oceanic oxygen minima in generating biodiversity in the deep-sea. Deep-Sea Res II 47:119-148

Rogers AD (2001) Molecular ecology and identification of marine invertebrate larvae. In: Atkinson D, Thorndyke M (eds) Environment and Animal Development, Genes, Life Histories and Plasticity. BIOS Scientific Publications Ltd, Oxford pp 29-69

Ruzzante DE, Taggart CT, Cook D (1996) Spatial and temporal; variation in the genetic composition of a larval cod (*Gadus morhua*) aggregation: Cohort contribution and genetic stability. Can J Fish Aquat Sci 53:2695-2705

Sedberry GR, Carlin JL, Chapman RW, Eleby B (1996) Population structure in the pan-oceanic wreckfish, *Polyprion americanus* (Teleostei: Polyprionidae), as indicated by mtDNA variation. J Fish Biol 49 (Suppl A.):318-329

Shaw PW, Pierce GJ, Boyle PR (1999) Subtle population structuring within a highly vagile marine invertebrate, the veined squid *Loligo forbesi*, demonstrated with microsatellite DNA markers. Mol Ecol 8:407-417

Siebenaller JF (1978) Genetic variation in deep-sea invertebrate populations: The bathyal gastropod *Bathybembix bairdii*. Mar Biol 47:265-275

Siebenaller JF, Somero GN (1989) Biochemical adaptation to the deep-sea. CRC Crit Revs Aquat Sci 1:1-25

Smith PJ, Benson PG (1997) Genetic diversity in orange roughy from the east of New Zealand. Fish Res 31: 197-213

Smith PJ, Francis RICC (1982) A glucosephate isomerase polymorphism in New Zealand ling *Genypterus blacoides*. Comp Biochem Physiol 73B:451-455

Smith PJ, Patchell PJ, Benson PG (1981) Genetic tags in the New Zealand hoki *Macruronus novaezelandiae*. Anim Blood Grps Biochem Genet 12:37-45

Smith PJ, McVeagh SM, Ede A (1996) Genetically isolated stocks of orange roughy (*Hoplostethus atlanticus*) but not of hoki (*Macruronus novaezealandiae*), in the Tasman Sea and south-west Pacific Ocean around New Zealand. Mar Biol 125:783-793

Stepien CA (1999) Phylogeographic structure of the Dover sole *Microstomus pacificus*: The larval retention hypothesis and genetic divergence along the deep continental slope of the northeastern Pacific Ocean. Mol Ecol 8:923-939

Thorpe JP (1982). The molecular clock hypothesis: Biochemical evolution, genetic differentiation and systematics. Ann Rev Ecol Syst 13:139-168

Tyler PA, Young CM (1998) Temperature and pressure tolerance in dispersal stages of the genus *Echinus* (Echinodermata: Echinoidea): Prerequisites for deep-sea invasion and speciation. Deep-Sea Res II 45:253-277

Vis ML, Carr SM, Bowering WR, Davidson WS (1997) Greenland halibut (*Reinhardtius hippoglossoides*) in the North Atlantic are genetically homogeneous. Can J Fish Aquat Sci 54:1813-1821

Ward RD, Elliot NG, Grewe PM, Last PR, Lowry PS, Innes BH, Yearsley GK (1998) Allozyme and mitochondrial DNA variation in three species of oreos (Teleostei: Oreosomatidae) from Australasian waters. New Zealand J Mar Freshwater Res 32:233-245

Whitehead PJP, Bauchot M-L, Hureau J-C, Nielsen J, Tortonese E (1986) Fishes of the Northeastern Atlantic and the Mediterranean, Volumes 1 and 2. UNESCO, Paris 1007p

Zeldis JR, Murdoch RC, Cordue PL, Page MJ (1998) Distribution of hoki (*Macruronus novaezelandiae*) eggs, larvae, and adults off Westland, New Zealand, and the design of an egg production survey to estimate hoki biomass. Can J Fish Aquat Sci 55:1682-1694

Larval and Reproductive Strategies on European Continental Margins

P. A. Tyler[*] and E. Ramirez-Llodra

School of Ocean and Earth Science, University of Southampton, SOC,
Southampton SO14 3ZH UK
* corresponding author (e-mail): pat8@soc.soton.ac.uk

Abstract: The European margin has given rise to a series of sampling programmes that have provided an insight into the reproductive biology of deep-sea species. From these observations it is apparent that no one reproductive pattern dominates. The observed life history pattern is a function of phylogenetic constraint, whilst the timing of reproduction in some species is determined by the local input of surface-derived material. Little is known of the larvae of deep-water species from any depth. A few larval types have been described, but we know little of the larval ecology. Experimental analysis of larval development has shown that embryos of certain taxa have wider temperature/pressure tolerances than the adult and, as a result, may allow a mechanism by which shallow-water species could invade the deep-sea. It is imperative that there is a better understanding of reproductive processes because anthropogenic impacts may have sub-lethal effects on the benthos by disrupting those processes without killing the adult. We need to understand the natural processes of reproduction in the deep-sea so that the effects of man on this environment can be modelled and predicted.

Introduction

The European continental margin is, arguably, the best-explored continental margin in the world. Although the first deep-sea animals were collected in Baffin Bay in 1818 (Tyler 1980), it was Forbes' "azoic theory" (Forbes 1844) that stimulated exploration in the deeper parts of the NE Atlantic and the Mediterranean. Wallich (1862) and Sars (1868) were the first to show scientifically the presence of a deep water fauna in the NE Atlantic. In the Mediterranean the solitary coral *Caryophylia borelis* was recovered from a submarine cable at 2400 m. In 1868, HMS *Lightning* carried out oceanographic observations to the NW of Scotland, and this was followed by HMS *Porcupine* in 1869/1870 that demonstrated there was a physical difference between the Norwegian Sea ("cold area") and the main NE Atlantic ("warm area") that had an influence on the distribution of the deep-water fauna. Subsequently, HMS *Porcupine* continued to explore deep water to the SW of the United Kingdom and on her last cruise sampled in the western Mediterranean (Wyville-Thomson 1873). These expeditions demonstrated that samples from as deep as 4000 m brought up

live animals. The impact of these observations was to give rise to the *Challenger* expedition of 1872 to 1876 that demonstrated unequivocally that the deep waters of the world's ocean were teeming with life, with the discovery of many taxa previously known only from the fossil record.

The *Challenger* expedition carried out very little sampling along the European margin (except off Portugal) and thus contributes relatively little to our review. However, one of the most perceptive observations of H.N. Moseley, a naturalist on *Challenger*, was that "..the flux of material to the deep-sea bed caused a little annual excitement amongst the inhabitants". This statement was recorded but had little impact until the observations of seasonal reproduction and vertical flux in the NE Atlantic (Tyler et al. 1982; Billett et al. 1983). An additional feature of the *Challenger* expedition is that some authors of the resulting monographs included observations of gonad morphology in their reports (e.g. Théel 1882).

Although *Caryophylia borealis* had been collected by cable ship from deep water in the Mediterranean, it was Prince Albert of Monaco who

From WEFER G, BILLETT D, HEBBELN D, JØRGENSEN BB, SCHLÜTER M, VAN WEERING T (eds), 2002,
Ocean Margin Systems. Springer-Verlag Berlin Heidelberg, pp 339-350

deployed the vessels *Hirondelle, Princess Alice I* and *II* for more systematic collection in the Mediterranean and the Bay of Biscay. These collections and observations appeared as a series of monographs of particular taxa, but little attention was paid to any aspects of reproduction. As a result of all these 19th century expeditions, it was established that the deeper water surrounding much of the European mainland supported a deep-water fauna and, although numbers in some areas were low, no area was azoic. As such the stage was set for the great oceanographic expeditions that sampled many other parts of the world ocean culminating in the circumnavigation by the *Galathea* and the sampling of the deepest trenches again showing that a fauna lived at these extreme depths.

In the early part of the 20th century, interest in deep-sea exploration centred on the more remote parts of the world's oceans and it was not until the latter part of the century that serious interest was revived in European waters. Driving the new approach was technological advancement. In the late 1960s manned submersibles were developed and these were used extensively in the Mediterranean (Pérès 1985). From surface vessels, quantitative remote seafloor sampling progressed rapidly with the invention of the USNEL spade box corer and quantitative deep-sea data are now available from many areas round the European margin (Gage and Tyler 1991).

Practically all the sampling programmes were spatial. To examine reproduction it is necessary to have a temporally-varying sampling programme and these are costly and time-consuming. In European waters the main temporal programmes have been in the Rockall Trough, the Porcupine Seabight and Abyssal Plain and in the NW Mediterranean.

Energy availability

For reproduction to be successful there are a number of criteria to be satisfied, the first of these being the successful production of gametes: eggs and sperm. These processes are energy-expensive and thus the availability of energy in the form of food at the seabed is critical. The deep-sea ecosystem, with some notable exceptions, is a heterotrophic system relying on the input of surface-derived primary production as its organic energy source. The original concept of a slow rain of particles to the deep-sea floor has been challenged in the last 20 years. Billett et al. (1983) were the first to show that, along the European margin of the NE Atlantic, there was an input of surface-derived phytodetrital material (essentially sinking phytoplankton composed of, *inter alia*, primarily diatoms and coccoliths) to the deep-sea bed. This input was seasonal and predictable (to within weeks), with production from the surface sinking at rates of ~100 m d^{-1}. Such a flux has now been observed at a number of locations in the NE Atlantic and in the Mediterranean (e.g. Monaco et al. 1990) although in the latter there is evidence that at certain sites the high organic input is derived from down-canyon transport. Tyler et al. (1982) suggested that such a seasonal input of phytodetritus would drive reproductive processes and correlation between gametogenic pattern and flux supports this concept but experimental evidence remains elusive (but see Tyler et al. 1993a; Campos-Creasey et al. 1994).

In addition to phytodetrital input, transfer of energy to the deep-sea bed may be in the form of larger packages (Gage and Tyler 1991). In the NE Atlantic there is strong evidence for a seasonal flux of dead mesopelagic fish to the deep-sea bed. The blue whiting (*Micromesistius poutassou*) dies after spawning in the spring and sinks to deep water in the Rockall Trough. This input is believed to have an inpact on the diet of both invertebrate and vertebrate populations at depth (Tyler at al. 1993a). Other potential large food falls observed in other parts of the world include whale carcasses and macrophytic debris (from macroalgae, seagrasses and wood) but direct observations of such inputs are lacking round the European margin (for details see Gage and Tyler 1991).

Organic matter may enter the deep-sea in the form of turbidity currents of down-slope and down-canyon transport. There is evidence that this is a significant source of material in the NW Mediterranean and evidence of past turbidity currents are found off Norway, but the contribution of organic matter to the deep-sea by such mechanisms is unquantified. Down-canyon currents have been observed transporting post-larval and juvenile ophiuroids into deep water in the Gollum Channel, east Porcupine Seabight (Tyler pers. obs. from the submersible *Cyana* 1986).

Last, but by no means least, is input by chemosynthetic production. No chemosynthetic environments are known to occur along the European margin (with the exception of the wreck *François Vieljeux* where a vestimentiferan community has been found to thrive in the reducing environment of the ship's rotting cargo), but with the discovery of hydrocarbon deposits to the NW of the UK it may be predicted such environments will eventually be found as in the Gulf of Mexico. There is also evidence of chemosynthetic communities along the Cretan Arc of the central Mediterranean (Corselli and Basso 1996). Chemosynthesic communities may also be found in the reducing sediments associated, particularly, with submarine canyons. Such environments are oxic at the surface of the sediment, but have a strong reducing environment below the surface derived from decomposing organic matter of various origins.

To put these environmental observations into a larval and reproductive context it is necessary to consider the reproductive pattern (Fig. 1) from initiation of gametogenesis to settlement of post-larvae. It is necessary to understand the sequence of life history events, their timing and their energy requirements and output. Most of the data we have from the European margin are observational, but there is a limited amount of experimental data.

Gametogenic patterns

Prior to 1980 the traditional concept was that reproduction in the deep-sea would conform to Orton's rule (see Gage and Tyler 1991) whereby in a "non-varying environment" gametogenesis (the production of eggs and sperm) would be quasi-continuous. To test such a paradigm requires a temporal sampling programme and the only long-term programme on the European margin dedicated to the interpretation of population variables is that at two stations in the Rockall area. These two stations,

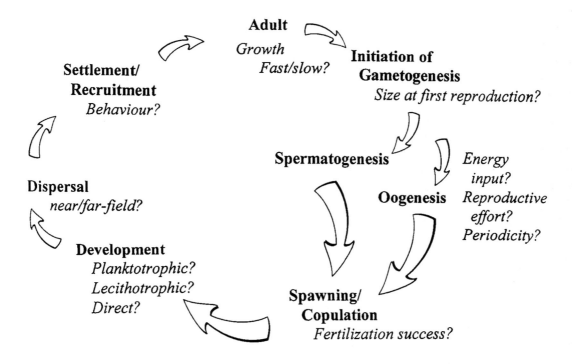

Fig. 1. The life history pattern of a marine invertebrate showing the stages, the points of energy input and the environmental control.

at 2200 m and 2900 m depth were sampled at different times of the year, over a number of years, and the data combined onto a single annual axis (Gage and Tyler 1991).

The early data were based on the analysis of gametogenic patterns in echinoderms. In the echinoids *Echinus affinis, E. elegans, E. alexandri* and *E. acutus*, the asteroids *Plutonaster bifrons* and *Dytaster insignis* and the ophiuroids *Ophiura ljungmani* and *Ophiocten gracilis* reproduction shows an inter-species annual synchrony of gamete development culminating in spawning in the early months of each year (Fig. 2). All these species also produce large numbers of small eggs (typically ~100 to120 μm diameter) that is interpreted as developing into a planktotrophic larva (larvae that feed in the water column during development).

However, only the larva of *Ophiocten gracilis* is known for sure (Tyler and Gage 1982). At upper bathyal depths the larva of the cidarid urchin *Cidaris cidaris* is now known to have been described for *Doriocidaris papillata* from a population off Banyuls sur Mer, in the Mediterranean. Although this list of "seasonal" breeders looks impressive, it should be borne in mind that this is only a very small subset of the deep-sea echinoderm fauna of the NE Atlantic (2200 m and 2900 m depth) and that the vast majority have a quasi-continuous gametogenic pattern producing a small number of large eggs throughout the year (Fig. 3) (see Gage and Tyler 1991). The diameter of the egg produced varies considerably among continuous-breeding species, with the smallest being ~600 μm in some asteroids to an enormous 4.4 mm

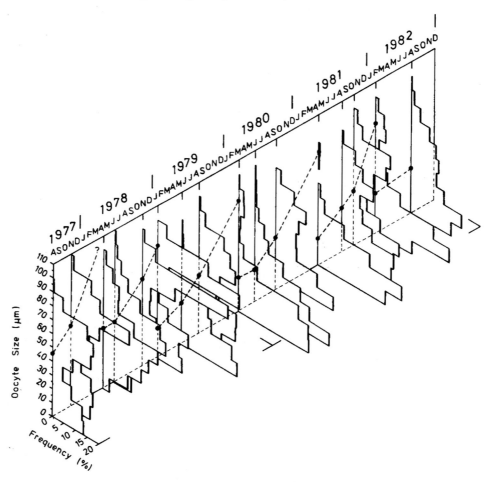

Fig. 2. Seasonal oocyte size/frequency in the oogenic pattern of reproductive pattern of *Echinus affinis*. Solid circles represent mean oocyte size for that sample. Reprinted from Deep-Sea Research, Vol. 31, Tyler and Gage, Seasonal reproduction of *Echinus affinis* in the Rockall Trough, NE Atlantic, 387-402, Copyright (1984), with permission from Elsevier Science

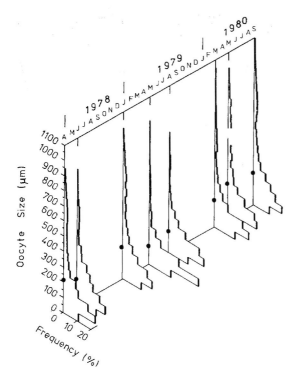

Fig. 3. "Continuous" oocyte size/frequency in the oogenic pattern of *Bathybiaster vexillifer*. Solid circles represent mean oocyte size for that sample.

in some species of holothurian. Such large lipid-rich eggs must be significant stores of energy in the deep-sea and such an egg size is indicative of direct development on the seabed (larval stage omitted) or of lecithotrophic development (reduced larval stage that develop by using their own nutrient reserves) in the water column. It is this pattern that may be considered typical of the deep-sea. Only the very common ophiuroid *Ophiomusium lymani* (from 2200 m) is known not to show either pattern. In this species the egg is intermediate (~400 µm diameter) and there is no seasonality in its production. However, recruitment is seasonal! In the Mediterranean, off Banyuls sur Mer, Ferrand et al. (1988) examined the reproductive seasonality of the irregular echinoid *Brissopsis lyrifera* from samples taken between 60 and 1000 m. Over the entire depth range the gametogenic pattern remained synchronous.

The echinoderms have given a clear steer to the patterns that may occur in the gamete production

in deep-sea invertebrates. It was seasonality that caught the imagination and the hunt was on to find species in other major taxa at different depths along the European margin that had seasonal processes. Such species were found in the Porifera (Witte 1996), Cnidaria (Bronsdon et al. 1993), Protobranchia (Bivalvia) (Tyler et al. 1993b) and Isopoda (Harrison 1988). However, all these taxa also had species that produced large gametes on a quasi-continuous basis. In some cases such gamete production was taxonomically constrained, e.g. in the octocoral family Pennatulida in which all the species examined to date have large eggs.

If we consider the diversity of Crustacea in the deep-sea it may be surprising that we know relatively little about their reproductive biology. The crab *Dorhynchus thomsoni* appears to be a seasonal species (Rice and Hartnoll 1983). Of considerable interest is the epizoitic stalked barnacle *Poecilasma kaempferi*. This barnacle is normally found on the carapace of stone crabs, but will settle on anthropogenic structures such as the time-lapse camera *Bathysnap* (Lampitt 1990). Since Darwin's monograph on stalked barnacles, all are known to grow quickly although it is with some surprise that *P. kaempferi*, which lives in cold water, grows rapidly (Lampitt 1990) and reproduces within 180 days of settling at 2000 m depth (Green et al. 1994). With very few exceptions, such high reproductive rates in cold water are known only for the meiofauna. The only other deep-sea taxon known to have such a rapid growth and reproduction is the wood-boring mollusc *Xylophaga* (at 550 m depth) that appears to be gametogenically active within 30 days of settlement on experimental wood blocks (Tyler and Young pers. obs). In the NW Mediterranean there has been extensive examination of the reproductive biology of commercially-important shrimp at bathyal depths (Demestre and Fortuno 1992; Company and Sardà 1997) and the scavenging isopod *Natatolana borealis* (Kaim-Malka 1997), whilst off Spitzbergen the reproductive output of the deep-water prawn *Pandalus borealis* has been determined (Clarke et al. 1991).

Although there is a greater movement of fishing activity into ever deeper water there remains

344   Tyler and Ramirez-Llodra

only limited data on the reproduction of deep-sea fish along the European margin although the effort in this direction worldwide is considerable.

From these data we can see that there is no "typical" gametogenic pattern in deep-sea invertebrates. There is a degree of phylogenetic constraint but there is a proximate effect of the environment in determining, at least, the timing of seasonal gametogensis. Although not experimentally shown, intuitively, it may be predicted that the deep-sea may have been invaded in the past by seasonally-breeding shallow-water species that have maintained their seasonal reproductive pattern by being entrained by the vertical flux from surface production. How long this could be maintained is subject to speculation.

A last variable on this proscenium is that the deeper a species lives on the continental margin, the earlier it appears to spawn. This has been observed in the shallow water and bathyal species of the genus *Echinus* found in the NE Atlantic (Fig. 4) (Tyler and Young 1998). Such a pattern is not easy to interpret, but may be related to the slower embryonic development in deeper colder waters. If this is the case embryos of planktotrophic species in the cooler deeper waters will develop more slowly and be at the feeding stage when the sinking phytodetritus arrives.

Controls on spawning

We have no evidence of any specific event that may stimulate spawning in the invertebrates of the European continental margin. When ripe many species will spawn naturally but other species require some stimulus. The only variable that could be correlated to spawning is increased eddy kinetic energy during the period when spawning occurred (Tyler and Gage 1984). We do not know if this increase in eddy kinetic energy acts as some form of mechanical stimulus or advects some pheromone that stimulates spawning in the population. It may be that the immediate presence of an individual of the opposite sex stimulates spawning. This may account for the observation, from submersibles in the Bahamas, of pairing in deep-sea echinoids (Young et al. 1992). No such observa-

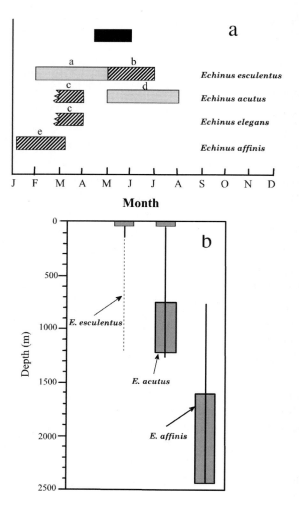

Fig. 4. a) Variation in the times of spawning with depth in the echinoid genus *Echinus* in the NE Atlantic. Solid represents main period of phytoplankton in the NE Atlantic. Depth of sampling: a=60 m; b=10 m; c=500 to 1271 m; d=60 m; e=2200 m. **b)** Depth distribution of species of *Echinus* (Reprinted from Deep-Sea Research II, Vol. 45, Tyler and Young, Temperature and pressure tolerances in dispersal stages of the genus Echinus: Prerequisites for deep-sea invasion and speciation? 253-402, Copyright (1998), with permission from Elsvier Science)

tions have been made in the NE Atlantic as video and photographic observation of the seabed has proved very difficult from surface vessels in the early months of the year when deep-sea seasonally breeding species spawn. However, there is indirect evidence that pairing may occur in the NE Atlantic bathyal hermaphroditic holothuroid *Paroriza pallens* (Tyler et al. 1992).

A feature of greater importance in deep-sea populations is fertilization success. Compared to many shallow-water populations the density of deep-sea populations can be very low. In free-spawning dioecious species it is necessary for a male and a female to be in close proximity at spawning to ensure fertilization. Sperm are rapidly diluted as they are advected from the male gonopore and it may be that just a couple of metres downstream the concentration of sperm is too low for fertilization success. The problem of low density is mitigated in motile species that may come together in pairs or herds during fertilization (as seen in the Bahamian *Stylocidaris lineata* (Young et al. 1992)) so that successful fertilization is achieved. In sessile dioecious species (on rock e.g. gorgonians or in sediment e.g. pennatulids), where the whole colony is a single sex, the density of colonies may be so low that the colony may never contribute genes to the next generation. Only when the population density is above a certain, as yet unknown level, will successful fertilization occur. There is some evidence from modelling that large egg size may contribute to fertilization success.

Under natural circumstances this variable has lead to extinction or adaptation in evolutionary time. More and more, parts of the deep-sea on the European margin are subject to anthropogenic impact, especially from deep water fishing and oil exploration, perhaps reducing population density artificially in sub-evolutionary time. As a result, the individual density may be reduced to a level where there is no successful fertilization in a population. Such events are already recognised in commercial shellfish populations and there is even evidence that at least one shallow-water species has become extinct (Roberts and Hawkins 1999). Such effects are being successfully modelled including the effect of turbulence on sperm dispersal and models of harvest refugia. These models will become particularly important in very sensitive long-lived species such as deep-water scleractinian corals.

Larval development

The first known deep-sea larva along the European continental margin was that of *Dorocidaris papillata*. Little else in known for the larval development other than some fully developed larvae. *Ophiopluteus ramosus* is now known to be the larva of the ophiuroid *Ophiocten gracilis* whilst *Ophiopluteus compressus* is the larval form of the ophiuroid *Ophiura carnea*. All three are found in the NE Atlantic and the last is found in the western Mediterranean. Off Villefranche sur Mer, in the northern Mediterranean, larvae of the shallow-water echinoid species *Paracentrotus lividus* and *Arbacia lixula* have been found alive as deep as 400 m, whereas the adult population rarely extends below 50 m (M.L. Pedrotti, pers. comm.). At the end of the 19th century, Chun (1887) noted the presence of a number of invertebrate larvae at 1400 m depth off Naples, although it is not clear whether these are truly deep-sea species (see below).

Adult gastropod and protobranch bivalves "keep" a record of their larval development on their larval shells. In the gastropods the prodissoconch I and II can be of different sizes in different species. If prodissoconch I is small and prodissoconch II large this is interpreted as evidence for a longer planktotrophic larval life, whereas the opposite suggests longer embryonic development employing lecithotrophy. In some species there is direct development in which the entire larval development occurs within the egg capsule. Such observations have been used to suggest the life history biology of gastropods and bivalves along the European margin as well as at hydrothermal vents (see Gage and Tyler 1991)

Experimental embryology

There has been considerable debate on the mechanisms by which animals have invaded the deep-sea. One pathway is the larval invasion of the deep-sea. To this end we (Young et al. 1997; Tyler and Young 1998) have conducted a series of experiments examining the temperature/pressure tolerance of early and late embryos of deep and shallow water echinoids. The main emphasis of this work in European waters has been on the species of the genus *Echinus* in the NE Atlantic (Tyler and Young 1998) and *Paracentrotus lividus, Arbacia lixula* and *Sphaerechinus granularis* from the northern Mediterranean. Figs 5 and 6 show the results for the temperature/pressure effects on the early embryos of different species of *Echinus* from different depths in the NE Atlantic. The outcome of these experiments is that larvae can tolerate much greater temperature/pressure regimes than the adult and would be able to penetrate the deep-sea along isotherms but across pressure gradients. In the NE Atlantic evolutionary time has allowed this to occur whereas the young age of the Mediterranean (~5 My) is too short for such processes (Young et al. 1997; Tyler and Young 1998).

Settlement and recruitment

There have been a number of experiments to determine recolonisation of defaunated sediments in the deep-sea (Gage and Tyler 1991), but there have been no experiments that we know of that have targeted the settlement and recruitment of juveniles to the adult population. The data available (Gage and Tyler 1981, 1991; Sumida et al. 2000) have been based on temporal sampling programmes and the identification of the periodicity of settlement and recruitment to the deep-sea benthos. Because of the fiscal and operating constraints, these observations have been fairly coarse but have identified to the nearest month when recruitment occurs in seasonally breeding populations. In species not reproducing seasonally there appears, from analy-

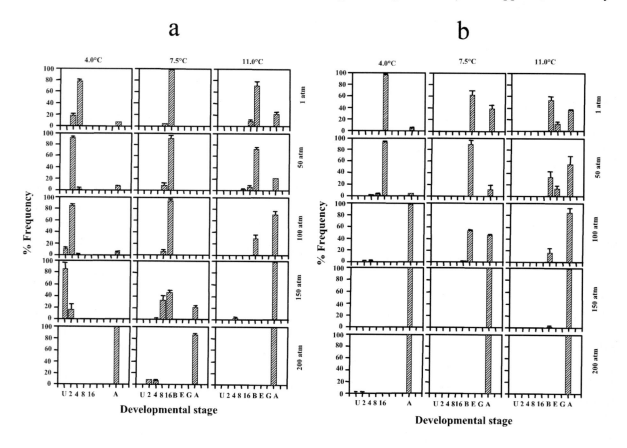

Fig. 5. Temperature/pressure tolerance of early embryos of *Echinus acutus* from shallow water, **a)** - at 24h; **b)** - at 48h. Development stage is 2 to 64 cell. U is uncleaved; B is blastula; E is early gastrula; G is gastrula and A is abnormal. Histograms are % mean and standard deviation. (Reprinted from Deep-Sea Research II, Vol. 45, Tyler and Young, Temperature and pressure tolerances in dispersal stages of the genus Echinus: Prerequisites for deep-sea invasion and speciation? 253-402, Copyright (1998), with permission from Elsvier Science)

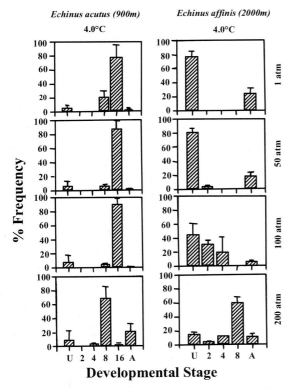

Fig. 6. Temperature/pressure tolerance of early embryos of *Echinus acutus* from 900 m and *E. affinis* from 2000 m cultured at 4°C at 1 to 200 atm. for 12h. Development stage is 2 to 16 cell. U is uncleaved and A is abnormal. Histograms are % mean and standard deviation. (Reprinted from Deep-Sea Research II, Vol. 45, Tyler and Young, Temperature and pressure tolerances in dispersal stages of the genus Echinus: Prerequisites for deep-sea invasion and speciation? 253-402, Copyright (1998), with permission from Elsvier Science)

sis of the population structure, that there is a low level of recruitment throughout the year.

Two species, the ophiuroid *Ophiocten gracilis* and the echinoid *Echinus affinis*, show very discrete periods of recruitment (Gage and Tyler 1981, 1985; Sumida et al. 2000). In the case of *Ophiocten gracilis*, sediment traps have collected settling post-larvae and it has been possible to determine the early growth rate in this species. Using preserved specimens from sediment traps it has also been possible to determine experimentally the settlement rate of this species (Sumida et al. 2000). Although the individuals used were preserved, a settlement rate of ~500 m d^{-1} was estimated (Fig. 7).

From the Rockall Trough sampling programme, the most obvious feature of juvenile recruitment has been the settlement of juveniles above and below the known depth range of the adult. In *Ophiocten gracilis* the juveniles settle as deep as 4000 m and as shallow as 150 m, although the adult range is around 700 to 1000 m. Some of these juveniles found outside the adult range even start developing gonads but all have died by the end of the first year (Gage and Tyler 1981; Sumida et al. 2000). It may be that the variation in larval tolerances between shallow and deep populations of invertebrates (see above) may, over evolutionary time, result in selection for increasing depth tolerance leading to the invasion and colonisation of the deep-sea.

Conclusions

In the last twenty years there has been considerable progress in our understanding of the reproductive biology of animals along the European continental margin, including the NE Atlantic and the Mediterranean but much remains to be understood. In light of the actual and potential anthropogenic impact from mineral exploration and exploitation, as well as fishing impact, it is an appropriate time for a renewed effort in understanding the basic processes, particularly reproduction, which maintain these ecosystems. It is possible to destroy a species by disrupting its reproductive pattern by sub-lethal effects. The adult may continue to live under such conditions and, especially in long-lived species, the effect may take years to manifest itself. It will be evident only when the adult dies without reproducing itself. We believe this is especially important in long-lived, slow growing species such as most cnidarians, especially the fragile deep-water scleractinian corals and gorgonians, but has application amongst all deep-water species. Finally, not only is fishing having an impact on the environment, but it is also having an impact on fish stocks. These impacts need to be assessed. The orange roughy fishery worldwide has already shown what uncontrolled fishing can do to a deep water population.

We believe the main areas of research that need addressing are:

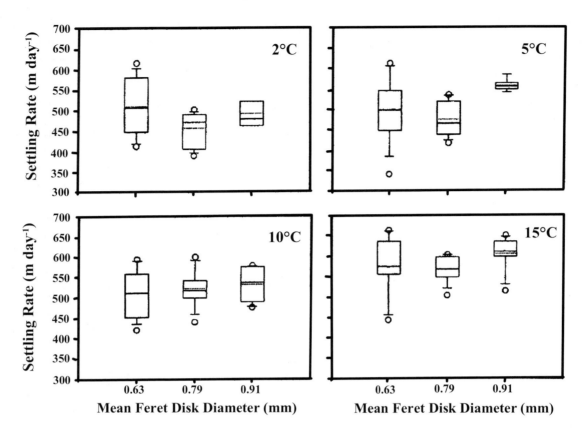

Fig. 7. Experimental settlement of the post-larvae of *Ophiocten gracilis* (from Sumida et al. 2000). Box and whisker plots show mean (dotted line); error bars (vertical bars) and outliers (circles).

- Timing of the initiation of gametogenesis both in real time (of year) and also in life history of adult.
- Timing and initiation of spawning.
- Fecundity variability in relation to environmental factors.
- Fertilization kinetics and ultimate success of embryos.
- Dispersal of larvae in the deep-sea. Planktotrophy vs. lecithotrophy: have we got it wrong up to now?
- Pressure/temperature tolerance of both shallow and deep living species.
- Vertical distribution of larvae in the water column.
- Settlement rates of postlarvae.
- Survival of juveniles once settled.
- Inter-annual variation in all of the above
- Is variation related to surface production?
- What are the proximate and ultimate controls? Is surface production proximate whilst phylogeny an ultimate control?

- What is the effect of reproduction in long-lived slow-growing species in relation to fast-growing short-lived species?
- How will the life history adaptations and tactics of all deep-sea species cope with potential exploitation within the deep-sea?

Answering such questions will give us an opportunity to understand the natural environment and from this understanding we should be able to produce models that will determine the impact arising from anthropogenic sources.

References

Billett DSM, Lampitt RS, Rice AL, Mantoura RFC (1983) Seasonal sedimentation of phytoplankton to the deep-sea benthos. Nature 302:520-522

Bronsdon SK, Tyler PA, Rice AL, Gage JD (1993) Reproductive biology of two epizoic anemones from bathyal and abyssal depths in the NE Atlantic Ocean. J Mar Biol Ass UK 73:531-541

Campos-Creasey LS, Tyler PA, Gage JD, John AWG

(1994) Vertical flux coupled to the diet and seasonal life history of the deep-sea echinoid *Echinus affinis*. Deep-Sea Res 41:369-388

Chun C (1887) Die pelagische Thierwelt in groesseren Meerestiefen und ihre Beziehungen zu der Ober-flaechenfauna. Biobliotheca Zoologica No. 1 Cassel: Theodor Fischer 72 p

Clarke A, Hopkins CCE, Nilssen EM (1991) Egg size and reproductive output in the deep-water prawn *Pandalus borealis* Krøyer 1838. Func Ecol 5:724-730

Company JB, Sardà F (1997) Reproductive patterns and population characteristics in five deep-water pandalid shrimps in the Western Mediterranean along a depth gradient (150-1100 m). Mar Ecol Prog Ser 148:49-58

Corselli C, Basso D (1996) First evidence of benthic communities based on chemosynthesis on the Napoli mud volcano (Eastern Mediterranean). Mar Geol 132:227-239

Demestre M, Fortuno J-F (1992) Reproduction of the deep-water shrimp *Aristeus antennatus* (Decapoda: Dendrobranchiata). Mar Ecol Prog Ser 84:41-51

Ferrand JG, Vadon C, Doumenc D, Guille A (1988) The effect of depth on the reproductive cycle of *Brissopsis lyrifera* (Echinoidea: Echinodermata) in the Gulf of Lions, Mediterranean Sea. Mar Biol 99: 387-392

Forbes E (1844) Report on the Mollusca and Radiata from the Aegean Sea. Report (1843) to the 13th meeting of the British Association for the Advancement of Science 30-193 p

Gage JD, Tyler PA (1981) Non-viable seasonal settlement of larvae of the upper bathyal brittlestar *Ophiocten gracilis* in the Rockall Trough abyssal. Mar Biol 64: 153-161

Gage JD, Tyler PA (1985) Growth and recruitment of the deep-sea urchin *Echinus affinis*. Mar Biol 90:41-54

Gage JD, Tyler PA (1991) Deep-Sea Biology: A Natural History of Organisms at the Deep-Sea Floor. Cambridge University Press 504 p

Green A, Tyler PA, Angel MV, Gage JD (1994) Gametogenesis in deep- and surface-dwelling stalked barnacles from the NE Atlantic Ocean. J Exp Mar Biol Ecol 184:143-158

Harrison K (1988) Seasonal reproduction in deep-sea Crustacea (Isopoda: Asellota). J Nat Hist 22:48-62

Kaim-Malka RA (1997) Biology and life cycle of *Natatolana borealis*. A scavenging isopod from the continental slope of the Mediterranean. Prog Oceanogr 44:2045-2067

Lampitt RS (1990) Directly measured rapid growth in a deep-sea barnacle. Nature 345:805-807

Monaco A, Biscaye P, Soyer J, Pocklington R, Heussner S (1990) Particle fluxes and ecosystem response on a continental margin: The 1985-1988 Mediterranean ECOMARGE experiment. Cont Shelf Res 10: 809-839

Pérès JM (1985) History of the Mediterranean biota and colonization of the depths. In: Margalef R (ed) Western Mediterranean. Pergamon Press, Oxford pp 198-232

Rice AL, Hartnoll RG (1983) Aspects of the biology of the deep-sea spider crab *Dorhynchus thomsoni* (Crustacea: Brachyura). J Zool Lond 201:417-431

Roberts C.M, Hawkins JP (1999) Extinction risk in the sea. TREE 14:241-246

Sars M (1868) Fortsatte Bemaerkninger over det dyriske Livs Udbreedning I Havets Dybder. Christiana. Videnskabs-Selskabs Fordandlinger for 1868

Sumida PYG, Tyler PA, Lampitt RS, Gage JD (2000) Reproduction, dispersal and settlement of the bathyal ophiuroid *Ophiocten gracilis* in the NE Atlantic. Mar Biol 137:623-630

Théel H (1882) Holothuroidea Part 1. Rep Sci Res Voy HMS Challenger Zoology 4:1-176

Tyler PA (1980) Deep-sea ophiuroids. Oceanography and Marine Biology: An Annual Review 18:125-153

Tyler PA, Gage JD (1982) *Ophiopluteus ramosus* the larval form of *Ophiocten gracilis* (Echinodermata: Ophiuroidea). J Mar Biol Ass UK 62:485-486

Tyler PA, Gage JD (1984) Seasonal reproduction of *Echinus affinis* (Echinodermata: Echinoidea) in the Rockall Trough, NE Atlantic. Deep-Sea Res 31: 387-402

Tyler PA, Young CM (1998) Temperature and pressure tolerances in dispersal stages of the genus *Echinus* (Echinodermata: Echinoidea): Prerequisites for deep-sea invasion and speciation? Deep-Sea Res II 45: 253-277

Tyler PA, Grant A, Pain SL, Gage JD (1982) Is annual reproduction in deep-sea echinoderms a response to variability in their environment? Nature 300: 747-750

Tyler PA, Young CM, Billett DSM, Giles LA (1992) Pairing behaviour, reproduction and diet in the deep-sea holothuria genus *Paroriza* (Holothuroidea: Synallactidae). J Mar Ass UK 72:447-462

Tyler PA, Gage JD, Paterson GJL, Rice AL (1993a) Dietary constraint on reproductive periodicity in two sympatric deep-sea astropectinid species. Mar Biol 115:267-277

Tyler PA, Harvey R, Giles LA, Gage JD (1993b) Reproductive strategies and diet in deep-sea nuculanid protobranchs (Bivalvia: Nuculoidea) from the Rockall Trough. Mar Biol 114:571-580

Wallich GC (1862) The North Atlantic Seabed. Comprising a Journey on HMS Bulldog in 1860 Van Voorst, London 160 p

Witte U (1996) Seasonal reproduction in deep-sea sponges- triggered by vertical particle flux? Mar Biol 124:571-581

Wyville-Thomson C (1873) Depths of the Sea. Macmillan and Co, London

Young CM, Tyler PA, Cameron JL, Rumrill SG (1992) Seasonal breeding aggregations in low-density populations of the bathyal echinoid *Stylocidaris lineata*. Mar Biol 113:603-612

Young CM, Tyler PA, Fenaux L (1997) Potential for deep-sea invasion by Mediterranean shallow water echinoids: Pressure and temperature as stage-specific dispersal barriers. Mar Ecol Prog Ser 154:197-209

Factors Controlling Soft Bottom Macrofauna Along and Across European Continental Margins

E.C. Flach

University of Stockholm, Department of Systems Ecology, 106 91 Stockholm, Sweden
corresponding author (e-mail): elsina@system.ecology.su.se

Abstract: Soft bottom macrofauna (benthos) in the deep-sea is influenced by a wide variety of environmental factors. In this review I will shortly evaluate the importance of some factors on different parameters of the benthos. Studies on macrobenthic communities from the Skagerrak in the north to the Iberian continental margin in the south are compared. Not one factor can be indicated as the factor controlling the benthic fauna at the European continental margin. Nearly all factors are coupled and directly or indirectly influencing the benthic community, indicating the complexity of this ecosystem. The overall pattern in decrease in total density and biomass with increasing water depth is mainly determined by the food input, but special topographic structures can highly influence this pattern. In canyons very high densities were found, whereas on very steep slopes and on sea mounts densities were very low. Not only quantity, but also quality, reliability and source of the food are important for the benthos. High flow velocities can resuspend the organic matter again and make it more available for suspension-feeders, whereas the fauna can change the flow and actively capture food that otherwise would pass. Extreme flow conditions can periodically disturb the fauna, allowing only deep living fauna to maintain and favor rapid colonizers, whereas a stable environment allows the development of a 'climax' community in which biotic interactions become very important. As we still know only very little about the biology of the deep-sea fauna, we also can say very little about the elasticity and carrying capacity of this remote, but probably very important ecosystem. The highly complex nature of the continental margins, with strong difference in appearance, steep slopes with rocky outcrops, smooth sedimental plateaus, bights, troughs, seamounts and canyons, results in high local variability. We need to know more about the actual life-history strategies, feeding-types and mobility of the deep-sea fauna. Long-term observations and experiments would give an idea about the reactions of the benthos on different environmental events, as changes in flow velocities, food falls, sedimentation, even pollution and disturbances, but also reactions on biological events, as predators, competitors, etc. DNA- analyses can give information about evolution and distribution of the fauna and probably allow us a better understanding of biodiversity.

Introduction

Soft bottom macrofauna (benthos) in the deep-sea is influenced by a wide variety of environmental factors. The most important factors are listed in Table 1. In this review I will evaluate briefly the influence of these factors on different parameters of the benthos. In contrast to shallow-water fauna, the deep-sea benthic fauna is dependent on organic matter produced from distant sources. Therefore, transport of food to the benthos is an important factor. The organic carbon flux to the benthos depends on primary production in the oceanic pelagic system, in adjacent shallow seas and on land. This production is related to radiation and temperature, and thus latitude, and nutrients. Nutrient conditions depend on the dregree of winter mixing, upwelling, outwelling, distance from land, depth and the steepness of slope. Transport is affected by current velocity, flow direction and topography. Current velocity, distance from land, depth and steepness of slope also determine sediment structure. The carbon requirements of the benthos are related to temperature. Temperature is related to latitude and depth and depth is coupled with pressure. No one factor alone controls the benthic

From WEFER G, BILLETT D, HEBBELN D, JØRGENSEN BB, SCHLÜTER M, VAN WEERING T (eds), 2002,
Ocean Margin Systems. Springer-Verlag Berlin Heidelberg, pp 351-363

Factor	along	across
food	++	++
flow	++	++
topography	++	++
sedimentation rate	++	++
grain-size	++	+
% calcium carbonate	+	+
temperature	+	+
biotic interactions	+	+
distance from land, outwelling	++	
upwelling	+	
pressure		+

Table 1. Some of the (most important) factors controlling soft bottom fauna along and across continental margins. + important, ++ very important

community. Nearly all the factors are coupled and their relative importance at different sites determines benthic community structure.

Factors

Food

Food limitation is the major factor controlling benthic standing stocks (Galéron et al. 2000). As food is so important a good estimate is needed. However, this has been proved to be quite difficult to resolve (Carney 1989). An often used, and relatively easy, method is to measure the amount of organic carbon and nitrogen in the sediment. However, Flach and Thomsen (1998) and Flach et al. (in press) found no relationship between macrofauna density and biomass and the amount of organic carbon and nitrogen in the sediment at a number of stations on five different transects across the European continental slope. In addition, in a deep trench in the Skagerrak there was no relationship between the amount of carbon and nitrogen and macrofauna density and biomass (Rosenberg et al. 1996). Sibuet et al. (1989) concluded from a comparison of different regions that the % POC and %N within the sediment were not good estimates of organic input. They found that ~ 85% of the sedimenting organic C was utilized before burial and that the organic C

within the surface sediments represented mainly what the fauna did not consume. Generally an exponential decrease in the abundance of benthic fauna with water depth is observed and this is usually explained by trophic input (Sibuet et al. 1989). Carney (1989) proposed that depth is a major determinant of the level of food input experienced by deep-sea fauna in most regions of the world ocean because of the exponential relationship of carbon flux with depth. On the northeastern margin of the Atlantic Ocean an overall decrease in macrofauna density was found with increasing depth (Fig. 1), but there were very large regional differences. At a water depth of about 400 m, macrofauna densities differed by a factor 10 between the Nazaré canyon, the Skagerrak and the Hebridean slope. Such differences may also be found at greater depth (Fig. 1), for example between the eutrophic EUMELI site and the La Coruña station at about 1500 m or between sites in the two Portuguese canyons at about 3400 m. Unfortunately different sieve mesh-sizes were used in the different studies making a comparison somewhat tricky. A standardized sieve-size for deep-sea research would be desirable. Biomass may fluctuate even more than density (Flach and Heip 1996a; Flach et al. in press). Biomass is directly linked to the carbon requirements (community respiration) of the benthos. Sibuet et al. (1989) concluded

Macrofauna density

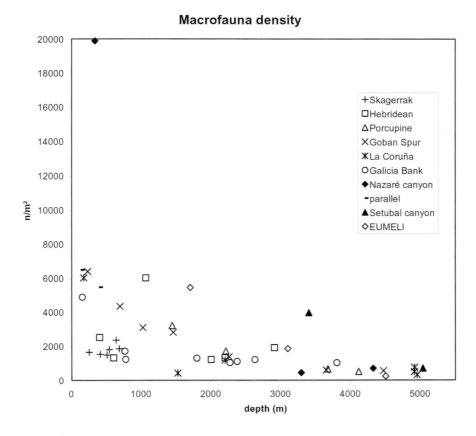

Fig. 1. Macrofauna densities (n/m^2) in relation to water depth in the NE Atlantic from the northern Skagerrak to the tropical EUMELI-sites. Positions (roughly) and sieve-size used: Skagerrak 58°N 9°E, 1 mm-sieve; Hebridean 56°N 9°W, 0.42 mm-sieve; Porcupine 50°N 13°W, 0.5 mm-sieve; Goban Spur 49°N 13°W, 0.5 mm-sieve; La Coruña 44°N 9°W, 0.5 mm-sieve; Galicia Bank 42°N 10°W, 0.5 mm-sieve; Nazaré canyon and parallel 39°N 9°W, 0.5 mm-sieve; Sebutal canyon 38°N 10°W, 0.3 mm-sieve; EUMELI 20°N 20°W, 0.25 mm-sieve. Data from the Skagerrak from Rosenberg et al. (1996), Hebridean slope and Sebutal canyon from J.D. Gage (pers. comm.), EUMELI from Galéron et al. (2000).

that the settling carbon seems to be the main factor controlling the biomass distribution at various trophic levels. On the Iberian margin a significant positive linear relationship was found between the organic carbon flux (data from Epping and Helder in press) and macrofauna total biomass (Flach et al. in press) (Fig. 2), but either factor was related significantly with depth. Depth, therefore, cannot be regarded as a "synonym" for food input/carbon flux.

Total macrofauna biomass is related to the carbon influx estimated over the whole year. However, inputs are often calculated from short-term measurements. Carbon influx can be calculated from *in situ* oxygen profile measurements (Epping and Helder in press) or from sediment traps. Comparison of sediment trap data with sediment oxygen consumption data of the Goban Spur shows a deficit of organic input as measured from the sediment trap (Fig. 3). We know, however, that carbon input does not occur as a steady slow detrital rain, but as seasonal and short-term pulses of labile material sinking rapidly from the photic zone (Billett et al. 1983; Lampitt 1985). Especially the small benthic size classes are found to react quickly to this food input (Pfannkuche et al. 1999). Phytoplankton blooms vary not only seasonally, but also between years (Pfannkuche et al. 1999) and thus

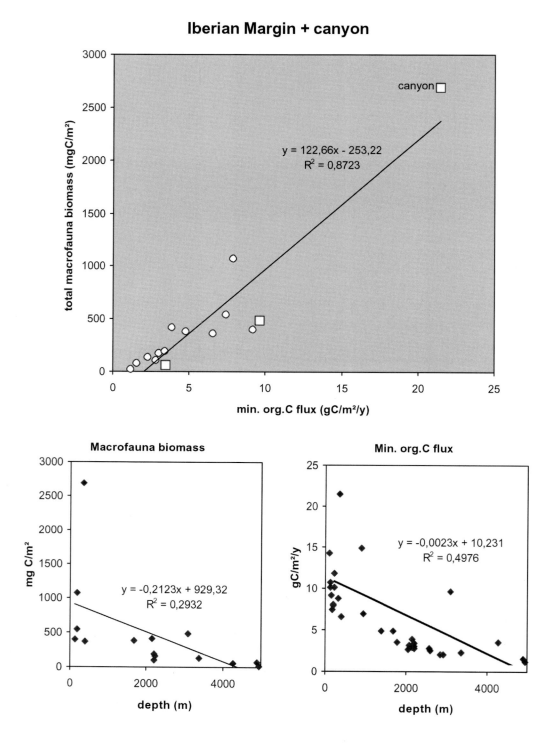

Fig. 2. Total macrofauna biomass (mgC/m²) in relation to the minimum organic carbon flux (gC/m²/y) and both parameters in relation to water depth at the Iberian margin (circles) and in the Nazaré canyon (squares). Organic carbon flux data from Epping and Helder (in press, pers. comm.)

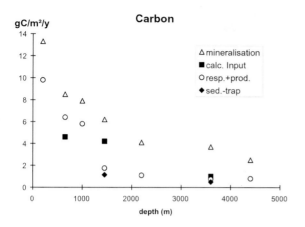

Carbon

gC/m²/y

△ mineralisation
■ calc. Input
○ resp.+prod.
◆ sed.-trap

depth (m)

Fig. 3. Carbon fluxes (gC/m²/y) across the Goban Spur, input (black symbols) calculated from the sediment-trap data (Antia, pers. comm.) and using BBL-measurements and results from flume-experiments (Thomsen, pers. comm.) and required by the benthos (open symbols) calculated from SCOC (sediment community oxygen consumption) measurements (Duineveld, pers. comm.) and respiration plus production (p + r) of the metazoan fauna (megafauna data from Lavaleye (pers. comm.) and meiofauna from J. Vanaverbeke (pers. comm.)).

short-term measurements are not accurate estimates of food input for long lived animals, such as most macro- and megafauna species. Some macro- and megafauna species reproduce seasonally and there are differences in reproductive success between years (Tyler et al. 1992, 1993).

Apart from seasonal and year-to-year variability in the quantity of organic matter, the quality of the food is also of great importance for benthic fauna (Carney 1989). Dauwe and Middelburg (1998) found that systems with similar organic loading might contain organic matter with widely varying degradability and therefore different food value to benthic organisms. They found strong differences in amino acid composition between four stations in the North Sea, which was reflected in a strong difference in benthic community structure (Dauwe et al. 1998). At a station in the German Bight receiving a large quantity of high qualtity organic matter they found mainly interface- and suspension-feeding animals feeding on freshly deposited or (re)suspended material, whereas at a

station in the Skagerrak receiving a large quantity of refractory material they found mainly deep-living deposit-feeders. The highest diversity of trophic groups and the largest individuals were found in sediment with intermediate quantity and quality of organic matter at the Frisian Front (Dauwe et al. 1998). On the continental shelf off Galicia López-Jamar et al. (1992) found mainly small, surface feeding, fast growing polychaetes in a zone dominated by seasonal upwelling, whereas larger subsurface deposit-feeding polychaetes were found in an area influenced by outwelling of organic rich matter from the Rías Bajas as well as upwelling. They concluded that the enrichment from upwelling production reached the seabed in pulses favouring small opportunistic species specialized in exploiting such irregular events, whereas the more regular outwelling from the Rías promote fauna of a higher successional state. On the continental margin of the Goban Spur the macrofauna was dominated by small opportunistic polychaetes at the shelf edge, whereas at abyssal depths larger, subsurface, deposit-feeding polychaetes dominated the fauna (Flach et al. 1998). This overall pattern was also found on the Iberian margin and at the Hebridean slope (Fig. 4). These deep-living, large, subsurface, deposit-feeders are favoured by a steady, slow sedimentation of refractory organic material (Rice and Rhoads 1989). Jumars et al. (1990) supposed that bacteria, which convert refractory material that is not directly available for deposit feeders, might be an important food source for subsurface deposit feeders. Pfannkuche and Soltwedel (1998) found indeed an increase in relative importance of bacteria with depth. However, a comparison of the relative importance of the small size class biomass (SSCB, Pfannkuche and Soltwedel 1998) with the relative importance of the subsurface deposit-feeding polychaetes (Flach et al. 1998) showed no good correlation (Fig. 5). On the Iberian shelf López-Jamar et al. (1992) found the highest bacterial production in the area which was mainly influenced by seasonal upwelling. The highest densities of small opportunistic species also occured a this area. In the area dominated by large subsurface deposit feeders bacterial production was much lower. So, although bacteria might be an important

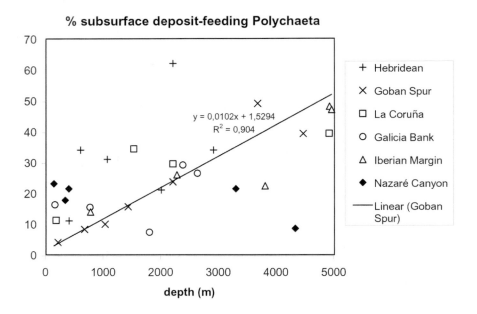

Fig. 4. The relative abundance (%) of subsurface deposit-feeders within the Polychaeta in relation to water depth. Hebridean slope data from J.D. Gage (pers. comm.).

food source for subsurface deposit-feeders no direct link between these two groups seems to exist. At station OMEX-II at 1450 m relatively few subsurface deposit-feeding polychaetes occured. At this station high numbers of filter- and interface-feeders were found (Flach et al. 1998) which might be related to the next very important factor in structuring the benthic fauna.

Flow

At the OMEX station at 1450 m on the Goban Spur high current velocities were measured (Thomsen and Van Weering 1998). Currents were influenced by the tidal cycle, but the maximum flow velocities were directed dominantly down-slope often exceeding 15 cm/s causing resuspension and transport of organic matter (Thomsen and Van Weering 1998). This resuspension makes food more available to suspension-feeders (Lampitt 1985). A zone of sessile suspension-feeders was found between 1100-1400 m in the Porcupine Seabight (Lampitt et al. 1986) coinciding with an area with strong tidal currents although rarely

exceeding 15 cm/s. At the Goban Spur a similar zone of sessile suspension-feeders (mainly the sponge *Pheronema carpenteri* and astrorhizid foraminifera) were found at about 1400 m (Flach et al. 1998). However, a direct comparison with the abundance of filter- and interface-feeders within the macrofauna did not show a good correlation (Fig. 6a), despite a significant positive correlation with the concentration of POC within the benthic boundary layer (BBL) (Fig. 6b). This shows that it is not the flow velocity itself that is important for the animals, but its indirect effect, on food availability within the water column just above the seabed.

Rosenberg (1995) reported more or less the same pattern in a trench in the Skagerrak. On the slopes of this trench, where strong bottom currents transport suspended matter, high densities of suspension- and interface-feeders occurred. However, at the bottom of the trench low current velocities and high sedimentation rates coincided with a dominance of subsurface deposit-feeders. In the deep Norwegian Trench a different pattern was

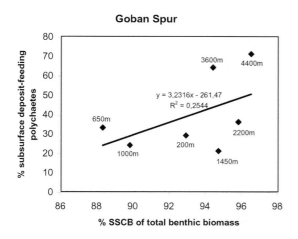

Fig. 5. The relative abundance (%) of subsurface deposit-feeders within the Polychaeta in relation to the relative (%) importance of the small size class biomass (SSCB) of total benthic biomass. SSCB data from Pfannkuche and Soltwedel (1998).

higher rate than sediment of similar bottom roughness without macrofauna. It would appear that oganic carbon, which might normally bypass this mid slope areas during high flow conditions, is trapped, accumulated and/or remineralized by the macrofauna. Carbon budgets calculated for the OMEX-stations B and III showed a deficit at station B, indicating an additional (horizontal) input at ~1000 m, whereas the vertical flux at ~4000 m was sufficient to fuel the benthic community at this depth (Flach and Heip 1996b).

found. In the deepest part of this trench the benthic community was dominated by suspension feeders, whereas subsurface deposit feeders were dominant at the shallow-slope stations (Rosenberg et al. 1996). These authors suggested that this pattern could be related to the hydrodynamic regime. On the slope sediment accumulation rates were higher than in the deeper areas, whereas in the deeper part of the trench near-bottom horizontal transport and resuspension processes were observed. The high densities of suspension-feeders in the deepest parts were mainly due to the polychaete *Spiochaeto-pterus bergensis*. This species is supposed to be able to switch from suspension feeding to surface deposit feeding under conditions of low particle concentration and could thus be classified as an interface-feeder. During mesocosm experiments (Thomsen and Flach 1997) under three different flow conditions video observations showed a shift from deposit feeding at low flow velocities to suspension feeding at intermediate (~6-7 cm/s) flow velocities in two interface-feeding species. These experiments also showed that during summer even under super critical flow conditions with erosion, the macrofauna was able to reduce phytodetritus and POC from the BBL with a 50% and 60%

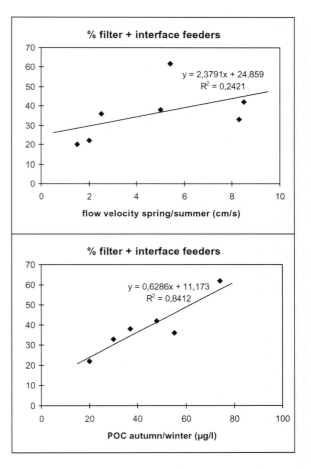

Fig. 6. The relative abundance (%) of filter plus interface-feeders within the macrofauna in relation to the flow velocities and the concentration of particulate organic carbon (POC) in the BBL (benthic boundary layer) 5 cm above the bottom. BBL data from L. Thomsen (pers. comm.).

358 Flach

At a station with high flow velocities in the Sebutal canyon, Gage et al. (1995) found high faunal densities. In areas with high flow velocities, faunal densities might be relatively high due to resuspension and horizontal transport (Jumars et al. 1990). At the Iberian margin, however, very low densities were found at a station with high flow velocities at about 1500 m water depth. Again the fauna was characterized by a relatively high density of suspension feeders, mainly sponges, *Bryozoas* and small colonial sea anemones (Flach et al. in press; Duineveld et al. in press). This station was situated on a very steep part of the slope and was characterized by current ripples parallel to the coastline and terraces, thus indicating the importance of topography and sediment structure.

Topography and sediment structure

The Iberian margin is so steep in many places, that it is impossible to sample with a box corer. Video images showed bare rocks and sediment ripples associated with strong currents. At these places sessile suspension-feeding megafauna could be observed, e.g. the cold water coral *Lophelia* (Duineveld et al. in press). This coral was also found on top of the Galicia Bank at about 700 m. Here, infaunal densities were very low and the macrofauna had a very special community structure (Flach et al. in press), with high numbers of carnivores polychaetes and suspension feeders. On the sedimented steep slopes of the Iberian margin and on the Galicia Bank, median grain-size was high as was the % $CaCO_3$, whereas the % org. C and % N were low. Sedimentation rates are likely to be very low at these sites, as found on the Goban Spur at the 1425 m station (Van Weering et al. 1998). However, unlike the Iberian margin stations, the median grain-size and % $CaCO_3$ at the 1425 m station were not extreme compared to the other Goban Spur stations. The topography of the Goban Spur is also much smoother than at the Iberian margin. Both, topography and flow velocities are important in structuring the sediment. High sedimentation rates were found at the deepest part of the Goban Spur transect, where median grain-size was low and the % $CaCO_3$ high, indicating pelagic sedimentation. At the deepest part of the

Iberian margin, median grain-size was similar to the Goban Spur, but the % $CaCO_3$ was lower whereas the % org. C and % N were higher (Flach et al. in press). As these stations were situated at the bottom of the steep slope close to the land, a greater input of terrestrial sediment would be expected and this could have an impact on the fauna. However, no differences in the composition of the macrofaunal community was found, with a dominance of subsurface deposit-feeders (Fig. 4). Meiofauna densities, on the other hand, were higher at the Iberian margin and a closer relationship with sediment composition of the meiofauna was found compared to the macrofauna (Flach et al. in press). Very high densities of meiofauna were found in the Nazaré canyon at all depths, whereas macrofauna densities were only high in the shallow part of the canyon. Macrofauna community structure was very different within the canyon compared to the other stations at similar depths. The number of subsurface deposit-feeders (Fig. 4) was very low. The macrofauna community was dominated by small opportunistic polychaetes (*Spionidae* and *Paraonidae*) and it could be expected that the special topography of the canyon initiated occasional sediment slides disturbing the fauna and that only small fast growing animals could live under these special circumstances. At a deep site (3400 m) in the Setubal canyon, Gage et al. (1995) found very high densities for all faunal groups retained on a 300 μm sieve (Fig. 1), although differences in relative abundances were found. Species diversity was low at this site. The authors suggested this was the result of occasional thinning of the benthos because of hydrodynamic disturbance. Especially fragile surface feeders were absent and deep burrowing species were abundant. They related this to the sediment structure, which showed sharp ripples indicating vigorous bottom currents. Again topography and flow structured the sediment and this had a strong impact on the fauna. Gage et al. (1995) compared the situation with that of the deep part of the Rockall Trough. In the Rockall Trough also a special flow regime, with occasional high current speeds (up to 20 cm/s), was found, but the fauna seemed to be less influenced. The number of subsurface deposit-feeders is close to the regression line of the Goban

Spur (Fig. 4) and total density only a little higher than at the Goban Spur and similar to the intermediate EUMELI-site (Fig. 1). In the Norwegian Trench in the Skagerrak Rosenberg et al. (1996) found high sedimentation rates on the slopes and high currents at the bottom of the trench and this was reflected in the faunal composition with relatively high densities of suspension-feeders at the bottom and high numbers of subsurface deposit-feeders on the slopes. Also relatively high amounts of subsurface deposit-feeding polychaetes were found on the slope of the Rockall Trough (Fig. 4) compared to the smooth continental slope of the Goban Spur. Thus, special features as seamounts, trenches/troughs and canyons, and steepness of slope have a strong impact on the structure of the benthic fauna.

Upwelling, outwelling and distance from land

Figure 1 shows strong differences in densities between sites sampled at similar depths. These differences can be explained partly by differences in topography and flow velocities, but high density in the Setubal canyon could be related to another factor as well. Lamont et al. (1995) suggest that the patchiness found at this site might be related to organic enrichment by seagrass, trapped in the current ripples. Seagrass can only grow in light, thus close to land, and must be transported to deep sites by downslope currents. The shelf of the Iberian margin is very narrow and influences of organic enrichment from land and shallow coastal areas could be expected here. However, densities at all Iberian margin stations, except the canyon stations, were not higher than stations on the Goban Spur, an area which is situated far from land (Fig. 1). Two stations situated close to the Nazaré canyon on the Portuguese margin agree nicely with the Goban Spur stations, and the same was found for the stations at the northern Iberian margin (La Coruña) transect. Densities at the other Iberian margin transect, situated close to the Rías Bajas, were also not higher, but were somewhat lower than on the Goban Spur. The Rías Bajas are known for a high production and outwelling of organic rich matter (López-Jamar et al. 1992) and the northern

Iberian margin (La Coruña) transect is situated close to land in an area with strong seasonal upwelling and high primary production (López-Jamar et al. 1992). However, neither has high macrofauna abundance. The high density at the shallowest EUMELI-site could be related to upwelling (Galéron et al. 2000). High primary production occurs at this "eutrophic" site, but this station was also situated relatively close to land and was characterized by high flow velocities. Very low densities were found in the Skagerrak and at the shallow Hebridean stations. These stations were situated close to land and the Skagerrak especially is known to be influenced by eutrophication.

Temperature and pressure

As both the Skagerrak and the Hebridean stations are situated at high latitudes (~58 and 56°N) and the EUMELI-site is set in the tropics (~20°N), temperature could be a factor in causing the differences in macrofauna densities. The mesotrophic EUMELI-site had a similar faunal density as the Rockall Trough (Hebridean) station at ~3000 m. Bottom-water temperature is similar for all depths greater than 2000 m (2-4°C), and therefore is only likely to be an important factor at the shallower depths. Indeed, the largest differences in macrofauna densities were found at shallow bathyal depths, although bottom-water temperature at the shallow, tropic, high density, EUMELI-site was ~4°C compared with ~6°C in the Skagerrak. Bottom-water temperature was about similar for all upper slope stations (Goban Spur-Iberian margin) (~12-13°C), and the shallow (334 m) canyon station (~11°C). On top of the Galicia Bank bottom-water temperature was also ~11°C, whereas densities were more than 10 times less than in the canyon. So, it appears that macrofauna densities are not related directly to bottom-water temperature.

However, as respiration is correlated to bottom-water temperature (higher carbon requirements at higher temperatures) the amount of found biomass that can be sustained by a given quantity of food is less at higher temperatures. Also, respiration is negatively correlated to body size. Large animals therefore use relatively less food. No significant pattern in mean individual weight of the macrofauna with water depth was found along six transects in the NE Atlantic (Flach and Thomsen 1998; Flach et al. in press), but the overall pattern suggested an increase in body size with increasing depth. This would suggest that under cold, food-poor conditions a larger body size is energetically more profitable. Jumars et al. (1990), however, found that under very poor feeding conditions small size-classes prevail. They suggested a bottleneck in size at about 1 mm and concluded that as juvenile survival rate is highly variable and dependent on pulses of labile food a long adult live and few offspring would be a better strategy in the deep sea. There seems, indeed, to be a change in dominance from r-selected species (high numbers of small pelagic larvae) to K-selected species (low number of large larvae and brood care) from the shelf/upper slope to the lower slope abyssal plain (Flach and De Bruin 1999). Larval development, however, is dependent not only on food supply, but also pressure can be important as well. Young et al. (1996) found in an experimental set-up that embryonic pressure tolerances could determine both the upper and lower bathymetric limits of the sea star *Plutonaster bifrons*. Pressure is directly correlated with depth and the general zonation pattern often found across continental margins, could be the result of differences in pressure tolerances between species/groups. However, many species are able to withstand pressures much greater than those found near the lower limits of their vertical ranges and some species can be found over very large depths ranges from the shelf to the abyssal plain. Within species that occurred over wide depth ranges no change in mean size could be found.

There is, however, a shift in relative importance from a more macrofauna dominated community to a more meiofauna dominated community with increasing water depth (Flach et al. in press). Galéron et al. (2000) found a correlation with food input and dominance of size-classes, with megafauna dominating under eutrophic and meiofauna under oligotrophic conditions, whereas there was hardly any difference in temperature. But they also found that within a single size group some taxa did not follow this trend of decreasing biomass with decreasing food input and they suggest that other both abiotic and biotic factors might be important.

Biotic interaction

Size is often determining the outcome of biotic interactions. Large animals are often better competitors and less susceptible to predation. Juveniles, especially, are very vulnerable to both competition and predation. Under stable physical conditions biotic interactions are important and K-selection is thus more profitable. Under highly variable, unpredictable and unstable conditions the importance of biotic interactions is limited and small, fast growing r-selected, opportunistic species can dominate. At two water depths (1800 m and 3600 m) a colonization experiment was carried out by Grassle and Morse-Porteous (1987) in the NW Atlantic Ocean. They found that even after five years the densities were not as high as in the surrounding sediment and that small opportunistic polychaetes were the main colonists. Highest densities were found in the treatments with greater organic enrichment. A few species increased in abundance under screens and they suggested that predation on the juveniles of these species normally prevents these species from becoming abundant. However, our knowledge on interactions in the deep-sea is still very limited and we can often only interpret deep-sea community structure from our knowledge using shallow water experiments.

Besides these direct interactions, indirect interactions can be important in structuring the benthic fauna. Tubes, burrows, mounds and feeding-pits can change the flow conditions and sedimentation rates locally, with a strong impact on other animals (Aller and Aller 1986). Bioturbation and other sediment disturbing activities of some species can also have a major impact on the benthic community structure.

Conclusions and future research

As stated in the introduction no single factor can be indicated as the one factor controlling the distribution of the benthic fauna at the European continental margin. Nearly all factors are coupled directly or indirectly, indicating the complexity of this ecosystem. The overall pattern in the decrease in total density and biomass with increasing water depth is mainly determined by the food input, but special topographic structures can highly influence this pattern. The quantity, quality, reliability and sources of food are important for the benthos. High flow velocities can resuspend the organic matter again and make it more available for suspension-feeders. Some fauna can change the flow regime and capture food that otherwise would not be available. Extreme flow conditions can disturb the fauna periodically, allowing only deep-living fauna to maintain their position and favouring rapid colonizers. In contrast, a stable environment allow the development of a 'climax' community in which biotic interactions become very important.

As we still know only very little about the biology of the deep-sea fauna, we also can say very little about the elasticity and carrying capacity of this remote, but important, ecosystem. The highly complex nature of the continental margins, with strong differences in regime, from steep slopes with rocky outcrops to smooth sedimental plateaus, to bights, troughs, seamounts and canyons, results in high local variability. As the environmental conditions are locally and temporally highly variable, future research should not be restricked to one place and to short time-scales. If we want to get a better knowledge about the processes that are important on continental margins, we should select a variety of locations. At these locations all possible parameters should be measured at different scales and over long time-scales. Besides these extensive monitoring programmes, high emphasize should be laid on experiments, both *in situ* and *in vitro*. Modern equipment, e.g. benthic landers, submersibles, ROVs, AUVs and video, allow us to study the benthic community in their natural environment. Long-term observations would give an idea about the reactions of the benthos not only to different environmental events, such as changes in flow velocities, food falls, sedimentation, even pollution and disturbances, but also to interaction between fauna, e.g. predation, competition, etc. These events should be simulated in experiments. The colonization-experiment of Grassle and Morse-Porteous (1987) showed that,

such experiments can enhance our knowledge of deep-sea processes. Also laboratory-based experiments, simulating deep-sea conditions, can give much information, such as flume-tank experiments (Thomsen and Flach 1997). These authors showed that not only does the environment influence the benthos, but the benthos also influences its environment. The activities of the benthos can have a strong impact on the chemical and geological processes and are therefore important for our understanding of biogeochemical cycles.

Our knowledge of the biology of the deep-sea benthos is very restricted and mainly based on comparison with shallow-water relatives. We need to know more about the actual life-history strategies, feeding-types and mobility of the deep-sea fauna. Already detailed study of morphological features as gut-length and feeding-apparatus of related species can give much information about the feeding conditions in the deep-sea, as is shown for the bivalve *Abra profundorum* by Allen and Sanders (1966). Morphological examination of the gonads, number and size of eggs and larvae and time of reproduction, will give us more information about life-history strategies. DNA-analyses can give information about evolution and distribution of the fauna and probably allow us a better understanding of biodiversity. Preliminary small-scale experiments showed that whole box-core samples from about 2000 m water depth could be kept in good condition in the lab for weeks. The cold-water coral *Lophelia* and its associated fauna from depths of about 700 m, can be maintained for months. This allows us to do experiments under controlled conditions and thus to study which factors in which way structure the benthic fauna.

References

Allen JA, Sanders HL (1966) Adaptations to abyssal life as shown by the bivalve *Abra profundorum* (Smith). Deep-Sea Res 13:1175-1184

Aller JY, Aller RC (1986) Evidence for localized enhancement of biological activity associated with tube and burrow structures in deep-sea sediments at the HEBBLE site, western North Atlantic. Deep-Sea Res 33:755-790

Billet DSM, Lampitt RS, Rice AL, Mantoura RF (1983) Seasonal sedimentation of phytoplankton to the deep-sea benthos. Nature 302:520-522

Carney RS (1989) Examining relationships between organic carbon flux and deep-sea deposit feeding. In: Lopez GR, Taghon GL, Levinton J (eds) Ecology of Marine Deposit Feeders. Springer, New York 24-58

Dauwe B, Middelburg JJ (1998) Amino acids and hexosamines as indicators of organic matter degradation state in North Sea sediments. Limnol Oceanol 43:782-798

Dauwe B, Herman PMJ, Heip CHR (1998) Community structure and bioturbation potential of macrofauna at four North Sea stations with contrasting food supply. Mar Ecol Prog Ser 173:67-83

Duineveld GCA, Lavaleye MSS, Berghuis EM, Kok A, Witbaard R (in press) Patterns of benthic megafauna on the NW Iberian Continental Margin in relation to the distribution of phytodetritus and RNA; a comparison with the Celtic continental margin. Prog Oceanog, Special Vol OMEX II

Epping E, Helder W (in press) Organic carbon mineralization in NE Atlantic margin sediments: A comparison between Goban Spur and the Iberian Margin. Prog Oceanog, Special Vol OMEX II

Flach E, Bruin W de (1999) Diversity patterns in macrobenthos across a continental slope in the NE Atlantic. J Sea Res 42:303-323

Flach E, Heip C (1996a) Vertical distribution of macrozoobenthos within the sediments on the continental slope of the Goban Spur area (NE Atlantic). Mar Ecol Prog Ser 141:55–66

Flach E, Heip C (1996b) Seasonal variations in faunal distribution and activity across the continental slope of the Goban Spur area (NE Atlantic). J Sea Res 36:203–215

Flach E, Thomsen L (1998) Do physical and chemical factors structure the macrobenthic community at a continental slope in the NE Atlantic? Hydrobiologia 375/376:265-285

Flach E, Lavaleye M, de Stigter H, Thomsen L (1998) Feeding types of the benthic community and particle transport across the continental slope of the Goban Spur. Progr Oceanog 42:209-231

Flach E, Muthumbi A, Heip C (in press) Meiofauna and macrofauna community structure in relation to sediment composition at the Iberian Margin compared to the Goban spur (NE Atlantic). Progr Oceanog Special Vol OMEX II

Gage JD, Lamont PA, Tyler PA (1995) Deep-sea macrobenthic communities at contrasting sites off Portugal, preliminary results: I Introduction and diversity comparisons. Int Rev ges Hydrobiol 80:235-250

Galéron J, Sibuet M, Mahaut M-L, Dinet A (2000) Variation in structure and biomass of the benthic communities at three contrasting sites in the tropical Northeast Atlantic. Mar Ecol Prog Ser 197:121-137

Grassle JF, Morse-Porteous LS (1987) Macrofaunal colonization of disturbed deep-sea environments and the structure of deep-sea benthic communities. Deep-Sea Res 34:1911-1950

Jumars PA, Mayer LM, Deming JW, Baross JA, Wheatcroft RA (1990) Deep-sea deposit-feeding strategies suggested by environmental and feeding constraints. Phil Trans Royal Soc London A 331:85-101

Lamont PA, Gage JD, Tyler PA (1995) Deep-sea macrobenthic communities at contrasting sites off Portugal, preliminary results: II Spatial dispersion. Int Rev ges Hydrobiol 80:251-265

Lampitt RS (1985) Evidence for the seasonal deposition of detritus to the deep-sea floor and its subsequent resuspension. Deep-Sea Res 32:885-897

Lampitt RS, Billett DSM, Rice AL (1986) Biomass of the invertebrate megabenthos from 500 to 4100 m in the northeast Atlantic Ocean. Mar Biol 93:69-81

López-Jamar E, Cal RM, González G, Hanson RB, Rey J, Santiago G, Tenore KR (1992) Upwelling and outwelling effects on the benthic regime of the continental shelf off Galicia, NW Spain. J Mar Res 50:465-488

Pfannkuche O, Soltwedel T (1998) Small benthic size classes along the NW European Continental Margin: Spatial and temporal variability in activity and biomass. Progr Oceanogr 42:189-207

Pfannkuche O, Boetius A, Lochte K, Lundgren U, Thiel H (1999) Responses of deep-sea benthos to sedimentation patterns in the North-East Atlantic in 1992. Deep-Sea Res 46:573-596

Rice DL, Rhoads DC (1989) Early diagenesis of organic matter and the nutritional value of sediment. In: Lopez GR, Taghon GL, Levinton JS (eds) Ecology of Marine Deposit Feeders. Springer, New York 59-79

Rosenberg R (1995) Benthic marine fauna structured by hydrodynamic processes and food availability. Neth J Sea Res 34:303-317

Rosenberg R, Hellman B, Lundberg A (1996) Benthic macrofaunal community structure in the Norwegian Trench, deep Skagerrak. J Sea Res 35:181-188

Sibuet M, Lambert CE, Chesselet R, Laubier L (1989) Density of the major size groups of benthic fauna and trophic input in deep basins of the Atlantic Ocean. J Mar Res 47:851-867

Tyler PA, Harvey R, Giles LA, Gage JD (1992) Reproductive strategies and diet in deep-sea nuculanid protobranchs (Bivalvia: Nuculoidea) from the Rockall Trough. Mar Biol 114:571-580

Tyler PA, Gage JD, Paterson GJL, Rice AL (1993) Dietary constraints on reproductive periodicity in two sympatric deep-sea astropectinid seastars. Mar Biol 115:267-277

Thomsen L, Flach E (1997) Mesocosm observations of fluxes of particulate matter within the benthic boundary layer. J Sea Res 37:67-79

Thomsen L, Weering TjCE van (1998) Spatial and temporal variability of particulate matter in the benthic boundary layer at the North East Atlantic continental margin (Celtic Sea). Progr Oceanogr 42:61-76

Van Weering TjCE, Hall IR, de Stigter H, McCave IN, Thomsen L (1998) Recent sediments, sediment accumulation and carbon burial at Goban Spur, NW European continental margin (47-50°N). Progr Oceanogr 42:5-35

Young GM, Tyler PA, Gage JD (1996) Vertical distribution correlates with pressure tolerances of early embryos in the deep-sea asteroid *Plutonaster bifrons*. J Mar Biol Ass UK 76:749-757

Reef-Forming Cold-Water Corals

A. Freiwald

*Institut für Geologie und Paläontologie, Universität Tübingen, Herrenberger Str. 51,
72070 Tübingen, Germany
corresponding author (e-mail): andre.freiwald@uni-tuebingen.de*

Abstract: Coral reefs are something we usually associate with warm, tropical waters and exotic fish, but not with the cold, deep and dark waters of the North Atlantic, where corals were regarded as oddities on the seafloor. It is now known that cold-water coral species also produce reefs which rival their tropical cousins in terms of their species richness and diversity. Increasing commercial operations in deep waters, and the use of advanced offshore technology have slowly revealed the true extent of Europe's hidden coral ecosystems. This article reviews current knowledge about the reef-forming potential and the environmental controls of the scleractinian *Lophelia pertusa* along different deep-shelf and continental margin settings with special reference to NE Atlantic occurrences.

Introduction

Scleractinian coral reefs thriving in cold and deep waters are widely distributed ecosystems along continental margin settings, but our understanding of the major reef-forming (and destructive) processes is fragmentary. Spectacular examples of deep-water coral locations are:

• the numerous deep shelf reefs off Norway with the Sula Reef as the best known (Freiwald et al. 1999),
• the giant mounds in the Rockall Trough and Porcupine Seabight, where the participation of deep-water corals is substantial but mound formational processes are yet not understood because of the lack of drilling (Henriet et al. 1998),
• the so-called Darwin Mounds on the Wyville-Thompson-Ridge, which may represent incipient mound stages (Bett 1999).

Existing data, however, indicate that the existence of the largest coral reef province is not confined to the shallow-water subtropical to tropical climatic belt as is conventionally cited in textbooks. Instead, the largest coral reef province thrives in mid-depths from the high latitudes to the low latitudes of both hemispheres along continental margins and seamounts. Along the NE-Atlantic margin which is focused in this review, the domi-

nant reef-forming scleractinan is the caryophyllid *Lophelia pertusa* and the oculinid *Madrepora oculata* (Fig. 1a, b). A common third species that abundantly occurs either on dead *Lophelia* or *Madrepora* colonies is *Desmophyllum cristagalli* (Fig. 1c). To date, a total of 868 species have been identified from *Lophelia pertusa* habitats in the North Atlantic (Rogers 1999). However, the data set is strongly biased as a result of different sampling methods and techniques in the past. Recently launched research programmes on deep-water coral communities currently alter our traditional views on the biodiversity and dynamics of deeper water ecosystems.

On a large-scale assessment, the distribution pattern of deep-water coral communities along continental margins seems to be best explained in terms of physical oceanography. Although *Lophelia pertusa* and *Madrepora oculata* have been known since pre-Linnean times to scientists and may be even much earlier to fishermen, we are only on the edge of understanding their biology, dispersal, longevity and environmental requirements. This was clearly demonstrated by the recent unexpected finding of thriving *Lophelia* colonies at the moorings of the Beryl Alpha oil platform in the North Sea

From WEFER G, BILLETT D, HEBBELN D, JØRGENSEN BB, SCHLÜTER M, VAN WEERING T (eds), 2002,
Ocean Margin Systems. Springer-Verlag Berlin Heidelberg, pp 365-385

Fig. 1. Major scleractinian corals that are commonly found in deep-water reefal frameworks: (A) *Lophelia pertusa*, Sula Reef, Norway, 276 m water depth; (B) *Madrepora oculata*, rapid growing colony on a plastic rope from Galicia Bank, 900 m water depth; (C) a rapid growing *Desmophyllum cristagalli* with parental corallite on a plastic rope from Galicia Bank, 900 m water depth.

production at the surface waters or by a redistribution of suspended particles in the bottom mixed layer (Frederiksen et al. 1992), and (2) the distribution of hydrocarbon seepage areas (Hovland et al. 1998).

The purpose of this study is to provide an assessment of the state-of-the-art knowledge in coral-supported accumulations in deep, mostly aphotic environments, emphasising the facts and ideas needed for a discussion of biosphere processes of continental margins and their sustainable use or conservation.

Unravelling the deeper waters: a dynamic environment

With increasing scientific and commerical efforts in the investigation of deep marine environments on the continental shelf, slope and deep-sea settings, a number of key discoveries demonstrated our limited knowledge of the fundamental processes and resources of the "Earth's inner space". Among these, the spectacular hydrothermal vents with their chemosymbiotic communities along seafloor spreading zones, the occurrences of gas hydrates and methane vents or seeps along the continental margins and the unravelling of enigmatic aphotic coral reefs in the North Atlantic which are discussed here, have become prime targets in offshore research. But it is not only the deep benthic environment that has offered spectacular findings, the hydrography below storm wave base is also much more dynamic than had been previously thought. Features such as internal waves, benthic storms, or topographically-guided currents in the form of Taylor columns, eddies and vortices, barocline tides and boundary currents, all dissipate energy at the sediment-water interface which can: (1) generate current-induced, mobile bedforms at great depths, and (2) can concentrate digestable detritus for consumption by heterotrophic organisms. These processes often transfer pulses of particle-laden waters to the deep-sea floor which, in turn, stimulate benthic communities to secure their metabolic demands and control the variations in the spatial distribution and aggregation of invertebrate communities. There is increasing evidence that these so-called benthic-pelagic coupling processes are

(Roberts 2000), and during the decommissioning of the infamous Brent Spar (Bell and Smith 1999). Nevertheless, even without a basic understanding of the biology of *Lophelia* reefs, they have been the subject of elaborate theories that explain their occurrence as being controlled by: (1) an increase of food availability either through higher primary

more prominent at the higher latitude seas where a strong seasonality in the annual evolution of the plankton food-web is developed. This fuels the megabenthic communities in the deep-sea and may also result in reproductive synchrony in deep-sea populations. Basically, the settings of azooxanthellate coral buildups in deep waters also generally coincide with seasonally fertile surface waters.

Scleractinian corals in cold and deep environments

In the following context the term "deep-water" is not meant to describe corals which live in oceanic deep-water masses. Deep-water is used here to delineate the ability of corals to thrive at great water depths under cold and dark environmental conditions usually below the storm wave base. The occurrences of modern deep-water coral buildups are governed by: (1) attachment on a hard substrate, (2) association with permanently or episodically strong currents that prevent the settling of fine-grained detritus, which foster the food particle flux rates and the exchange of waters and possibly larvae and gametes.

All scleractinian corals which exist in cold, deep and – seasonally – eutrophicated environments lack photosymbionts in their soft tissue. Consequently, these corals fall into the azooxanthellate or non-zooxanthellate group. Azooxanthellate corals as a whole are found within a wide bathymetric range, from 0 - 6200 m water depth and at a temperature range from -1° to 29° C (Stanley and Cairns 1988). In contrast to zooxanthellate corals, the azooxanthellate group is cosmopolitan.

Colonial azooxanthellate corals and their different framework-constructing capacity

Seventy-five percent of the 121 known azooxanthellate coral species are solitary while the remaining are colonial – a major prerequisite for framework construction. The available data on the preservation of deep-water coral buildups is still limited and coring through these structures has rarely been attempted. Among the colonial azooxanthellate corals, only a few have any great potential to produce frameworks over the surrounding seafloor. Most of them create low-relief thickets which generate large quantities of coral rubble rather than *in situ* frameworks after death. The coral rubble grounds may become a wackestone, packstone or more likely a rudstone in the geological record while the *in situ* buildups may be preserved as framestones.

Species such as *Madrepora oculata* predominantly form anastomosing, bushy colonies with predominantly outward-growing branches by continuing budding of polyps (Fig. 1b). This mode of growth results in 30 - 50 cm high, often fan-shaped colonies. The potential for framework aggradation or progradation of this fragile growth habit is, however, rather limited. Only thickets and/ or rubble facies rather than frameworks are known from *Madrepora*-dominated coral buildups. The same limited framework potential was reported by Stetson et al. (1962) for the Blake Plateau occurrences of *Dendrophyllia profunda*. In contrast, most ecotypes of *Lophelia pertusa* also produce 30 - 100 cm (sometimes up to 150 cm) high massive, dendroid and bushy skeletal colonies but the neighbouring branches commonly grow together, thus increasing the architectural stability of the framework considerably. It should be kept in mind, that aside from the colony growth habit, several other factors including the various types of sediment-infill and the diagenetic regime exert contemporary and significant controls on the preservational style of deep-water coral frameworks. Within environments which do not support rapid early diagenetic cementation, the flux of baffled detrital particles from the suspension-load in the water, the growth rate of the coral frameworks and the bioerosive processes form a sensitive balance between framework-supported and/or coral rubble facies, or mud-supported seafloor elevations (Freiwald and Wilson 1998).

Deep-water coral frameworks

The habitat requirements for framework-producing azooxanthellate corals such as *Lophelia pertusa* are reasonably well known. The *Planula* larvae – which yet has never been found – need a hard substrate for settling and metamorphosis. Strong bottom currents prevent the deposition of fine-grained deposits and therefore, *Lophelia* is preferentially found on various kinds of topographic highs such as submerged moraine ridges, clay ridges, iceberg ploughmark levees, spurs, outcropping rocks and artificial substrates. Since the beginning of this century, an increasing number of terms are used to describe deep-water coral buildups: coral patch (Wilson 1979a), coral bank (Teichert 1958), coral reef (Dons 1944), coral reef mound (Freiwald et al. 1997), coral mound (Mullins et al. 1981), carbonate mound (Hovland et al. 1994) and bioherm (Mortensen et al. 1995). This wide spectrum of terms mirrors the advances in offshore technologies. Earlier workers who obtained their results solely on the basis of dredging and echosoundings either used the term bank or reef. The diversification in the terminology arose with the increased operation of higher resolution acoustic systems and cameras which are operated from towed systems, remotely operated vehicles and submersibles. The deep-water coral buildups can form considerable topographic elevations above the surrounding seafloor. Again, because of the lack of coring, the information on the interior of these coral structures is still poorly known.

Coral patches

This term was introduced by Wilson (1979a) to describe the *Lophelia pertusa* settlements on Rockall Bank. The mature patches measure 10 m to 50 m in diameter and are 100 m to 200 m apart and reach elevations above the surrounding seafloor of about 1.5 m (Fig. 1). Similar *Lophelia* patches were documented from Porcupine Bank at 200 m to 500 m water depth (Scoffin and Bowes 1988) and from Gollum Channel, Porcupine Seabight, in 725 m to 920 m water depth (Tudhope and Scoffin 1995). Based on the genetic approach of Squires (1964) and Wilson (1979a), coral patches start with single colonies which arose from the settling of a *Planula* larva. Under ideal growth conditions, an outer ring of younger satellite colonies develops around the initial colony predominantly by means of vegetative reproduction. The maximum size of a single *Lophelia* colony seems to be in the range of up to 1 m in height and up to 1.5 - 2 m across and has a circular outline. With increasing growth, the lower part of the initial colony becomes vulnerable to boring sponge infestation as a result of which mechanical breakage of parts of the colony (as a result of weakening by bioerosion, especially sponge excavation) occurs and portions of the colony fall onto the sediment adjacent to the colony (Wilson 1979a). This leads to *in situ* collapse structures of coral skeletons forming a coral rubble layer at the base while the outer satellite colonies continue to grow. Coral *thickets* of this type rarely exceed 6 - 8 m across. The next stage is rep-resented by the *coppice* at which coral debris begins to accumulate considerably about the base of the thicket (Squires 1964; Wilson 1979a). Colonies, thickets and coppices represent progressive stages in the coral patch evolution. If Wilson's (1979a) model on the progressive development from colonies to patches could be substantiated by comparative genetic studies, this would mean that all members of a patch represent clones of a once sexually produced individual that settled down on a hard substrate as a *Planula* larvae and became the founder of a coral colony after metamorphosis. Following this line of thought, a coral patch represents a clonal organism which has a considerable lifespan of probably thousands of years.

Coral banks and reefs

In descriptions of modern accumulations in the marine realm, the term "bank" is used to describe *any* topographic feature on the seafloor. A widely used term for deep-water coral buildups in the North Atlantic is that of a "coral bank" (Teichert 1958; Stetson et al. 1962; Squires 1964) for topographic elevations predominantly consisting of corals. The coral bank represents the final evolutionary step in the formation of deep-water coral buildups and shows greatest similarities to shallow-water coral reefs. Principally, a coral bank is characterised by three distinct units: (1) a cap of living coral colonies which, (2) rest on an open-spaced but dead coral framework and debris zone, and (3) a zone of coral framework and debris that is clogged solidly with sediment. However, recolonisation of corals takes place on each of these three delineated zones. The greatest density of living coral colonies is observed on the bank top and upper flanks (Mortensen et al. 1995).

Except from the philosophical point of view, there is virtually no difference between a coral bank and a deep-water coral reef (Rogers 1999). In general, coral reefs form topographic highs on the basis of skeletal framework accretion which is enhanced by biological encrustation over dead surfaces, prolific sediment production and concurrent inorganic or biologically-controlled cementation. Destructive processes are dominated by rasping, grazing and boring organisms, and by high-energy hydrodynamics. In the strict ecological sense, the *Lophelia* banks represent coral reefs with respect to their framework-building capacity which lead to highly elevated structures on the seafloor, which support a highly diverse associated fauna with biodiversity indices similar to those found in tropical shallow-water reefs (Jensen and Frederiksen 1992; Rogers 1999). Moreover, like shallow-water counterparts, the deep-water reefs have to withstand severe physical disturbances. These disturbances do not occur in the form of tropical storms, but as oceanic internal waves (Frederiksen et al. 1992), deep barotrophic tides and shock waves produced by submarine slide events (Freiwald et al. 1999), but most likely at a reduced frequency. One of the largest framework-supported *Lophelia* reef complexes that is intensively studied on the Mid-Norwegian Shelf in 250 to 350m water depth is the Sula Reef (Fig. 2; Freiwald et al. 1999; Fosså et al. 2000). The length of the reef chain is 13 km across with an average framework height of 15 m but many individual reefs are 20 and 30 m thick. Existing ^{14}C-data provided by Hovland et al. (1998) reveal an Early Holocene (8060 ± 130 YBP) initial growth after the retreat and final deglaciation of the Fennoscandian icesheet.

Carbonate mounds

The now famous clusters of apparent seafloor elevations which are grouped in distinct mound provinces in the Rockall Trough and Procupine Seabight are the scientific subject of three EU-FP 5 Projects (ACES, ECOMOUND, GEOMOUND). These often high-relief mounds have a dimension of 5 by 1 km and a thickness of up to 100 m above the surrounding seafloor (Fig. 3). There is much debate on their genesis and internal structure but many of the surveyed mounds show a high density of living and dead deep-water corals on the flanks (Chachkine and Akhmetzhanov 1998). There are several lines of evidence that the whole structure forms a build-up consisting of coral growth periods that alternate with increased deposition of drift sediments. The enhanced baffling potential of the coral framework gave way to higher and focused accumulation rates compared to the off-mound areas where even erosion in the form of moats and contourites are known. Alternative hypothesis prefer a microbially-triggered link to deep biosphere-biosphere coupling and hydrocarbon emanation (Hovland et al. 1994; Henriet et al. 1998). However, all the theories remain speculative as long as there is no deep drilling survey into these coral-bearing mounds operating.

The geological record of *Lophelia*

Since the Cretaceous, the caryophylliid corals have evolved more genera than any other family and most of them are azooxanthellate (Veron 1995). The earliest fossil documentation of *Lophelia* (and *Madrepora*) dates from the Paleocene of Seymour Island (Filkorn 1994) and from the Eocene Gulf Coast (Squires 1957). The oldest known deep-water *Lophelia* buildup is preserved in a mudstone near Hinakura, North Island, New Zealand, from the Middle Tongaporutuan Stage, Late Miocene (Squires 1964; Vella 1964). The coral thickets were formed entirely by *Lophelia parvisepta* that developed at least 3.4 m-high and 36.6 m-wide patches. However, taking the outcrop situation into account, the full dimension of the coral thickets is estimated to about 75 m in diameter (Squires 1964). Nearby, *Lophelia parvisepta* thickets of the same dimension were exposed in mudstones of the Mangapanian Stage, Late Pliocene at Lake Ferry. According to the associated foraminifer fauna, the bathymetric ranges indicate a 1500 - 2500 m water depth for the Late Miocene thicket and 200 - 300 m for the Late Pliocene thicket (Vella 1964). Both thickets existed near the tectonic fault zone

8°10'00"E

64°08'00"N

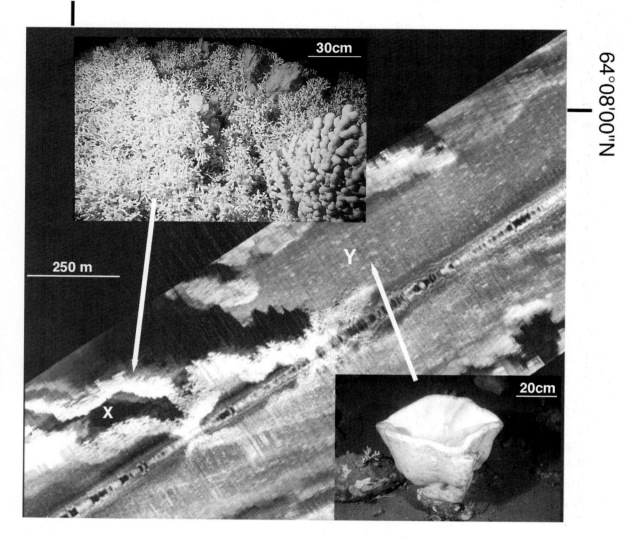

Fig. 2. Sula Reef: Side-scan sonography showing a Late Pleistocene iceberg ploughmark (x) whose boulder levees provide hard substrate for the Holocene to Recent *Lophelia* reef complex. The transition from reef to off reef settings is abrupt. The off reef environment is characterised by ice-rafted boulder fields (Y) which are intensely colonised by sponges and octocorals (RV *Poseidon* Cruise 254).

separating the North Island from the South Island at Palliser Bay, Cook Strait. This environmental setting favours severe currents through the palaeo-strait and high rates of tectonic uplift.

The next fossil hot spot is the Mediterranean Sea. After the Messinian Event, azooxanthellate corals invaded the Mediterranean deep-sea through the Strait of Gibraltar since the Early or Middle Pliocene. Pliocene *Lophelia* records in the Mediterranean region are concentrated in regions of intense tectonic uplift. Such a region is the Messina

Strait, where numerous Pliocene and Early Pleistocene *Lophelia* occurs in outcrops along both flanks of the strait. The tectonic subsidence of the Messina Strait commenced in the Miocene passing from shallow-water to bathyal conditions (Montenat et al. 1991). The bathymetric window for *Lophelia* in the Messina Strait was open at about 2 Mio. years ago. Starting in the Early to Middle Pleistocene, intense uplifting and block faulting was active until present. Bathyal marls deposited at more than 600 m water depth are now raised to

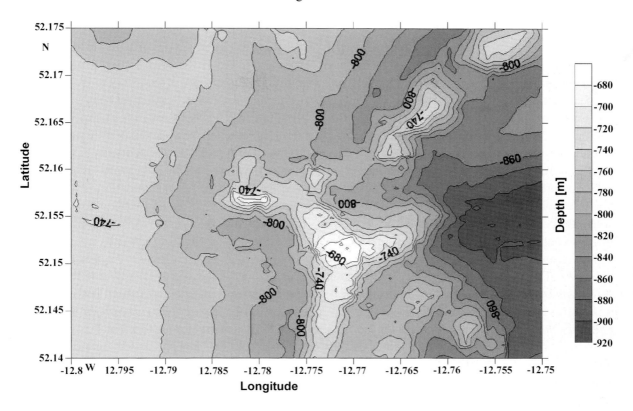

Fig. 3. Bathymetric map from the Propeller Mound, Hovland Mound Province, northern Porcupine Seabight, surveyed during RV *Poseidon* Cruise 265.

400 m along the Calabrian Margin (Montenat et al. 1991). *Lophelia pertusa* and other deep-water corals are generally preserved in bathyal marls in close vicinity to palaeo-escarpments - their initial settling substrate.

Late Pliocene to Early Pleistocene *Lophelia pertusa* and *Madrepora oculata* thickets outcrop on Rhodes, southern Aegean Sea (Hanken et al. 1996). Rhodes as part of the Dodekanese Island Arc is located at the uplifted front facing the Hellenic Trench subduction zone. The corals are preserved as *in situ* colonies or broken colony fragments in a limestone or bathyal marl.

Late Pleistocene deep-water corals are frequently dredged from deep-sea hardgrounds and bathyal marls in the Mediterranean basins. The specimens are often chalky, covered by ferro-mangenese oxides and encrusted and filled by biomicrite (Delibrias and Taviani 1985). In the Atlantic, Pleistocene coral buildups are incorporated in current-swept and early diagenetically cemented lithoherms which are described

from the Florida Strait (Messing et al. 1990) and the outer Blake Plateau (Paull et al. 2000).

There appears to be a sudden increase in occurrences of *Lophelia pertusa* since the Late Pliocene with particular concentrations in the Mediterranean Sea. Strikingly, the occurrence of *Lophelia pertusa* in Mediterranean deposits often coincides with glacial periods (Delibrias and Taviani 1985). At present, living *Lophelia* is known to exist in only a few places in the western Mediterranean Sea.

Diagnostic features in comparison with shallow-water coral reefs

Aside from the different bathymetric range, cold- and deep-water coral reefs differ in the following aspects from warm- and shallow-water reefs:

Large-scale nutritional setting – The strong relationship between low nutrient supply and photo-synthesising epibenthic invertebrates has been demonstrated for zooxanthellate corals (Cowen 1983) and for symbiont-bearing large foraminifers

(Hallock and Schlager 1986). Low nutrient settings promote photosymbiosis because it allows internal recycling of phosphorous, nitrogen, carbon dioxide and oxygen between host and symbiont (Brasier 1995). The dominant trophic strategies in coralgal reefs are the auto- and mixotrophic modes. Recycling is coordinated by grazers which give way to the photozoan association of the warm-water carbonate realm (James 1997). Deep-water coral buildups predominate below the pathways of at least seasonally eutrophicated surface waters where the coral communities benefit from benthic-pelagic coupling. There are many speculations as to how *Lophelia pertusa* obtains its metabolic demands. The lack of direct observations made scientists believe that *Lophelia* predominantly utilises dissolved organic matter. However, *Lophelia pertusa* was observed *in situ* on the Sula Ridge in May 1997 during a submersible operation and was seen to prey voraciously on zooplankton such as copepods and cumacean crustaceans (Freiwald, video documentation). As these zooplankton communities follow a pronounced seasonal cyclicity which is controlled by the appearance of the phytoplankton, the benthic-pelagic coupling is clearly the dominant process in the nourishing of the *Lophelia* reef ecosystem. Oceanic circulation and hydrography are closely linked with the trophic state of a given water mass. While the warm-water coral reefs thrive within the *oligotrophic* oceanic gyres generally *above* the thermocline, the deep-water coral reefs exist under seasonally *eutrophicated* surface water conditions *beneath* the thermocline.

Zonation patterns – Shallow-water coral-algal reefs thriving in the trade-wind belts exhibit a clear energy zonation. This windward-leeward zonation is expressed by the wind-exposed reef crest with its characteristic encrusting community. Leeward, a reef flat with pavements of reworked coralgal clasts, rhodoliths and skeletal sands is developed, while the back reef is the depocenter for carbonate muds (James and Bourque 1992). As a result of the paucity of visual inspections, energy zonations in deep-water buildups are rarely documented. The few examples point to a preferentially upcurrent growth of colonial corals as has been documented from the Blake Plateau (Stetson et al.

1962), the Strait of Florida (Messing et al. 1990) and from the Stjernsund, northern Norway (Freiwald et al. 1997). In deeper shelf areas, where semidiurnal shifting currents driven by tides can occur, such a preferred settlement behaviour is less pronounced. The depth zonation in shallow-water coral-algal reefs is expressed mostly along the reef front and fore reef with the ecotypes of the corals changing with decreasing light levels. In the aphotic zone such a depth-relation zonation is developed predominantly in terms of a taphonomic gradient (Freiwald and Wilson 1998). An ideal gradient following downslope from a healthy *Lophelia* reef shows an intensely microbially-triggered Fe-Mn-coated zone which seems to stimulate the attachment of sessile invertebrates (high portion of the hemichordate *Rhabdopleura normannii*). In this zone, the dead coral framework is not clogged with baffled sediment. Deeper downslope long-living sponges such as *Pachastrella*, *Geodia* and *Plakortis* in concert with octocorals become dominant and are attached to a now sediment-clogged degraded coral framework. In places, where vigorous bottom currents find their way through the reefal seafloor elevations, the baffled material seems to be winnowed thus creating a coral rubble facies while in more sheltered settings a coral-rich hemipelagic mud is formed.

Diversity of primary framework builder – In the warm-water realm, reef construction is supported predominantly by a diverse suite of zooxanthellate corals. In contrast, deep-water reefs are formed by only one or two main framework constructing azooxanthellate corals (Squires 1964; Mullins et al. 1981).

Functional groups of biologic framework stabilisation – In shallow-water reefs encrusting coralline algae are the most important biological stabilisers. They form the wave-resistant algal ridges along the windward faces of reefs. In the deep environments, the dominant stabilising role is passed to encrusting sponges such as *Hymedesmia* spp. and *Plakortis* spp. (Freiwald and Wilson 1998).

Bioerosion – The bioerosional impact is high on both shallow-water coral-algal reefs and deep-water coral reefs. Shallow-water reefs or carbonate buildups have been affected by intense infestation of cyanobacteria and thallophytes since the Precambrian, and are sites of deeply excavating grazer communities of sea-urchins and asteroids. In the aphotic zone, the corals are also infested by endoliths but primarily by heterotrophs such as boring bacteria, fungi (*Dodgella priscus*), bryozoans, sponges (*Aka labyrinthica*, *Alectona millari*), and sabellids (*Perkinsiana socialis*). In places clusters of the parasitic and boring foraminifer *Hyrrokkin sarcophaga* are common on near living coral polyps and associated molluscs (Freiwald and Schönfeld 1996). Traces of excavating grazers are rarely observed.

Sedimentary infill and diagenetic regime – Shallow-water reefs both modern and ancient are predominantly not preserved as *in situ* frameworks. Mostly reworked and bioeroded piles of carbonate grains which are derived from the coral reef community itself are rapidly lithified through processes of meteoric and/or vadose diagenesis. Deep-water reefs however are different. The sedimentary infill is dominated by imported pelagic muds consisting of either calcareous plankton (foraminifers, coccolithophorids, pteropods), or by suspended terrigenous muds and silt-grade deposits. The internally produced reef debris dominates in places such as in the coral rubble zone (Mortensen et al. 1995; Freiwald et al. 1997) but is generally diluted. Other important constituents of deep-water reef sediment infill are: sponge spicules, benthic foraminifers, crinoid ossicles, octocoral and ascidian sclerites. Locally bioeroded sponge-chips are abundant (Freiwald and Wilson 1998). The diagnetic regime is poorly studied. Microbial micritisation carried out by bacteria and fungi creates a characteristic micritic envelope which is generally regarded as a shallow-water indicator. Submarine cementation is best described by Wilbur and Neumann (1993) from the coral-bearing and current-swept lithoherms from the Florida Strait. The cementation started with crypto-crystalline precipitation of magnesium calcite which converts a primarily grain-supported fabric into a mud-supported fabric rich in peloids with microspar rims and micritic limestones. Another early diagenetic process seems to be very common in deep-water coral reefs in the dysaerobic zone above the sediment-water interface. This is where dead coral skeletons become infested by an iron and manganese precipitating microbial consortium consisting of *Siderocapsa*-like bacteria, fungi and undetermined protists such as stalked peritrichs. The product is a faint metastable iron-manganese coat on exposed skeletal surfaces and unfilled borings into the skeleton (Freiwald and Wilson 1998). Evidently, this 10 – 15 µm-thick biofilm seems to attract pelagic larvae to settle and stimulate metamorphosis of sessile organisms such as sponges, bryozoans, brachiopods, crinoids, serpulids, cirripeds, hydroids, octocorals and ascidians.

Global distribution of deep-water coral buildups

On plotting global distribution, a concentration of deep-water coral patches, reefs and mounds in the North Atlantic seem to exist, while records from the South Atlantic, Indian Ocean and Pacific Ocean are scarce (Fig. 4). This distribution, however, mirrors the intense scientific, economic and military investigations along the North Atlantic shelves. It also demonstrates the coincidence between fertile surface water masses and the occurrence of deep-water reefs in the higher latitudes.

Deep-water coral buildups were first discovered in Norwegian waters (Dons 1944). These *Lophelia pertusa* occurrences which preferentially occur in 250 to 500 m water depth have been the subject of a great number of research projects up to the present (Freiwald et al. 1997, 1999; Fosså et al. 2000). Deep-water coral buildups of various dimensions seem to follow the northwest European continental margin around the British Isles (Wilson 1979a), the Faroer Islands (Frederiksen et al. 1992), Rockall Bank (Wilson 1979b), the Porcupine Bank and Seabight area (Scoffin and Bowes 1988), the Western Approaches and offshore Brittany (LeDanois 1948) and along the northern and northwestern shelves off the Iberian Peninsula (LeDanois 1948) in 250 to 1000 m water depth. In addition to *Lophelia pertusa*, other corals, such as *Madrepora oculata* and *Desmophyllum cristagalli*, contribute to, or even dominate the coral buildups south of Norway (Rogers 1999). Information of the occurrence of deep-water corals

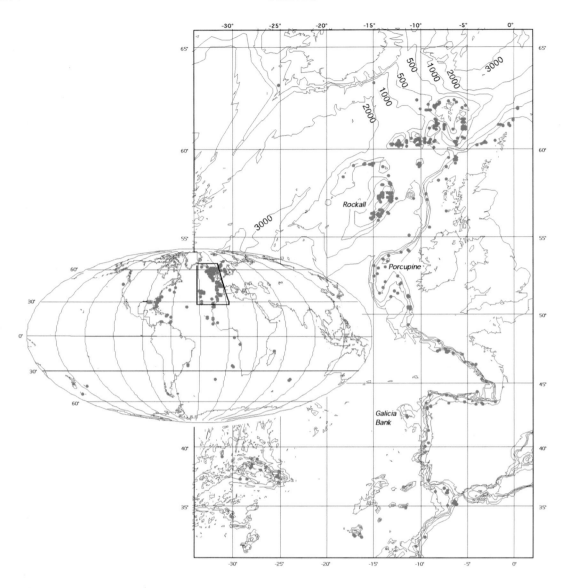

Fig. 4. Compilation of deep-water coral occurrences on a global scale (insert map) and focused in the NE Atlantic (Data sources: see Rogers 1999).

along the Mid-Atlantic Ridge are scanty except for the Reykjanes Ridge, where *Lophelia* and *Madrepora* have been collected between 500 and 900m water depth (Copley et al. 1996).

In the Northwest Atlantic, the Blake Plateau is a well-known location harbouring large numbers of deep-water coral constructions formed by *Enallospammia* (*Dendrophyllia*) *profunda* and less abundantly, *Lophelia pertusa* growing on top of the mounds, at 640 - 870 m water depth (Stetson et al. 1962; Squires 1964). Over 200 coral mounds up to 146 m in height occur within an area

of 3240 km² on the Blake Plateau (Reed 1992). Coral mounds consisting predominantly of *Solenosmilia variabilis* and solitary corals are common along the northern lower slope of the Little Bahama Bank at water depths of 1000 - 1300 m (Mullins et al. 1981). Elongated coral-bearing mounds – "lithoherms" – are present along the western margin of the Little Bahama Bank at water depths of 500 - 700 m (Messing et al. 1990). Although not consisting entirely of azooxanthellate corals, *Lophelia pertusa* occurs abundantly at upcurrent positions on these mounds (Messing et al. 1990). Recent studies by Paull et al. (2000) investigated more than 400 m wide and more than 4400 m long and 150 m high lithoherms along the Florida - Hatteras Slope below 440 m water depth. Extant *Lophelia* buildups and fossil coral mounds were found in the Gulf of Mexico in 400 - 500 m water depth (Newton et al. 1987).

Although, azooxanthellate corals occur on most if not all seamounts and offshore banks in the South Atlantic (Rogers 1999), only a few deep-water coral buildups have been located so far. Extensive Holocene coral banks were reported from the Campos Basin slope, off southwestern Brazil (Viana et al. 1997).

In the subantarctic South Pacific, *Solenosmilia variabilis* forms coral buildups on a seamount in 550 - 915 m water depth (Cairns 1982). In New Zealand waters coral buildups of *Goniocorella dumosa* are known from Palliser Bay, Cook Strait, at water depths of 448 - 512 m (Beu and Climo 1974) and from the Campbell Plateau in 312 m water depth (Squires 1965).

The oceanic boundary conditions of *Lophelia pertusa* in the North Atlantic

The North Atlantic receives water from several sources (Hopkins 1988). The South Atlantic and the Caribbean Sea are the sources of the North Atlantic surface water which form the Gulf Stream or North Atlantic Current (NAC) that flows across the central North Atlantic along the northwestern European continental margins to as far north as Svalbard. Polar water masses flow southward along the western margins of the ocean with the East Greenland Current and the Labrador Current. The mid-depths in the North Atlantic are fed by a Southern Ocean component water at approximately 900 m water depth (Reid 1996). Another input at mid-depths is the Mediterranean Outflow Water (MOW) that transports relatively warm and saline water to as far north as the Norwegian Sea (Reid 1979). The MOW appears in the North Atlantic as a salinity maximum at about 900 - 1100 m water depth off the Gibraltar Strait. The MOW extends as tongues of relatively warm and saline water both northward along the northwestern European continental margin as an eastern bound-

ary current and westward across the Atlantic. As the MOW reaches the western boundary it provides warm water of high salinity to the Gulf Stream at 700 - 2500 m water depth and to the southward flow off Cape Hatteras (Reid 1994). North Atlantic Deep-Water (NADW) is formed on a large scale in convection cells in the Labrador Sea and the Norwegian-Greenland Sea (Hopkins 1988). These convection cells fuel the deep-sea with cold, oxygen-enriched and saline water. The North Atlantic deep-sea receives old, corrosive and low-salinity water as Antarctic Bottom Water from the south where this dense water mass underflows the NADW. The North Atlantic in effect trades cold deep-water for warm shallow-water (Berger and Wefer 1996). This complex circulation pattern in the North Atlantic is documented in an east-west hydrographic transect at 42° northern latitude (Fig. 5) and a north-south transect at 12° western longitude (Fig. 6) where the mean annual temperature, salinity and the dissolved oxygen content are plotted against depth. Focusing on the east to west transect, all three parameters reveal the same trend: bathymetrically crowded isolines at the western margin of the North Atlantic and widely spaced isolines at the eastern margin (Fig. 5). This apparent asymmetry is largely caused by the lateral inflow of dense, warm and saline MOW through the Gibraltar Strait from the west and from the southward flow of the dense NADW layer that is thicker along the western boundary due to the Coriolis force.

The large scale assessment of oceanographic boundary conditions of living *Lophelia pertusa* occurrences (data compiled from Rogers 1999) are discussed below in respect of the complex oceanic regime of the North Atlantic. The emphasis is laid on selected structural highs such as seamounts, continental slopes and shelves in the Northeast Atlantic where sufficient information of living *Lophelia* occurrences exist. This selection of locations includes a transect from Cap St. Vincent, Portugal, to the Azores (centred at 37° N), northwestern Iberian Margin to Galicia Bank (43° N), Banc de la Chapelle area (47° N), Porcupine Seabight (50.3° N) and the Norwegian Sea (64° N) and is meant as a physical oceanographic assessment to understand the environmental con-

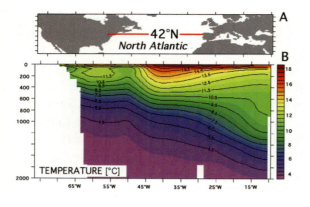

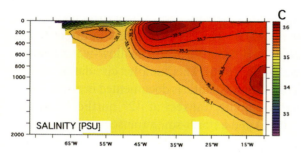

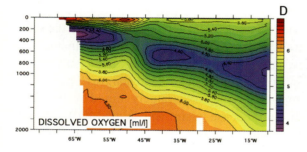

Fig. 5. a) Oceanographic parameters along a latitudinal gradient (42° N) in the North Atlantic: **b)** Temperature, **c)** Salinity, and **d)** Dissolved oxygen (data sources are cited in the text).

trols of *Lophelia*- and *Madrepora*-dominated buildups. The bathymetric range of living *Lophelia* has been plotted against hydrographic tracers: mean annual temperature, salinity and dissolved oxygen. The data based on Levitus and Boyer (1994a, b) and Levitus et al. (1994) can be aquired online from the NOAA-databank (http://ferret.wrc.noaa.gov/las/main.html):

Mean annual temperature

The MOW brings warm water into the Northeastern Atlantic and the eastern boundary current

flow of the MOW is detectable by elevated seawater temperatures at all selected locations south of the Scotland-Faroer Ridge (Fig. 7). The 4 °C-isotherm as the lower temperature limit of *Lophelia pertusa* is about 2000 m deep off the Iberian Peninsula and it shallows up to 1600 m along the Celtic Slope. In the Norwegian Sea, however, this critical isotherm is developed at about 400 m water depth. The upper temperature limit of *Lophelia pertusa* is reported with 12° C (Rogers 1999). The corresponding mean annual

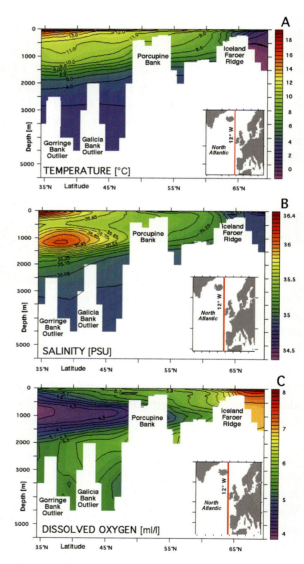

Fig. 6. Oceanographic parameters along a longitudinal gradient (12° W) in the North Atlantic: **a)** Temperature, **b)** Salinity, and **c)** Dissolved oxygen (data sources are cited in the text).

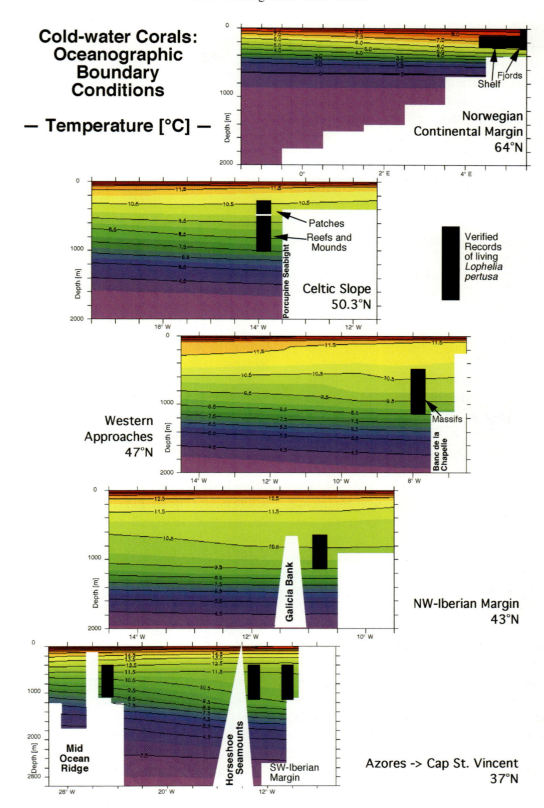

Fig. 7. Mean annual temperature regimes from the surface to 2000 m water depth plotted from selected *Lophelia*-rich areas from 37° N to 64° N along the northwest European continental margin. The bathymetric range of verified records of live *Lophelia* is indicated (data sources are cited in the text).

isotherm lies at 500 m water depth off the south-western Iberian Peninsula but crops out off the northwestern Iberian Peninsula and as far north as the Celtic Slope. The mean annual temperatures in the Norwegian Sea do not reach the 12° C-isotherm (Fig. 7). The verified records of living *Lophelia pertusa* plot within the:

• 8.5° - 12.5° C range along the Cap St. Vincent - Azores transect at 1200 to 500 m depth,
• 9.5° - 11° C range off the northwestern Iberian Peninsula and the Western Approaches at 1200 to 400 m depth,
• 7.5° - 10.5° C range off the Celtic Slope at 1000 to 180 m depth.
• 6° - 8° C range in Norwegian waters at 300 to 250 m depth but even much shallower in some Scandinavian fjords.

Mean annual salinity

Cores of elevated salinity in the North Atlantic at 32° N occur in both shallow and mid-depth water down to 2000 m water depth (Fig. 5c). The elevated salinity in the deeper water levels delineates the MOW intrusion into the Northeast Atlantic with values of up to 36 ‰ and it forms a distinctive core as an eastern boundary current along the northwestern European continental margin south of the Iceland-Faroer-Shetland Ridge (Fig. 8). The MOW is well developed as an intersecting core between 1400 - 600 m as far north as the Porcupine Seabight, Celtic Sea, at 50° N where the saline waters crop out (Figs. 6b and 8).

The *Lophelia* occurrences plot in the 35 - 37 ‰-range in the North Atlantic but the figures show a preference torwards higher salinities in the centre of the MOW boundary current (Fig. 8). However, salinity is slightly reduced in many Norwegian fjords with values around 33 - 34.5 ‰ and it is strongly elevated in the Mediterranean, where today only scattered records of live *Lophelia* are documented, with values >37 ‰. North of the Iceland-Faroer-Shetland Ridge the salinity window for *Lophelia* is confined to the NAC water mass in the upper 400 m (Fig. 9). At the Norwegian continental slope, this salinity window is narrowed by (1) the lower saline Norwegian Coastal Current

(NCC) surface water on top and (2) by the lesser saline NADW beneath. To conclude, *Lophelia* or *Lophelia*-dominated buildups show a preference for the highest salinity water in any given location.

Mean annual dissolved oxygen

An important factor to determine the distribution of heterotrophic life in the oceans is the oxygen content of the water which is expressed as dissolved oxygen per millilitre. Focusing on the North Atlantic, the vertical profiles of dissolved oxygen reveal the following trends: (1) High oxygen content in the upper water column by phytoplankton photosynthesis and by trapped air through breaking waves. (2) Below, the level of dissolved oxygen decreases because no plants or phytoplankton add oxygen to the water. Instead, oxygen consumption through the decomposition of sinking organic matter which is derived from productive surface water is the dominant process which leads to the formation of a distinct oxygen minimum zone. (3) Oxygen content increases towards greater depths as deep-water carries oxygen into the deep ocean basins.

The degree of oxygen depletion in the North Atlantic does not reach levels of <1 ml/l which would induce anoxia. In the eastern North Atlantic, the depleted dissolved oxygen zone or oxygen minimum zone (OMZ) is characterized by 3 - 5 ml/l at 1600 - 500 m (Fig. 6c) and is underlain by the more oxygenated NADW (Levitus et al. 1994). Toward the western margin of the North Atlantic, the OMZ shallows to 600 - 200 m at 42° N and to 700 - 500 m at the Blake Plateau, 32° N (Fig. 6d). Off Norway, the OMZ occurs in much shallower water depths but beneath the thermocline depth as a consequence of seasonal eutrophication through the rich plankton communities of the higher latitudes (Fig. 9).

In areas where the OMZ intersects with seamounts, offshore banks and the continental slope, the depth distribution of *Lophelia pertusa* strikingly coincides with the OMZ layer in the North Atlantic (Fig. 9). The *Lophelia* reefs on the Norwegian shelf occur in 250 - 500 m (Mortensen et al. 1995; Freiwald et al. 1999), the *Lophelia*-bearing reefs and lithoherms on the Blake Plateau and Florida -

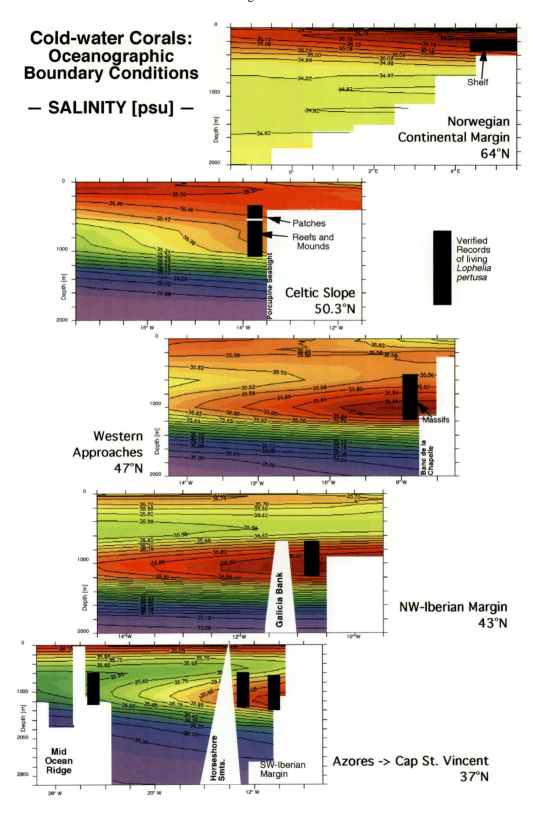

Fig. 8. Mean annual salinity regimes from the surface to 2000 m water depth plotted from selected *Lophelia*-rich areas from 37° N to 64° N along the northwest European continental margin. The bathymetric range of verified records of live *Lophelia* is indicated (data sources are cited in the text).

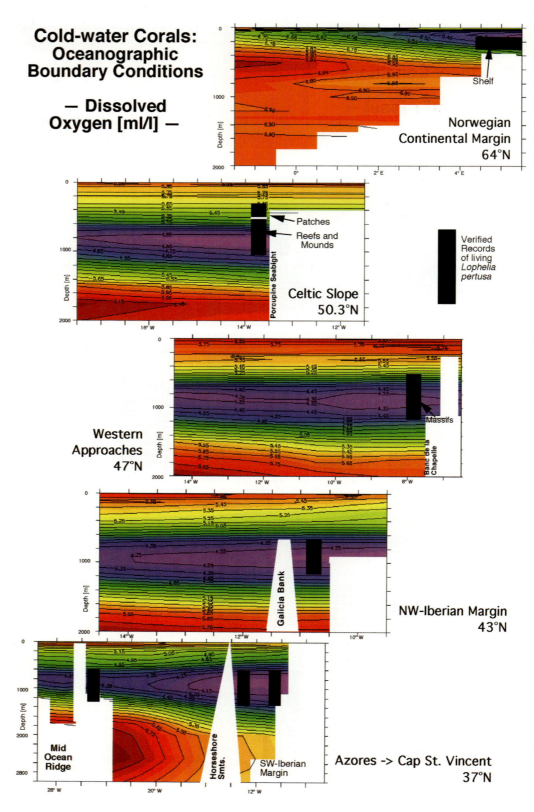

Fig. 9. Mean annual dissolved oxygen regimes from the surface to 2000 m water depth plotted from selected *Lophelia*-rich areas from 37° N to 64° N along the northwest European continental margin. The bathymetric range of verified records of live *Lophelia* is indicated (data sources are cited in the text).

Hatteras Slope in 500 - 700 m (Stetson et al. 1962; Paull et al. 2000) and the *Lophelia* occurrences along the Celtic Slope, Banc de la Chapelle and off the Iberian Peninsula form buildups of unverified size and geometry between 500 - 1000 m water depth (Fig. 9; LeDanois 1948). The underlying reason for this apparent coincidence may be an increased availability of digestable food for *Lophelia* at these levels. However, it should be kept in mind, that *Lophelia* also occurs in shallower depths than the OMZ layer such as on Rockall Bank where it forms patches (Wilson 1979b) rather than mature reefs as in the deeper waters.

For each given location or transect, there are many more records of dead *Lophelia pertusa* findings and sometimes the indication of fossil *Lophelia* is mentioned in the station lists. Such findings may become a useful tool to reconstruct the properties of water masses, changes in the circulation and ventilation pattern and the residence time of water masses in the oceans. A basis for such an effort will be the dating of *Lophelia pertusa* with the $^{238}U/^{230}Th$ method (Adkins et al. 1998; Mangini et al. 1998).

Thermocline depth, storm wave base and light

The minimum depth of occurrence is considered to be the vertical force of wave action that could induce collapse of the coral frameworks (Frederiksen et al. 1992) and the thermocline depth (Dons 1944). In the northeastern Atlantic, the mean annual depth of the thermocline shifts with the intensity of wind- and current-driven mixing processes. In the Canary Island region, the thermocline depth is located between 100 - 150 m water depth. It increases to 250 m off the Iberian Peninsula and to 500 m in the Bay of Biscay, Porcupine Seabight and Celtic Sea (Rice et al. 1994). *Lophelia pertusa* occurs in aphotic and subphotic depths. Light *per se* is not regarded as a limiting factor for *Lophelia*, as this coral has been kept alive for several years in an aquarium under normal illumination conditions (Mortensen, pers. comm.). Owing to the lack of data, the maximum depth occurrence for *Lophelia pertusa* is unknown but the hydrographic plots already indicate that this species occurs above the cold core of the North Atlantic Deep Water.

Deep-water corals as environmental recorder

As in the case of skeletons of other azooxanthellate corals, *Lophelia pertusa* does not secrete skeletal aragonite in isotopic equilibrium with the ambient seawater. This disequilibrium leads to a depletion of both the stable oxygen and carbon isotopic composition whereby a strong positive $\delta^{18}O/\delta^{13}C$ correlation is characteristic (McConnaughey 1989a; Freiwald et al. 1997). The simultaneous depletion of oxygen and carbon isotope ratios is explained as a combined kinetic and metabolic effect (McConnaughey 1989a, b). Kinetic isotopic effects result from discrimination against the heavy isotopes of carbon and oxygen during carbon dioxide hydration and hydroxylation (McConnaughey 1989b). Despite the kinetic disequilibrium of the stable isotope ratios in azooxanthellate corals, there are reasonable expectations that it will be possible to use *Lophelia pertusa* and other azooxanthellate corals as tools to age deep waters and changes in the ventilation of deep waters (Smith et al. 1997, 1999).

Existing data on the average annual growth rates of *Lophelia pertusa* expressed by the linear extension of the corallite varies from 4.1 - 25 mm yr^{-1} (Fig. 10; Mikkelsen et al. 1982; Freiwald et al. 1997; Mortensen and Rapp 1998). The higher value of the average growth of *Lophelia* falls in the range as for most of the massive hermatypic and zooxanthellate corals of the subtropical - tropical climates. This unexpected finding explains the enormous reef-forming potential of *Lophelia pertusa* along Europe's continental margin.

Perspectives

Our knowledge of the biology, distribution and environmental controls of *Lophelia pertusa* as a biospecies and as a reef constructor is quite poor. We are just beginning to realise how widely this species is distributed along the northwest European continental margin and elsewhere and about its role as the key species of a distinct deep-water eco-

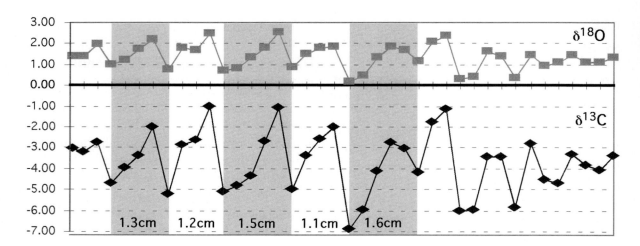

Fig. 10. Stable isotope study (oxygen and carbon) of a *Lophelia pertusa* branch from Sula Ridge, 270m water depth. The growth direction is from left to right (data were measured in the Stable Isotope Laboratory of the Alfred-Wegner-Institute, Bremerhaven by Dr. A. Mackensen).

system that supports an extremely rich associated biodiversity. Moreover, our understanding of the formation of coral reefs is challenged by the potential to produce large reef tracts of the same dimensions and at a similar rate to those found in shallow-water reefs in warm climates. This also draws attention to the fact, that hitherto unknown quantities of calcium carbonate are produced on non-tropical shelves whose impact is still not yet acknowledged in any global carbon-cycling model. The types of biological buildups provide clues for the understanding of the widespread but still enigmatic formation of mud mounds during the Phanerozoic which also occurred in deeper water settings.

The presence of coral reef ecosystems in European waters is as yet unknown to many of the stakeholders. Europe's coral reefs mostly lie within the Exclusive Economic Zones and are, therefore subject to potential threats from increasing commercial activities in Europe's deeper waters. For the first time, these ecosystems now face the prospect of serious anthropogenic impact. It is likely that the most serious impacts will result from deep-sea trawling which is already operational to depths of at least 2,000 m on the European margin - spanning the full bathymetric range of deep-water coral ecosystems. The second key potential threat

is related to the expanding hydrocarbon exploration and production industry which, like deep-water trawling, is already active to depths of at least 2,000 m around Europe and elsewhere in the world. Oil production has already started in the deep waters to the west of Shetland in the Faeroe-Shetland Channel where deep-water coral ecosystems are known to occur. This potential conflict between industry and environment has already caused concern; witness the recent controversial debates between conservationists and industrialists concerning the potentially irreversible damage to reefs (i.e. the Greenpace Court Case in the U.K.). This debate remains unresolved and could present a barrier to future economic development in this sector until sufficient independant research is carried out to: (1) ascertain the potential sensitivities and vulnerability of the coral ecosystem, (2) identify the most significant conservation issues involved, and (3) meet the needs of environmental managers and regulatory authorities by recommending the measures necessary to permit sustainable resource development alongside effective conservation in the vicinity of these enigmatic cold reef communities. Therefore, the results obtained through the ongoing ACES project will have a strong strategic importance in the future sustainable use of biological offshore resources. To close that gap, a joint research effort embracing various disciplines of bio- and geosciences is now on its way – ACES (Atlantic Coral Ecosystem Study; Freiwald 2000).

Acknowledgements

This review based upon the authors participation in several projects which are/were funded by the Deutsche Forschungsgemeinschaft (Grant He-1671/5 and Heisenberg Grant Fr-1134/1), the Bundesministerium für Bildung und Forschung (BOSMAN Project No. 03F0256A-D) and the 5th European Framework Programme (ACES, Contract No. EVK3-CT1999-00008).

References

Adkins JF, Cheng H, Boyle EA, Druffel ERM, Edwards RL (1998) Deep-sea coral evidence for rapid change in ventilation of deep North Atlantic 15,400 years ago. Science 280:725-728

Bell N, Smith J (1999) Coral growing on North Sea oil rigs. Nature 402:601

Berger WH, Wefer G (1996) Central themes of South Atlantic circulation. In: Wefer G, Berger WH, Siedler G, Webb DJ (eds) The South Atlantic: Present and Past Circulation. Springer, Berlin pp 1-11

Bett BJ (1999) RRS Charles Darwin Cruise 112C, 19 May – 24 Jun 1998. Atlantic Margin environmental survey: Seabed survey of deep-water areas (17th round Trances) to the north and west of Scotland. Southampton Oceanography Centre Cruise Report #25: 171p

Beu AG, Climo FM (1974) Mollusca from a recent coral community in Palliser Bay, Cook Strait. New Zealand J Mar Freshwater Res 8:302-332

Brasier MD (1995) Fossil indicators of nutrient levels. 1: Eutrophication and climate change. In: Bosence DWJ, Allison PA (eds) Marine Palaeoenvironmental Analysis from Fossils. Geol Soc SpecPubl 83:113-132

Cairns SD (1982) Antarctic and subantarctic scleractinia. In: Kornicker LS (ed) Biology of the Antarctic Seas. American Geophysical Union, Washington pp 1-74

Chachkine P, Akmetzhanov A (1998) Subbottom currents on the Porcupine Margin study by side-scan sonars. Intergovernmental Oceanographic Commission, Workshop Report 143:30

Copley JTP, Tyler PA, Sheade, M, Murton J, German R (1996) Megafauna from sublittoral to abyssal depths along the Mid-Atlantic Ridge south of Iceand. Oceanol Acta 19:549-559

Cowen R (1983) Algal symbiosis and its recognition in the fossil record. In: Tevesz MJS, McCall PW (eds) Biotic Interactions in Recent and Fossil Benthic Communities. Plenum Press, New York pp 431-479

Delibrias G, Taviani M (1985) Dating the death of Mediterranean deep-sea scleractinian corals. Mar Geol 62: 175-180

Dons C (1944) Norges korallrev. Det Kongelige Norske Videnskabers Selskab, Forhandlinger 16:37-82

Filkorn HF (1994) Fossil scleractinian corals from James Ross Basin, Antarctica. Antarctic Research Series 65: 1-96

Fosså JH, Mortensen PB, Furevik DM (2000) Lophelia-korallrev langs norskekysten forekomst og tilstand. Fisken og Havet 2:1-94

Frederiksen R, Jensen A, Westerberg H (1992) The distribution of the scleractinian coral Lophelia pertusa around the Faroe Islands and the relation to internal tidal mixing. Sarsia 77:157-171

Freiwald A (2000) The Atlantic Coral Ecosystem Study (ACES): A margin-wide assessment of corals and their environmental sensitivities in Europe's deep waters. EurOCEAN 2000, Project Synopses Vol. 1: 313-317

Freiwald A, Schönfeld J (1996) Substrate pitting and boring pattern of Hyrrokkin sarcophaga Cedhagen, 1994 (Foraminifera) in a modern deep-water coral reef mound. Mar Micropal 28:199-207

Freiwald A, Wilson JB (1998) Taphonomy of modern, deep, cold-temperate water coral reefs. Historical Biol 13:37-52

Freiwald A, Henrich R, Pätzold J (1997) Anatomy of a deep-water coral reef mound from Stjernsund, West-Finnmark, northern Norway. In: James NP, Clarke JAD (eds) Cool-Water Carbonates. SEPM Spec Publ 56:141-161

Freiwald A, Wilson JB, Henrich R (1999) Grounding icebergs shape deep-water coral reefs. Sed Geol 125:1-8

Hallock P, Schlager W (1986) Nutrient excess and the demise of coral reefs and carbonate platforms. Palaios 1:389-398

Hanken NM, Bromley RG, Miller J (1996) Plio-Pleistocene sedimentation in coastal grabens, northeast Rhodes, Greece. Geol J 31:393-418

Henriet JP, De Mol B, Pillen S, Vanneste M, Van Rooij D, Versteeg W, Crocker PF, Shannon PM, Unnithan V, Bouriak S, Chachkine P (1998) Gas hydrate crystals may help build reefs. Nature 391:647-649

Hopkins TS (1988) The GIN Sea. SACLANTCEN Report SR-124:1-190

Hovland M, Croker PF, Martin M (1994) Fault-associated seabed mounds (carbonate knolls?) off western Ireland and north-west Australia. Mar Petrol Geol 11:232-246

Hovland M, Mortensen PB, Brattegard T, Strass P, Rokoengen K (1998) Ahermatypic coral banks off mid-Norway: evidence for a link with seepage of light hydrocarbons. Palaios 13:189-200

James NP (1997) The cool-water carbonate depositional realm. In: James NP, Clarke JAD (eds) Cool-Water Carbonates, SEPM Spec Publ 56:1-20

James NP, Bourque PA (1992) Reefs and mounds. In: Walker RG, James NP (eds) Facies Models – Response to Sea Level Change. Geological Association of Canada pp 323-347

Jensen A, Frederiksen R (1992) The fauna associated with the bank-forming deepwater coral *Lophelia pertusa* (Scleractinia) on the Faroe shelf. Sarsia 77: 53-69

LeDanois E (1948) Les profondeurs de la mer. Payot, Paris 303 p

Levitus S, Boyer, TP (1994a) World Ocean atlas 1994, Volume 2: Oyxgen. NOAA Atlas NESDIS 2, US Department of Commerce, Washington DC 186 p

Levitus S, Boyer TP (1994b) World Ocean atlas 1994, Volume 4: Temperature. NOAA Atlas NESDIS 4, US Department of Commerce, Washington DC 117 p

Levitus S, Burgett R, Boyer TP (1994) World Ocean atlas 1994, Volume 3: Salinity. NOAA Atlas NESDIS 3, US Department of Commerce, Washington DC 99 p

Mangini A, Lomitschka M, Eichstädter R, Frank N, Vogler S, Bonani G, Hajdas I, Pätzold J (1998) Coral provides way to age deep water. Nature 392:347-348

McConnaughey T (1989a) ^{13}C and ^{18}O isotopic disequilibrium in biological carbonates: I. Patterns. Geochim Cosmochim Acta 53:151-162

McConnaughey T (1989b) ^{13}C and ^{18}O isotopic disequilibrium in biological carbonates: II. *In vitro* simulation of kinetic isotope effects. Geochim Cosmochim Acta 53:163-171

Messing CG, Neumann AC, Lang JC (1990) Biozonation of deep-water lithoherms and associated hardgrounds in the northeastern Straits of Florida. Palaios 5:15-33

Mikkelsen N, Erlenkeuser H, Killingley JS, Berger WH (1982) Norwegian corals: radiocarbon and stable isotopes in *Lophelia pertusa*. Boreas 11:163-171

Montenat C, Barrier P, Ott d'Estevou P (1991) Some aspects of the recent tectonics in the Strait of Messina, Italy. Tectonophys 194:203-215

Mortensen PB, Rapp HT (1998) Oxygen and carbon isotope ratios related to growth line patterns in skeletons of *Lophelia pertusa* (L) (*Anthozoa,*

Scleractinia): Implications for determination of linear extension rates. Sarsia 83:433-446

Mortensen PB, Hovland M, Brattegard T, Farestveit R (1995) Deep water bioherms of the scleractinian coral *Lophelia pertusa* (L.) at 64°N on the Norwegian shelf: Structure and associated megafauna. Sarsia 80:145-158

Mullins HT, Newton CR, Heath K, Vanburen HM (1981) Modern deep-water coral mounds north of Little Bahama Bank: Criteria for recognition of deep-water coral bioherms in the rock record. J Sed Petrol 51: 999-1013

Newton CR, Mullins HT, Gardulski AF, Hine AC, Dix GR (1987) Coral mounds on the West Florida slope: unanswered questions regarding the development of deep-water banks. Palaios 2:359-367

Paull CK, Neumann AC, am Ende BA, Ussler W, Rodriguez NM (2000) Lithoherms on the Florida – Hatteras Slope. Mar Geol 166:83-101

Reed JK (1992) Submersible studies of deep-water *Oculina* and *Lophelia* coral banks off southeastern USA. In: Cahoon LB (ed) Diving for Science. University of North Carolina, Wilmington pp 143-151

Reid JL (1979) On the Mediterranean Sea outflow to the Norwegian-Greenland Sea. Deep-Sea Res 26:1199-1223

Reid JL (1994) On the total geostrophic circulation of the North Atlantic Ocean: Flow patterns, tracers, and transports. Progr Oceanogr 33:1-92

Reid JL (1996) On the circulation of the South Atlantic Ocean. In: Wefer G. Berger WH, Siedler G, Webb DJ (eds) The South Atlantic: Present and Past Circulation. Springer, Berlin pp 13-44

Rice AL, Thurston MH, Bett BJ (1994) IOSDL DEEPSEAS programme: Introduction and photographic evidence for the presence and absence of a seasonal input of phytodetritus at contrasting abyssal sites in the northeastern Atlantic. Deep-Sea Res 41:1305-1320

Roberts M (2000) Coral colonies make a home on North Sea oil rigs. Reef Encounter 27:17-18

Rogers AD (1999) The biology of *Lophelia pertusa* (Linnaeus 1758). Internationale Revue der gesamten Hydrobiologie 84:315-406

Scoffin TP, Bowes GE (1988) The facies distribution of carbonate sediments on Porcupine Bank, northeast Atlantic. Sed Geol 60:125-134

Smith JE, Risk MJ, Schwarcz HP, McConnaughey TA (1997) Rapid climate change in the North Atlantic during the Younger Dryas recorded by deep-sea corals. Nature 386:818-820

Smith JE, Schwarcz HP, Risk MJ, McConnaughey TA,

Keller N (1999) Paleotemperatures from deep-sea corals: Overcoming "vital effects". Palaios 15:25-32

Squires DF (1957) New species of caryophylliid corals from the Gulf Coast Tertiary. J Paleontol 31:992-996

Squires DF (1964) Fossil coral thickets in Wairarapa, New Zealand. J Paleontol 38:904-915

Squires DF (1965) Deep-water coral structure on the Campbell Plateau, New Zealand. Deep-Sea Res 12: 785-788

Stanley GD, Cairns SD (1988) Constructional azoo-xanthellate coral communities: An overview with implications for the fossil record. Palaios 3:233-242

Stetson TR, Squires DF, Pratt RM (1962) Coral banks occurring in deep water on the Blake Plateau. American Museum Novitates 2114:1-39

Teichert C (1958) Cold- and deep-water coral banks. Amer Ass Petrol Geol Bull 42:1064-1082

Tudhope AW, Scoffin TP (1995) Processes of sedimentation in Gollum Channel, Porcupine Seabight: Submersible observations and sediment analysis. Trans Roy Soc Edinburgh Earth Sci 86:9-55

Vella P (1964) Foraminifera and other fossils from Late Tertiary deep-water coral thickets, Wairarapa, New Zealand. J Paleontol 38:916-928

Veron JEN (1995) Corals in Space and Time. Comstock/ Cornell, Ithaca 321 p

Viana AR, Faugéres JC. Kowsmann RO, Lima JAM, Caddah LFG, Rizzo JG (1997) Hydrology, morphology and sedimentology of the Campos continental margin, offshore Brazil. Sed Geol 115:133-157

Wilber RJ, Neumann AC (1993) Effects of submarine cementation on microfabrics and physical properties of carbonate slope deposits, northern Bahamas. In: Rezak R, Lavoie DL (eds) Carbonate Microfacies. Springer, New York pp 79-94

Wilson JB (1979a) "Patch" development of the deep-water coral *Lophelia pertusa* (L.) on Rockall Bank. J Mar Biol Ass UK 59:165-177

Wilson JB (1979b) Biogenic carbonate sediments on the Scottish continental shelf and on Rockall Bank. Mar Geol 33:M85-M93

Life at the Edge:
Achieving Prediction from Environmental Variability and Biological Variety

A. Rogers[1*], D. Billett[2], W. Berger[3], E. Flach[4], A. Freiwald[5], J. Gage[6], D. Hebbeln[7], C. Heip[8], O. Pfannkuche[9], E. Ramirez-Llodra[1], L. Medlin[10], M. Sibuet[11], K. Soetaert[8], O. Tendal[12], A. Vanreusel[13], M. Wlodarska-Kowalczuk[14]

[1]*School of Ocean & Earth Science, University of Southampton, Southampton Oceanography Centre, European Way, Southampton, SO14 3ZH, United Kingdom*
[2]*George Deacon Division for Ocean Processes, Southampton Oceanography Centre, European Way, Southampton, SO14 3ZH, United Kingdom*
[3]*Scripps Institution of Oceanography, UCLA, San Diego, La Jolla, CA 92093-0215, USA*
[4]*Department of Systems Ecology, University of Stockholm, 106 91 Stockholm, Sweden*
[5]*Universität Tübingen, Institut und Museum für Geologie und Paläontologie, Herrenberger Str. 51, D-72070 Tübingen, Germany*
[6]*Scottish Association for Marine Science, Dunstaffnage Marine Laboratory, P.O. Box 3, Oban, Argyll, PA34 4AD, United Kingdom*
[7]*Universität Bremen, Fachbereich Geowissenschaften, 28359 Bremen, Germany*
[8]*NIOO-CEMO, PO Box 140, 4400 AC Yerseke, The Netherlands*
[9]*GEOMAR, Wischofstraße 1 – 3, 24148 Kiel, Germany*
[10]*Alfred Wegener Institut, Am Handelshafen 12, 27570 Bremerhaven, Germany*
[11]*DRO/Dépt. Environment Profond, IFREMER, Centre de Brest, BP 70, 29280 Plouzane Cedex, France*
[12]*University of Copenhagen, Zoological Museum, Universitetsparken 15, 2100 Copenhagen, Denmark*
[13]*Marine Biology Section, University of Gent, Ledeganckstraat 35, 9000 Gent, Belgium*
[14]*Marine Ecology Department, Polish Academy of Sciences, Institute of Oceanography, ul. Powstancow Warszawy 55, 81 – 712 Sopot, Poland*
* *corresponding author (e-mail): a.d.rogers@soc.soton.ac.uk*

Abstract: The ocean margins contain a great variety of habitats and biological communities. Recent discoveries, such as deep-water coral reefs, show that these communities are poorly described and understood. However, observations have already indicated that benthic communities on ocean margins show high levels of spatial and temporal variation at all scales. The European continental margin is increasingly exploited for both biological resources (fisheries) and non-biological resources (oil, gas, minerals). Environmental management of the exploitation of continental margins requires an understanding of natural levels of variation inherent in biological communities that are potentially impacted by such activities. This paper presents a synthesis of the present knowledge of the spatial and temporal variation of slope communities. Priorities for future research and its technological development are discussed. The aim of this research is to provide a scientific basis for the environmental management of the continental slopes of Europe.

From WEFER G, BILLETT D, HEBBELN D, JØRGENSEN BB, SCHLÜTER M, VAN WEERING T (eds), 2002, *Ocean Margin Systems.* Springer-Verlag Berlin Heidelberg, pp 387-404

Introduction

Ocean margin systems contain a wide variety of habitats including coral mounds, mud volcanoes, deep canyons, sediment slopes, gravel banks and submarine mountains (Heezen and Hollister 1971). If the habitats were on land we would be amazed by their spectacular variety. As the exploration of the ocean margin proceeds, the rate of fundamental discoveries appears to be gathering pace. Each new technological development leads to major new findings. Within the last ten years hundreds of giant mounds more than 100 m high and approximately 1km in diameter have been found along the European continental margin (depths of 600 – 1000 m). Dense and diverse coral communities cover many of them (Rogers 1999). That such basic discoveries should still be made in the present day is a measure of our ignorance of ocean margin systems.

The commercial exploitation of living and non-living deep-water resources on the ocean margin is increasing rapidly. Deep-sea fisheries and oil and gas production may be having significant effects on wide areas of the ocean margin. Direct scientific sampling of the seabed on the ocean margin covers at the most only a few tens of km², depending on the type of technology used, and the density of coring of the seabed is inadequate in most regions to describe biological communities (Gage 1996). The discrepancy between available scientific information and the level of commercial exploitation presents a severe challenge to the development of environmental management plans. How do we plan the sustainable development of the European margin from Spain to Norway based on knowledge gained from an area the size of a few football fields?

Any future programme of work must couple the excitement of the exploration of the ocean margin with the management requirements of the European marine estate. It must utilise new technological advances in underwater vehicles (AUVs, manned submersibles and ROVs) and underwater remote sensing techniques to provide a synoptic view of natural variability of ocean margin ecosystems, in both space and time. Scientific research will also provide an accurate assessment of changes wrought by human impact.

It is known that there is a wide variety of habitats found along the ocean margin that change with increasing depth from the shelf break to the abyssal plain. However, there is a growing appreciation that certain communities are fluid with time. Changes in the dominance and diversity of species can occur on a variety of timescales from days to tens of years. Measuring the scale of human impacts will only be useful if we understand how significant external impacts relate to natural system variability.

Natural environmental variation

Along slope

Benthic communities show strong variations between sites situated at a similar depth along the continental margin, but very little is known about the scales of spatial variability and the factors that influence them (see review in Gage and Tyler 1991). Benthic communities also show changes with time. Temporal variation in community structure is difficult to observe because of the remoteness of the deep-sea environment, but is critical in assessing the real impacts of human activity.

Along the European continental margin differences in abundances, biomass and community structure have been observed among the relatively few sites investigated to date. These differences can be partly explained by differences in the physical environment such as differences in slope steepness, flow velocities, sediment structure, organic matter input, distance from land etc. The largest variations are found on the upper part of the slope, the area most impacted by human activities. Some of the apparent spatial variation may actually represent communities at different successional stages. Such a scenario would suggest that benthic community structure is deterministic and that stochastic factors, such as variations in recruitment or the occurrence of events such as benthic storms, are less important.

To understand along-slope natural variability, a large number of contrasting sites along the European continental margins need to be investigated over timescales that identify natural environmental change. At all sites, the physical parameters that might influence benthic community structure should be measured using standardised methods in relation to differences in abundance, biomass, size classes and community structure, including species and genetic diversity, feeding types and life-history characteristics. The information gathered would be used to determine, for instance, whether the abundance of species along the margin can be predicted from local environment and depth or whether there is a strong stochastic element in continental slope communities. To learn what factors naturally influence the characteristics of the benthic community, multivariate analyses of controlling factors and elements of community structure will be applied across all size classes of the benthic community. It will also help to determine the importance and spatial variability of rare species (e.g. Grassle and Maciolek 1992). Deep-sea communities are very diverse but many species exist in low numbers. The apparent contribution rare species make to species diversity and ecosystem function may be very important especially when communities are considered over a number of years. Resulting hypotheses regarding relationships between community structure and environmental variables need to be tested by means of *in situ* experiments.

Vertical patterns of spatial variation

Continental margins are regions where strong differences in depths occur over relatively short distances. With increasing depth a number of environmental factors change markedly, with a corresponding impact on the benthic community. General patterns include a decrease in abundance and biomass with increasing depth, a change that affects larger size classes more than smaller ones (Gage and Tyler 1991). This pattern is generally ascribed to a decrease in food supply, with the largest size classes suffering more from food depletion. Anomalies from this general pattern can

be found and related to special environmental conditions, such as high flow velocities. Together with density and biomass, community structure changes with depth. Species diversity in the macrofauna may show a parabolic pattern, with the highest diversity at mid-slope depths (e.g. Paterson and Lambshead 1995). In places (e.g. the Goban Spur), species richness (number of species per unit area) shows a parabolic pattern with increasing depth but there is an increase in species diversity (number of species per number of individuals sampled) with depth (e.g. Flach and de Bruin 1999). We know very little about the factors that induce these diversity patterns. What are the factors driving speciation and which are important for the maintenance of high diversity? As the pattern can be different for different taxonomic groups, are life-history strategies important as well?

The dominant feeding types may change with increasing depth from filter-feeding and surface feeding species on the upper slope to surface and subsurface deposit feeders on the lower slope (Flach et al. 1998). To explain this change we need to improve our understanding of the processes determining the flux of organic matter to the seabed, including seasonal and interannual variations. In addition, more information is needed on the way in which deep-sea species feed. What are the adaptations to the special feeding conditions occurring at different depths?

Seasonal and interannual variability in deep-sea benthos

The deep-sea, with the exception of hydrothermal vents and cold seeps, is a heterotrophic environment relying on the input of surface-derived primary production as its organic energy source. Seasonal variations of primary production in surface waters leads to the seasonal deposition on the seabed of organic matter rich in labile material (Billett et al. 1983; Rice et al. 1994). Variations in the quantity and quality of this phytodetritus occur both within and between years. On the continental margins, the input of organic matter to the benthic system can also be related to down-slope transport often through canyon systems. However, the relative importance of phytodetrital input direct from the sea surface and down-slope transport processes are still unknown.

The response of the different benthic groups to the arrival of phytodetritus depends on their feeding strategies and their ability to use specific com-pounds within the organic flux for growth and reproduction. Rapid increases in the biomass of bacteria and flagellates occur on timescales of hours to days in the surface sediments and in the phytodetritus (Yayanos 1998). Some foraminifera rapidly colonise the organic particles. There is also evidence of differential abundance and distribution of foraminifera and nematodes in relation to phytodetrital inputs (Vanreusel et al. 1995; Smart and Gooday 1997). The macro- and megafauna also show higher biomass and abundance at abyssal sites affected by seasonal deposition of surface-derived organic matter compared to the populations at sites that do not receive such inputs (Rice et al. 1994).

Seasonal, pulsed food inputs into the deep-sea raise many important scientific questions. Generally, organic particles are extensively decomposed before they reach the seabed, leaving a large proportion of less labile compounds. It is clear that members of the benthic community, especially microorganisms, can break less labile organic compounds down into simpler molecules, making them available to other members of the benthic community (see review in Turley 2000). Discovering the molecular mechanisms of these processes are important for understanding the community responses to the arrival of phytodetritus and may reveal important new enzyme systems for biotechnological applications.

Critical questions are: How does the quality of the organic matter reaching the seabed influence the benthic community? How do members of the benthic community survive after the phytodetritus has been used up? There is evidence that bacteria may switch to alternative metabolic pathways or may become more efficient at scavenging nutrients. How do members of the macrofauna and megafauna survive for months when no phytodetritus is available?

Seasonal and interannual variability and life history strategies

To understand the forces driving an ecosystem, it is imperative to understand the life history strategies of the species composing the communities. The reproductive patterns of a species integrate information on their phylogeny (the gametogenic pathways are constrained by the ancestry of the species) and on environmental factors such as food availability and habitat stability. A good knowledge of the life history of a species (age at first maturity, gametogenic pattern, egg size, size- and age-specific fecundity, larval development and dispersal potential, and the population dynamics including post larval and adult growth and survivorship) and the relationships between these parameters is necessary to understand the functioning of the populations within an ecosystem, and the functioning of the ecosystem itself.

Although the main reproductive pattern of deep-sea megafauna is the quasi-continuous production of large eggs that develop into lecithotrophic larvae, there is evidence now for seasonal reproduction coupled to the arrival of phytodetritus in several species of echinoderms (echinoids, ophiuroids and asteroids), sponges, cnidarians and bivalves (e.g. Tyler et al. 1982; Witte 1996; Bronsdon et al. 1993). These species produce a large number of small eggs that develop into planktotrophic larvae. The phytodetritus deposition in the deep-sea is used to fuel gametogenesis and is thought to feed the larvae in the water column.

Evidence from studies in shallow water has shown that larval survival may be low and that the contribution of an individual reproducing adult to the next generation is highly variable (i.e. small effective population size; Hedgecock 1994). Differential survival of larvae may reflect the nutritional status of the breeding adult and/or the timing of larval release (i.e. whether it coincides with suitable environmental conditions such as high food availability, temperature etc.). This has obvious implications for the deep-sea. The timing, quantity and quality of phytodetritus arriving at the seafloor is likely to have a strong influence on the reproductive success of populations through their impact on gametogenesis and larval survival.

The deep-sea is highly food-limited and it is likely that such factors are important in understanding the maintenance of deep-sea populations. The development of new technologies for *in situ* observation and experimentation, such as ROVs, and advances in molecular biology, are likely to provide a better understanding of the feeding, growth and reproduction of the deep-sea fauna and the relationship of these to temporal variation of environmental factors. Such an understanding is crucial in order to assess the impacts of large-scale environmental changes on benthic populations. It is also important when considering the impacts of smaller-scale anthropogenic activities that may disrupt the reproductive cycle of species by sub-lethal effects. Such impacts are well known in shallow-water environments, especially through the actions of compounds such as TBT or those that mimic hormones such as oestrogen.

Decadal change

There is a growing appreciation of decadal-scale changes in primary production in response to environmental forcing. Large-scale changes have occurred, for instance, in relation to the El Niño-Southern Oscillation (ENSO), and the North Atlantic Oscillation (NAO). What is not known is how such decadal change might affect seabed communities that are dependent on organic flux from the sea surface. Recent evidence from abyssal plain ecosystems indicated that dominance within benthic communities, and presumably their functioning can vary significantly on the scale of several years (Billett et al. in press). Similar variations should be expected for ocean margin systems, but remain to be documented. As such, natural changes are critical in the design of environmental impact monitoring programmes. Studies are needed on representative benthic communities of the ocean margin over a timescale that allows the effects of decadal variation to be identified and modelled.

Measuring long-term climatic changes

Long-term observational time series of benthic communities along the ocean margins do not exist.

However, there are a few examples of where successive sampling of deep-sea sites has produced data running over several years (e.g. Dunstaffnage Marine Laboratory, Permanent Station, Rockall Trough). Generally, the only way at present to assess the natural variability and the effects of climate change on such communities is to use the sedimentary record. This record is affected by bioturbation almost everywhere, leaving only a smoothed record of any changes. However, the sedimentary record provides the best archive for study of the responses of communities to a changing environment. This will always be true for time scales of glacial-interglacial variability up to decadal changes, but is also true, for decadal to interannual scales, using finely laminated sediments deposited in dysoxic settings.

Traditionally, when using the sedimentary record, any assessment of former benthic populations has to rely on those groups leaving hard parts in the sediment (i.e. shells, skeletons), which are preserved over geological time scales. The most important groups in the analysis of such fossil assemblages have been benthic foraminifera, molluscs and corals. Their taxonomic compositions, for example, can provide clues to temperature and oxygen concentrations at the seabed when such organisms were deposited (e.g. Cronin and Raymo 1997). A number of geochemical proxy data can also be extracted from these fossils as well as from the sediments, and these can be used to describe former environmental settings. The most important are stable isotopes that provide clues about water temperatures, oxygen utilization (and "age" of deep waters), and the quality of organic matter. In addition, sedimentological proxies (e.g. grain size distributions) provide information on other physical conditions (e.g. current strength at the bottom). Significant progress in applying molecular techniques to such records may open the possibility of also gathering information about animals that leave no obvious remains once they have died. On shorter time scales the extraction of trace elements and stable isotopes from the hard parts of long-living corals, molluscs, sclerosponges and similar biological recorders can provide continuous data on time scales of 30 to 60 years. However, the study of subfossil occurrences of such skeletons can provide information over longer time-spans, possibly analogous to tree-ring work on the land.

The sedimentary record also stores information concerning the surface ocean system, which probably experiences much greater environmental change on short time scales than the more stable deep ocean. Any changes observed at the ocean margin floor are most likely initiated by changes in the food supply derived from the surface ocean. Thus, to study the impact of climate change on benthic communities one needs to reconstruct both surface and deep-water environments.

Climatic changes influence the distributions and movements of water masses. Current patterns suggest that major shifts in currents in near-bottom layers are followed by changes in faunal composition, species abundance and distribution. These kinds of changes are most obvious at boundaries of ecological regimes and will be detected there first. Thus in the outermost branches of the North Atlantic Current, analysis of shifts in the distribution of a number of selected benthic species have been detected at the scale of decades (1878-1931, 1949-1959, 1984), following shifts in the boundary between Boreal and Arctic water masses (Blacker 1957, 1965; Dyer et al. 1984). Such boundary areas may act as useful "early warning systems" of environmental change on the deep-sea and may also provide useful models of the impact of such perturbations on slope communities. Three target areas for future investigations of this kind are available in European waters. One is the outer shelf of Northern Norway, another is the margin of the Kola Peninsula and a third is the continental margin from Bjørnoya to West Svalbard.

The impacts of long-term climatic changes

Margins are important areas in the global carbon cycle because they are close to regions of high production. In Europe the carbon storage function of the margin has been studied in a few areas along the Atlantic and in the Mediterranean, but there is, as yet, no quantitative understanding of the role of margins in the carbon cycle. The main consequences of global climate change on ocean margin systems are in the changes of quality and quantity of productivity in the pelagic system caused by changes in mixing and nutrient supply in the upwelling water, from where they are propagated to the benthic system. Additional complications are induced by the deep hydrodynamic setting, affecting the temperature of bottom waters and their current characteristics, which in turn will change species distributions, sedimentation and burial efficiency of organic matter. Warming of the surface waters of the open ocean and warming of the entire shelf water mass will change pelagic food webs in terms of species and productivity, as well as the level of activity in the microbial loop. This in turn will change the amount and nature of the organic matter exported to the margin sediments.

Changes in the deep thermohaline circulation and the Mediterranean outflow can affect the current system along the European margin. The existence of a strong along-slope current is crucial for the distribution of many benthic communities and therefore benthic metabolism. Changes in the position or the strength of that current may have dramatic effects on productivity, benthic-pelagic coupling and burial of organic matter. A significant weakening of the current may enhance the development of an oxygen minimum and denitrification maximum zone on the slope. In addition, stratified shelves will become more stable and mixed shelves will become warmer. As a result the density gradient between the deep-sea and the shelf will increase. This may result in less upwelling and mixing, increased seasonality in productivity, increased internal wave-action and consequent fragmentation (localization) of productivity. Changes in the frequency and intensity of storms are not likely to produce changes on the continental slope itself, but increased resuspension of organic matter and sediments on shelves may change the export of organic matter to the continental slope system.

Changes on land (land use and vegetation cover) will be regionally different but effects on the oceans may occur through changes in wind action and supply of sediments and dissolved matter through rivers. Examples of possible impacts are a postulated partial greening of the Sahara and lower rainfall in the circum-Mediterranean region that may potentially change primary production, as a result of changes in influx of bio-nitrogen and iron to the Atlantic or of phosphorus to the Mediterranean. Coastal phenomena are likely to have an impact on margins where the shelf is narrow. Changes at the coastlines (e.g. from increased frequency of storms and effects on near-shore communities such as kelp, mangrove forests and on coastal morphology) may have ramifications in the food webs offshore.

Direct human impacts

This section addresses human affects on the continental margin at a more immediate scale than those from climate change, which is dealt with above under Temporal Variation. For the most part the impacts listed below are of recent origin and have accompanied an extension of commercial exploitation from coastal waters into the deep water of the continental margin.

Fishing

Unsustainable pressures on inshore fish stocks, caused by increased fishing effort and improved fish-location technology, have severely depleted many traditional fisheries in European waters. New fisheries have been developed by multinational fishing fleets in the deep water along the margin that target non-quota unregulated stocks. Stock size, recruitment, growth rates and migration patterns are partially understood for some exploited deep-water fish populations. Elsewhere in the world, many deep-water fisheries have rapidly collapsed following exploitation (e.g. the orange roughy fishery off New Zealand). Along the European margin, stocks of deep-water species, such as roundnose grenadier, are showing signs of depletion that indicate the urgent need for fisheries management (Koslow et al. 2000). Research has shown that deep-water species often share a number of features including, slow growth, late maturity, high longevity, intermittent spawning and the habit of forming aggregations that are easily targeted by fishing vessels (Rogers 1994). Some of these issues are now actively being investigated by DGXIV FAIR projects, but the lack of knowledge is still an important factor in the

management of deep-sea species. It is even more striking that virtually nothing is known as yet about the ecosystem impacts of the new deep-water fisheries. In fact, the successful expansion of fishing effort into the deep-water ecosystem challenges our present view of deep-ocean productivity as an essentially food-limited environment, where top-down predation plays a subordinate role.

The impact can be categorized into two fairly distinct areas:

(i) Food web impacts: intense fisheries, by removing top predators, may change patterns in food availability to the rest of the ecosystem; whereas the large amount of discarded biomass (up to 50% of trawled biomass in the mainly French roundnose grenadier fishery off the northern British Isles) may radically alter patterns in packaged supply of food falls.

(ii) The physical impact of large trawls towed by powerful trawlers over the seabed results in the crushing of delicate epifauna, such as glass sponges and xenophyophores, and the severe damage of coral ecosystems. Furthermore, it creates a diffuse, smothering impact from the settling of clouds of disturbed sediment over a wider area. In addition, the effects of repeated ploughing of superficial sediment, on natural biogeochemical fluxes are unknown, not to mention damage to submarine cables and *in situ* scientific hardware and moorings. For example, a recent cable break off Australia caused by deep-water trawling, caused massive internet communications breakdown throughout Australia (BBC World Service, October 21st, 2000).

Research on the biology of deep-water fish species targeted by commercial fisheries is urgently required so that this resource can be managed in a sustainable fashion. The conclusion of such research may be that deep-sea populations can only be exploited at a very low level. This strategy should be preferred to the "gold-mine fishing" practiced by some nations. Investigation of wider ecosystem level impacts of deep-sea fishing is urgently required. There is evidence already of wide spread damage to some habitats from trawl impact. Removal of deep-sea fish species by trawling may also have an impact on their predators, such as toothed whales, though at present little is known about trophic interactions in communities associated with the continental slope (Koslow 1997).

Oil/gas and other minerals

The extension of oil prospecting beyond the continental shelf has led to substantial investment and innovative engineering. Along the European Margin productive wells occur in a number of deep-sea areas including the Faeroe-Shetland Channel, Northern North Sea and off the coast of Norway. Exploration is at present taking place on the continental margin to the west of the United Kingdom and Ireland. Deep-water hydrocarbon exploration and production is also taking place elsewhere in the world (e.g. Campos Basin – Brazil, off West Africa).

A general lack of knowledge complicates impact assessment, but some progress has been made in understanding of the main influences of offshore oil production on benthic communities. Large-scale environmental surveys have been conducted in a scientifically rigorous way on the European Atlantic Margin. The aim is to minimize long-term impacts. The chief uncertainties in this management process derive from the poorly known dynamics of the deep-water ecosystem, particularly the species-rich habitat associated with soft sediments. These sensitivities (some of which relate to chemical communication) remain largely speculative, and methods of assaying sublethal effects of contaminants at the species level will be dependent on new technology for experimentally maintaining deep-sea biota in pressurized aquaria. They will also rely on the applications of new molecular methods for assessment of gene expression (e.g. microarrays) to investigate the responses of organisms to contaminants or pollutants at the molecular level. Such methodologies will enable biomonitoring of organisms *in situ* on the slope when used in conjunction with direct intervention technologies such as ROVs. Until such methodologies are developed the best means to detect impacts remains, as on the continental shelf, the analysis of

the species-related structure of the benthic community (e.g. Olsgard and Gray 1995). New studies are needed to test for species redundancy in order to see whether time and cost savings are possible by addressing community status at higher taxonomic levels. The importance of monitoring of small size classes, such as meio- and microfauna, in environmental impact assessment also needs to be understood. Larger size classes, including non-motile species, such as corals, xenophyophores and glass sponges, provide a ready focus for study and may be easier to monitor from seafloor imaging from towed camera systems, AUV or ROVs. Such (apparent) dominant species form visually spectacular communities that also readily acquire a growing public profile.

Substantial impacts on the seabed community at the continental margin may materialize from mining clathrates (gas hydrates) in deep water. Huge natural deposits of gas hydrates are thought to exist along many continental margins, especially in the Arctic. It is essential that informed scientific input is involved in planning the modes of any exploitation of these resources, both because of implications for atmospheric pollution (methane is a major greenhouse gas) and for the health of marine ecosystems. A treaty-based regulatory framework, when and if it becomes necessary, must be based on science from the continental margin environment, rather than on extrapolation from better known areas in shallow water.

Mining of metals from the deep-sea has been discussed for many years. The possibilities of gathering manganese nodules, or mining manganese crusts or metalliferous sediments from hydrothermal settings have now become technologically possible. Some of these sources of metals exist on the European Margin (e.g. manganese crusts on Marsili Seamount in the Tyrrhennian Sea) and may play host to specialised communities of benthic organisms.

Pollution

Although not yet realized along the European continental margin, pressures for using the deep-water environment beyond the shelf edge as repositories for bulk waste, both inert and low toxicity, will mount as landfill areas become exhausted. Also, in the context of the Kyoto Protocol there are several exploratory initiatives regarding emplacement of frozen CO_2 (as gas hydrate) in deep water (Ormerod et al. 1993). However, at present the chief ongoing concern is from an ever-rising burden of exotic chemical species of anthropogenic origin found in marine biota, especially those subject to biological amplification along food chains. The role of deep-sea benthic biota along the continental margin in the re-cycling and mobilisation of such chemical species remains unclear. This issue is closely linked to export fluxes of organic carbon over the continental shelf edge, the existence of a carbon 'depocentre' on the continental slope, and possibly from enhanced particle flux from shelf-edge blooms, such as coccolithophores. The re-distribution of contaminants along the continental slope by means of slope currents, and the role of physical forcing, such as that from internal wave energy and from downward propagated eddy kinetic energy creating periodic benthic storms needs to be investigated in relation to dispersal of both dissolved and solid-phase contaminants, particularly those associated with living biomass. Again the sublethal effects of such chemicals on the slope fauna should be investigated using an experimental approach based on the use of pressurised aquaria, molecular biology and *in situ* studies.

Other impacts

Other anthropogenic impacts include military activities that are largely unknown to the academic science community. Sunken shipping vessels, both from accidental and deliberate action may result in the release of contaminants. Although deliberate disposal into the deeper waters of continental margin of large man-made objects, such as oil production platforms, may not constitute an ongoing threat (because of the London Dumping Convention), the past legacy from sunken ships provides a large range of uncertainties. The deliberate scuttling of several ships carrying munitions and chemical and biological weapons off Scotland after

WW2 -provides an example of such risk. These wrecks lie in an area whose seabed has been intensively trawled for several years.

Marine protected areas for research and conservation

The high public profile of cold-water coral ecosystems, in part through activities by NGOs, such as Greenpeace, provide an opportunity to secure extension of accepted concepts for conservation on land and in shallow water to deep-water ecosystems. Although certain anthropogenic impacts are now subject to a regulatory framework that may be sufficient to assess impacts from reports obtained, this is not the case for the ongoing deep-water trawling for non-quota fish stocks. At present, no large area can be regarded as sufficiently free of fishing effort so that impact can be studied by comparison between affected and non-affected areas. To provide this means of assessment, areas closed to fishing must be declared and adequately policed. Choice of areas should not be left to the fishing industry, and should be decided on the basis of (1) the highest risk to the natural ecosystem and (2) our present knowledge. Although pristine areas are desirable, such locations may not now be available. Areas closed to fishing should include both near-pristine conditions and areas already experiencing high levels of fishing impact, so that at least *post hoc* recovery may be monitored.

Special habitats

Corals

Coral reefs are something we usually associate with warm, tropical waters, but not with the cold, deep, dark waters of the North Atlantic. It is now known that cold-water coral species also produce reefs that rival their tropical counterparts in terms of their associated species richness and diversity (Rogers 1999). Increasing commercial operations in deep waters, and the use of advanced offshore technology have slowly revealed the true extent of Europe's hidden coral ecosystems. In the

Northeast Atlantic, the geographic distribution of deeper water coral ecosystems can be traced from the continental slopes and banks off the Iberian Peninsula as far north as the Scandinavian Shelf. Apart from the European margins, deep-water coral build-ups are known from the waters off New Zealand and Australia and have recently been detected off western Africa and SE Brazil. Hence, it is evident that the deep-water coral reef province is globally much more widely distributed than the well-known shallow-water reef provinces in low latitudes.

Along the NW European Margins, extraordinary 10km-long chains of reef-building corals, with the key species *Lophelia pertusa* and *Madrepora oculata*, have been found in 300 to 500 m water depth on the deep Norwegian Shelf and upper margin (e.g. Freiwald et al. 1999). These coral structures have profoundly challenged conventional views. This same coral assemblage is also found associated with large unique seabed structures in the Porcupine Seabight and Rockall Trough, where they are so abundant that their skeletal remains have contributed to mound structures in 700 – 1200 m water depth. These high relief mounds have a dimension of 5 by 1 km and a height of up to more than 100m above the surrounding seafloor (Freiwald this volume). There is much debate on their genesis and internal structure, but many of the surveyed mounds show a high density of living and dead deep-water corals on the flanks. There are several lines of evidence that show the whole coral structure forms from a build-up consisting of coral growth periods that alternate with increased deposition of drift sediments. This suggests that they are potentially useful palaeo-environmental recorders of global change and might be used to detect both long-term and short-term variations in deep circulation. This potential, however, has not yet been exploited for the understanding of ocean margin processes. Reef habitat mapping through ACES (Atlantic Coral Ecosystems Study, FP 5 project) has clearly demonstrated the unexpectedly large biodiversity of the reef-dwelling community and the high degree of space partitioning within various reef habitats. However, visual observations have also demonstrated that human impact, resulting from

deep-water trawling, has already severely damaged or even wiped out many deep-water coral reefs along Europe's margin.

Canyons and gullies

The continental margin is dissected in many places by canyons, formed as erosional features in lithified sedimentary bed forms of Meso- and Cenozoic ages. During these periods there were repeated sea level low stands and rivers reached, or nearly reached, the shelf edge. On formerly glaciated shelves, such as the Norwegian Shelf, unlithified Cenozoic deposits were deposited through the activity of glaciers. Consequently, erosive gullies, rather than canyons, were carved out from the action of seasonally cascading water and slope failures. Canyons vary greatly in size and shape but some are very large, exceeding 320 km in length and 37 km in width.

Canyons are characterized by very special environmental conditions, often with high rates of accumulation of organic matter transported from the shelf to the deep-sea and high levels of secondary production (e.g. Vetter and Dayton 1998). This is facilitated by the funnelling of currents that are driven by internal waves and upwelling. These processes potentially produce high levels of physical instability. There have been relatively few studies of canyons in part, because of the instability of these systems and in part because the technology has not been available for in-depth studies in these topographically complex environments. It has been known for some time that the faunal communities within canyons are distinct from the communities on the adjacent continental shelf (e.g. Rowe 1971). Often the abundance and biomass of species within canyons can be extremely high (e.g. La Jolla Canyon, benthic amphipods and leptostracan Crustacea reach densities of more than 3 million individuals and a biomass of more than 1kg per m^2; Vetter 1994). The distribution of this fauna is extremely patchy and largely governed by substrate, sediment characteristics, local currents and whether a particular area is subject to erosion or deposition of material. In the Lydonia Canyon, for example, off the United States, some areas are dominated by high abundances of sessile filter feeding organisms such as pennatulids, tube-dwelling polychaetes, corals (gorgonians up to 15 feet high), sponges, hydroids and brittlestars. Other areas, characterised by sediment deposition, may have huge densities of brittlestars (Hecker 1989). High densities of fish have been observed in canyons and several commercially fished species are associated with canyons (e.g. hoki, *Macruronus novaezelandiae*).

No individual canyon is identical to another and this is reflected by differences in fauna between canyons even located along the same stretch of shelf (e.g. Hecker 1989). Along the European margin, observations within the Nazare Canyon have indicated the presence of high densities of small, shallow living, opportunistic species. In the Sebutal Canyon, on the other hand, high densities of all faunal groups have been observed. Here, especially fragile surface feeders are absent and deep burrowing species are more abundant. Benthic communities within canyons are clearly adapted to the special conditions within them.

There is, therefore, a requirement to investigate more canyons and to try to relate the benthic communities found in them to the special conditions occurring within individual canyons. These canyons should preferably be studied over sufficient timescales to study the effects of periodic slides and slumps on the benthos. Canyon systems may act as barriers for along-slope processes, such as dispersal of benthic organisms. It is therefore necessary that together with studies within canyons, isobathic stations located outside and on both sides of the canyons should also be sampled, in order to examine the differences between canyon and adjacent areas. If canyons can act as barriers to shelf populations, or are the location of special canyon-adapted species, which by definition would have to have a discontinuous distribution, then this would provide evidence that these habitats can play a role in speciation in the deep-sea.

Cold seeps and associated chemosynthetic communities

Seep biology and ecology is a new field of research that has followed the discovery of fluid emissions at the sea floor. Whereas detritus-based benthic communities are fed indirectly by seasonal photosynthetic production derived from the upper layer of the ocean, chemosynthetic communities are based on bacterial growth on redox gradients (chemosynthesis) *in situ*. Many more locations of cold seeps have been identified on active margins than on passive margins (Sibuet and Olu 1998; this volume). However, future exploration should favour new discovery of cold (i.e. methane) seeps on passive margins, linked to special geological settings and various structures, such as faults, mud volcanoes (formed by mud extrusion) and pockmarks (from fluid emission).

A major characteristic of seep communities is the remarkable visual impact of symbiont-containing metazoans, such as large bivalves and giant tube-worms. These live in dense clusters (biomass of several kg/m²) and cover areas up to 1,000,000 m² (i.e. in the passive margin, Gulf of Mexico continental slope).

Many fundamental questions still remain to be answered with regards to the biology of chemosynthetic communities. For example: how variable are the types of biological response by microorganisms, meiofauna, macrofauna and megafauna to fluid escape at the sea floor? How do biological systems adapt to fragmented and variable habitats in time? How is the diversity of symbiotic and non-symbiotic communities controlled? What are the consequences on the surrounding detritus-based communities of such rich but localised biological production? Do the organisms found at seeps harbour enzyme systems with biotechnological applications, such as hydrocarbon bioremediation?

Seamounts (banks), knolls and hills

Seamounts, knolls and hills are topographic highs with a limited extent at the summit. They are distinguished by elevation (seamounts > 1000m, knolls 200-1,000m, hills <200m). Whereas these features play host to important concentrations of biological resources, particularly in European waters (i.e. fisheries for black scabbard fish, sea breams, alfonsino, rockfish), they are poorly studied and understood (Rogers 1994). Seamounts represent highly dynamic systems in which current-topography interactions play a crucial role in shaping unusual benthic communities and exert a strong influence on the physical structure and biological communities of the overlying water column (Boehlert and Genin 1987). On geological time-scales, seamounts may act as important stepping stones for trans-oceanic dispersal of marine taxa and as centres for speciation through reproductive isolation of populations (e.g. Shaw et al. 1999).

To date, fisheries targeted on seamount-associated fin-fish have taken place with little knowledge of the biology of target species (Rogers 1994). As a result, such fisheries have been poorly managed (or not managed at all), resulting in the rapid (< 10 years) commercial extinction of many populations of valuable species (i.e. orange roughy in the southern hemisphere, pelagic armorhead in the Pacific). Very recently it has also become recognised that modern trawling methods have a dramatic impact on fragile communities of sessile marine invertebrates located on seamounts (Koslow and Gowlett-Jones 1998). These communities are formed by corals, sponges, bryozoans and calcareous algae, and have a high diversity of associated species. High levels of endemism are exhibited by species inhabiting individual seamounts or seamount provinces. Yet the benthic communities of seamounts are almost unstudied in European waters. Benthic surveys on these habitats that have been subject to trawl fisheries in other parts of the world (i.e. Tasman Seamounts, south of Australia) have shown that such communities maybe entirely removed by trawling and are unlikely ever to recover, as there are no refuges from impact.

There is an urgent requirement for research on all aspects of seamount biology to provide the data on which to base management practices for these rich habitats. Such research must address fundamental questions relating to bentho-pelagic interactions on seamounts, benthic species diversity, life-history and stock structure of commercially important species. Because of interactions between seamounts and the overlying water column, it is also likely that they may exert a significant influence on the surrounding ocean in terms of exporting nutrients and elevating productivity

(e.g. through vortex shedding and attendant downstream mixing). It is also likely that the impingement of diurnally migrating plankton populations on the summit and flanks of seamounts provide a food source for concentrations of demersal fish species at these localities. Such processes are also likely to occur on continental slopes and may explain why the biomass of demersal fish species on the slope maybe higher than expected from benthic biomass estimates (e.g. Mauchline and Gordon 1991). Various groups of seamounts exist within European waters that are suitable for future investigations, including those in the Eolian Arc, Tyrhennian Sea, the Horseshoe Seamounts to the west of Spain and Portugal and those around the Azores, many of which are locations of valuable fisheries.

Technology

Instrumentation

The definition of margins considered here includes the outer shelf seas down the continental slopes on to the adjacent abyssal plain, covering a depth range of typically 100 m to 4,500 m. The physiography of the margin creates narrowly banded physical, sedimentological and biological gradients that must be assessed by adequate sampling and observational and analytical techniques. Development of appropriate techniques represents a substantial challenge and will impact the growth of SMEs in this market sector.

Margin research requires vessels that are equipped with multibeam echo sounders, side-scan sonars and fibre optic cables (typically 6000 to 7000 m in length). Dynamic positioning and ultra-short baseline navigation is a prerequisite for placing instruments with precision on narrowly defined sites. All sampling gear, such as trawls, plankton nets, grabs and coring devices, should be equipped with a transponder for ultra-short baseline navigation and be controlled by online video. The latter requires the use of a fibre optic cable for the purpose of collecting high-quality images. Coring techniques have to be changed to enable the autoclaving of sampling gear and the retrieval of undecompressed sediment samples for transfer into pressure chambers. Such chambers should allow manipulation of seabed samples for experiments under *in situ* pressures.

Manned submersibles (if available) will play an important role in the investigation of special habitats such as hot vents, seeps, and rocky substrates. Conventional submersibles can be replaced by ROVs for many purposes. These have several advantages; for example, they can be used for extended missions on a variety of ships, they can place instrument packages on the sea floor and they can service long-term observatories. ROVs and submersibles will also be required for collection of high quality samples from the slope especially where small-scale studies are undertaken. This may require the further development of sampling equipment for such vehicles, especially for the collection of undecompressed samples for shipboard experimentation.

A major step in enlarging the spatial coverage of benthic investigations will be the use of near-bed driven AUVs and of mobile vehicles (rovers) for motoring about on the seafloor. AUVs should be capable of navigating about 2 m above the seafloor for continuous imaging and nepheloid layer measurements. Typical missions will involve running several cross-slope profiles in a lawnmower pattern. Rovers will have a more restricted radius (ca. 1 km) but will enable direct measurements on the seafloor (e.g. sediment profiling, electrodes, optrodes). Again the development of these vehicles, to extend operational capabilities, and of instrumentation will be a major requirement of future programmes of research on the continental slope.

Besides conventional ship expeditions, automated long-term observatories are needed to monitor processes and to sample *in situ* both the seabed and the water column. Bi-directional communication between such observatories and land-based laboratories will be essential. Stationary long-term observatories will be used at selected key sites for long-term measurements, particularly in observing seasonal change. Lander-type instrumentation maybe placed on specific features of the seafloor. Long-term platforms should carry sampling, sensing, imaging and *in situ* analytical systems. They may work either in a pre-pro-

grammed mode or, if autonomously operated, should be able to recognize environmental events and modify their observational routines in an appropriate fashion. There will then also be a requirement for new stable sensors. Special attention should be given to *in situ* analytical instrumentation for sediment and water properties, such as nutrient concentrations or chlorophyll abundance. Instrumented moorings with profiling capacity may also be used for the investigation of the water column.

The installation of long-term platforms on continental margins is constantly threatened by fisheries activities, especially otter trawling. Such platforms, therefore, will have to be designed to withstand this kind of attack. Permanent fibre optic cables can interlink a network of devices, deployed over a long time. Such a link has the advantage of constant power supply and data-transfer. Such a system will also allow efforts involving public education and outreach, which are viewed as an important and integrated part of environmental research.

Molecular biology

Despite the study of biodiversity over the last 20 years there are still many fundamental questions that need to be addressed. This is especially so for the marine environment where biodiversity is severely underestimated (Grassle 1991). The lack of information is caused partially by the expense and effort required for exploring the deep-sea seabed. However, it has also arisen because of difficulties in applying morphological criteria to the species concept for many groups of marine organisms (e.g. sponges, nematodes, nemerteans, flat-worms etc.). Even for animals with obvious morphological characteristics, such as exoskeletons, the occurrence of cryptic species, that are morphologically indistinguishable but are genetically isolated from each other, has proved to be a problem. Why cryptic species exist, and the extent of their occurrence, are interesting questions in themselves. Physical forces and grazing pressures can explain some convergence of form (Sournia 1988). Tilman et al. (1982) predicted that phenotypic diversity would decline in populations where there was a high cost for obtaining non-limiting nutrients. Similar requirements for morphological adaptations may result in the evolution of 'super species' that can exploit a wide variety of environmental conditions. These 'super species' maybe composed of many cryptic species.

The recognition of a species is highly controversial and many definitions (concepts) of a species exist (e.g. Cracraft 1989; Manhart and McCourt 1992; Gosling 1994). The biological species concept is the most widely used (species are reproductively isolated populations). However, it is often difficult to document especially in species where little is known about their life history and reproductive biology. The phylogenetic species concept represents monophyletic groupings and introduces an element of time into the species concept (Gosling 1994). Molecular data can be most easily incorporated into this species concept. Notably, species may be differentiated based on molecular divergence, but may not be reproductively isolated (syngens).

Adoption of the phylogenetic species concept means that rapid identification of environmental samples of animals from the slope maybe aided, in the future, by the application of molecular methods. For organisms that are particularly difficult to identify by morphology, because of the requirement of a specialist taxonomist, or because of time-consuming preparation of samples, such molecular approaches may revolutionise our understanding of the distributions and ecology of slope species. Such methods are already applied to microorganisms, through the application of PCR amplification of 16S rDNA, followed by separation of amplification products by denaturing gradient gel electrophoresis. 16S products may be sequenced subsequently and compared to DNA sequence databases to identify species. This approach may be used for identification of other metazoan taxa through the use of Molecular Operational Taxonomic Units (M-OTUs). The advent of microarray or DNA chip technology, however, raises possibilities of alternative means for even more rapid and accurate identification of taxa, and one that may be largely automated. Development of such technologies is crucial to extend investigations of the benthic slope fauna to include many groups that are currently not considered in investigations and which maybe crucial to ecosystem structure and function.

Spatial structure of genetic diversity is an important element of biodiversity below the species level. It is important to examine the interaction of species with environmental parameters that structure ecosystems. Until recently, it was assumed that marine organisms with high dispersal capaci-ties would show similar genotype frequencies over their entire range thus preventing establishment of genetically differentiated, geographically separated populations (Rogers this volume). Support for this convention has come mainly from phenotypic comparisons and isozyme variation studies. With the advent of nucleic acid methods, however, the view that spatial genetic structure is related directly to life history for all marine species has been challenged. Past history of populations, behaviour of larvae and a multitude of physical environmental factors have all been found to be powerful influences on spatial genetic structure. The concept of a single globally distributed species, very apparent in much of the deep-sea literature, is in great doubt. In addition, it has recently been found that temporal genetic changes in populations, caused by low effective population size, apparently can be greater than spatial change (e.g. Hedgecock 1994; Ruzzante et al. 1996). Temporal genetic variation within marine populations are the highest that have been observed in the natural world and may play a role in determining how local adaptation and speciation can occur in apparently homogenous populations. Further studies of processes driving spatial and temporal genetic variation will lead to a greater understanding of how biodiversity is related to climate change, ecosystem structure, dynamics and resilience.

Molecular biology is partially driven by the development of new technologies. At the present time developments are taking place at an astonishing rate, especially in the fields of genomics and microarray or DNA chip technologies. The availability of genome sequence data provides a new level of resolution for investigations by marine molecular ecologists. The next 10 years will see the application of such data to the study of specific interactions between genes and the environment.

Microarray technology means that expression of selected genes or multiple genes within the entire genome may be studied in experimental or *in situ* investigations. The relationships between benthic population dynamics in the deep-sea and external overriding physical and biological factors (i.e. seasonal changes in nutrient availability) will only be fully understood through such molecular approaches. As life evolved in the oceans, it is likely that an appreciation of sequence diversity at the level of genes and genomes, from marine taxa, will be significant in understanding the molecular evolution of all life.

Modelling

The deposition of organic matter on the seafloor triggers a set of responses leading to changes in benthic activity and biomass, as well as changes in the geochemical conditions of the sediment. At present the interactions between external forcing, and the biological responses and the biogeochemical processes in the sediment are incompletely understood. Most commonly, models of the sediment are based on the biogeochemical approach (so-called diagenetic models) in which the activity of infaunal animals is greatly simplified. More complex modelling approaches that include the combined description of biological and biogeochemical processes in the sediment will enhance our understanding of the mutual interaction between biological and geochemical processes. Such models will greatly benefit from time-series measurements of biological and geochemical processes and quantities.

The complexity of the deep-sea benthic food web may be explored by means of inverse modelling, where biomass distributions and rate measurements are used to constrain the assumed flows of energy between the different groups (Vézina and Platt 1988). However, such models are subject to problems arising from combining different species with very different biological characteristics into trophic functional groups.

Statistical models, including artificial neural network models, are tools that maybe used to handle multivariate data. These types of models are increasingly being applied to a number of real-world problems of considerable complexity. They are good pattern recognition engines and robust classifiers with the ability to generalise in making decisions about relationships between sets of input data with severe sampling problems.

New approaches to taxonomy and benthic fauna analysis

Databases provide access to large compilations of a rich variety of data in a manner convenient to users including non-specialists. Large interdisciplinary databases facilitate recognition of relationships between different types of data and provide a basis for extrapolations of observations to larger spatial scales. Data from several large-scale programs have been incorporated into certain existing databases (e.g. OMEX, Fishbase). Such databases should be identified whenever planning relevant observations so that additional data can be incorporated in a systematic fashion. Databases are also an invaluable tool for model-versus-data comparison or as a source for initial conditions and constraints required in modelling exercises. When coupled to a taxonomic key, databases may provide guidance to non-taxonomic researchers for future work.

Development of databases

There is a growing need for monitoring ocean margins in order to assess environmental changes in time and space. To understand these changes, we shall need to expand our knowledge of the present biological communities. Inevitably the task of monitoring the deep-sea margins will be confronted with the need for reliable indicators for environmental change. Such indicators include change in physical processes, life history parameters and community structure, e.g. the distribution of key species or functional groups.

Deep-sea ecological research is confronted with an enormous diversity of taxa and habitats. For larger groups (megafauna) the taxonomic groups are relatively well known, which allows reasonable estimates of biodiversity and community structure at different levels of inventory. One aspect, as yet poorly understood, is the genetic structure of populations that can only be explored by the combination of molecular data and morphological characterisation. For smaller benthic groups (macro- and meiofauna) information on the present taxonomic groups is widely scattered. It needs to be brought together in databases. With the use of the Internet and the development of E-databases, there is no reason to let information go unused. This information is precious and must be collected, maintained and made available for ongoing and future research.

Databases should include:
- morphological characterisation (e.g. pictures, diagnostic features) of taxa.
- molecular differentiation of populations.
- geographical distribution of taxa.
- ecological information on habitat, life history, trophic position, etc. of taxa.
- a facility for requesting specific information from participating experts.

The construction and use of similar databases will allow:
- estimation of biodiversity on local, regional and wider geographical scales (alpha and gamma).
- estimation of species turn-over between habitats and over environmental and geographical gradients (beta).
- will improve further identification with the implementation of queries (interactive electronic keys).
- will allow the selection of indicator groups for environmental changes.
- will allow assignment of key species for particular habitats.
- will contribute to assignment of MPA's and other recommendations for ocean margin management.

Standardization of techniques

In order to facilitate the integration of data over larger time and spatial scales for further and better understanding of large scale processes (like climate changes, human impacts, natural changes) there is a strong need for standardization of data collection. Recommendation (standardization) on data collection should be made within and between scientific projects (programmes). This information can be collected in databases.

Public awareness of scientific issues

Ocean margin research is a major challenge in terms of the scientific and technological endeavour. Therefore, co-ordinated measures to increase public awareness and the education of a wider public audience should be set up from the beginning of projects. Such measures will add weight to explaining the socio-economic impact of deep and offshore studies. These measures will consist of a consortium between PIs and media groups to generate and maintain dynamic websites with interactive discussion forums between scientists and the public (schools, environmental stakeholders, etc.). Special effort must be invested in the organisation of live reports during cruises with real-time video facilities. Permanent real-time video from selected deep-margin sites, such as a spectacular reef front in water of 800m depth will be a technological challenge in itself. In the future, 3D-visualisation will become an instructive tool to explain complex processes at great depths. Standard press releases and popular science articles will underpin such activities.

References

Billett DSM, Bett DJ, Rice AL, Thurston MH, Galéron J, Sibuet M, Wolff GA (in press) Long term changes in the megabenthos of the Porcupine Abyssal Plain (NE Atlantic). Prog Oceanogr
Billett DSM, Lampitt RS, Rice AL, Mantoura RFC (1983) Seasonal sedimentation of phytoplankton to the deep-sea benthos. Nature 302:520-522
Blacker RW (1957) Benthic animals as indicators of hydrographic conditions and climatic change in Svalbard waters. Fish Invest MAFF (ser 2) 20 (10): 49 pp
Blacker RW (1965) Recent changes in the benthos of the West Spitsbergen fishing grounds. Spec Publ Int Comm NW Atlantic Fish 6:791-794
Boehlert GW, Genin A (1987) A review of the effects of seamounts on biological processes. In: Keating BH, Fryer P, Batiza R, Boehlert GW (eds) Seamounts, Islands and Atolls. Geophysical Monograph 43, American Geophysical Union, Washington DC pp 319-334
Bronsdon SK, Tyler PA, Rice AL, Gage JD (1993)

Reproductive biology of two epizoic anemones from bathyal and abyssal depths in the NE Atlantic Ocean. J Mar Biol Ass UK 73:531-541

Cracraft J (1989) Speciation and its ontology, the empirical consequences of alternative species concepts for understanding patterns and process of differentiation. In: Otte D, Endler JA (eds) Speciation and its Consequences. Sinauer Associates, Sunderland MA pp 28-59

Cronin TM, Raymo ME (1997) Orbital forcing of deep-sea benthic species diversity. Nature 385:624-627

Dyer MF, Cranmer GJ, Fry PD, Fry WG (1984) The distribution of benthic hydrographic indicator species in Svalbard waters 1978-1981. J Mar Biol Ass UK 64:667-677

Flach E, de Bruin W (1999) Diversity patterns in macrobenthos across a continental slope in the NE Atlantic. J Sea Res 42:303-323

Flach E, Lavalaye M, de Stiger H, Thomsen L (1998) Feeding types of the benthic community and particulate transport across the slope of the NW European continental margin (Goban Spur). Prog Oceanogr 42:209-231

Freiwald A, Wilson JB, Henrich R (1999) Grounding Pleistocene icebergs shape Recent deep-water reefs. Sed Geol 125 (1-2): 1-8

Freiwald A (2002) Reef-forming cold-water corals. In: Wefer et al. (eds) Ocean Margin Systems. Springer, Berlin pp 365-385

Gage JD (1996) Why are there so many species in deep-sea sediments? J Exp Mar Biol Ecol 200:257-286

Gage JD, Tyler PA (1991) Deep-Sea Biology: A Natural History of Organisms at the Deep-Sea Floor. Cambridge University Press, Cambridge, UK 504 p

Gosling EM (1994) Speciation and species concepts in the marine environment. In: Beaumont AR (ed) Genetics and Evolution of Aquatic Organisms. Chapman & Hall, London pp 1-15

Grassle JF (1991) Deep-sea benthic biodiversity. Biosciences 41:464-469

Grassle JF, Maciolek NJ (1992) Deep-sea species richness: Regional and local diversity estimates from quantitative bottom samples. Am Nat 139:313-341

Hecker B (1989) Megafaunal populations in Lydonia canyon with notes on three other Atlantic canyons. Proceedings of the North Atlantic Submarine Canyons Workshop, Feb 7-9, 1989 Vol 11:63-66

Hedgecock D (1994) Does variance in reproductive success limit effective population sizes of marine organisms. In: Beaumont AR (ed) Genetics and Evo-lution of Aquatic Organisms. Chapman & Hall, London pp 122-134

Heezen BC, Hollister CD (1971) The Face of the Deep. Oxford University Press, New York 659 p

Koslow JA (1997) Seamounts and the ecology of deep-sea fisheries. Am Sci 85:168-176

Koslow JA, Boehlert GW, Gordon JDM, Haedrich RL, Lorance P, Parin N (2000) Continental slope and deep-sea fisheries: Implications for a fragile ecosystem. ICES J Mar Sci 57:548-557

Koslow JA, Gowlett-Jones K (1998) The seamount fauna off Southern Tasmania: Benthic communities, their conservation and impacts of trawling. Final Report to Environment Australia and the Fisheries Research Development Coorperation. Fisheries Research and Development Coorperation, Australia 104p

Manhart JR, McCourt RM (1992) Molecular data and the species concept in the algae. J Phycol 28:730-737

Mauchline J, Gordon JDM (1991) Oceanic pelagic prey of benthopelagic fish in the benthic boundary layer of a marginal oceanic region. Mar Ecol Prog Ser 74:109-115

Olsgard F, Gray JS (1995) A comprehensive analysis of the effects of offshore oil and gas exploration and production on the benthic communities of the Norwegian continental shelf. Mar Ecol Prog Ser 122:277-306

Ormerod WG, Webster IC, Audus H, Riemer PWF (1993) An overview of large scale CO_2 disposal options. In: Riemer PWF (ed) Proceedings of the International Energy Agency Carbon Dioxide Disposal Symposium, 1993. Energy Convers Manage 34 (9-11):833-840

Paterson GLJ, Lambshead PJD (1995) Bathymetric patterns of polychaete diversity in the Rockall Trough, northeast Atlantic. Deep-Sea Res 42:1199-1214

Rice AL, Thurston MH, Bett BJ (1994) The IOSDL program – introduction and photographic evidence for the presence and absence of a seasonal input of phytodetritus at contrasting abyssal sites in the northeastern Atlantic. Deep-Sea Res 41:1305-1320

Rogers AD (1994) The biology of seamounts. Adv Mar Biol 30:305-350

Rogers AD (1999) The biology of *Lophelia pertusa* (Linnaeus 1758) and other deep-water reef-forming corals and impacts from human activities. Int Rev Hydrobiol 84:315-406

Rogers AD (2002) Molecular ecology and evolution of slope species. In: Wefer et al. (eds) Ocean Margin Systems. Springer, Berlin pp 323-337

Rowe GT (1971) Observations on bottom currents and epibenthic populations in Hatteras Submarine Canyon. Deep-Sea Res 18:569-581

Ruzzante DE, Taggart CT and Cook D (1996) Spatial and temporal; variation in the genetic composition of a larval cod (*Gadus morhua*) aggregation: Cohort contribution and genetic stability. Can J Fish Aquat Sci 53:2695-2705

Shaw PW, Pierce GJ, Boyle PR (1999) Subtle population structuring within a highly vagile marine invertebrate, the veined squid *Loligo forbesi*, demonstrated with microsatellite DNA markers. Mol Ecol 8:407-417

Sibuet M, Olu K (1998) Biogeography, biodiversity and fluid dependence of deep-sea cold-seep communities at active and passive margins. Deep-Sea Res II 45:517-567

Sibuet M, Olu K (2002) Cold seep communities on continental margins. In: Wefer et al. (eds) Ocean Margin Systems. Springer, Berlin pp 235-251

Smart CW, Gooday AJ (1997) Recent benthic foraminifera in the abyssal northeast Atlantic Ocean: Relation to phytodetrital inputs. J Foram Res 27:85-92

Sournia A (1988) *Phaeocystis* (*Prymnesiophyceae*) – How many species? Nova Hedwigia 47:211-217

Tilman D, Kilham SS, Kilham P (1982) Phytoplankton community ecology – the role of limiting nutrients. Ann Rev Ecol Syst 13:349-372

Tyler PA, Grant A, Pain SL, Gage JD (1982) Is annual reproduction in deep-sea echinoderms a response to variability in their environment? Nature 300:747-750

Turley C (2000) Bacteria in the cold deep-sea benthic boundary layer and sediment-water interface of the NE Atlantic. FEMS Microbiol Ecol 33:89-99

Vanreusel A, Vincx M, Bett BJ, Rice AL (1995) Nematode biomass spectra at 2 abyssal sites in the NE Atlantic with a contrasting food supply. Int Rev Ges Hydrobiol 80:287-296

Vetter EW (1994) Hotspots of benthic production. Nature 372:47

Vetter EW, Dayton PK (1998) Macrofaunal communities within and adjacent to a detritus-rich submarine canyon system. Deep-Sea Res II 45:25-54

Vézina AF, Platt T (1988) Food web dynamics in the ocean. 1. Best-estimates of flow networks using inverse methods. Mar Ecol Prog Ser 42:269-287

Witte U (1996) Seasonal reproduction in deep-sea sponges triggered by vertical particle flux? Mar Biol 124:571-581

Yayanos AA (1998) Empirical and theoretical aspects of life at high pressure in the deep-sea. In: Horikoshi K, Grant WD (eds) Extremophiles: Microbial Life in Extreme Environments. J Wiley & Sons, Chichester pp 47-92

Processes driven by the Small Sized Organisms at the Water-Sediment Interface

K. Lochte[1*] and O. Pfannkuche[2]

[1]Institut für Meereskunde, Düsternbrooker Weg 20, 24105 Kiel, Germany
[2]GEOMAR, Forschungszentrum für Marine Geowissenschaften, Wischhofstraße 1-3, 24148 Kiel, Germany
*corresponding author (e-mail): klochte@ifm.uni-kiel.de

Abstract: The small sized organisms including prokaryotes (bacteria and archaea), protozoa and metazoan meiofauna (< 250 μm) are the driving forces for biogeochemical fluxes in surficial deep-sea sediments under oxic conditions. The relative proportion of small sized organisms increases along trophic gradients from eutrophy to oligotrophy or from the continental margin towards the mid oceanic deep-sea. They can consume up to 10% of freshly sedimented organic matter per day. The small sized fauna consumes and respires the largest part of organic matter, while macrofauna is instrumental in incorporating fresh detritus into the sediment, structuring the environment and thus facilitating microbial processes. Small organisms, in particular prokaryotes, can adapt to amount and quality of organic matter input. Under nutrient starvation probably a large proportion of the prokaryotic community is dormant and is reactivated during sedimentation events. On time scales of 7-10 days (metabolism) to 2-3 weeks (biomass increase) they can react to pulses of deposition of organic material. However, the history of food supply influences the speed of adaptation and effectiveness of growth. At stations close to continental margins estimates of organic matter input from sediment traps largely disagree with measurements of benthic respiration, carbon turnover or estimates obtained from geochemical modelling. This discrepancy is much smaller at mid-oceanic stations. Lateral inputs from productive shelf seas into the deep-sea are suspected to cause this discrepancy.

Introduction

In this contribution, we describe the role of small sized organisms in the recycling of organic matter at the deep-sea floor. These organisms react rapidly to sedimentation of detritus and they are the main consumers of organic matter reaching the sea floor. The organisms considered here include prokaryotes (bacteria and archaea), protozoa and metazoan meiofauna (< 250 μm). In many cases the analytical methods do not allow to distinguish between these groups. Biochemical determinations of bulk parameters typically include the whole size spectrum from prokaryotes to meiofauna. Similarly, common processes such as community respiration include all organisms in a sample, while more specific reactions, e.g. sulfate reduction, may be attributed more precisely. Therefore, the small sized biota has to be treated in many cases

as a functional unit. Whenever it is possible to distinguish prokaryotes from meiofauna (including protozoa and metazoa) this will be indicated. This analysis deals with the microbial processes at the water-sediment interface in the deep-sea in general and does not focus on continental margins in particular. Where applicable, references to the conditions at margins are made.

As long as oxygen is present in marine sediments, the processes mediated by small sized organisms are in the first place governed by the availability of biologically labile organic material. The deposition of particulate organic carbon (POC) at the sea floor is coupled to seasonal productivity cycles of the phytoplankton. In temperate, polar and monsoon regions the benthos experiences strongly pulsed deposition caused by the highly seasonal

From WEFER G, BILLETT D, HEBBELN D, JØRGENSEN BB, SCHLÜTER M, VAN WEERING T (eds), 2002, *Ocean Margin Systems.* Springer-Verlag Berlin Heidelberg, pp 405-418

export production in the surface water, while in the central oceanic gyres the sedimentation is relatively uniform and low. The magnitude of the POC export is determined by both, input of nutrients into the productive upper layer by winter mixing or upwelling and the balance between autotrophic and heterotrophic processes.

Individual sedimentation events featuring significantly higher sedimentation rates relative to background sedimentation are usually restricted to a short period of a few days only and lead to a dramatic increase in phytodetrital material at the sea floor. Phytoplankton pigments (especially chlorophyll a) are used as tracers for the input of relatively fresh plankton material. A mass sedimentation of phytodetritus observed in the NE-Atlantic in July/August 1986 showed a more than 10-fold increase in concentrations of chlorophyll a and 6- to 8-fold increase in phaeopigments in surface sediments (Fig. 1). In comparison, in August 1992 there was no distinct pulse of phytodetritus deposition and no rise in pigments (Pfannkuche et al. 1999). Fresh phytodetritus is readily utilized by benthic organisms. However, there are no field

observations so far which follow the whole cycle of biological responses after the deposition of organic material. The observations available are only spotlights illuminating certain stages of the reactions of benthic organisms to food pulses.

Continental margins present very special settings in respect to food supply. They act as transgression belts between neritic and oceanic provinces and are much more influenced by lateral transport processes than open ocean sites. Depending on the morphology of the slope the transport processes may differ strongly and hydrographical, geological and biogeochemical features show considerable variations. Deposition centres with high accumulation rates of fine grained, organic rich material are found in blocked-off canyons and other depressions on the slope and along the base of the continental rise, while areas stripped of organic rich particles by high current speeds in exposed locations may be in close vicinity. Particularly high material fluxes occur in canyons which are recognised as major conduits for transporting material from the shelves to the deep-sea, but the quantification of their role in the carbon

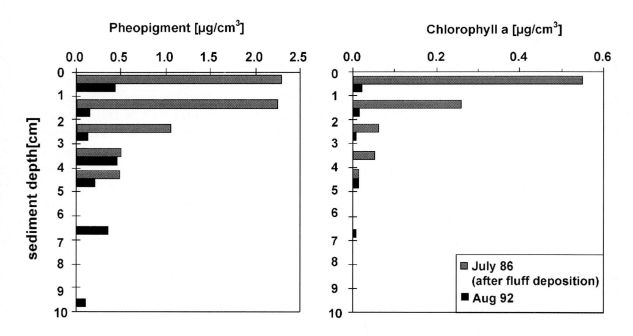

Fig. 1. Increase in chloroplastic pigment concentrations (Chlorophyll a, pheopigments) after a phytodetritus (=fluff) sedimentation pulse in the abyssal NE-Atlantic in 1986 compared to a year without distinct seasonal phytodetritus sedimentation in 1992 (modified after Pfannkuche et al. 1999).

budget is still insufficient and a challenge for future research. Jahnke and Jackson (1987) and Jahnke (1996) estimated that around 50% of all deep-sea benthic carbon consumption occur within 500 km of the continental margins. This highlights the importance of transport from the shelf towards the deep-sea plains. It is also an expression of the increased productivity along the continental margins. Major upwelling regions are found along continental margins and provide a large input of organic matter to the slope and adjacent deep-sea regions. This causes the development of extensive anoxic regions in relatively deep waters, which do not provide a suitable habitat for most eukaryotes. Prokaryotic organisms and anaerobic processes dominate such anoxic sediments. Since these regions differ in many aspects from the "normal" habitats found at continental margins and the deep-sea, they are not considered here. However, they are an important part of continental margin systems, and they are challenging for future research in respect to novel prokaryotic organisms and pathways.

Trophic gradients and the biomass of small sized organisms

Sea floor trophic gradients at continental margins are caused by decline in surface water productivity from the shelf towards the open ocean and by the progressive POC degradation with increasing water depth. Decreasing nutrient supplies downslope continental margins are reflected by decreasing biomasses of benthic organisms. In a global analysis of prokaryotic biomass and growth, Deming and Baross (1993) found a significant positive correlation between biomass and concentration of organic carbon in the sediments and vertical particulate organic carbon flux. The abundance of metazoan meiofauna was also found to decline significantly with increasing water depth and with decreasing phytoplankton pigments in the sediments at various North Atlantic continental margins (review Soltwedel 2000). The relationship between pigments and abundance of meiofauna demonstrates a general connection between food availability and benthic standing stocks. However, the slopes of the regressions between meiofauna

standing stock and water depth differ significantly in various climatic regions. These differences are reflecting the specific depositional regimes at the respective margins.

Studies at the Celtic continental margin (European Union OMEX-Programme) showed that at all investigated stations from 135 m to 4500 m the benthic biomass is dominated by the small sized biota (organisms typically < 250 µm) amounting to 90% on the shelf and 97-98% in the bathyal and abyssal parts (Pfannkuche and Soltwedel 1998). Downslope, the biomasses of all size groups decrease, but the gradient of decline of macrofauna (organisms > ca. 1 mm) is significantly steeper than that of small sized biota (Fig. 2). At the NW-African margin the ratio between meiofauna and macrofauna increased significantly between a station influenced by upwelling (eutrophic) and an oligotrophic station (Table 1, Galéron et al 2000). Thiel (1975) already postulated such a general tendency to smaller sized organisms with increasing water depth and oligotrophy.

A similar tendency is also obvious in a trophic gradient in the Arabian Sea when comparing the biomasses of macrofauna and prokaryotes. The ratio of prokaryotes to macrofauna is lower at stations receiving high POC input from monsoon

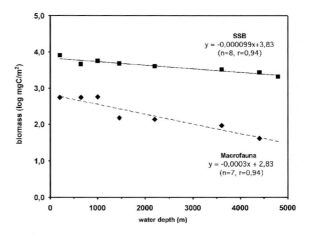

Fig. 2. Decline in macrofauna vs. small sized biota (SSB) biomass along the NW-European Margin, Goban Spur (modified after Pfannkuche and Soltwedel 1998).

Stations	(mgC/m²)	(mgC/m²)	ratio
	meiofauna	macrofauna	ratio meio./macro.
NW Africa (EUMELI):[1]			
eutrophic (1600-2100 m)	96.3	251.0	0.4
oligotrophic (4480-4640 m)	11.2	4.3	2.6
	prokaryotes	macrofauna	ratio prokar./macro.
Arabian Sea:[2]			
WAST (4030 m, eutrophic)	200	106	1.9
SAST (4410 m, oligotrophic)	100	21	4.8

[1](Galéron et al. 2000)

[2](Boetius et al. 2000, Witte 2000)

Table 1. Changes in the biomasses of meiofauna or prokaryotes relative to macrofauna at stations with different trophic status.

driven productivity in the western part (WAST) compared to the southern oligotrophic region (SAST, Table 2). This comparison also shows that prokaryotic biomass changes relatively little while metazoa are more affected by changes in food supply.

Since prokaryotic numbers are determined by direct microscopic counts, the activity level of the cells cannot be assessed and inactive or dead ones are included in the counted numbers. While prokaryotic biomass decreases by a factor of around 2 between the eutrophic and the oligotrophic station in the Arabian Sea (Table 2), remineralisation rates decrease by a factor of 3.3, enzyme activity by 3.2 and production rates by 13 (see also below). At the continental slope of the Laptev Sea, pro-

karyotic biomass decreases from the shelf (ca. 25 m) to the deep-sea (ca. 3500 m) by a factor of 2 to 3, but the activity of their enzymes (chitinase) decreases 14- to 35-fold (Boetius and Damm 1998). Hence in a trophic gradient, the activity of prokaryotes is reduced more strongly than their biomass. This indicates that in oligotrophic deep-sea regions a higher proportion of the prokaryotic community has a very low metabolic activity or is inactive compared to nutrient enriched sediments. Since prokaryotes have a much greater capacity to reduce their metabolism and to remain dormant than metazoa, they can maintain higher standing stocks at low food supply. However, such a prokaryotic community can be activated rapidly when adequate food substrates arrive.

Adaptation to food supply

Activation of metabolic activity by pulses of organic matter in the deep-sea has been studied in benthic prokaryotes. The key process in the breakdown of macromolecular organic material by these microorganisms is extracellular hydrolysis by enzymes. The turnover of some model substrates by extracellular enzymes is measured easily and, hence, provides a good opportunity to investigate the adaptation and regulation in response to food supply.

Within 10 days, the activity of enzymes involved in the degradation of β-polysaccharides and chitin increases when organic material is experimentally added to sediment samples from multiple corer hauls under deep-sea conditions (Boetius and Lochte 1996a). Meyer-Reil and Köster (1992) even found immediate stimulation of enyzme activity when sediments from the Norwegian Sea (1400 - 2000 m) were amended with organic matter. The increase is directly proportional to the amount of added specific substrate, e.g enrichment with chitin triggers an increase of the relevant hydrolytic enzyme chitobiase depending on the added quantity (Boetius and Lochte 1996a). Obviously, once a certain type of organic material becomes available it induces the production of new enzymes. Such regulation of the enzyme production is an important survival mechanism under starvation conditions, since the production of new enzymes demands additional energy from the cells.

The results of these experimental studies are useful for the interpretation of trophic conditions in deep-sea sediments. In sediments sampled on a transect down the continental slope in the Laptev Sea, eastern Arctic, enzymes degrading chitin and other polysaccharides decline with water depth (Fig. 3). This is positively correlated with the decline of available organic material and reflects the limitation of enzyme production under nutrient poor conditions (Boetius and Lochte 1996b; Boetius and Damm 1998). Similarly, in the Arabian Sea these enzymes show very high activities at station WAST with high primary productivity and sedimentation during monsoonal upwelling, but low activities at the oligotrophic station SAST (Fig. 4) (Boetius et al. 2000). In both cases, the numbers or biomasses of prokaryotes did not show similar

variations indicating that the changes in enzyme activity are independent from shifts in abundance. A mass deposition of 1 dead swimming crab m^{-2} at the deep-sea floor caused very high activities of chitin degrading enzymes in the surrounding sediment (Christiansen and Boetius 2000). Both, laboratory experiments and field studies, demonstrate that enzymes involved in degradation of chitin and other β-polysaccharide compounds respond directly and rapidly to inputs of the relevant organic material and are useful indicators of the nutritional status of the investigated site. Not all of the dissolved organic carbon (DOC) produced by the enzymatic hydrolysis is utilized directly by the benthic microorganisms, but may diffuse out of the sediment. Hence, changes in extracellular enzyme activity may impact DOC fluxes from the sediment. The DOC efflux can account for a relatively large fraction of the POC input and could vary according to the rates of extracellular hydrolysis mediated by microbial enzymes.

In contrast to the above discussed enzymes β-glucosidase and chitobiase, peptidases catalysing the hydrolysis of proteins are reacting very differently (Fig. 3 and 4). Their activity immediately dropped when free amino acids were added in experiments (Boetius and Lochte 1996a, b). This indicates that peptidases are strongly inhibited when free end products (amino acids or peptides) are present. In sediments, peptidase activity increases with water depth and with declining availability of organic material. At the slope of the Laptev Sea, peptidases were at 3500 m more active than at 25 m (Fig.3). This trend was inversely related to the concentration of dissolved free amino acids in the pore water. Vetter and Deming (1994) also found a negative relationship between supply of particulate organic matter and peptidase activity in deep-sea sediments of the Northeast Atlantic. Poremba and Hoppe (1995) observed increased peptidase activity at the lower continental slope of the Celtic Sea. In the Arabian Sea, peptidase activity remained high although the vertical flux of organic matter decreased from the productive region in the northwest (station WAST) towards the relatively oligotrophic region in the south (station SAST) (Fig. 4). These surprising results demonstrate that peptidase activity is one

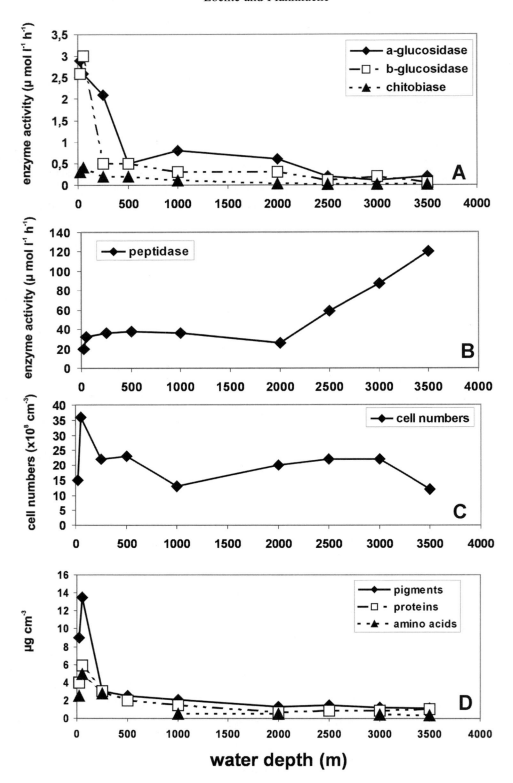

Fig. 3. Enzyme activity, numbers of prokaryotes and organic compounds in sediments from increasing water depth at the slope of the Laptev Sea. **a)** Enzyme activity of chitobiase, a- and b-glucosidase. **b)** Enzyme activity of peptidase. **c)** Numbers of prokaryotes. **d)** Concentration of organic components: pigments given in chloroplastic pigment equivalents, proteins given in globuline equivalents, aminoacids given in glycine equivalents (modified after Boetius and Lochte 1996b, Boetius and Damm 1998).

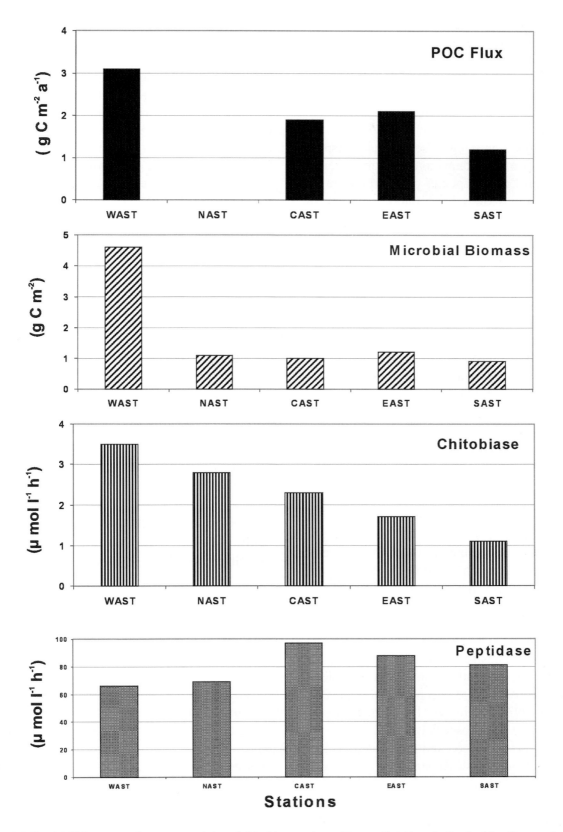

Fig. 4. Microbial biomass and enzyme activity of chitobiase and peptidase at 5 stations in the Arabian Sea receiving different inputs of particulate organic carbon (POC flux) (modified after Boetius et al. 2000).

of the few biochemical variables which increase in the deep-sea, and this is probably related to poor availability of nitrogen compounds. The protein degrading enzymes seem to be regulated differently in sedimentary prokaryotes compared to ones from the water column (Boetius and Lochte 1996a).

There are only few observations in respect to responses of meiofauna to deposition of organic material. Certain species of foraminifera were found in extremely high numbers in deposited aggregates of phytodetritus. They seem to be specialist feeders that develop opportunistically when phytodetritus becomes available, feeding either on the detrital material or on the associated microorganisms, while other species react negatively or not at all to the organic influx (Gooday and Lambshead 1989). Species of bacterivorous flagellates were found to multiply in phytodetritus (Turley et al. 1988), but in other meiofauna taxa, e.g. nematodes, such reactions were not observed (Gooday et al. 1996). In their review, Gooday and Turley (1990) state that the intermittant supply of phytodetritus favours primarily the temporal development of opportunistic species. Thus, apart from an increase in biomass, a shift in species composition of the small sized biota seems to be a response to pulsed deposition of organic matter compared to regions with more uniform sedimentation patterns.

Effects of sedimentation pulses on benthic biota have been observed to date only in a few regions of the deep-sea (Martin and Bender 1988; Graf 1989; Pfannkuche 1993; Soetaert et al. 1996; Smith and Druffel 1998; Pfannkuche and Lochte 2000). The results generally indicate that the small sized organisms respond most strongly to such pulses. They degrade 40-80% of a phytodetritus pulse deposited on the deep-sea sea floor (Pfannkuche 1993; Boetius et al. 2000). However, a competition for deposited organic material between prokaryotes, protozoa and macrofauna is likely. Macrofaunal deposit feeders were shown to take up and bury large amounts of the freshly deposited organic material in the sediments (Bett and Rice 1993). Such rapid transport of fresh organic material via bioturbation into deeper layers of abyssal sediments was detected by analysis of chloro-

phyll (Graf 1989) or natural radionuclides (Smith et al. 1986; Wheatcroft and Martin 1996; Levin et al. 1997). Although the small sized organisms are the major consumers of organic matter, the larger organisms have an important function by transporting the organic material into deeper layers of the sediment and structuring the environment (Witte 2000).

Unfortunately, the present approach has been limited to the determination of bulk parameters of benthic prokaryotes. Hence, at present little information about differences in species composition or succession of species is available. To some degree the same is true for meiofauna, with the exception of foraminifera which have received much more attention than other groups. With molecular biological methods it is now possible to identify spatial or temporal changes in species composition in relation to changing environmental conditions or to detect specific taxonomic or functional groups. It is an important task of future studies to link the occurrence of functional groups or species with specific biogeochemical processes. This new methodological development in ecological studies will provide major new insights into the regulation of prokaryotic communities and their function in the environment.

Temporal and quantitative links between sedimentation of organic matter and activity of small sized biota

The temporal and quantitative links between deposition of organic material and benthic response of natural communities are very difficult to determine, because the observations are usually too short and too infrequent to observe the whole cycle of reactions. During the North Atlantic Bloom Experiment (NABE) in 1989, the development of the spring phytoplankton bloom and its sedimentation were observed continuously at the BIOTRANS station (Lochte et al. 1993). The benthic reaction, however, was compiled from measurements performed over several years (Fig. 5). Assuming that this compilation represents an average course of development, a link between the surface water productivity and the sea floor in 4500 m depth can be demonstrated. Concen-

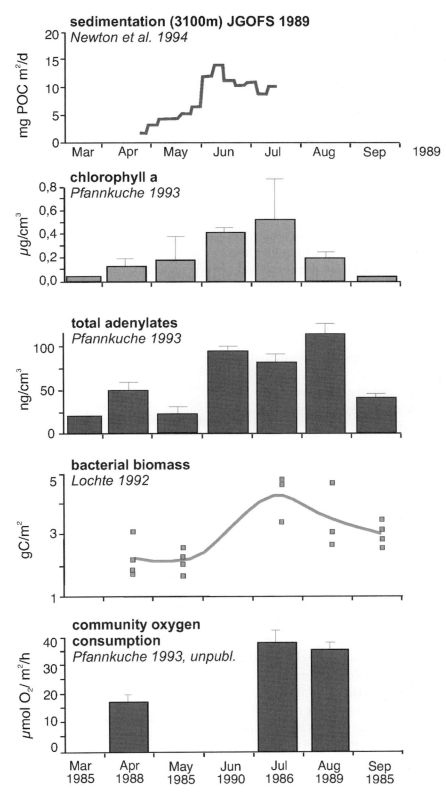

Fig. 5. Seasonal pattern of sedimentation of particulate organic carbon (POC) during the NABE study in 1989, and seasonal fluctuations in the sediment at the same station of chlorophyll *a*, total adenylates, prokaryotic biomass and oxygen consumption of the sediment community. The sediment data are compiled from several years. The sources of data are indicated with each variable.

trations of organic compounds, like phytoplankton pigments and adenylates, as well as respiration of the whole benthic community and prokaryotic biomass increase approximately 4 weeks after the sedimentation peak measured in deep moored traps and approximately 6 - 8 weeks after the spring phytoplankton bloom. In contrast, in a similar study in 1992 high storm frequency prevented the build up of a typical phytoplankton bloom and sedimentation patterns were changed. During that year no seasonal signal in the benthos could be detected (Pfannkuche et al. 1999). Due to such interannual variability of the surface water productivity cycle, assessments of benthic seasonal patterns have a high uncertainty.

In a recent *in situ* experiment, chitin was deposited on the surface of deep-sea sediments in chambers of a benthic lander at 4800 m depth in the Porcupine Abyssal Plain, North Atlantic. After recovery of the lander 9 days later, the activity of chitinases had doubled (Fig. 6). This was the first time such a rapid metabolic response to the deposition of organic matter by undisturbed microbial communities in deep-sea sediments could be shown directly. The results of the laboratory experiments discussed in the previous chapter where, thus, supported by the *in situ* studies. In particular the fast response was verified and an increase in enzyme activity can be expected in about 7-10 days after deposition of organic material at deep-sea temperature and pressure.

Biomass production could not be observed on such short time scales. From laboratory experiments, growth response of prokaryotes was found after 2 - 3 weeks; meiofauna growth is even more retarded. Biomass production seems possible only, when the input of organic material is high and the material is biologically labile (Boetius and Lochte 1996a). Amount and reactivity of the deposited organic matter determine the extent of metabolic reactions and at low vertical flux no biomass production can be expected. A temporally shifted and dampened reaction of the benthos to sedimentation of phytodetritus (Fig. 7) is therefore well supported.

The history of organic matter supply seems to influence microbial processes as well. At the relatively nutrient poor station BIOTRANS in the

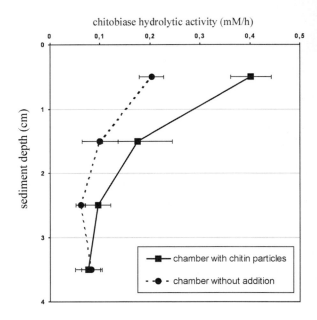

Fig. 6. Increase of chitobiase activity in sediments after 9 days in an *in situ* experiment carried out at station BENGAL (4800 m) in the Northeast Atlantic. Organic material (chitin and phytodetritus) was added and incubated in a benthic lander chamber (Boetius, unpublished).

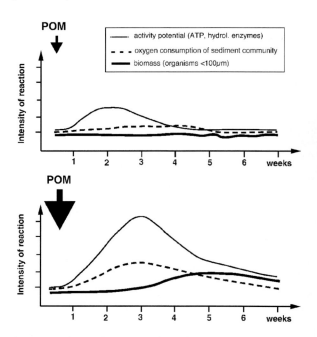

Fig. 7. Schematic curve of benthic reactions to sedimentation of particulate organic matter (POM). Reprinted from Deep-Sea Research, 46, Pfannkuche et al., Responses of deep-sea benthos to unusual sedimentation patterns in the North-East Atlantic in 1992, 573-596, 1999, with permission from Elsevier Science.

North Atlantic addition of organic matter resulted in stimulation of enzyme production after 10 days, while at the nutrient rich station in the western Arabian Sea no new enzymes were produced. Instead the added organic matter was directly channelled into new biomass production of pro-karyotes since an elevated level of enzymes was already present as a result of relatively high or frequent organic matter supply (Boetius et al. 2000). This study estimated that up to 10% of freshly sedimented detritus was consumed per day; the prokaryotes turned 35% of the ingested material into new biomass. Benthic foraminifera consumed up to 1.8% of the detrital input per day. Such pre-conditioning seems to determine the state of activity of the microbial community and influences the

efficiency with which consumed organic carbon is turned to biomass production.

In Table 2 the amount of organic matter input is compared to rates of benthic turnover in the Arabian Sea. Vertical flux (eight years average) measured by sediment traps moored in deep water indicate the sedimentation of organic material at different stations. This vertical flux covers on average only about 20% of the benthic carbon demand determined by *in situ* respiration rates. Carbon demands were also estimated from pro-karyotic remineralisation of ^{14}C-labelled phyto-plankton, from biomass and an average literature value of respiration rate as well as from a geo-chemical model. These values are in between the vertical flux and the respiration rates measured *in*

Stations	WAST	CAST	SAST	BIOTRANS
	16°N 60°E	15°N 65°E	10°N 65°E	47°N 20°E
	4030 m	3960 m	4410 m	4500 m
POC flux[1]	3.2	2.0	1.8	1.4
SCOC[2]	20.0	13.0	3.5	3.1
Microb. remin.[3]	10.0	7.7	3.0	2.7
SSB C_{org} demand[4]	15.0	2.2	1.5	1.4
Model C-turnover[5]	8.8	1.9	1.5	

[1] sediment traps 1000 m above bottom (Haake et al. 1993)
[2] sediment community oxygen consumption (SCOC) is measured *in situ* by benthic landers and converted to C (Witte and Pfannkuche 2000)
[3] microbial remineralisation is estimated from $^{14}CO_2$ release from labelled algal material and input of organic matter (Boetius et al. 2000)
[4] SSB C_{org} demand is estimated from biomass of small sized biota and literature values of respiration of this biomass (Boetius et al. 2000)
[5] Model estimates of benthic C-turnover are estimated from pore water profiles (Luff et al. 2000)

Table 2. Comparison of different stations in the Arabian Sea and the BIOTRANS station in the North Atlantic in respect to POC flux and turnover of organic carbon in the sediment determined by different approaches. All values are in gC m^{-2} a^{-1}.

situ by landers. Such discrepancies between vertical flux rates and benthic carbon demand are known from a number of investigations (e.g. Smith and Kaufmann 1999). The discrepancy is particularly large near continental margins (stations WAST and CAST) and it is usually attributed to additional input of organic matter via lateral transport from the shelf. However, lack of adequate quantification of lateral transport renders such assumptions still speculative.

At present there are only few intensely studied deep-sea benthic regions in the Arabian Sea, the North Atlantic and the NE Pacific (Smith and Druffel 1998) for which sufficient chemical and biological data are available for geochemical numerical modelling of element cycles. The model results for the Arabian Sea indicate the occurrence of a very labile organic carbon fraction with a degradation constant of 15 - 30 a^{-1} equivalent to a half live of about 30 days at stations with high vertical flux (Luff et al. 2000). Moderately labile fractions with degradation constants of 0.2 - 0.6 a^{-1} dominate the POC input at more oligotrophic stations. This supports the notion that high export systems not only provide the sea floor with more organic matter, but also with a higher proportion of biologically labile material. Phytoplankton production and formation of aggregates in the surface water, but also transformation of this material in the water column beneath the upper mixed layer ("twilight" zone) determine the level of export and loss of labile organic matter from the sinking POC. In the Arabian Sea, a high proportion of POC reaches the sea floor, which may be attributed to faster sinking speed of aggregates due to ballast by dust particles or to slower degradation rates in the extensive oxygen minimum zone. The geochemical model also indicates that benthic oxygen fluxes closely follow the seasonal variations in surface water export production and vertical flux and are, therefore, not in a steady state. Since the temporal changes after sedimentation events are so difficult to observe directly, such modelling techniques are a useful tool to assess benthic responses.

Outlook

We have now obtained a general understanding of the biogeochemical processes and their rates driven by the small benthic biota of the deep-sea. This knowledge has been pieced together from small and sometimes inadequate data sets. There is still a severe lack of sufficient data of good quality which is mainly caused by technical difficulties. In order to follow the temporal dynamics of biologically driven carbon and nutrient fluxes in the deep-sea, detailed multidisciplinary studies as well as long-term observations need to be carried out at key sites/provinces. Only such studies provide adequate data for a mechanistic understanding, a quantitative assessment and for biogeochemical modelling.

There are two important new methodologies which promise major progress in deep-sea research: *in situ* instrumentation (remotely operated or autonomous) and molecular biological methods. *In situ* measurements and experimentation with advanced lander techniques or benthic observatories are essential to observe processes and measure rates of turnover of material under deep-sea conditions correctly. Such *in situ* experiments are also needed to investigate the physiological adaptations of organisms and changes in species composition in response to disturbances. Reactions to natural disturbances, such as pulses of detritus deposition and turbidites, or to anthropogenic impacts need to be understood in order to assess potential consequences of environmental change for biogeochemical fluxes and biodiversity in the deep-sea. With new molecular biological methods it is now possible to link biogeochemical processes and the occurrence of functional groups of microorganisms. This important next step in environmental microbiology will help to understand the function of microbial communities in the environment and in the regulation of biogeochemical cycles.

The transport mechanisms between shelf and deep-sea and the deposition of organic material along continental margins are at present still poorly quantified. We know that major fluxes of organic material from shallow waters enter the deep-sea, but their magnitude needs to be determined much more precisely in order to assess the relative importance of margins for global ocean fluxes. Furthermore, the highly productive upwelling systems at continental margins often create very special anoxic habitats in the deep-sea. These specific settings need further research in respect to unusual microorganisms and biogeochemical pathways.

References

Bett BJ, Rice AL (1993) The feeding behaviour of an abyssal echiuran revealed by *in situ* time-lapse photography. Deep-Sea Res 40: 1767-1779

Boetius A, Damm E (1998) Benthic oxygen uptake, hydrolytic potentials and microbial biomass at the Arctic continental slope. Deep-Sea Res I 45: 239-275

Boetius A, Lochte K (1996a) Effect of organic enrichment on hydrolytic potentials and growth of bacteria in deep-sea sediments. Mar Ecol Prog Se 140: 239-250

Boetius A, Lochte K (1996b) High proteolytic activities of deep-sea bacteria from oligotrophic polar sediments. Arch. Hydrobiol. Spec. Issues Advanc Limnol 48: 269-276

Boetius A, Ferdelmann T, Lochte K (2000) Bacterial activity in sediments of the deep Arabian Sea in relation to vertical flux. Deep-Sea Res II 47: 2835-2875

Christiansen C, Boetius A (2000) Mass sedimentation of the swimming crab *Charybdis smithii* (Crustacea: Decapoda) in the deep Arabian Sea. Deep-Sea Res II 47: 2673-2685

Deming JW, Baross JA (1993) The early diagenesis of organic matter: Bacterial activity. In: Engel H, Macko SA (eds) Organic Geochemistry. Plenum Press, New York. pp 119-144

Galéron J, Sibuet M, Mahaut M-L, Dinet A (2000) Variation in structure and biomass of the benthic communities at three contrasting sites in the tropical Northeast Atlantic. Mar Ecol Prog Ser 197: 121-137

Gooday AJ, Lambshead PJD (1989) Influence of seasonally deposited phytodetritus on benthic foraminiferal populations in the bathyal northeast Atlantic: The species response. Mar Ecol Prog Ser 58: 53-67

Gooday AJ, Turley CM (1990) Responses by benthic organisms to inputs of organic material to the ocean floor: A review. Phil Trans R Soc Lond A 331: 119-138

Gooday AJ, Pfannkuche O, Lambshead PJD (1996) An apparent lack of response by metazoan meiofauna to phytodetritus deposition in the bathyal North-eastern Atlantic. J Mar Biol Assoc UK 76: 297-310

Graf G (1989) Benthic-pelagic coupling in a deep-sea benthic community. Nature 341: 437-439

Haake B, Ittekkot V, Rixen T, Ramaswamy V, Nair RR, Curry WB (1993) Seasonality and interannual variability of particle fluxes to the deep Arabian Sea. Deep-Sea Res I 407: 1323-1344

Jahnke RA (1996) The global ocean flux of particulate organic carbon: Areal distribution and magnitude. Glob Biogeochem Cycl 10: 71-88

Jahnke RA, Jackson GA (1987) Role of sea floor organisms in oxygen consumption in the deep North Pacific Ocean. Nature 329: 621-623

Levin L, Blair N, DeMaster D, Plaia G, Fornes W, Martin C, Thomas C (1997) Rapid subduction of organic matter by maldanid polychaetes on the North Carolina slope. J Mar Res 55: 595-611

Lochte K (1992) Bacterial standing stock and consumption of organic carbon in the benthic boundary layer of the abyssal North Atlantic. In: Rowe GT and Pariente V (eds) Deep-Sea Food Chains and the Global Carbon Cycle, NATO ASI Series, Vol. C360, Kluwer Academic Publishers, Dordrecht, pp 1-10

Lochte K, Ducklow HW, Fasham MJR, Stienen C (1993) Plankton succession and carbon cycling at 47°N, 20°W during the JGOFS North Atlantic Bloom Experiment. Deep-Sea Res 40: 91-114

Luff R, Wallmann K, Grandel S, Schlüter M (2000) Numerical modeling of benthic processes in the deep Arabian Sea. Deep-Sea Res II 47: 3039-3072

Martin WR, Bender ML (1988) The variability of benthic fluxes and sedimentary remineralization rates in resonse to seasonally variable organic carbon rain rates in the deep-sea: A modelling study. American J Sci 288: 561-574

Meyer-Reil L-A, Köster M (1992) Microbial life in pelagic sediments: The impact of environmental parameters on enzymatic degradation of organic material. Mar Ecol Prog Ser 81: 65-72

Newton PP, Lampitt RS, Jickells TD, King P, Boutle C (1994) Temporal and spatial variability of biogenic particle fluxes during the JGOFS northeast Atlantic process studies at 47°N, 20°W (1989-1990). Deep-Sea Res I 41: 1617-1642

Pfannkuche O (1993) Benthic response to the sedimentation of particulate organic matter at the

BIOTRANS station, 47°N, 20°W. Deep-Sea Res II 40: 135-149

Pfannkuche O, Lochte K (2000) The biogeochemistry of the deep Arabian Sea: Overview. Deep-Sea Res II 47: 2615-2628

Pfannkuche O, Soltwedel T (1998) Small benthic size classes along the NW European continental margin: Spatial and temporale variability in activity and biomass. Prog Oceanog 42: 189-207

Pfannkuche O, Boetius A, Lundgreen U, Lochte K, Thiel H (1999) Responses of deep-sea benthos to unusual sedimentation patterns in the North-East Atlantic in 1992. Deep-Sea Res I 46: 573-596

Poremba K, Hoppe H-G (1995) Spatial variation of benthic microbial production and hydrolytic enzymatic activity down the continental slope of the Celtic Sea. Mar Ecol Prog Ser 118: 237-245

Smith CR, Jumars PA, DeMaster DJ (1986) *In situ* studies of megafaunal mounds indicate rapid sediment turnover and community response at the deep-sea floor. Nature 323: 251-253

Smith KLjr, Druffel ERM (1998) Long time-series monitoring of an abyssal site in the NE Pacific: An introduction. Deep-Sea Res II 45: 573-586

Smith KLjr, Kaufmann RS (1999) Long-term discrepancy between food supply and demand in the deep eastern North Pacific. Science 284: 1174-1177

Soetaert K, Herman PMJ, Middelburg JJ (1996) A model of early diagenetic processes from the shelf to abyssal depths. Geochim Cosmochim Acta 60: 1019-1040

Soltwedel T (2000) Metazoan meiobenthos along continental margins: A review. Prog Oceanog 46: 59-84

Thiel H (1975) The size structures of the deep-sea benthos. Int Rev Gesamten Hydrobiol 60: 575-606

Turley CM, Lochte K, Patterson DJ (1988) A barophilic flagellate isolated from 4500 m in the mid-North Atlantic. Deep-Sea Res 35: 1079-1092

Vetter YA, Deming J (1994) Extracellular enzyme activity in the Arctic Northeast Water polynya. Mar Ecol Prog Ser 114: 23-34

Wheatcroft RA, Martin WR (1996) Spatial variation in short-term (234-Th) sediment bioturbation intensity along an organic-carbon gradient. J Mar Res 54: 763-792

Witte U (2000) Vertical distribution of metazoan macrofauna within the sediment at four sites with contrasting food supply in the Arabian Sea. Deep-Sea Res II 47: 2979-2997

Witte U, Pfannkuche O (2000) High rates of benthic carbon remineralisation in the abyssal Arabian Sea. Deep-Sea Res II 47: 2785-2804

Nucleic Acid-Based Techniques for Analyzing the Diversity, Structure, and Function of Microbial Communities in Marine Waters and Sediments

B. J. MacGregor[*], K. Ravenschlag, R. Amann

Max Planck Institut für Marine Mikrobiologie, Celsiusstrasse 1, 28359 Bremen, Germany
* corresponding author (e-mail): bmacgreg@mpi-bremen.de

Abstract: Many of the biogeochemical reactions that occur in marine sediments are catalyzed by the complex communities of bacteria and archaea living there. Linking specific microorganisms to specific chemical transformations has been a challenge for microbiologists, because microorganisms generally lack morphological detail and are therefore much more difficult to identify than macroorganisms. Identification has traditionally required pure-culture isolation, followed by often time-consuming chemotaxonomic characterization. In contrast to their narrow range of morphologies, microorganisms are genetically very diverse. This genetic diversity has recently been exploited for the *in situ* identification of individual microbial cells, and even of their biochemical activities. This paper is intended to give scientists of neighboring disciplines some insight into how nucleic acid-based tools such as cloning, sequencing and hybridization are used by microbiologists to analyze the diversity, structure and function of microbial communities.

Introduction

Marine sediments are complex ecosystems that play a prominent role in global biogeochemical cycles. Prokaryotes are the main catalysts of the biogenic reactions, e.g. the mineralization of organic matter. They are highly abundant (up to $> 10^9$ cells cm^{-3}) in the sediments as complex communities of very different Bacteria and Archaea. Whatever is measured in terms of a biogenic transformation in marine sediments should therefore be viewed in the light of the responsible microorganism(s). In an ideal world, the identification and quantification of defined transformation would be linked to the identification and quantification of the microbial population that catalyzes it. Just imagine that, based on solid knowledge of the relevant prokaryotes and their metabolism, good predictions could be made on the influence of natural and anthropogenic perturbations on the functioning of ecosystems such as marine sediments.

Unfortunately, microorganisms are much more difficult to identify than macroorganisms, since except in a few special cases (e.g. the giant sulfur bacteria) they lack morphological detail. For a long time, microbial identification started with the often time-consuming isolation of pure cultures for laboratory-based physiological and chemotaxonomic characterization. It can be difficult to mimic all the environmental requirements of a bacterium of unknown metabolism on an agar plate. Specific enrichments are often necessary to adapt the wild microbes to the lab, complicating quantification attempts, and all the information on interactions that in macroecology can be obtained from stable spatial arrangements is lost. Therefore, for the sake of speed and precision, there was an urgent demand for *in situ* identification methods.

Quite in contrast to their narrow range of morphologies, but in line with their amazing biochemical potential, microorganisms are genetically very diverse. In fact, when a functionally conserved macromolecule such as the small-subunit RNA of the ribosome (the central protein factory found in all cells) is compared among species, most of the diversity resides within the microorganisms. This genetic diversity, expressed in every cell on the

From WEFER G, BILLETT D, HEBBELN D, JØRGENSEN BB, SCHLÜTER M, VAN WEERING T (eds), 2002, *Ocean Margin Systems*. Springer-Verlag Berlin Heidelberg, pp 419-438

levels of DNA, RNA and protein, has recently been exploited for the *in situ* identification of individual microbial cells in such complex systems as marine sediments. This contribution intends to give scientists of neighboring disciplines some insight as to how molecular biologists nowadays use nucleic acid-based tools such as cloning, sequencing and hybridization to analyze the diversity, structure and function of microbial communities.

Background

Molecular biology basics

To interpret the results of molecular microbiological studies, it is important to understand how cellular levels of the molecules of interest are regulated, and how they are extracted from samples and measured. This review deals with RNA, DNA, and protein, whose roles are summarized by the "central dogma" of biochemistry. The enzyme RNA polymerase transcribes DNA-encoded genes to messenger RNA (mRNA), which is then translated to protein by ribosomes (Fig. 1).

DNA is a double-stranded polynucleic acid, with complementary "sense" and "antisense" strands joined by hydrogen bonds between adenine-thymine (A·T) and guanine-cytosine (G·C) base pairs. Prokaryotes (Bacteria and Archaea) typically contain a single circular DNA chromosome with on the order of several thousand genes, although multiple chromosomes, linear DNA, and extrachromosomal DNA are found in some species.

RNA is a single-stranded polynucleic acid, transcribed as a complement to the sense strand of DNA, with uracil (U) replacing thymine. The enzyme RNA polymerase recognizes specific promoter sequences preceding genes. Regulatory proteins produced or activated in response to environmental conditions (e.g. nitrogen limitation) can modulate promoter recognition. RNA polymerase moves along the DNA and produces an RNA molecule complementary to the coding strand. Transcription continues until the polymerase reaches a terminator sequence, and dissociates from the DNA.

There are three major classes of RNA: ribosomal RNA (rRNA), transfer RNA (tRNA),

and protein-encoding messenger RNA (mRNA). rRNAs are components of ribosomes, which are two-subunit rRNA/protein complexes responsible for the translation of mRNA into proteins. The genes encoding rRNA are referred to as rDNA. They are present in from 1 to as many as 14 copies per cell. In prokaryotes, the small ribosomal subunit (SSU) includes a 16S rRNA, and the large subunit includes 23S and 5S rRNAs. Eukaryotes have 18S (SSU), 28S, 5S, and 5.8S rRNAs. Production and assembly of ribosomal components is tightly regulated in accordance with cellular protein requirements. tRNAs are small RNA molecules which carry specific amino acids on one end and have a 3-nucleotide RNA sequence (anticodon) complementary to the corresponding 3-nucleotide codon on the mRNA.

Ribosomes bind to specific ribosome-binding sequences in mRNA, complementary to a sequence in SSU rRNA, and begin translation at an initiation codon (most often AUG). The ribosome matches mRNA codons to tRNA anticodons, and forms peptide bonds between sequential amino acids. Translation continues until the termination codon (UAG, UAA, or UGA) is reached. Newly produced proteins must then fold into their active configuration, and may require additional processing, cofactors, or localization to become functional.

From this overview, it can be seen that DNA-, RNA-, and protein-based methods measure different aspects of community activity. Both active and inactive microbes contain DNA, but metabolically active ones contain proportionally more RNA. Therefore DNA concentrations are somewhat analogous to cell numbers, and RNA concentrations to cellular activity. However, DNA can be stable extracellularly, so that the DNA in a sample may not always derive from the current population. Extracellular RNA, by contrast, is notoriously labile to both chemical and enzymatic degradation. mRNA and protein stability may differ for a given gene, and proteins may have both active and inactive forms, so that mRNA levels measured for a given gene cannot always be equated with enzymatic activity. Furthermore, it is difficult to be sure that nucleic acids (or any other biomarkers) are extracted with equal efficiency

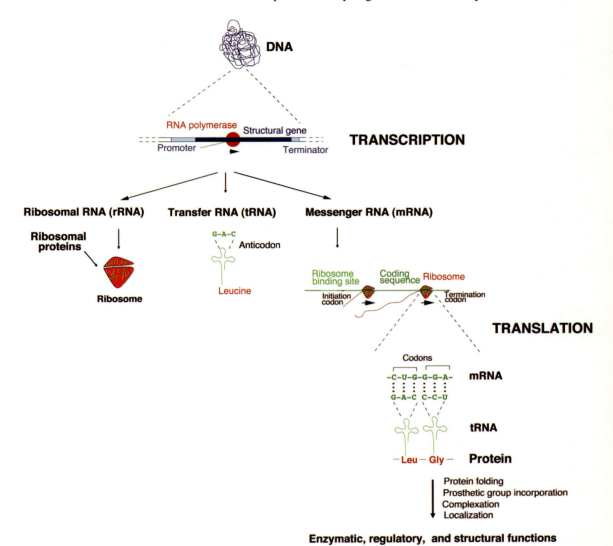

Fig. 1. Transcription and translation.

from all species in a sample, which may contain a complex mixture of mostly uncharacterized species. Finally, the appropriate sampling scale can be difficult to determine. Environmental patchiness (temporal or physical), the amount of sample needed to extract sufficient material for analysis, and available time and money must all be taken into consideration.

Phylogenetic reconstruction

Nucleic acid sequences can be used to infer phylogenetic relationships, and to identify unknown microbes by database comparisons. Steps in the construction of a phylogenetic tree are outlined in Figure 2. A collection of DNA sequences (in this case rDNA) is aligned. The dashes indicate positions where other species (not shown) have additional bases. Mutations tend to occur less frequently in sequences encoding essential functions such as substrate binding or secondary structure formation, so alignments generally reveal more- and less-conserved regions.

From the alignment, the number of base changes that would be required to change one sequence to another is computed, for all pairs of species. Allowance is made for the presumed rate of back mutation. The resulting distance matrix can be shown as a tree, here in two different formats. In the radial tree, the evolutionary distance between species is represented as the sum of the lengths of the line segments connecting them. In the linear

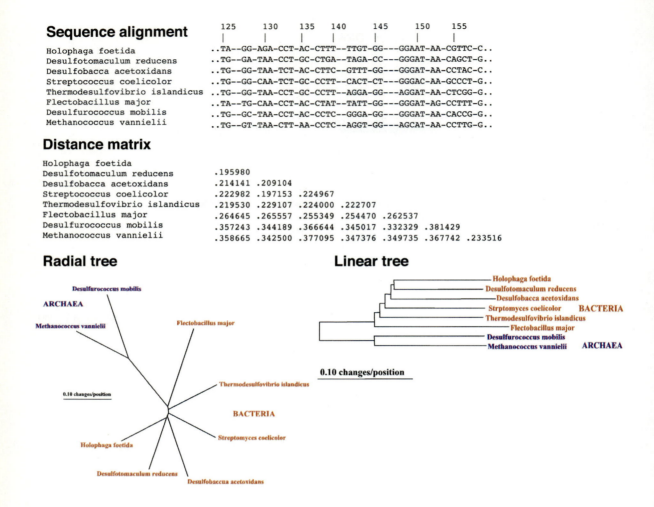

Sequence alignment

```
             125      130     135     140       145      150      155
              |        |       |       |         |        |        |
Holophaga foetida            ..TA--GG-AGA-CCT-AC-CTTT--TTGT-GG---GGAAT-AA-CGTTC-C..
Desulfotomaculum reducens    ..TG--GA-TAA-CCT-GC-CTGA--TAGA-CC---GGGAT-AA-CAGCT-G..
Desulfobacca acetoxidans     ..TG--GG-TAA-CCT-GC-CTTC--GTTT-GG---GGGAT-AA-CCTAC-C..
Streptococcus coelicolor     ..TG--GG-CAA-TCT-GC-CCTT--CACT-CT---GGGAC-AA-GCCCT-G..
Thermodesulfovibrio islandicus ..TG--GG-TAA-CCT-GC-CCTT--AGGA-GG---AGGAT-AA-CTCGG-G..
Flectobacillus major         ..TA--TG-CAA-CCT-AC-CTAT--TATT-GG---GGGAT-AG-CCTTT-G..
Desulfurococcus mobilis      ..TG--GC-TAA-CCT-AC-CCTC--GGGA-GG---GGGAT-AA-CACCG-G..
Methanococcus vannielii      ..TG--GT-TAA-CTT-AA-CCTC--AGGT-GG---AGCAT-AA-CCTTG-G..
```

Distance matrix

```
Holophaga foetida
Desulfotomaculum reducens       .195980
Desulfobacca acetoxidans        .214141 .209104
Streptococcus coelicolor        .222982 .197153 .224967
Thermodesulfovibrio islandicus  .219530 .229107 .224000 .222707
Flectobacillus major            .264645 .265557 .255349 .254470 .262537
Desulfurococcus mobilis         .357243 .344189 .366644 .345017 .332329 .381429
Methanococcus vannielii         .358665 .342500 .377095 .347376 .349735 .367742 .233516
```

Radial tree   ## Linear tree

Fig. 2. Construction of a phylogenetic tree.

representation, distances are shown by the horizontal segments only. In both types of trees, nodes between branches represent inferred common ancestors.

The distance between two species is related to the time elapsed since their divergence, but cannot be considered a direct measure of time. Mutation rates may vary among species, and between genes in the same species. For eukaryotic organisms the molecular "clock" can be calibrated by comparison with the fossil record. For prokaryotes, with only a limited fossil record, calibration relies on evidence such as microbial molecules preserved in geological formations.

DNA sequences alignments are the basis for designing short oligonucleotide probes and primers for the detection of nucleic acids derived

from particular species, as illustrated in Figure 3. Oligonucleotide probes were designed to target SSU rDNA from sulfate-reducing bacteria, at several levels of phylogenetic resolution. Probe DSS658 encompasses several genera, while probe c181-644 targets two specific sequences isolated from Svalbard sediments. Ideally, nested sets of probes can be designed to measure the proportions of different groups in the population.

Molecular measures of microbial community structure

Small-subunit rDNA or rRNA is currently the most common basis for molecular microbial ecology studies. It is an essential gene, found in all known organisms. It includes both quickly- and

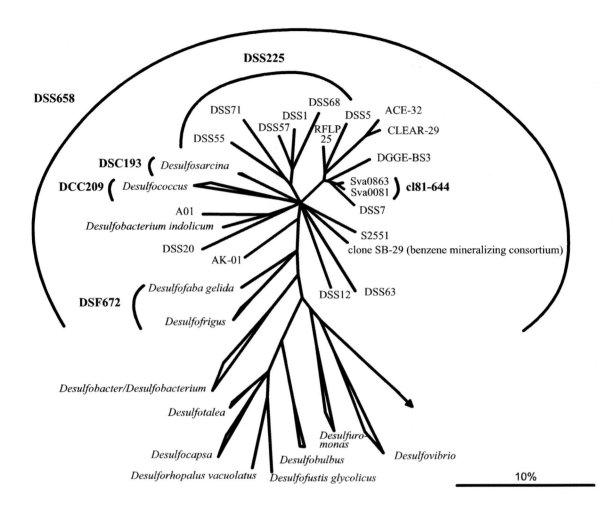

Fig. 3. Phylogenetically-based probes for the delta proteobacteria.

slowly-evolving regions, so that comparisons across both long and short evolutionary distances are possible. An extensive database of SSU rDNA sequences has accumulated, so newly acquired sequences may quickly be identified, at least in relative terms. For closely related species or strains, however, longer or more quickly evolving sequences may give better resolution.

rDNA-based methods

The polymerase chain reaction (PCR) is the basis of many methods in molecular microbiology because (with the proper controls) it allows specific detection of trace amounts of nucleic acid. For sediments, the DNA extracted from less than one

gram of sample is often sufficient for the detection of dozens of microbial species, so sampling can be done on a quite detailed vertical or horizontal scale.

PCR employs oligonucleotide primers and thermostable DNA polymerase (e.g. *Taq* polymerase, derived from the thermophilic bacterium *Thermus aquaticus*) to amplify target DNA sequences by temperature-controlled cycles of strand separation, primer annealing, and primer extension (Fig. 4). DNA isolated from an environmental sample is mixed with DNA polymerase, dideoxy nucleotide triphosphates (dNTPs), and oligonucleotide primers complementary to the upstream and downstream ends of the gene to be amplified. Reactions are performed in a thermal cycler capable of rapid cycles of heating and cooling. The DNA mixture is heated to separate it into single strands, then cooled to a temperature that allows stable but specific binding of the primers to their target sequences. DNA polymerase binds the resulting short double-stranded regions and elongates them. Repeated cycles result in numerous copies of the target sequence. Amplification proceeds exponentially until primer, dNTP, or enzyme concentration becomes limiting.

The primers used for PCR can be designed to target phylogenetic groups from the "universal" to the subspecies level, based on analysis of the available sequence database. Annealing temperature, buffer composition, and number of cycles can be adjusted to control specificity; optimum conditions must be determined empirically for each primer pair. Extreme care must be used in the preparation and handling of plasticware and reagents, especially with low-biomass samples and broadly-targeted primers, to avoid amplification of contaminating DNA.

PCR amplification is prone to several types of artifacts that must be considered when designing experiments and interpreting results. Some DNA sequences are more readily amplified than others, for reasons not yet completely understood. Because of the exponential nature of PCR, small differences in reaction rates can result in large differences in final product concentration.

Chimeric sequences may be also produced by PCR, joining partial sequences amplified from two different templates (e.g. Kopczynski et al. 1994). This is especially true if short elongation times or too-numerous cycles allow partially amplified sequences to accumulate. Chimeras may form between species, or between different *rrn* operons of the same species. Computer algorithms are available to help detect these mixed sequences by comparing different segments of amplified products to the sequence database separately. If the two ends of a sequence have different closest relatives, the sequence is probably chimeric. The effectiveness of this check depends on the completeness of the database, however – and may be compromised if the database includes unrecognized chimeric sequences. Mutations in the form of insertions, deletions, and base changes may also be introduced during the elongation step of PCR,

but with a high-fidelity DNA polymerase these should be found in only a very small proportion of product molecules.

Further problems are associated with the isolation of DNA from environmental samples. Organic and inorganic compounds, in particular humic acids, co-purify with DNA and can inhibit *Taq* polymerase. No single method of removing these seems applicable to all sample types. This presents difficulties for the development of quantitative PCR-based methods, discussed further below.

Clone libraries

Cloning is a method of separating the mixture of fragments obtained by PCR amplification, restriction enzyme digestion, or physical fragmentation of a DNA sample for characterization of the different products (Fig. 5). The fragments are ligated with a plasmid, a small circular DNA molecule that can be maintained by bacteria as an extrachromosomal element. In the example shown, the plasmid encodes an antibiotic resistance enzyme and a lethal gene. The plasmid is linearized with a restriction enzyme that cuts it once, within the lethal gene, creating a site for insertion of exogenous DNA. This insertion prevents production of the lethal gene product. Ligated molecules are introduced into *Escherichia coli* cells made competent for direct DNA uptake by (for example) calcium chloride treatment, and the cells spread on antibiotic-containing agar plates. Only cells carrying active antibiotic resistance genes, but not expressing the lethal gene, will divide and give rise to colonies; the majority of such cells should contain plasmids with a lethal gene interrupted by a fragment of environmental DNA. Individual colonies are purified by repeated streaking on antibiotic-containing agar plates, and the cloned fragments identified by DNA sequencing or fingerprinting (discussed below).

Methods of DNA sequence collection and analysis are evolving rapidly, driven by genome-sequencing efforts, and it has become fairly routine to collect large SSU rDNA sequence libraries. These generally include few or no sequences identical to those already in the database, implying the

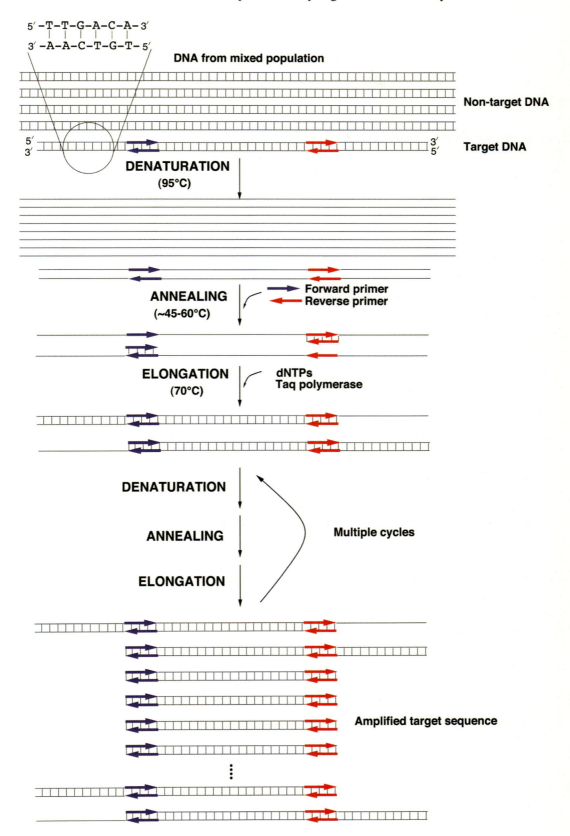

Fig. 4. The polymerase chain reaction (PCR).

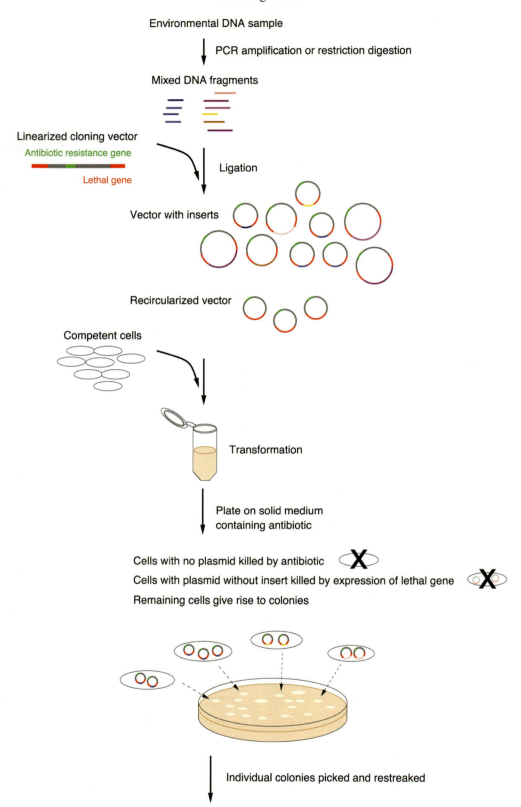

Fig. 5. Cloning of DNA fragments from environmental samples.

existence of an enormous diversity of microbial species. Examples from ocean-margin environments are listed in Table 1. Such sequence collections cannot be considered quantitative representations of microbial community structure, given the possibilities of PCR bias and chimera formation discussed above. Comparisons between collections may be misleading, even when identical primers are used, due to differences in DNA extraction and PCR methods. However, these collections are essential as a basis for probe and primer development.

DGGE

PCR products (amplificates) of similar length can be screened for sequence heterogeneity by several methods, sometimes referred to as "fingerprinting" techniques because community diversity is revealed as banding patterns upon fragment separation by gel electrophoresis. Denaturing gradient gel electrophoresis (DGGE; Fig. 6) allows separation of molecules of similar length but different nucleotide composition (Muyzer and Ramsing 1995). A·T base pairs, with two hydrogen bonds, are more easily denatured than G·C pairs, with three. DNA from a mixed population is PCR-amplified with primers targeting the gene of interest (e.g. bacterial SSU rRNA). One of the primers has a GC-rich "tail", which is amplified as one end of the PCR product molecules. The mixture of product molecules is then electrophoresed on an acrylamide gel prepared with an increasing gradient of denaturant. DNA is negatively charged, due to the phosphate groups in its sugar backbone, and will move towards the positive electrode. The double-stranded molecules run through the gel until they reach a denaturant concentration that separates all but the high-GC clamp into single strands. The resulting Y-shaped molecule essentially stops migrating, forming a band which can later be visualized by staining with DNA-binding dyes such as ethidium bromide. Bands may be identified by comparison with known standards; hybridization to labeled oligonucleotide probes specific for particular groups; or DNA sequencing. The patterns of amplificates obtained from different environments, or from the same environment

at different times, can be compared to detect changes in microbial community composition. A variation of DGGE, TGGE (temperature gradient gel electrophoresis), uses a heat gradient to separate molecules with different melting temperatures (Muyzer 1999).

For example, Sievert et al. (1999) used DGGE to study the diversity of the bacterial population in the sediments surrounding a hydrothermal vent in the Aegean Sea. DNA was extracted from sediment samples collected between 10 and 235 cm from the vent center, and amplified with primers targeting the domain Bacteria. For samples collected in June, the number of different bands detected increased with distance from the center (between 10 cm and 235 cm), suggesting a more diverse population at moderate temperature and pH. No such gradient was detected in September, however, which the authors suggest may be due to disturbance by a storm a few days before sampling.

RFLP and T-RFLP

Restriction fragment length polymorphism (RFLP) analysis employs sequence-specific DNA-cutting enzymes (restriction enzymes), which will yield fragments of different sizes from PCR products with different sequences. These are separated by size by gel electrophoresis and the number of different sequences estimated from the number of different banding patterns. Moyer et al. (1994) used RFLP of 16S rDNA amplificates to study the population structure of a microbial mat at a hydrothermal vent near Hawaii. By rarefaction analysis they estimated that their library of 48 clones, which yielded 12 different banding patterns, covered most of the diversity of the vent community. As the authors discuss, this assumes there was no significant bias in DNA extraction, PCR, or cloning.

T-RFLP is a method for estimating the number of different sequences in a mixture of PCR products (Liu et al. 1997), which unlike RFLP does not require a cloning or other separation step. One of the two PCR primers is fluorescently labeled. After amplification, the PCR products are digested with a restriction enzyme chosen to cut at different

Sampling site	Water depth	Clone name	Accession- no.	PCR primers	Reference
Puget Sound, USA coastal sediment 47°37'N, 122°30'W	13 m	EH-*	U43630-U43651	**8FB**:GAGTTTGATCCTGGC TCAG **1492RU**:GGTTACCTTGTTA CGACTT	Gray and Herwig 1996
Hornsund, Svalbard coastal arctic sediment 76°58'N, 15°34'E	155 m	Sva*	AJ240966-AJ241022; AJ297453-AJ297473	**EUB008**:AGAGTTTGATCM TGGC **EUB1492**:TACCTTGTTACG ACTT	Ravenschlag et al. 1999
Bassin d'Archachon seagrass-colonized coastal sediment (n.g.)	n.g.	B2M*	AF223253-AF223307	**27f**: **1492r**:	Cifuentes et al. 2000
Brackish water coastal sediment (n.g.)	n.g.	CE* MT*	AF211258-AF211331		Tanner et al. 2000
Black Sea shelf sediment Enrichments for Mn reducers	n.g.	A3b3, A3b2, B4b1, D1Mn, D1a1	AJ271653-AJ271657	**GM3F**:AGAGTTTGATCMT GGC **GM4R**:TACCTTGTTACGAC TT	Thamdrup et al. 2000
Black Sea shelf sediment 43°53'N, 29°58'E	77 m	BS*	AJ011657-AJ011668	**341F**:CCTACGGGAGGCAG CAG **907R**:CCGTCAATTCCTTTR AGTTT	Rosselló-Mora et al. 1999
Tokyo Bay coastal sediment 35°21,0'N, 139°47,3'E	15 m	TIHP302-* 0-2 cm depth TIHP368-* 6-8 cm depth	AB031590-AB031662		Urakawa et al. unpubl.
Taynaya Bay, Antarctic 68°46'S, 78°29'E	32 m	TAYNAYA*	AF142950-AF142975	**530f**:GTGCCAGCMGCCGC GG **1492r**:TACGGYTACCTTGT TACGACTT	Bowman et al. 2000
Papua, New Guinea estuarine sediment (8°09'S, 144°80'E)	50 m	n* a*	AF194185-AF194214; AF193560-AF193570	**530F**:GTGCCAGCMGCCGC GG **1494R**:GGYTACCTTGTTAC GACTT	Todorov et al. 2000
Sagami Bay SB: 35°4,0'N, 139°14,5'E SA: 35°0,2'N, 139°20,5'E Tokyo Bay 35°20,8'N, 139°47,1'E	1159 m 1516 m 43 m	SA* SB* TK*	AB022607-AB022642	**519f**:CAGCMGCCGCGGTA ATWC **1492r**:GGTTACCTTGTTAC GACTT	Urakawa et al. 1999
Santa Rosa Sound, USA coastal sediment (n.g.)	n.g.	A* (SRB only)	U08385-U08397	First round: **fD1**:AGAGTTTGATCCTGGC TCAG **rP2**:ACGGCTACCTTGTTAC GACTT Second round: **385F**:CGGCGTCGCTGCGTC AGG **907R**:TCTAGAAGCTTCCCC GTCAATTCCTTTGAGTTT	Devereux and Mundfrom 1994

Table 1. Bacterial 16S rDNA clone libraries from marine coastal sediments (n.g.: not given).

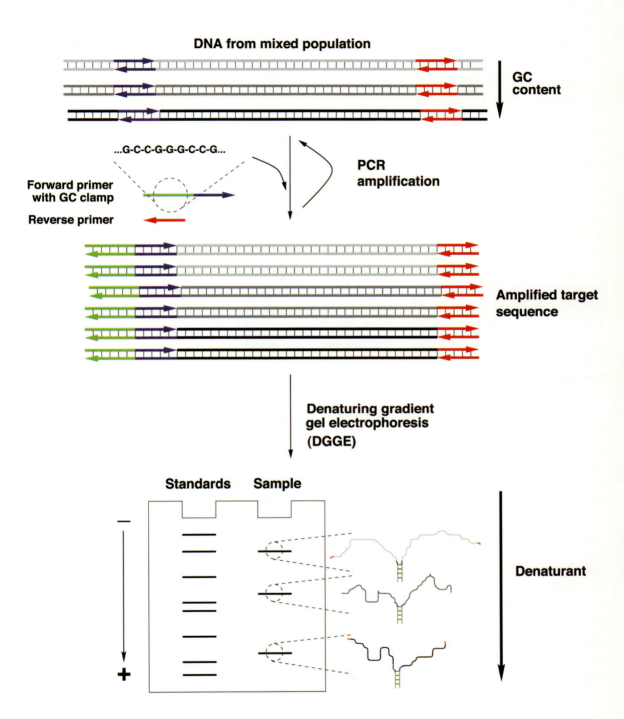

Fig 6. Denaturing gradient gel electrophoresis (DGGE).

positions in different product sequences. The digested DNA is then separated by gel electrophoresis on a DNA sequencing machine that can detect the fluorescent label. Fragment lengths are determined by comparison with fragments of known length. Because only the terminal fragments of each PCR product molecule are labeled, ideally a unique band is detected for each type of sequence. As currently used, this method requires that the sequences of all molecules likely to be present in the sample are already known, or are separately determined.

DGGE and T-RFLP fingerprinting were compared in a study of the bacterioplankton populations of the Aegean and Adriatic Seas (Moesender et al. 1999). Between-site comparisons by statistical analysis of DNA banding patterns obtained by the two methods were similar, although more different bands were detected by T-RFLP. By T-RFLP, samples taken from different depths at individual sites were more similar than samples from different sites, and samples were clustered geographically.

Quantitative PCR

Most-probable-number PCR (MPN-PCR) is conceptually similar to the MPN method of bacterial enumeration, in which serial dilutions of a sample are made in media specific for the group of interest. Cell numbers can be estimated statistically from the lowest dilution at which growth is detected. In MPN-PCR, the number of copies of a target sequence in a sample (which may or may not be equivalent to the number of cells carrying the gene) is estimated by amplification of serial dilutions of extracted DNA. One difficulty in the case of natural samples is that if enzyme inhibitors are present in a sample, amplification efficiency may increase with dilution.

In competitive PCR (cPCR), a competitor template is designed which is amplifiable by the same primer pair as the target sequence, but yields a product of different length, so that target and competitor amplificates can be distinguished by gel electrophoresis. A known amount of competitor template is mixed with the sample DNA. Assuming competitor and target are amplified at the same rate, the ratio of the two products at the end of the reaction will be the same as their ratio at the beginning, and the initial target concentration can be calculated from the known initial competitor concentration.

Relatively few studies of marine environments have employed competitive PCR to date, perhaps in part because of the difficulty of designing proper controls. Michotey et al. (2000) tested two primer sets for the quantification of the gene encoding cytochrome cd_1-type nitrite reductase in DNA from marine samples. Both showed decreased efficiency when the proportion of DNA from target species was low (1%). They estimate that denitrifying bacteria may be between 0.1 and 10% of natural populations, so even the less-affected primer set might lead to underestimates, interfering with comparisons among samples.

rRNA-based methods

Cellular rRNA content generally increases with growth rate, so rRNA levels reflect activity. Oligonucleotide probes can be designed to target the RNA of specific microbial groups, from the subspecies to the domain level, so that changes in their relative activities can be followed over time or compared between sites. It should be borne in mind, however, that the relationship between rRNA content and growth rate is not necessarily linear, and may differ among species and strains. Some microbial species may maintain a ribosomal reserve at very low growth rates, perhaps to allow quick response to improved conditions. Thus the precise relationship between rRNA concentration and metabolic activity in a mixed population depends on community composition.

RNA-based methods not including a PCR step may require larger samples than DNA-based methods, sometimes on the order of tens of grams. Because cellular RNA content can change rapidly in response to perturbation, and because RNA released by lysed cells can be quickly degraded chemically or enzymatically, samples must be handled carefully and preserved quickly.

Membrane hybridization

Membrane hybridization, illustrated here for SSU rRNA (Fig. 7), is an example of an *in vitro* method. Total RNA is extracted and blotted in triplicate onto a positively-charged nylon membrane (RNA is negatively charged), along with a range of concentrations of a reference RNA. The membranes are hybridized with oligonucleotides targeting particular phylogenetic groups - for example, archaeal or bacterial SSU rRNA. The concentration of target RNA in the samples is calculated by comparison with the reference RNA. To control for cross-hybridization, nested sets of probes are recommended - for example, universal probe hybridization should ideally equal the sum of bacterial, archaeal, and eukaryotic probe hybridization. Membrane hybridizations have been used, to study the distribution of sulfate-reducing bacteria in sediments from the Mariager Fjord, Denmark (Sahm et al. 1999b). Sulfate reduction rates were measured in the same sediments. Cell numbers estimated from the membrane hybridizations and direct cell counts yielded a more reasonable rate per cell (relative to pure-culture rates) than did cell numbers estimated by most-probable-number enrichments, suggesting that cultivation methods significantly underestimate population size.

Fluorescence in situ hybridization (FISH)

Individual microbial cells can be identified and quantified in their natural microhabitats by fluorescence *in situ* hybridization (FISH). FISH detects whole cells by a fluorescently labeled rRNA-targeted probe that hybridizes specifically to its target sequence within the cell. In contrast to the quantitative rRNA slot blot hybridization, FISH allows not only quantification of the relative abundance of different bacterial groups but also determination of absolute cell numbers, their morphology and spatial distribution, and microbial interactions (Table 2). Automated cell counting methods may allow it to be used to estimate cellular ribosome contents as well, although differential cell permeability presents a problem.

The general protocol includes the following steps (Fig. 8; for a detailed protocol see Snaidr et al. 1997): (i) fixation and permeabilization of the cells; (ii) hybridization with a rRNA-targeted probe and washing steps to remove unbound probes; (iii) visualization of the signals by epifluorescence microscopy. An overview of commonly used probes for the characterization of sediments is given in Table 3. Most commonly, oligonucleotide probes (15-20 base pairs in length) are used for FISH, but the sensitivity of FISH in environmental samples was significantly increased using polyribonucleotide probes labeled with several fluorescent dye molecules (DeLong et al. 1999). However, so far polyribonucleotide probes can only differentiate on the domain-specific level, not on the genus or species level.

FISH has demonstrated great power in the analysis of marine microbial communities (e.g. Boetius et al. 2000; Ravenschlag et al. 2000), although the autofluorescence of organic matter, and the relatively low rRNA content of cells in the often nutrient-poor deeper layers, can complicate *in situ* hybridizations of marine sediments. Llobet-Brossa et al. (1998) succeeded in FISH analysis of Wadden Sea sediments after the introduction of new fluorescent dyes and improved microscopic filter sets, and an optimization of the protocol (Fig. 9).

The simultaneous use of two (or more) probes, labeled with fluorescent dyes with different excitation and/or emission maxima, allows the detection of microbial interactions. For example, Boetius et al. (2000) identified a microbial consortium apparently mediating the anaerobic oxidation of methane above marine gas hydrates.

RT-PCR

There are exceptions to the central dogma: of particular relevance to this chapter, genetic information in retroviruses is carried by RNA rather than DNA, and transcribed to DNA by the viral enzyme reverse transcriptase. This enzyme has proven useful in molecular biological studies for the detection of rRNA and also of mRNA, which is present at much lower copy number per cell. The

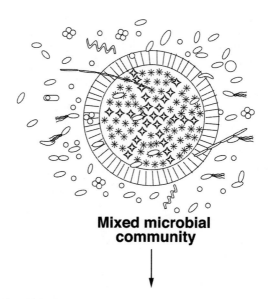

Mixed microbial community

- ● Extract total RNA
- ● Dilute, denature
- ● Blot on nylon membranes with appropriate standards
- ● Probe with labeled oligonucleotides

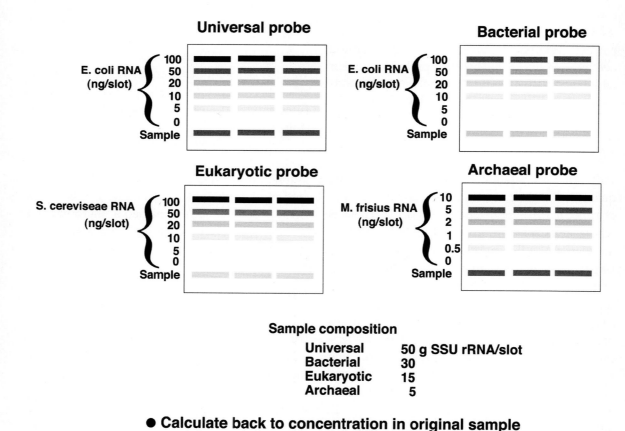

Sample composition

Universal	50 g SSU rRNA/slot
Bacterial	30
Eukaryotic	15
Archaeal	5

● Calculate back to concentration in original sample

Fig. 7. Membrane hybridization for the detection of rRNA.

	In situ-hybridization (FISH)	slot-blot hybridization
detection **depends on**	**single cell level** rRNA-content per cell	**rRNA-Pool** rRNA-content per cell + number of target cells
data	% of DAPI-stained cells or cells ml^{-1}	% prokaryotic rRNA or ng ml^{-1}
detection limit	**0,5% - 1%** of DAPI-stained cells	**0,5%** of prokaryotic rRNA **but**: specific rRNA **often still detectable** if cellular rRNA-content is below detection limit for FISH
cell numbers	+	not inferable
cell morphology	+	-
spatial distribution	+	limited
probe target site/ **probe accessibility**	**in situ accessibility** of probe target sites determines the fluorescence intensity of probes	**no restriction**
limitations	**autofluorescence** (e.g. of sediments) **impermeabiliy** of cell walls (e.g. of Gram-positives)	species-dependent differences in **cell lysis efficiency**

Table 2. Comparison of fluorescence in situ hybridization (FISH) and slot-blot hybridization.

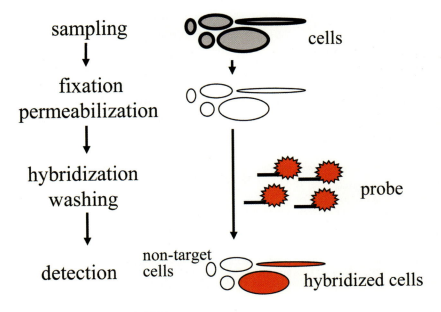

Fig. 8. Fluorescent in situ hybridization (FISH).

Probe	Specificity	Sequence (5'→ 3')	Position [a]	FISH [FA] [c]	slot blot T_d [°C]
Uni1390 [1]	Universal- all organisms	GACGGGCGGTGTGTACAA	1390-1407	--	44*
EUB338 [2]	*Bacteria*	GCTGCCTCCCGTAGGAGT	338-355	0-35	54*
EUB338-II [3]	supplement to EUB338: Planctomycetales	GCAGCCACCCGTAGGTGT	338-355		n.d.
EUB338-III [3]	supplement to EUB338: Verrucomicrobia	GCTGCCACCCGTAGGTGT	338-355		n.d.
NON338 [4]	negative control	ACTCCTACGGGAGGCAGC	338-355	10	--
ARCH915 [5]	*Archaea*	GTGCTCCCCCGCCAATTCCT	915-935	35	56*
EUK1379 [6]	*Eukarya*	TACAAAGGGCAGGGAC	1379-1394	--	42*
ALF968 [7]	α- proteobacteria, wide variety of δ- proteobacteria	GGTAAGGTTCTGCGCGTT	968-985	20	n.d.
ALF1b [8]	α- proteobacteria, wide variety of δ- proteobacteria, most spirochetes	CGTTCGYTCTGAGCCAG	19-35	20	n.d.
BET42a [8]	β- proteobacteria	GCCTTCCCACTTCGTTT	1027-1043 [b]	35	58
GAM42a [8]	γ- proteobacteria	GCCTTCCCACATCGTTT	1027-1043 [b]	35	60
PLA886 [9]	*Planctomycetales,* some *Eucarya*	GCCTTGCGACCATACTCCC	886-904	35	62
CF319a [10]	*Cytophaga/Flavobacterium*-cluster	TGGTCCGTGTCTCAGTAC	319-336	35	56
HGC69a [11]	Gram-positive bacteria with high DNA G+C content	TATAGTTACCACCGCCGT	1901-1918 [b]	35	n.d.
LGC354a,b,c [12]	Gram-positive bacteria with low DNA G+C content	YSGAAGAATCCCTACTGC	354-	20-35	n.d.
ARC94 [13]	*Arcobacter* spp.	TGCGCCACTTAGCTGACA	94-111	20	n.d.
SRB385 [2]	sulfate-reducing bacteria of the δ-proteobacteria; many Gram-positive bacteria	CGGCGTCGCTGCGTCAGG	385-402	35	n.d.
DSR651 [14]	*Desulforhopalus* spp.	CCCCCTCCAGTACTCAAG	651-668	35	62
DSS658 [14]	*Desulfosarcina* spp./*Desulfofaba* sp./ *Desulfococcus* spp./*Desulfofrigus* spp.	TCCACTTCCCTCTCCCAT	658-685	60	58
DSV698 [14]	*Desulfovibrio* spp.	GTTCCTCCAGATATCTACGG	698-717	35	58
DSV214 [14]	*Desulfomicrobium* spp.	CATCCTCGGACGAATGC	214-230	10	n.d.
DSMA488 [14]	*Desulfarculus* sp./*Desulfomonile* sp./ *Syntrophus* spp.	GCCGGTGCTTCCTTTGGCGG	488-507	60	n.d.
DNMA657 [15]	*Desulfonema* spp.	TTCCGCTTCCCTCTCCCATA	657-676	35	68
660 [16]	*Desulfobulbus* spp.	GAATTCCACTTTCCCCTCTG	660-679	60	59*
221 [16]	*Desulfobacterium* spp.	TGCGCGGACTCATCTTCAAA	221-240	35	57*
DSB985 [14]	*Desulfobacter* spp./*Desulfobacula* spp.	CACAGGATGTCAAACCCAG	985-1003	20	53
DTM229 [17]	*Desulfotomaculum* spp.	AATGGGACGCGGAXCCAT	229-246	15	56
Sval428 [18]	*Desulfotalea* spp./*Desulfofustis* sp.	CCATCTGACAGGATTTTAC	428-446	25	52*
DSF672 [19]	*Desulfofrigus* spp./*Desulfofaba* sp.	CCTCTACACCTGGAATTCC	672-690	45	n.d.
DSC193 [19]	*Desulfosarcina* spp.	AGGCCACCCTTGATCCAA	193-210	35	n.d.
DCC209 [19]	*Desulfococcus* spp.	CCCAAACGGTAGCTTCCT	209-226	25	n.d.
DRM432 [19]	*Desulfuromonas* spp./*Pelobacter* spp.	CTTCCCCTCTGACAGAGC	432-449	40	62

[a]   Position in the 16S rRNA of *E.coli*
[b]   Position in the 23S rRNA of *E.coli*
[c]   Formamide concentrations in the hybridization buffer in % (v/v)
*   Use of washing buffer containing 1xSSC, 1%SDS
n.d.   not determined

Probes BET42a, GAM42a, and PLA886 are used with competitor [8,9]
[1] Zheng et al. 1996; [2] Amann et al. 1990a; [3] Daims et al. 1999; [4] Wallner et al. 1993; [5] Amann et al. 1990b; [6] Hicks et al. 1992; [7] Neef 1997; [8] Manz et al. 1992; [9] Neef et al. 1998; [10] Manz et al. 1996; [11] Roller et al. 1994; [12] Meier et al. 1999; [13] Snaidr et al. 1997; [14] Manz et al. 1998; [15] Fukui et al. 1999; [16] Devereux et al. 1992; [17] Hristova et al. 2000; [18] Sahm et al. 1999a; [19] Ravenschlag et al. 2000

Table 3. Oligonucleotide probes commonly used in sediments.

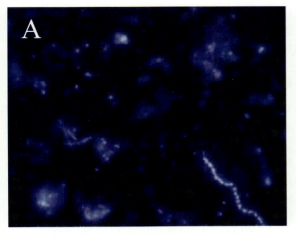

Fig. 9. a) DAPI staining and **b)** FISH.

advantage of detecting rRNA, as opposed to rDNA, is that it is more likely to derive from actively growing cells.

Methods include in situ reverse transcription (RT), and *in situ* or *in vitro* RT-PCR. In either case, RNA is hybridized with a specific oligonucleotide primer. Reverse transcriptase binds to the RNA/oligonucleotide hybrid and transcribes the RNA to DNA. The DNA may then be PCR-amplified, using labeled nucleotides to allow detection. Quantification is possible by means of competitor templates. RT-PCR has been used to detect the expression of prokaryotic genes in lakes (e.g. Zani et al. 2000), but we are not aware of any similar studies yet in marine systems.

Connecting the *in situ* identification of microorganisms with their *in situ* function

We have now shown, in some detail, how molecular biologists examine the diversity and community composition of marine sediments. The new information obtained will then add most to the understanding of ocean margin systems if it is linked to defined microbial transformations. What we know today about the physiology of microorganisms is mainly based on laboratory studies of pure cultures. In the simplest scenario, molecular techniques are used to quantify and localize well-characterized microorganisms with known functions. A good example in this respect is work done on sulfate-reducing biofilms (Ramsing et al. 1993) with a

combination of microsensors and FISH. Specific functional zones are first identified and quantified by means of microelectrodes or microoptodes. The very same biofilm is subsequently fixed, sliced, and used for *in situ* localization and quantification of populations known for their respective transformations. With the assumption that every cell is equally active, average transformation rates per cell and per day can be calculated.

The situation is much more difficult if no pure cultures of the population of interest are available. In this case, there is the option of perturbation studies, in which micro- or mesocosms are used to correlate changes in function with changes in the community composition. A good example for this type of a study is the one by Rossello-Mora et al. (1999). Cyanobacterial biomass was added in different amounts to anaerobic sediment in small plastic bags to simulate the natural input of complex organic substrate that occurs in nature after algal blooms. Molecular tools were used to follow community dynamics during incubation at different temperatures, and the anaerobic mineralization in the bags was monitored by analysis of total carbon mineralization, sulfate reduction and ammonium production rates. The addition of organic material resulted in significant changes in the composition of the microbial community at all temperatures tested. Sulfate reduction was the main mineralization process detected. However, not sulfate-reducing bacteria but members of another bacterial group, the *Cytophaga-Flavobacterium* cluster, showed the highest increase in the bacterial cells as detected by FISH. The authors hypothesize that this group catalyzes the initial hydrolysis and fermentation of the cyanobacterial biomass, which is the rate-limiting step in the anaerobic degradation of macromolecules. Their identification in high numbers in the field may indicate recent deposition events.

Yet another promising way to assign *in situ* functions to individual cells is the combination of microautoradiography to monitor the uptake of radiolabeled substrates and FISH for cell identification. It has been shown, for example, that members of the alpha subclass of *Proteobacteria* and of the *Cytophaga-Flavobacterium* group not only numerically dominated a marine bacterioplankton sample but also significantly participated in the uptake of dissolved amino acids (Ouverney & Fuhrman 1999). Stable isotopes may help in the assignment of specific processes to certain bacteria. ^{13}C-labeled substrates may be traced in lipids that are biomarkers for distinct groups of microorganisms (Boschker et al. 1998). Naturally occurring isotope signatures like the one found in biogenic methane may be helpful to link a process like the anaerobic oxidation of methane to certain prokaryotes (Hinrichs et al. 1999; Boetius et al. 2000).

There are additional nucleic acid based-techniques. These include the *in situ* visualization of functional genes or their mRNA by FISH or by *in situ* PCR. However, these methods are still very demanding experimentally, and a long way from routine application. An advanced strategy for predicting the function of hitherto uncultured bacteria originates in the new area of genome sequencing. Large DNA fragments, containing multiple genes, are obtained directly from the environment and cloned in suitable vectors (Stein et al. 1996), and screened to select those including taxonomically and/or functionally identifiable genes. For example, a DNA fragment from a marine sample was found to encode both 16S rRNA affiliated with the SAR86 cluster of marine bacteria, and a gene encoding a bacterial rhodopsin which functioned as a light-driven proton pump when expressed in *Escherichia coli* (Beja et al. 2000). This suggests a novel type of photosynthetic metabolism for this uncultured bacterial group, which by molecular evidence is widespread in the world's oceans.

Conclusions

We hope to have demonstrated the potential of molecular biological methods for the study of the biodiversity and function of marine sediments. Molecular microbial ecologists have just started to participate in the study of ocean margin processes. Like scientists from all other disciplines, they realize that oceanography is an interdisciplinary science that will develop best if new techniques are rapidly and smoothly integrated. This will require that geologists can understand results obtained by molecular biology and vice versa, and that they share a common language or at least have access to interpreters.

References

Amann RI, Binder BJ, Olson RJ, Chisholm SW, Devereux R, Stahl DA (1990a) Combination of 16S rRNA-targeted oligonucleotide probes with flow cytometry for analyzing mixed microbial populations. Appl Environ Microbiol 56:1919-1925

Amann RI, Krumholz L, Stahl DA (1990b) Fluorescent-oligonucleotide probing of whole cells for determinative, phylogenetic, environmental studies in microbiology. J Bacteriol 172:762-770

Beja O, Aravind L, Koonin EV, Suzuki MT, Hadd A, Nguyen LP, Jovanovich S, Gates CM, Feldman, RA, Spudich JL, Spudich EN, DeLong EF (2000) Bacterial rhodopsin: Evidence for a new type of phototrophy in the sea. Science 289:1902-1906

Boetius A, Ravenschlag K, Schubert CJ, Rickert D, Widdel F, Gieseke A, Amann R, Jørgensen BB, Witte U, Pfannkuche O (2000) Microscopic identification of a microbial consortium apparently mediating anaerobic methane oxidation above marine gas hydrates. Nature 407:623-626

Boschker HTS, Nold SC, Wellsbury P, Bos D, Graaf W, Rel R, Parkes RJ, and Cappenberg TE (1998) Direct linking of microbial populations to specific biogeochemical processes by ^{13}C-labelling of biomarkers. Nature 392:801-805

Bowman JP, Rea SM, McCammon SA, McMeekin TA (2000) Diversity and community structure within anoxic sediment from marine salinity meromictic lakes and a coastal meromictic marine basin, Vestfold

Hills, Eastern Antarctica. Environ Microbiol 2:227-237

Cifuentes A, Anton J, Benlloch S, Donnelly A, Herbert RA, Rodriguez-Valera F (2000) Prokaryotic diversity in *Zostera noltii*-colonized marine sediments. Appl Environ Microbiol 66:1715-1719

Daims H, Brühl A, Amann R, Schleifer K-H (1999) The domain-specific probe EUB338 is insufficient for the detection of all Bacteria: Development and evaluation of a more comprehensive probe set. Syst Appl Microbiol 22:434-444

DeLong EF, Taylor LT, Marsh TL, Preston CM (1999) Visualization and enumeration of marine planktonic Archaea and Bacteria by using polyribonucleotide probes and fluorescent *in situ* hybridization. Appl Environ Microbiol 65:5554-5563

Devereux R, Kane MD, Winfrey J, Stahl DA (1992) Genus- and group-specific hybridization probes for determinative and environmental studies of sulfate-reducing bacteria. Syst Appl Microbiol 15:601-609

Devereux R, Mundfrom GW (1994) A phylogenetic tree of 16S ribosomal-RNA sequences from sulfate-reducing bacteria. Appl Environ Microbiol 60:3437-3439

Gray JP, Herwig RP (1996) Phylogenetic analysis of the bacterial communities in marine sediments. Appl Environ Microbiol 62:4049-4059

Hicks RE, Amann RI, Stahl DA (1992) Dual staining of natural bacterioplancton with 4',6-diamidino-2-phenylindole and fluorescent oligonucleotide probes targeting kingdom-level 16S rRNA se-quences. Appl Environ Microbiol 58:2158-2163

Hinrichs K-U, Hayes JM, Sylva SP, Brewer PG, DeLong EF (1999) Methane-consuming archaebacteria in marine sediments. Nature 398:802-805

Hristova KR, Mau M, Zheng D, Aminov RI, Mackie RI, Gaskins HR, Raskin L (2000) *Desulfotomaculum* genus- and subgenus-specific 16S rRNA hybridization probes for environmental studies. Environ Microbiol 2:143-159

Kopczynski ED, Bateson MM, Ward DM (1994) Recognition of chimeric small-subunit ribosomal DNAs composed of genes from uncultivated organisms. Appl Environ Microbiol 60:746-748

Liu W-T, Marsh TL, Cheng H, Forney LJ (1997) Characterization of microbial diversity by determining terminal restriction fragment length polymorphisms of genes encoding 16S rRNA. Appl Environ Microbiol 63:4516-4522

Llobet-Brossa E, Rossello-Mora R, Amann R (1998) Microbial community composition of Wadden Sea sediments as revealed by fluorescence *in situ*

hybridization. Appl Environ Microbiol 64:2691-2696

Manz W, Amann R, Ludwig W, Wagner M, Schleifer K-H (1992) Phylogenetic oligodeoxynucleotide probes for the major subclasses of proteobacteria: problems and solutions. Syst Appl Microbiol 15:593-600

Manz W, Amann R, Ludwig W, Vancanneyt M, Schleifer K-H (1996) Application of a suite of 16S rRNA-specific oligonucleotide probes designed to investigate bacteria of the phylum *Cytophaga-Flavobacter-Bacteroides in* the natural environment. Microbiol 142:1097-1106

Manz W, Eisenbrecher M, Neu TR, Szewzyk U (1998) Abundance and spatial organization of gram negative sulfate-reducing bacteria in activated sludge investigated by *in situ* probing with specific 16S rRNA targeted oligonucleotides. FEMS Microbiol Ecol 25:43-61

Meier H, Amann R, Ludwig W, Schleifer K-H (1999) Specific oligonucleotide probes for *in situ* detection of a major group of Gram-positive bacteria with low DNA G+C content. Syst Appl Microbiol 22:186-196

Michotey V, Méjean V, Bonin P (2000) Comparison of methods for quantification of cytochrome cd_1-denitrifying bacteria in environmental marine samples. Appl Environ Microbiol 66: 1564-1571

Moeseneder MM, Arrieta JM, Muyzer G, Winter C, Herndl GJ (1999) Optimization of terminal-restriction fragment length polymorphism analysis for complex marine bacterioplankton communities and comparison with denaturing gradient gel electrophoresis. Appl Environ Microbiol 65: 3581-3525

Moyer CL, Dobbs FC, Karl DM (1994) Estimation of diversity and community structure through restriction fragment length polymorphism analysis of bacterial 16S rRNA genes from a microbial mat at an active, hydrothermal vent system, Loihi Seamount, Hawaii. Appl Environ Microbiol 60: 871-879

Muyzer G (1999) DGGE/TGGE a method for identifying genes from natural ecosystems. Curr. Opin Microbiol 2:317-322

Muyzer G, Ramsing NB (1995) Molecular methods to study the organization of microbial communities. Water Sci Techn 32:1-9

Neef A (1997) Anwendung der *in situ*-Einzelzell-Identifizierung von Bakterien zur Populations-Analyse in komplexen mikrobiellen Biozönosen, Fakultät für Chemie, Biologie und Geowissenschaften. Dissertation. Technische Universität München, München, pp. 142

Neef A, Amann R, Schlesner H, Schleifer K-H (1998) Monitoring a widespread bacterial group: *In situ* detection of planctomycetes with 16S rRNA-targeted probes. Microbiol 144:3257-3266

Ouverney CC, Fuhrman JA (1999) Combined microauto-radiography-16S rRNA probe technique for determination of radioisotope uptake by specific microbial cell types *in situ*. Appl Environ Microbiol 65:1746-1752

Ramsing NB, Kühl M, Jørgensen BB (1993) Distribution of sulfate-reducing bacteria, O_2 and H_2S in photosynthetic biofilms determined by oligonucleotide probes and microelectrodes. Appl Environ Microbiol 59:3820-3849

Ravenschlag K, Sahm K, Pernthaler J, Amann R (1999) High bacterial diversity in permanently cold marine sediments. Appl Environ Microbiol 65:3982-3989

Ravenschlag K, Sahm K, Knoblauch C, Jørgensen BB, Amann, R (2000) Community structure, cellular rRNA content and activity of sulfate-reducing bacteria in marine Arctic sediments. Appl Environ Microbiol 66:3592-3602

Roller C, Wagner M, Amann R, Ludwig W, Schleifer K-H (1994) *In situ* probing of gram-positive bacteria with high DNA G+C content using 23S rRNA-targeted oligonucleotides. Microbiol 140:2849-2858

Rossello-Mora R, Thamdrup B, Schäfer H, Weller R, Amann R (1999) The reponse of the microbial community of marine sediments to organic carbon input under anaerobic conditions. Syst Appl Microbiol 22:237-248

Sahm K, Knoblauch C, Amann R (1999a) Phylogenetic affiliation and quantification of psychrophilic sulfate-reducing isolates in marine arctic sediments. Appl Environ Microbiol 65:3976-3981

Sahm K, MacGregor BJ, Jørgensen BB, Stahl DA (1999b) Sulphate reduction and vertical distribution of sulphate-reducing bacteria quantified by rRNA slot-blot hybridization in a coastal marine sediment. Environ Microbiol 1:65-74

Sievert SM, Brinkhoff T, Muyzer G, Ziebis W, Kuever J (1999) Spatial heterogeneity of bacterial populations along an environmental gradient at a shallow submarine hydrothermal vent near Milos Island (Greece). Appl Environ Microbiol 65:3834-3842

Snaidr J, Amann R, Huber I, Ludwig W, Schleifer K-H (1997) Phylogenetic analysis and *in situ* identification of bacteria in activated sludge. Appl Environ Microbiol 63:2884-2896

Stein JL, Marsh TL, Wu YK, Shizuya H, DeLong EF (1996) Characterization of uncultivated prokaryotes: isolation and analysis of a 40-kilobase-pair genome fragment from a planktonic marine archaeon. J Bacteriol 178:591-599

Tanner MA, Everett CL, Coleman WJ, Yang MM, Youvan DC (2000) Complex microbial consortia inhabiting hydrogen sulfide-rich black mud from marine coastal environments. Biotechn 8:1-16

Teske A, Ramsing NB, Habicht K, Fukui M, Küver J, Jørgensen BB, Cohen Y (1998) Sulfate-reducing bacteria and their activities in cyanobacterial mats of Solar Lake (Sinai, Egypt). Appl Environ Microbiol 64:2943-2951

Thamdrup B, Rosselló-Mora R, Amann R (2000) Microbial manganese and sulfate reduction in Black Sea shelf sediments. Appl Environ Microbiol 66:2888-2897

Todorov JR, Chistoserdov AY, Aller JY (2000) Molecular analysis of microbial communities in mobile deltaic muds of Southeastern Papua New Guinea. FEMS Microbiol Ecol 33:147-155

Urakawa H, Kita-Tsukamoto K, Ohwada K (1999) Microbial diversity in marine sediments from Sagami Bay and Tokyo Bay, Japan, as determined by 16S rRNA gene analysis. Microbiol-UK 145:3305-3315

Wallner G, Amann R, Beisker W (1993) Optimizing fluorescent *in situ* hybridization with rRNA-targeted oligonucleotide probes for flow cytometric identification of microorganism. Cytometry 14:136-143

Zani S, Mellon MT, Collier JL, Zehr JP (2000) Expression of *nifH* genes in natural microbial assemblages in Lake George, New York, detected by reverse transcriptase PCR. Appl Environ Microbiol 66:3119-3124

Zheng D, Alm EW, Stahl DA, Raskin L (1996) Characterization of universal small subunit rRNA hybridization probes for quantitative molecular microbial ecology studies. Appl Environ Microbiol 62:4314-4317

Carbonate Mounds as a Possible Example for Microbial Activity in Geological Processes

J.P. Henriet[*], S. Guidard and the ODP "Proposal 573" Team

Renard Centre of Marine Geology, Krijgslaan 281, S8, 9000 Ghent, Belgium
* corresponding author (e-mail): jeanpierre.henriet@rug.ac.be

Abstract: Carbonate mounds from the geological record provide ample evidence of microbial mediation in mound buildup and stabilization. Advanced models argue for the prominent role which biofilms may have played at the interface between the fluid and mineral phases. While up to the early nineties, there was little evidence of mud-mound formation from Late Cretaceous times onwards, recent investigations have increasingly reported occurrences of large mound clusters on modern ocean margins, in particular in basins rich in hydrocarbons. Mound provinces are significant ocean margin systems, up to now largely overlooked. How do such recent mound provinces relate to the fossil examples, and do the modern mound provinces provide a new window on the microbiota that were instrumental in building giant mounds throughout Phanerozoic times? These are burning questions, and the answer will only come through a new dialogue between experts of the past, explorationists of the recent ocean, and microbiologists. An example is given of the power of new exploration tools, which can highlight controls on mound nucleation and patterns of early diagenesis – typically microbially driven processes. New insights can pave the way for new sampling opportunities, both by targeted surface sampling and controlled subsurface sampling through drilling.

Introduction: new views on old mounds

In his book "Life as a Geological Force", Peter Westbroek (1991) narrates his encounter with Claude Monty on the famous Devonian carbonate mounds near Couvin, in the southern part of Belgium (Fig. 1). He cites his guide: "We believe that the whole dome was a hill of slime, bacteria and carbonate mud. Originally it was a coherent mass, which explains the steep slopes. But the slightest earthquake could destabilize this jellyish mass, hence causing slumping, collapse and sedimentary dykes". Laminated limestone filling the cavities would argue for the role of bacterial mats. Corals on top of the mounds got cought in slumps, shaping rubble layers. These mounds are some 350 Ma old, but - geologically speaking - they would have grown in a relatively short time span, less than 1.5 Ma, most probably in periods of sediment starvation. Claude Monty considered fluxes of organic matter from slope transport or deep oceanic currents as the main source of energy, but Peter Westbroek believed "that the hill builders thrived on methane (or another nutrient gas), which seeped from deeper faults towards the seabed".

The debate between "believers" and "disbelievers" of the role of methane and of the key role played by bacteria in mound building in modern oceans has not yet settled, on the contrary. For instance, the recent discovery of deep-water corals settling on top of thick authigenic carbonate crusts (Fig. 2) capping mound-shaped mud volcanoes rich in methane and gas hydrates in the Bay of Cadiz (Ivanov et al. 2000) adds to fuel debates and a major scientific mobilisation in Europe and elsewhere. This should culminate in conclusive sampling and drilling experiments.

The - not casual - meeting between Peter Westbroek and Claude Monty has a near to symbolic significance. Peter Westbroek is no doubt one of the leading scientists (some would say gurus) of the turn of the century, who has added a major piece of oceanic evidence (*Emiliana huxleyi*) to the paradigm of "The Biosphere", as first defined by Vladimir Vernadsky (1926). As to Claude Monty and his fellow editors of the monograph "Carbonate Mud-Mounds - Their Origin and Evolution" (1995), they no doubt should be credited for a most sub-

From WEFER G, BILLETT D, HEBBELN D, JØRGENSEN BB, SCHLÜTER M, VAN WEERING T (eds), 2002, *Ocean Margin Systems.* Springer-Verlag Berlin Heidelberg, pp 439-455

Fig. 1. Frasnian (Upper-Devonian) mud mound, Beauchâteau quarry, Senzeille, Belgium. The mound is red coloured by microaerophilic iron bacteria Sphaerotilus-Leptothrix and Siderocapsa (Bourque and Boulvain 1993).

Fig. 2. Photograph of a piece of authigenic carbonate crust with pogonophores and a small coral as pioneer settler. This crust was sampled during the TTR 10 cruise of the Training Through Research programme of IOC-UNESCO on a mound of mud-volcanic origin in the Bay of Cadiz, off the Morocco margin.

stantial account of Life's role in the construction of carbonate mud-mounds, throughout geological times. A review of some highlights of microbial clues in the fossil record, summarized in the next chapter, largely drew from this study. Monty et al.'s monograph also includes case studies from recent shallow water carbonate environments, but at the time of redaction of their book, "ocean-truthing" models for deep-water carbonate mounds were still largely lacking.

In recent years, the seismic exploration and multibeam mapping of oceanic margins has increasingly reported evidence for large clusters of giant carbonate mounds, down to much greater water depths than any fossil evidence ever might have suggested. Deep-water carbonate mound systems are alive, down to 1000 m depth, or at least they were alive in recent geological times, and their size and extent suggests they are more than mere curios. Within the carbon cycle and the ocean/climate machine, they represent giant sinks of carbon, locked on the seabed at discrete moments of climatic cycles, and locked by Life.

Recent deep-water mounds become increasingly accessible through modern technologies of acoustic and seismic remote sensing and imaging, in 2D and 3D, through careful surface sampling and through deep drilling with adequate recovery techiques. Without overlooking the pitfalls of Actualism, we can build upon such evidence for gaining "new views on old mounds" and untangle the basic mechanisms by which bacterial life in particular can shape mounds and stabilize their slopes. A curiosity-driven search that might open a new window on "biotechnology" at the seabed, and insights in the nuts and bolts of some of those giant oceanic "bioreactors" (Boetius et al. this volume) which can yield more than fundamental perspectives.

There is however still a long way to go before the research worlds of ancient and modern mounds can be bridged. Petrologists and palaeontologists of ancient mounds on one side, and geophysicists and oceanographers studying recent mound settings on the other side form two communities, speaking a different language, and for whom in addition the dialogue with microbiologists is some kind of a challenge. Taking the challenge will involve the

definition of a common strategy, where the new exploratory power and holistic view of geophysics on full mound systems in their natural oceanic environment will pave the way to targeted sampling opportunities, and new analytical breakthroughs.

Three windows of opportunities towards synergy are briefly discussed, which potentially may shed a new light on the biota and processes at the interface between mound and ocean, on processes of mound nucleation and on the dynamics of diagenesis and lithification of the "core" of a mound. The potential role of gas seeps in mound nucleation, the pathways of fluid flow in a mound reservoir and the physical patterns of diagenesis and lithification can increasingly be approached by pushing geophysical techniques to their frontiers of very high spatial resolution.

Microbial mediation in carbonate mounds: evidence from the geological record

Bacteria have left clear evidence of their presence and activity in mound sediments in the form of fossils, fabrics and chemical signatures, from Archaean times onwards (Riding and Awramik 2000). Mud-mounds are exclusively marine buildups made essentiallly of muddy carbonate sediments. Although they may reach large sizes, most of them are not framework-supported. "Microbial" mounds are defined by Bosence and Bridges (1995) in a ternary classification jointly with biodetrital mounds and framework reefs, as those formed by the action of microbes which calcify, trap and bind sediments. Microbes in such context largely comprise bacteria, but also include small algae and fungi, and protozoans.

Monty (1995) further identified four types of microbial mounds based on their meso- and microscale structures, textures and biota:
• stromatolitic mud-mounds,
• non-laminated microbial (thrombolitic) mud-mounds with mottled or bioturbated fabrics,
• metazoan boundstones and framework reefs built through incorporation of metazoans into microbial fabrics,
• stromatactoid mud-mounds (stromatactis are still enigmatic crystalline fills of cavities in very

fine-grained carbonates, with a typical smoothly curved base and irregular top). Variations of such classification have been introduced, with few differences, by other authors (e.g. Pratt 1995).

According to Bosence and Bridges (1995), Late Precambrian mounds are constructed largely by non-calcified microbes, without higher fauna. Monty (1995) however reports that the decline in stromatolites in Late Precambrian times coincides with the rise of calcified cyanobacteria (Late Riphean, 800-700 Ma) which, in association with the upcoming metazoans, played an important role in the development of mud-mounds.

The true dawn of carbonate mud-mounds is in Cambrian times, in the Early Palaeozoic, when mounds suddenly feature a diversity in microbial and biodetrital fabrics with abundant mound-building calcified microbes, calcified coralline and green algae and a variety of Palaeozoic benthic invertebrates, which may have played an ancillary role in mound construction (Bosence and Bridges 1995; Pratt 2000). Phototrophic and heterotrophic bacteria degraded sponges and mucilages. Iron bacteria may have intervened in some cases. Fossil cyanobacterial filaments had an extensive bathymetric range, down to relatively deep waters (370 m, Monty 1995), which parallels observations in the Recent period. Living cyanobacteria may survive photosynthetically in substantial water depths, by modifying the quality and quantity of the pigments.

Photosynthesis would however not be essential for the precipitation of carbonates: carbonate precipitation increases for instance with depth in bacterial mats, and hence follows more complex processes. These are summarized in "active" bacterial precipitation, related to the properties of the bacterial membranes which ensure calcium transportation and binding, and "passive" precipitation. Active precipitation initiates the formation of carbonate nuclei which will ultimately serve as support for much of the passive precipitation. The latter process principally follows two metabolic pathways that alter the physico-chemical environment toward increased alkalinization: the nitrogen cycle and the sulphur cycle (Castanier et al. 2000).

Carbonate was precipitated in Palaeozoic mounds under the form of micrite (grains smaller than 4 µm) or microsparite (4 µm to 31µm). The role of cyanobacteria in the formation of micritic to microsparitic matrix was essential and related to the large amount of mucilage secreted by these microbes. The polysaccharides were heavily browsed by heterotrophic bacteria. Microbial micrites can be recognized by the morphology of grains and the presence of fossilized bacteria in their centre (*Girvanella, Epiphyton, Renalcis, Bacinella*). *Girvanella* is compared with the living filamentous cyanobacteria *Plectonema*.

In mid- to Late-Ordovician times, the dramatic rise of large skeletal metazoans such as stromatoporoids, corals (rugosa and tabulata) and bryozoans as well as higher algae paved the way for the strong development of reefs and typical stromatactoid mud-mounds (e.g. in western Newfoundland, Sweden). Thick Silurian microbial buildups, including slope sequences, have been reported in the Canadian Arctic, on Ellesmere Island and on Anticosta Island.

Lower Devonian (Emsian) conical carbonate mounds (kess-kess) of the Moroccan Anti-Atlas are related to precipitation from hydrothermal fluids, some of which interfered with a carbon source with low $\delta^{13}C$ values, suggesting a contribution of methane-bearing fluids (hydrocarbon) (Mounji et al. 1998). Such low $\delta^{13}C$ values were obtained on samples at the base of a mound.

The Belgian Frasnian (Upper Devonian) is famous for its carbonate mounds (Bourque and Boulvain 1993, Monty 1995), quarried as "red marble" (Fig. 1). In these mounds as well as in the Waulsortian (Early Carboniferous) banks of Belgium and Ireland (Lees and Miller 1995), the lowermost part of the mounds is generally deprived of, or very poor in calcimicrobes, and frequently dominated by sponges. One explanation might be that the deep-water initial bacterial population was not fossilized or that the bacteria were too small to be recognized as such. In the basal horizons of the red stromatactis mounds, some authors consider that it is the bacterial degradation of sponges and slimes that triggered limited precipitation of lime mud. In the middle and particularly in the upper part of the Frasnian mounds, a variety of micritic filaments, microbes and cyanobacterial microbial films or mats can generally be identified. Those biofilms may have induced and supported the colonization of metazoans such as bryozoans.

Some of the most impressive of Early Carboniferous bank aggregates, reaching up to 1 km in thickness, are those known as the Waulsortian (de Dorlodot 1909, Lees and Miller 1995). They are of Tournaisian or early Visean age and have been described from Belgium, Ireland and the English Midlands, but are also known in North America (e.g. New Mexico), central Asia and possibly North Africa. One of the most distinctive features of Waulsortian banks is the occurrence of a sequence of generations of mud development (polymuds), of bacterial origin. Macroskeletal components are mainly crinoids, bryozoans and brachiopods. There is no evidence for significant macrobioturbation. The importance of Waulsortian mounds for our insight in microbially mediated surface processes will be discussed in more detail in the next paragraph.

Waulsortian banks frequently developed around fault-controlled basinal highs or in deep ramp environments, well below the photic zone. No satisfactory explanation for their occurrence and distribution has been identified yet (Lees and Miller 1995).Upwelling is not considered as a significant factor. Methane seeps have been invoked as a locating factor (Hovland 1990), but Lees and Miller (1995) question this hypothesis as an adequate explanation for large stacked complexes such as that in Ireland, which covered tens of thousands of square kilometres. The Waulsortian mounds of New Mexico are related to the dynamics of internal waves in a stratified ocean, which would have causing mixing of an oxygen minimum zone, rich in organic matter, with better oxygenated water masses (Stanton et al. 2000).

The Permo-Triassic produced mud-mounds of different composition, including newly evolved calcisponges, microbial crusts and sessile foraminifera coated by microbial encrustations (Pratt 1995). The Early Permian carbonate buildups of

the Southern Urals (Russia) are important hydrocarbon reservoirs.

In full Mesozoic times, a decline in the abundance and diversity of microbial mounds is recorded from the Triassic to the Cretaceous (Bosence and Bridges 1995). Middle-Triassic mud-mounds with reefal caps and microbial crusts have been described in the Catalan ranges of north-eastern Spain. The Late Jurassic was a time of extensive formation of coral, siliceous sponge and microbial mounds (Schmid et al. 2000). Microbial micrite precipitation is also inferred in Albian carbonate mounds of the Basque-Cantabrian region in northern Spain (García-Mondéjar and Fernandez-Mendiola 1995). Turonian and Coniacian mounds in Tunisia (Camoin 1995) might represent the last known truly microbial (non-chemosynthetic) mud-mounds in the fossil record.

From the mid-Cretaceous onwards microbial fabrics are only known as components to metazoan framework reefs, as documented by Pratt (1995), and would not be found in mud-mounds (Bosence and Bridges 1995). Most Cenozoic mud-mounds would be of biodetrital origin (Monty 1995), though microbial components might have remained significant in deeper water (Pratt 1995). The chemosynthetic microbial mounds from the Miocene of Italy (Clari and Martire 2000) might form an exception.

In terms of environmental conditions, Monty (1995) and Lees and Miller (1995) are tempted to correlate the size of mud-mounds in general with raised global seawater temperatures. This was the case in the Lower Cambrian and the Upper Devonian, where cyanobacteria in particular could abundantly multiply and calcify, and for the tropical environments of Waulsortian banks. Schmid et al. (2000) however point to an additional factor which may have stimulated the growth in particular of Upper Jurassic mounds, namely the competition at a certain development stage between pulsating sedimentation rates and the growth responses of sediment-baffling organisms. Sedimentation stimulated the growth of those organisms, as a mere quest for survival, but if the rates of sediment deposition exceeded their capacity to grow, they got buried and died.

Debates on mounds and microbes

In the wake of the publication of the International Association of Sedimentologists Special Publication on "Carbonate Mud-Mounds", Claude Monty and several colleagues launched in October 1995 in Paris project 380 of UNESCO's International Geological Correlation Programme (IGCP): "Biosedimentology of Microbial Buildups". It broadened the objectives of IGCP project 261 which had been concerned with stromatolites. IGCP 380 focused on detailed studies of microbial mediation in sedimentary buildups, and on "life as a geological force". Apparently, the dialogue between Monty and Westbroek had been sustained.

Particular attention went to mud mounds and their evolutionary paths, from the lowermost Cambrian to the Late Cretaceous. To understand micro- and macromorphologies of microbial buildups, it was deemed necessary to elucidate the processes acting between living microbial mats/biofilms and the host sediment, to find out "who" is doing "what", "where" and "what for". "Why" was maybe left for a next project. But the proponents left the reader with a big "why": why stop at the Late Cretaceous? If mound-building microbes had been zealous - with ups and downs – over more than half a billion years, why not a few million years more? Have the microbial builders of giant mounds been wiped off the planet's surface, jointly with the dinosaurs, or are we simply assisting to one of the interludes in the mounds' symphony?

Large was consequently the surprise of the (first) author, when – about the same time, in 1995 - he came across the picture of a strange, giant mound on the ocean floor: a seismogram from Hovland et al. (1994) of a modern mound in a presumed seep setting, proposed as a site for drilling under the EU concerted action CORSAIRES. The workshop on "Gas Hydrates – Relevance to World Margin Stability and Climatic Change" in Ghent in September 1996 (Henriet and Mienert 1998) and a joint action between CORSAIRES

and the EU project ENAM II (European North Atlantic Margin) set the stage for the cruises of R/V *Belgica* and R/V *Prof. Logachev* to the Porcupine Basin, off southwest Ireland, in May and July 1997. This led to the discovery – next to the "Hovland" mounds – of a buried field of more than a thousand of "Magellan" mounds, and of the impressive range of "Belgica" mounds (Henriet et al. 1998, Kenyon et al. 1998).

The CORSAIRES-TTR (Training Trough Research, IOC-UNESCO) post-cruise conference on "Carbonate Mud Mounds and Cold Water Reefs", organized in Ghent in February 1998 (De Mol 1998) and a subsequent meeting hosted by the Irish Marine Institute in Dublin one year later set the stage for the preparation of a triptych of EU projects under the 5th Framework Programme: GEOMOUND, ECOMOUND and ACES, focusing respectively on the internal controls on mound buildup, on the external controls and on the Atlantic coral ecosystems, including the role of microbiota. These projects teamed up with the microbiology-driven EU project Deep-Bug under the OMARC umbrella (Ocean Margin Deep-Water Consortium) to propose a major Ocean Drilling Programme action: the "Modern Carbonate Mounds: Porcupine Drilling" project.

The surface processes and populations

As correctly stated in the terms of reference of the IGCP project 380, it is necessary – if we want to understand mounds - to elucidate the processes acting between living microbial mats/biofilms and the host sediment. This interface should become the first priority in modern mound research. But what do we know about such microbial systems, what had been inferred from the fossil record, and how to access this interface?

The surface transfer system: the role of microbial mats and biofilms

Microbial processes can not only create chemical environments that favour mineral precipitation, but they also can stabilize mineral frameworks. Mobile sediments and moving water encourage microbes to produce adhesive polysaccharide secretions for anchorage and protection (Riding and Awramik 2000). These stabilize grains and also provide sites for mineral nucleation.

The microbial communities - predominantly populated by prokaryotes - that colonize the surface of sediments can be either microbial mats or biofilms. Their structure are reviewed by Stolz (2000). Microbial mats such as those invoked by Monty in cavities of Devonian mud-mounds may display a predominance of cyanobacterial species and are known for their role in the formation of laminated sedimentary structures. They are characterized by steep chemical gradients, the abundance of phototrophic microorganisms, and stratification of the microbial populations into distinct layers. In many cases, such complex communities may fulfill the definition of an ecosystem – or even an organized society - in that all necessary trophic levels are present: e.g. the three basic functional groups: the primary producers, consumers, decomposers.

Biofilms have been described as a population of microorganisms and their extracellular products bound to a solid surface. The bacterial colonization of surfaces is dependent on the formation of extracellular polymeric substance (EPS), which can trap inorganic and abiotic components, as well as immobilize water. Biofilms may be structured and partitioned through the formation of exopolymer-mediated microdomains, which have the potential to segregate extracellular activities at molecular scales and to facilitate e.g. the sequestering of nutrients. Biofilms may also be involved in the formation of mineral precipitates. In this respect, they are similar to microbial mats. Biofilms however may vary greatly in species composition, from a single species to complex communities. Microbial mats could in a way be regarded as complex biofilms (Stolz 2000).

The slope of mound flanks: clue for rheological modifications and early lithification?

In general, the flanks of Palaeozoic mud-mounds are very steep, up to 30-50°, which argues for the cohesivity of the gel-like muds (Lees and Miller 1995). But in addition to modifications of the rheological properties of the mud surface, there seems to be a rapid passage from the off-mound silic-

iclastic sediments to the carbonate of the mound. And as a whole, the micrites of mudmounds are made of very pure carbonates, even when they grow on or are surrounded by an argillaceous sea-floor. This means that the mud of Palaeozoic mud-mounds could not have been imported detrital sediments, but also that the carbonate lithified quite rapidly.

Another observation regarding slopes is that Palaeozoic mounds do not generally show asymmetrical flanks, as would be expected if the flanks had been hydraulic build-ups (Monty 1995). Some Waulsortian banks have aggraded with a strong lateral component, but the incompleteness of the outcrops did not allow to infer any asymmetry (Lees and Miller 1995). There are consequently few arguments that the Palaeozoic mud-mounds would have formed from sediment baffled by fenestellids, crinoids or other metazoa (Monty et al. 1995). The production of micritic sediments must have been a local process, which took place in the mound itself. The microbial origin of such rapid lithification of the carbonate mud is recognized by many authors (Garcia-Mondéjar and Fernandez-Mendiola 1995; Lees and Miller 1995; Monty 1995).

Lees and Miller's process-response model for an active surface biofilm

In order to investigate the implications of microbial mud generation on the surface systems of Waulsortian banks, Lees and Miller (1995) developed a process-response model, represented in two parts, one for the surface layer and one for the early burial history. The model relates known microbial mechanisms to Waulsortian fabrics, textures and biota.

It is proposed that bacterial cells, together with fungal/cyanobacterial filaments and bacterial extracellular polymeric material (BEPM) produced a mucus-like biofilm which induced precipitation of lime mud at the surface of Waulsortian mounds (Fig. 3, top). The highly anionic nature of BEPM enables it to scavenge Ca^{2+} and Mg^{2+} from seawater and porewaters, with other sites on the polymer able to bind CO_3^{2-}. Bacteriogenic calcites can thus be Mg-rich. The presence of the muci-

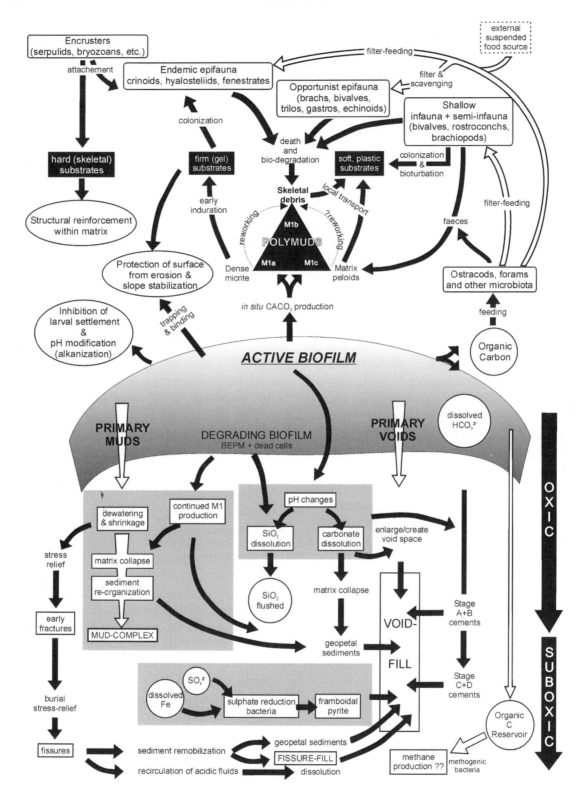

Fig. 3. Process-response models for the surface layer of Waulsortian banks (top of the figure), and the shallow-burial diagenesis of such banks (modified after Lees and Miller 1995). These models illustrate the complexity of the surface and subsurface processes potentially able to stabilize the soft muds and to support the steep slopes, as well as the key role attributed to active or degrading biofilms.

laginous biofilm would have discouraged the early settlement of many organisms and also contributed to the binding of the surface and the prevention of significant erosion. Early diagenesis which accompanied shallow burial resulted in micritization and cementation (Fig. 3, bottom).

This model is based on purely circumstancial evidence and is necessarily speculative, but it is a useful guide for research on surface processes on mounds in the Recent ocean – with due care considering the changes in biota and differences in environment.

First microbiological analyses on Hovland mounds

The adequate sampling of the intricate surface bacterial world on mounds is not even in its infancy, but some very exploratory results can be reported. Gravity cores taken on the Training Through Research cruise TTR7 from R/V *Prof. Logachev* on Hovland mounds in water depths of 780-795m have been sampled immediately after recovery and analyzed at the Laboratory of Microbiology of the University of Ghent (Heyrman et al. 1997). The main objective was to determine first of all the occurrence of aerobic heterotrophic bacteria at different locations and soil depths (0-0.4m). The Hovland mounds display an upper surface layer of foraminiferal sand rich in azooxanthellate corals (*Madreporia, Lophelia*), molluscs and echinoderm spines, covering a silty clay (De Mol et al. in press).

Colonies were counted and analysed for their morphology after an incubation period of 5 days at 20 °C. The density of heterotrophic bacteria ranged from 10^6 to 10^7 colony forming units per gram dry weight of sediment, an order of magnitude larger than references for such deep-water environments from literature (10^5 - 10^6 CFU/g).

In terms of taxonomic diversity among heterotrophs, the dominant gram-positive bacteria was *Micrococcus lylae*, the dominant gram-negative bacteria was *Pseudomonas stutzeri* and – remarkably if corroborated - the samples contained the phototrophic bacteria *Rhodococcus capsulatus*. *Micrococcus* bacteria are strict aerobes, *Pseudomonads* are a major group of chemoorganotrophic

aerobes (using organic chemicals as energy sources), and *Rhodococcus* bacteria are common soil saprophytes, utilizing hydrocarbon. Care should however be taken about these very preliminary results, which have mainly an exploratory value. Neither the site nor the sampling strategy could be relevant for testing the model of Lees and Miller, if this ever could be tested in modern oceans. But it is a first step.

Mound nucleation and lithification in the present ocean: evidence from the Australian and North Atlantic margins?

An important observation made by Monty (1995) is that mud-mounds do not seem to grow everywhere. Their development often appears to be related to particular situations such as seeps, halokinesis, argilokinesis, faults and vents, upwells and particular isobaths. This results in the fact that they generally occur in groups or clusters.

Such observation, made on fossil settings, is to some extent corroborated in the modern ocean, or at least large mound clusters are increasingly discovered in settings also characterized by shallow gas, seeps and gas hydrates – prime sites for prolific microbial life. While causal links are not yet established, the association deserves attention. We discuss below some interesting mound settings, reported recently.

Australian margins: cool-water carbonate "reefs" and hydrogen sulphide, methane and brines (ODP Leg 182)

Microbial buildups rich in bryozoans and sponges have been reported at depths greater than 100m off the shelf of south-eastern Australia. Their high growth rate (1 m per 1000 years) has been related to active upwelling. Clusters of large knolls rich in *Halimeda* coral algae and bryozoa have also been described in the Timor sea, off north-west Australia (Hovland et al. 1994), in the hydrocarbon province of the Vulcan Subbasin. Their emplacement has been correlated with faults, which may have provided conduits for upward migrating hydrocarbons.

During drilling on the southern Australian continental margin (Leg 182 of the Ocean Drilling Programme) in the Great Australian Bight, fluids with unusually high salinities, high concentrations of H_2S, CH_4 and CO_2 have been encountered in Miocene to Pleistocene sediments, which host sequences of bryozoan mounds (Feary et al. 1999). The mounds might be considered the first modern analogues for e.g. the Late Cretaceous to Danian cool-water carbonate ramp with bryozoan mounds in the Danish basin (Surlyk 1997). It is postulated that the gases in the host sediments drilled in Leg 182 are associated with H_2S-dominated hydrates (Swart et al. 2000). Such environments are known as sites of very prolific bacterial activity, part of the newly discovered "Deep Biosphere".

Blake Plateau and Orphan Knoll

A large cluster of over 200 mounds, described as coral banks (Stetson et al. 1962), has been mapped on the outer rim of Blake Plateau, off the North American East coast, in water depths of 700-900 m. The dominant corals are *Lophelia* and *Dendrophyllia*. They are located exactly upslope of what would later be identified as probably one of the largest methane hydrate provinces of our planet, below Blake Ridge (Paull et al. 1998). The gas hydrate province stretches from depths of 4000 m up to less than 1000 m. In glacial times, the hydrate stability zone must have migrated upslope, encompassing the mound province. While no direct causal link has hitherto been evidenced, such coincidence between mound provinces and the upper slope regions which have witnessed the waxing and waning of a hydrate stability zone in response to glacial/interglacial oscillations deserves further attention (Henriet et al. 1998; Henriet et al. in press).

A large cluster of mounds has also been described on Orphan Knoll, north of the Grand Banks off Labrador (Canada) in water depths of about 2000 m, but their nature and origin (modern structures or updoming fossil mounds) is still debated (Van Hinte et al. 1995). Remarkably, in mid-Cretaceous times, the conjugate position of Orphan Knoll was directly adjacent to Porcupine Bank, in such way that the bedrock of the Orphan Knoll mound cluster and that of an extensive carbonate mound range now surveyed on the south-eastern slope of Rockall Basin once formed a continuum.

The Porcupine and Rockall mounds: elements towards a model?

The largest recent carbonate mounds discovered up to now are found on the slopes of Rockall and Porcupine Basin, off south-west Ireland (Hovland et al. 1994; Croker and O'Loughlin 1998; De Mol 1998; Henriet et al. 1998; Kenyon et al. 1998; De Mol et al. in press). Some mounds in Rockall Basin exceed 200 m in height (Tj. Van Weering, pers. com.).

Recent very-high resolution seismic investigations (R/V *Belgica* cruise 2000) provide an exceptionally detailed view on the sedimentary facies below the mounds and on the potential of seismic attribute analysis, in particular in the Belgica mound range. The *Challenger* mound, proposed as an ODP target (Fig. 4), roots on a sharp slope break of an ancient erosional surface. Below this surface, an enigmatic sequence of sigmoidal, apparently high-energy slope deposits features reflectors with negative polarity which - jointly with other seismic attributes - might betray the presence of (faint) concentrations of gas. In terms of mound nucleation controls, two hypotheses might be considered: a morphological one, impacting on hydraulics, and a seep hypothesis, involving the advection of energy from the subsurface. In either case, microbial mediation in crust formation (hardground on erosional surface or authigenic carbonates) might have enhanced the chances for settlement of metazoans and for triggering mound growth. Such a root zone is consequently a prime target for drilling.

Reflectors below the core of the mound display an apparent upwarping (a "velocity pull-up", on a section which has a vertical "two-way time" dimension): the core would have a significantly higher seismic velocity, compared to the flanks and foot deposits, and to the embedding sedimentary sequences. The small buried mound section to the left of the *Challenger* mound is a profile through the foot of a neighbouring large mound (cf. the projection of the track on a digital terrain model,

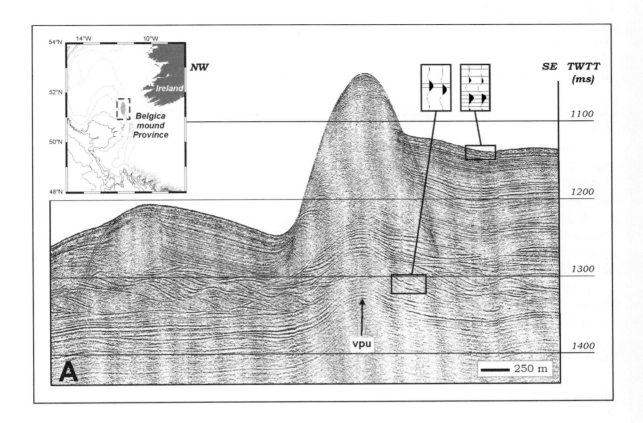

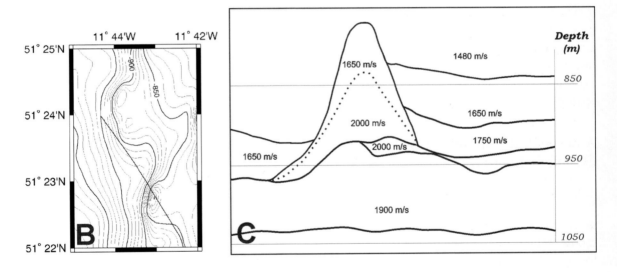

Fig. 4. Very high resolution reflection seismic profile through the Challenger mound in the Belgica mound province, southwest of Ireland. The profile shows a strong "velocity pull-up" effect indicating a high-velocity core. Reflectors below the base of the mound display reversals of signal polarity, a possible clue for faint quantities of gas (methane?). A velocity-controlled depth model restores the real structure below the mound.

insert b on Fig. 4). As it does not feature a velocity pull-up, one can conclude that the velocity of these foot deposits does not significantly differ from that of adjacent sediments.

One velocity (step) model which can account for the velocity effect is projected on a depth section through the *Challenger* mound on Fig. 4 (insert c). The absence of any internal reflector in the mound however may argue for a continuous velocity gradient, leading to basal velocities up to 2200 m/s or more. Such a velocity in carbonates suggests a beginning of diagenesis (Anselmetti and Eberli 1997), most probably controlled by microbial processes. The core of such a mound is consequently a prime drilling target for assessing how "early" "early diagenesis" may be.

A third observation is the (asymmetric) burial by a tail-shaped sequence of drift sediments, which apparently have moulded the flank of a near to fully grown-up mound. This suggests a long period of non-deposition and/or a rapid initial mound growth, again most probably boosted and supported by microbial processes of slope consolidation. The "coral banks" we now observe at the surface (De Mol et al. in press), where cold water corals apparently baffle current-transported sediments, may represent a final stage of development, where corals and other metazoans compete with oscillating sedimentary fluxes in a struggle for survival of the mound ecosystem – a plot where the terminal burying of the mound is looming large.

A development model for deep-water carbonate mounds

The above observations on a section through a Belgica mound should in principle not lead to the definition of an archetype. Even in the Belgica mound range, some mounds do not display a velocity pull-up at all. There are obviously many ways to make a mound, and up to now each newly dis-covered giant mound shares mainly one feature with the former ones: a giant question mark. But the general morphological coherence of mound structures within so-called mound "provinces", such as for instance described in Porcupine Basin, calls for some possible overruling scenario. Scenarios may have changed through geological times, in pace with the evolution of species and of their habitats, but mound formation has remained a universal phenomenon throughout the history of Life, from Archaean times onwards.

A common thread linking mounds throughout the Phanerozoic leads to a common key player, with as most probable candidate the one who has taken a prominent role on the sedimentary stage from the Archaean onwards: microbial life. Other actors are not overlooked, and corals or other metazoa may have been instrumental in mound building, but there is few evidence that they ever would have played first fiddle. We however still largely face a bacterial "default" conviction: we found the "corpse", we may have identified circumstances for the "crime", but we are left to gather conclusive evidence for convicting those elusive bacteria, and for eliciting their role and complicity in mound genesis, buildup and lithification. And when bacterial remains are found on the site, can they truly point to the key player, or rather to the most easily fossilized guest?

A major handicap in our quest is the fragmentary nature of our observations, in a multi-dimensional world. To our present knowledge, the build-up processes and the fate of mounds may be controlled by:
• climate and ocean dynamics in space and time ("external" fluid controls),
• sub-seabed fluid dynamics in space and time ("internal" fluid controls within a sedimentary basin: fluid migration from the depths, or discharge and upwelling of fluids from an emerged shelf or continental range); such pulses are by nature of their own controls transient, rather than steady-state,
• sedimentary fluxes, steady-state or pulsating,
• biological speciation and ecosystem dynamics in space and time: the intrinsic dynamics of "Life as a geological force".

One strategy to cope with the fragmentary nature of our observations is to take different angles of perspective, combining and confronting – with due care, as suggested above – the fragmentary data from various disciplines: geophysics, biogeochemistry, petrology of fossil and recent mounds, micropalaeontology, palaeoecology, isotope chemistry, etc.

To further blur the picture, various stages in development of one single mound may also call for subsequent and different controls, as suggested for the *Challenger* mound: the genesis (nucleation) and the development stages (sustained growth) are probably controlled by different processes, and might mobilise different players. Process relays are important to understand mounds.

These comments being made, we nevertheless wish to venture towards a model – whatever debatable – to pave the way towards targeted experiments, in first instance by drilling, and to trigger new insights. In what follows we will consequently formulate a tentative model for carbonate mound settings in the Recent ocean and in deeper waters, off the shelf edge and on the slope. Typical water depths for such settings range between 600 and 1000 m, well below the photic zone. Our basic observations primarily build upon research in the Belgica province of the Porcupine Basin and in the Gulf of Cadiz, but we endeavour to search for ground rules, with a broader outlook.

We postulate that mound buildup is rarely a homogenous process, and that it most probably involves a suite of relaying development stages – three to four in this case (Fig. 5) - which conceptually should be analyzed distinctly for their dominant external forcing and relevant microbiological players.

The trigger stage (Fig. 5, lower case) would imply the prime building of a solid substratum, as base for settlement of coral larvae, sponges or other metazoa. Possible scenario's – both bacterially mediated - are the development of a hardground in long periods of non-sedimentation on seabed highs, or the construction of thick authigenic carbonate crusts, linked to a fluid pulse of internal origin (CH_4 and/or H_2S). In the latter case, the spatial confinement of a mound cluster may reflect the dispersion area of the seeps. Fossil models could be found in the basal horizons of some Monferrato (Clari and Martire 2000) or Hamar-Lakhad Ridge mounds (Mounji et al. 1998). As mentioned in the introduction, modern evidence might come from the Bay of Cadiz, where methane-derived carbonate crusts with worm tubes and attached cold water corals (Fig.2) have been recovered on juvenile

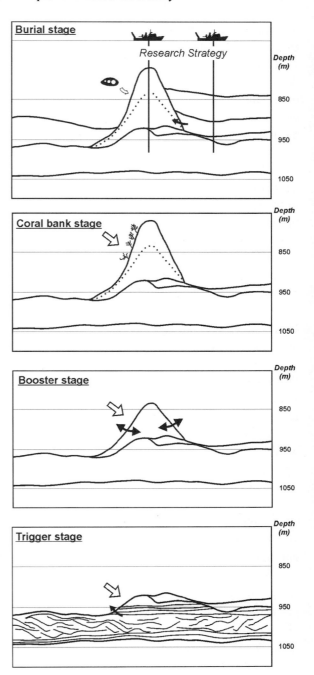

Fig. 5. Development model of a mound such as the Challenger mound in the Belgica province, southwest of Ireland. Successive stages involve relays of main controls. The initial stage (trigger) may be related to fluid venting and the development of a sole of authigenic carbonates, substratum for further colonization of the site and fast growth of the mound (booster stage). The coral cap is the presently observed surficial facies, also verified by coring. Asymmetric burial by drift sediments may impact on internal hydraulic gradients and flushing of the mounds by ground water.

mound-shaped structures of mud-volcanic origin (Ivanov et al. 2000).

The booster stage involves the catalyzation of the mound growth by the settling of metazoans on the hardgrounds or crusts, in conjunction with microbially-mediated carbonate precipitation, and it heralds the relay of internal controls - wherever applicable - by external fluxes. This implies the spatial confinement of mounds to slope sites and depths with high fluxes of organic matter and high mixing rates (internal waves?). Early lithification through bacterially induced micrite precipitation (role of biofilms) can account for the stabilization of such mud build-ups and for the relatively steep slopes. Quite a few Palaeozoic or Mesozoic mounds (Monty 1995; Pratt 1995; García-Mondéjar and Fernandez-Mendiola 1995) may bear witness of such development. A modern model, if any is to be found, might be searched in the inferred partly lithified core of some Belgica mounds in the Porcupine Basin. The flat base of such mounds argues for a relatively fast growth towards near-full size, as evoqued in the "booster" concept.

The coral bank stage implies the full relay by benthic biota and colonizing filter feeders (corals, etc.). The spatial zonation of habitats is controlled by the morphology of the mounds and their inter-action with strong currents. Pelagic ooze and current-transported siliciclastic sediments get baffled by the coral framework and rubble patches. Heterotrophic bacteria activity might dominate. Such processes would not imply early lithification, as is suggested by the seismic velocities which do not significantly differ from off-mound sediments. Slumping would occur frequently on the mound flanks and contribute to rubble layers. A modern example might be the "reef" to "coral bank" facies (De Mol et al. in press) characterizing the surface of the Belgica and Hovland mounds, Porcupine Basin.

In addition, we might consider – where relevant – the possible impact of a fourth stage: the incipient burial, which in general will shape an asymmetry in exposure of the flanks to oceanic and intergranular flows, and hence a possible hydraulic gradient driven by compaction in the adjacent sediments. Such hydraulic gradient may drive internal flow within the mound (Fig. 5, top

case). This potentially important process, high-lighted by the setting illustrated by the high-resolution section through the *Challenger* mound, may impact on the patterns of internal diagenesis.

Conclusions and outlook

Messages from the fossil record

The fossil record has been up to now our principal source of information about deep-water carbonate mounds, and has led to outstanding studies. Most fossil mounds have left clues of microbial mediation, but hypotheses about the key processes lack "ocean-truthing", and we are not sure that identifiable microbial remains in the fossil record lead to the key players, or rather to the most easily fossilized guests.

Where energy fluxes converge

Modern mounds recently documented in various places of the ocean are prominent and dynamic ocean margin systems, beats of Life at the cross-roads of energy fluxes from internal basinal origin, from oceanic origin, from sedimentary origin and from ecosystem dynamics and evolution.

The clustered structure of mound swarms already observed by Monty (1995) for Palaeozoic mounds might suggest that the "crossroads" concept may be taken literally. Where transient fluxes of methane in a hydrocarbon basin seep at the seabed within an ocean depth range displaying negligible fluxes and mixing rates of nutrients, features like pockmarks and authigenic carbonates may form (Hovland and Judd 1988), but they might make little chance for initiating a relaying mound buildup stage.

On the other hand, the prolific coral and sponges belt of the Atlantic margin of Europe may twist as a ribbon over thousands of kilometers in a narrow bathymetric window (Le Danois 1948; Freiwald 1998). However, it comes across giant mound clusters only locally, in discrete, sharply confined patches, in general in hydrocarbon-bearing basins: "hot spots", transient meeting places with a flux of energy from the depths?

And where the initial trigger involves chemo-synthetic processes in a seep environment, the relay by a subsequent complex woven of microbial processes and colonization by metazoans will fully reflect the evolution, extinction, competition and ecological success of various groups of prokaryotes and invertebrates in the considered oceanic window, and at the relevant moment in Earth and Life history.

The way ahead

Whether the above preliminary model – or any other model - could be substantiated or killed will only depend upon comprehensive, multidisciplinary studies, in which the drilling of some full mound structures – down to their roots - and of their embedding sequences will stand centrally. Of critical importance will be the uncontaminated sampling of sub-seabed biota, their preservation at ambient pressure and temperature conditions up to the surface, and their preliminary study in autoclaves.

But in addition, we should search for the key sites which – by adequate surface observation and sampling – will provide insight in the ways microbial processes may consolidate "slopes of mud". In a province like the Belgica mounds off southwest Ireland, we would advise to hit the steepest slopes, now easily tracked by multibeam bathymetry, side scan sonar and 2D/3D very high-resolution seismic investigations. Submarines and ROV's will be essential tools in this quest, and the development of advanced sampling techniques is a prerequisite. Maybe the most promising sites for such studies will be in provinces where mound-building is in an incipient stage, such as probably in the Gulf of Cadiz.

Last but not least, such exercise bears the potential of an exciting international educational exercise, which can reveal to students (of any age) how Earth and Life, Geosphere and Biosphere may have interacted throughout Phanerozoic times, and still interact in a most spectacular way.

Acknowledgements

ODP Proposal 573 has been submitted by J.P. Henriet, B. De Mol, W.C. Dullo, A. Freiwald, B. Jørgensen, J. Parkes and J. Patching on behalf of the EU projects ACES, ECOMOUND, Deep-Bug and GEOMOUND, within the 5th Framework Programme. The merits of this large-scale joint effort are shared by the whole team. The errors are the privilege of the first author. Preparatory research has been supported by various national agencies, through national programmes and PhD grants. In Belgium, the study of the Porcupine Basin is supported by the "Porcupine-Belgica" project (University of Ghent, Special Research Fund), by the Federal Office for Scientific, Technical and Cultural Affairs, by the Management Unit of the Mathematical Model of the North Sea and Scheldt Estuary, by the Flemish Fund for Scientific Research and the Flemish Institute for the Promotion of the Scientific-Technological Research in the Industry. IOC-UNESCO is acknowledged for its support granted to the Training Through Research cruises.

References

Anselmetti FS, Eberli GP (1997) Sonic velocity in Carbonate Sediments and Rocks. Society of Exploration Geophysicists, Geophys Develop Ser 6: 53-74

Bosence DWJ, Bridges PH (1995) A review of the origin and evolution of carbonate mud-mounds. In: Monty CLV, Bosence DWJ, Bridges PH, Pratt BR (eds) Carbonate Mud-Mounds. Their Origin and Evolution. Blackwell Science, Oxford, International Association of Sedimentologists, Spec Publ 23: 3-9

Bourque PA, Boulvain F (1993) A model for the origin and petrogenesis of the red stromatactis limestone of Palaeozoic carbonate mounds. J Sed Petr 63: 607-619

Camoin GF (1995) Nature and origin of Late Cretaceous mud-mounds, north Africa. In: Monty CLV, Bosence DWJ, Bridges PH, Pratt BR (eds) Carbonate Mud-Mounds. Their Origin and Evolution. Blackwell Science, Oxford, International Association of Sedimentologists, Spec Publ 23: 385-400

Castanier S, Le Métayer-Levrel G, Perthuisot JP (2000) Bacterial roles in the precipitation of carbonate minerals. In: Riding RE, Awramik M (eds) Microbial Sediments. Springer, Berlin pp 32-39

Clari PA, Martire L (2000) Cold seep carbonates in the Tertiary of Northwest Italy: Evidence of bacterial degradation of methane. In: Riding RE, Awramik M (eds) Microbial Sediments. Springer, Berlin pp 261-269

Croker PF, O'Loughlin O (1998) A catalogue of Irish offshore carbonate mud mounds. In: De Mol B (ed)

Geosphere-biosphere coupling: Carbonate Mud Mounds and Cold Water Reefs. IOC-UNESCO Workshop Report 143: 11

De Dorlodot H (1909) Les faunes du Dinantien et leur signification stratigraphique. Bull Soc Belge Géol XXIII: 153-174

De Mol B (ed) (1998) Geosphere-Biosphere coupling: Carbonate Mud Mounds and Cold Water Reefs. IOC-UNESCO Workshop Report 143

De Mol B, Van Rensbergen P, Pillen S, Van Herreweghe K, Van Rooij D, McDonnell A, Huvenne V, Ivanov M, Swennen R, Henriet JP (in press) Large deep-water coral banks in the Porcupine Basin, southwest of Ireland. Mar Geol

Feary D, Hine A, Malone M and the Leg 182 Scientific Party (1999) Cool-water 'reefs', possible hydrogen sulphide / methane clathrates, and brine circulation – preliminary results of Leg 182 drilling in the Great Australian Bight. Joides Journal 25, 2:4-7

Freiwald A (1998) Geobiology of *Lophelia pertusa* (Scleractinia) Reefs in the North Atlantic. Habil Thesis, University of Bremen

Garcia-Mondéjar J, Fernandez-Mendiola PA (1995) Albian carbonate mounds: Comparative study in the context of sea-level variations (Soba, northern Spain). In: Monty CLV, Bosence DWJ, Bridges PH, Pratt BR (eds) Carbonate Mud-Mounds. Their Origin and Evolution. Blackwell Science, Oxford, International Association of Sedimentologists Spec Publ 23: 359-384

Henriet JP, Mienert J (eds) (1998) Gas Hydrates. Relevance to World Margin Stability and Climatic Change. Geol Soc London Spec Publ 137, 338p

Henriet JP, De Mol B, Pillen S, Vanneste M, Van Rooij D, Versteeg W, Croker PF, Shannon PM, Unnithan V, Bouriak S, Chachkine P (1998) Gas hydrate crystals may help build reefs. Nature 391: 648-649

Henriet JP, De Mol B, Vanneste M, Huvenne V, Van Rooij D and the Porcupine-Belgica 97, 98 and 99 Shipboard Parties (in press) Carbonate mounds and slope failures in the Porcupine Basin: A development model involving fluid venting. Geol Soc London Spec Publ

Heyrman J, Huys G, Swings J (1997) - Bacteriological analysis of soil samples from the Porcupine Basin. Internal Report, Laboratory of Microbiology, University of Ghent

Hovland M (1990) Do carbonate reefs form due to fluid seeps? Terra Nova 2: 8-18

Hovland M, Judd AG (1988) Seabed Pockmarks and Seepages. Graham and Trotman, London

Hovland M, Croker PF, Martin M (1994) Fault-associated seabed mounds (carbonate knolls?) off western Ireland and north-west Australia. Mar Petrol Geol 11: 232-246

Ivanov M, Pinheiro L, Henriet JP, Gardner J, Akhmanov G (2000) Some evidences of the relationship between carbonate mound formation and cold seepage in the Gulf of Cadiz. AAPG Fall Session, Special Session Geosphere-Biosphere Coupling: Cold Seep Related Carbonate and Mound Formation and Ecology, abstract

Kenyon NH, Ivanov MK, Akhmetanov AM (1998) Cold water carbonate mounds and sediment transport on the Northeast Atlantic margin. IOC-UNESCO Technical Series, 52, 178 pp

Le Danois E (1948) Les profondeurs de la mer. Payot, Paris

Lees A, Miller J (1995) Waulsortian banks. In: Monty CLV, Bosence DWJ, Bridges PH, Pratt BR (eds) Carbonate Mud-Mounds. Their Origin and Evolution. Blackwell Science, Oxford, International Association of Sedimentologists Spec Publ 23: 191-271

Monty CLV (1995) The rise and nature of carbonate mud-mounds: an introductory actualistic approach. In: Monty CLV, Bosence DWJ, Bridges PH, Pratt BR (eds) (1995) Carbonate Mud-Mounds. Their Origin and Evolution. Blackwell Science, Oxford, International Association of Sedimentologists Spec Publ 23: 11-48

Mounji D, Bourque P-A, Savard MM (1998) Hydrothermal origin of Devonian conical mounds (kess-kess) of Hamar-Lakhad Ridge, Anti-Atlas, Morocco. Geol 26 : 1123-1126

Paull CK, Borowski, WS, Rodriguez NM and the ODP Leg 164 Shipboard Scientific Party (1998) Marine gas hydrate inventory: Preliminary results of ODP Leg 164 and implications for gas venting and slumping associated with the Blake Ridge gas hydrate field. In: Henriet JP, Mienert J (eds) Gas Hydrates. Relevance to World Margin Stability and Climatic Change. Geol Soc London Spec Publ 137: 153-160

Pratt BR (1995) The origin, biota and evolution of deep-water mud-mounds. In: Monty CLV, Bosence DWJ, Bridges PH, Pratt BR (eds) (1995) Carbonate Mud-Mounds. Their Origin and Evolution. Blackwell Science, Oxford, International Association of Sedimentologists Spec Publ 23: 49-123

Pratt BR (2000) Microbial contribution to reefal mud-mounds in ancient deep-water settings: Evidence from the Cambrian. In: Riding RE, Awramik M (eds) Microbial Sediments. Springer, Berlin pp 282-288

Riding RE, Awramik M (eds) (2000) Microbial Sediments. Springer, Berlin

Schmid DU, Leinfelder RR, Nose M (2000) Growth dynamics and ecology of Upper Jurasic mounds, with comparison to Mid-Palaeozoic mounds. In: Neuweiler F, Boulvain F and Bourque P A (eds) (2000) Palaeoceanography of Carbonate Mud-Mounds, Workshop Abstract volume, IGCP 380 Biosedimentology of Microbial Buildups, Liège, pp 29-30

Stanton R J, Jeffery D L, Guillemette RN (2000) Oxygen minimum zone and internal waves as potential controls on location and growth of Waulsortian Mounds (Mississippian, Sacramento Mountains, New Mexico). Facies 42 : 161-176

Stetson TR, Squires DF, Pratt RM (1962) Coral Banks Occurring in Deep Water on the Blake Plateau. Novitates, Amer Mus of Nat Hist, New-York, 2114 : 1-39

Stolz JF (2000) Structure of microbial mats and films. In: Riding RE, Awramik M (eds) Microbial Sediments.

Springer, Berlin pp 1-8

Surlyk F (1997) A cool-water carbonate ramp with bryozoan mounds : Late Cretaceous – Danian of the Danish Basin. In: James NP, Clarke JAD (eds) Cool-Water Carbonates. SEPM Spec Publ 56:293-308

Swart PK, Wortmann UG, Mitterer RM, Malone MJ, Smart PL, Feary DA, Hine AC (2000) Hydrogen sulfide-rich hydrates and saline fluids in the continental margin of South Australia. Geol 28:1039-1042

van Hinte JE, Ruffman A, van den Boogaard M, Jansonius J, van Kempen TMG, Melchin MJ, Miller TH (1995) Palaeozoic microfossils from Orphan Knoll, NW Atlantic Ocean. Scripta Geologica, Leiden, pp 1-64

Vernadsky VI (1926) The Biosphere. Engl. Translation : Langmuir DB (1997). Nevraumont Publ Cy, New York

Westbroek P (1991) Life as a Geological Force. WW Norton and Company, New York

The Anaerobic Oxidation of Methane: New Insights in Microbial Ecology and Biogeochemistry

K.-U. Hinrichs[1,2*], A. Boetius[3]

[1] *Woods Hole Oceanographic Institution, Woods Hole, MA 02543, USA*
[2] *Hanse - Wissenschaftskolleg, 27753 Delmenhorst, Germany*
[3] *Max Planck Institute for Marine Microbiology,Celsiusstr.1, 28359 Bremen, Germany*
*corresponding author (e-mail): khinrichs@whoi.edu

Abstract: As the major biological sink of methane in marine sediments, the microbially mediated anaerobic oxidation of methane (AOM) is crucial in its role of maintaining a sensitive balance of our atmosphere's greenhouse gas content. Although there is now sufficient geochemical evidence to exactly locate the "hot spots" of AOM, and to crudely estimate its contribution to the methane cycle, a fundamental understanding of the associated biology is still lacking, consequently preventing a thorough biogeochemical understanding of an integral process in the global carbon cycle. Earlier microbiological work trying to resolve the enigma of AOM mostly failed because it was largely focussed on the simulation of AOM under laboratory conditions using cultivable candidate organisms. Now again, understanding the biological and biochemical details of AOM is the declared goal of several international research groups, but this time in a combined effort of biogeochemists and microbiologists using novel analytical tools tailored for the study of unknown microbes and habitats. This review gives an overview on very recent progress in the study of AOM that dramatically advanced this ~ 30-yr-old field. New insights on the quantitative significance of AOM are combined to refine older estimates.

Introduction

Since pioneering reports by Reeburgh (1976) and Barnes and Goldberg (1976), subsequent studies employing stable isotopes, radiotracers, modeling, and inhibition techniques have established that methane in marine sediments is oxidized biologically under anoxic conditions (see review of Valentine and Reeburgh 2000, and references therein). Despite the compelling evidence for the anaerobic oxidation of methane (AOM), details of the related biochemical mechanisms and organisms are still unknown. Zehnder and Brock (1979) showed that incubation of active methanogenic cultures with $^{14}CH_4$ would result in formation of some $^{14}CO_2$, thus demonstrating the oxidation of a small fraction of methane under anaerobic conditions. Although no anaerobic methanotroph has ever been isolated, biogeochemical studies have shown that the overall process involves a transfer of electrons from methane to sulfate (Iversen and Jørgensen 1985; Hoehler et al. 1994). The isotopic

and genetic signature of the microbial biomass in environments enriched with methane shows that this transfer is probably mediated by a microbial consortium that includes archaea and sulfate-reducing bacteria (Hinrichs et al. 1999; Boetius et al. 2000). The distributions of abundant lipid products apparently derived from members of AOM consortia indicate a substantial diversity among the microbial players in different methane-rich environments. However, the intermediate substrate (e.g. $H_2 + CO_2$, acetate, formate), which is exchanged between archaeal and bacterial members of the consortium and thereby coupling methane oxidation and sulfate reduction remains unidentified. In several sedimentary environments, AOM can be the dominant sulfate-consuming process, e.g. in sediments from Carolina Rise and Blake Ridge (Borowski et al. 1996) or at Hydrate Ridge (Boetius et al. 2000). Also, AOM may prove to be the oxidative process that penetrates most deeply

From WEFER G, BILLETT D, HEBBELN D, JØRGENSEN BB, SCHLÜTER M, VAN WEERING T (eds), 2002,
Ocean Margin Systems. Springer-Verlag Berlin Heidelberg, pp 457-477

into anoxic environments wherever sulfate is available. In the methane budget proposed by Reeburgh (1996), more than 80% of the methane produced annually in marine sediments is consumed mostly in anoxic environments before it can reach the atmosphere. The previously estimated 75 Tg/yr methane consumption is nearly twice the annual increase in the atmospheric inventory of CH_4 (40 Tg/yr). However, our updated compilation of published AOM rates shows that the consumption of methane in anoxic sediments is probably several times higher than previously estimated, implying that methane production estimates are probably too low as well. In contrast to earlier assessments, AOM may have been an important biogeochemical process in earlier stages of Earth history.

The Process

The metabolic process of AOM is still unknown and all reactions discussed in the literature remain speculative as long as the elusive microorganisms involved in AOM are not available for physiological investigations. However, different lines of evidence, including pore water concentration profiles, radioisotope measurements, biomarker studies, phylogenetic analyses, and thermodynamic models serve as a basis to examine the likelihood of the various proposed pathways.

In the mid 70's, it was established that in certain horizons in anoxic sediments and waters at continental margins net consumption of methane occurs (Barnes and Goldberg 1976, Reeburgh 1976). It was observed that AOM peaks coincide with increased sulfate reduction. Thus, Barnes and Goldberg (1976) suggested sulfate as the terminal electron acceptor for this process according to reaction (1).

$$CH_4 + SO_4^{2-} \rightarrow HCO_3^- + HS^- + H_2O \qquad (1)$$

Thermodynamic models show that this reaction can become favorable at in situ conditions in marine sediments, however, only with a low free energy yield of -25 kJ per mol methane consumed (Hoehler et al. 1994). This is only approximately half the energy required for the formation of ATP. Until

today it has been discussed controversially whether AOM can support microbial growth. However, recently, extremely high amounts of biomass of aggregated archaea and sulfate reducing bacteria involved in AOM were found in surface sediments above marine gas hydrate (Boetius et al. 2000). The biomass of the consortia in the zone of AOM (1-10 cm sediment depth) exceeded the total microbial biomass of surrounding sediments by an order of magnitude and even served as food to some members of the macrofauna community, as indicated by stable isotope analysis (Levin, unpubl. data). Furthermore, the cell-specific rates of methane based sulfate reduction were similar to those of cultivated SRB under optimal culture conditions. These field data indicate that the process of AOM can support significant cell growth and activity.

Counterintuitive to thermodynamic considerations predicting an extremely low energy gain of AOM, the involvement of two or more microorganisms as syntrophic partners was discussed already at the beginning of microbiological research on this process (see review of Valentine and Reeburgh 2000, and literature therein). It was proposed that a consortium of microorganisms involving both methanogens and bacteria may mediate AOM, with the latter oxidizing intermediate products derived from methane. In field experiments conducted by Hoehler et al. (1994), AOM was possible as long as hydrogen concentrations were kept at extremely low levels. The authors concluded that AOM is mediated by two syntrophic partners, which rely on interspecies hydrogen transfer: methanogenic archaea mediating the oxidation of methane with water (reaction 2), and sulfate reducing bacteria scavenging the intermediate hydrogen (reaction 3).

$$CH_4 + 2 H_2O \rightarrow CO_2 + 4 H_2 \qquad (2)$$
$$SO_4^{2-} + 4 H_2 + H^+ \rightarrow HS^- + 4 H_2O \qquad (3)$$

The free energy gain of the reversal of methanogenesis as shown in reaction (2) depends to a large degree on the concentration of dissolved hydrogen in pore waters. For example, under reactant/product concentrations and temperatures typical for Cape Lookout Bight sediments, AOM becomes favorable at H_2 concentrations below 0.3 nM (Hoehler et al. 1994). The involvement of methanogenic archaea in the first part of the reaction is supported by the occurrence of strongly ^{13}C-depleted archaeal biomarkers in environments with evidence for high rates of AOM (Hinrichs et al. 1999, Elvert et al. 1999, Thiel et al. 2001). Extreme depletions of ^{13}C were also observed in bacterial biomarker lipids, consistent with a derivation from methane-C (Boetius et al. 2000, Hinrichs et al. 2000, Elvert et al. 2000). To explain the low $\delta^{13}C$ in the lipids of the SRB by reaction (3), one would have to assume a direct transfer and assimilation of the $\delta^{13}C$-depleted CO_2 produced from methane within the archaea/SRB-consortium. This mechanism requires formation of microenvironments with strong physicochemical gradients, so that utilization of CO_2 from reaction (2) would be much faster than diffusion of pore water CO_2 to the SRB partner. According to a diffusion model, this is only possible if the distance between both partners is smaller than 10 µm, which agrees with dimensions observed in aggregates of archaea/SRB in sediments at Hydrate Ridge (Boetius et al. 2000 and literature therein). However, uncertainty remains with respect to the fractionation factors for autotrophic consumption of CO_2 by the SRB, which could also cause significant depletion of ^{13}C in their lipids. In principle, biosynthesis from inorganic carbon in the SRB appears feasible and is consistent with recent observations at Hydrate Ridge, where many relatives of the sulfate reducers associated with archaea are facultative autotrophs.

According to the biomarker and 16S rRNA analyses, the dominating archaea in zones of AOM at methane seeps and above methane hydrate (Hinrichs et al. 1999; Boetius et al. 2000; Orphan et al. in press) are closely related to the *Methanosarcinales*. These methanogens include the known major producers of hydroxy-archeols and typically use acetate and reduced C_1-compounds for methanogenesis. Hoehler et al. (1994) discussed acetate as an intermediate in AOM. The reversal of acetoclastic methanogenesis would involve the formation of acetate according to reaction (4), which could serve as a carbon as well as an energy source to the sulfate re-

ducing bacteria. Also, cultivated bacteria of the *Desulfosarcina/Desulfococcus* group of the deltaproteobacteria are known as complete acetate oxidizers, able to mediate reaction (5).

$$CH_4 + HCO_3^- \rightarrow CH_3COO^- + H_2O \qquad (4)$$
$$SO_4^{2-} + CH_3COO^- \rightarrow HS^- + 2HCO_3^- \qquad (5)$$

However, to make AOM with acetate as intermediate thermodynamically attractive, acetate concentrations would have to be lower than 2 nM to gain at least −10 kJ of free energy (Hoehler et al. 1994; Boetius et al. 2000). Such low concentrations are uncommon for most marine sediments and would be far below the *Michaelis Menten constant* for acetate uptake by known SRB.

Valentine and Reeburgh (2000) recently proposed an alternative pathway involving acetate that yields a higher change of free energy (-50 kJ) and could explain isotopic depletions in biosynthetic products of the SRB partner more satisfactorily than reactions (2) and (4). They suggest a formation of acetate solely from methane, and a transfer of both acetate and hydrogen to the SRB partners (reaction 6-8). However, the biochemical pathways of reaction (6) are without example in cultured organisms.

$$2CH_4 + 2H_2O \rightarrow CH_3COOH + 4H_2 \qquad (6)$$
$$4 H_2 + SO_4^{2-} + H^+ \rightarrow HS^- + 4H_2O \qquad (7)$$
$$CH_3COOH + SO_4^{2-} \rightarrow 2HCO_3^- + HS^- + H^+ \quad (8)$$
$$2(CH_4 + SO_4^{2-} \rightarrow HCO_3^- + HS^- + H_2O)$$
$$\text{(net reaction)}$$

The *Methanosarcinales* uniquely include the methylotrophic methanogens. These organisms are capable of disproportionating methanol, methylamines, or methyl sulfides to methane and carbon dioxide, indicative of an enzyme system to oxidize $R–CH_3$ to CO_2 anaerobically. Hence, such C_1 compounds could also be potential intermediates according to reactions (9/10) with methanol as an example. However, only very few known SRB are able to use methanol as their sole carbon substrate, and no SRB have been isolated yet which can grow on methylamine (Hansen pers. comm. 2000).

$$4CH_4 + 4H_2O \rightarrow 4CH_3OH + 4H_2 \qquad (9)$$
$$3SO_4^{2-} + 4CH_3OH \rightarrow 3HS^- + 4HCO_3^- + 4H_2O + H^+ \qquad (10)$$

A thermodynamic model considering the difference in concentrations between the producing and consuming partners by Sørensen et al. (in press) tests the likelihood of the different possible elec-tron shuttles hydrogen, acetate, methanol, and formate. Only transfer of formate according to reactions (11/12) resulted in free energy gain. Formate is used by members of the order *Methanobacteriales* and *Methanococcales* for methanogenesis but no member of the order *Methanosarcinales* is known to utilize formate.

$$CH_4 + 3\,HCO_3^- \rightarrow 4HCOO^- + H^+ + H_2O \qquad (11)$$
$$SO_4^{2-} + 4HCOO^- + H^+ \rightarrow HS^- + 4HCO_3^- \qquad (12)$$

In the attempt of predicting the metabolic pathways of AOM on the basis of our current knowledge, a reversal of methanogenesis on the basis of known enzymes appears attractive. In the case of methanogenesis, some of the enzymes of methanogens used for the reduction of CO_2 to CH_4 are operating in reverse in the oxidative pathway of the sulfate-reducing archaeon *Archaeoglobus*. However, the final enzymatic step in methane production involves the protonation of methyl-nickel by the enzyme methyl-CoM reductase and is considered irreversible (see review of Hoehler and Alperin 1996 and literature therein). Hence, it is likely that other, yet unknown enzymes are involved. The complete sequencing of the genomes of the methanogenic archaea *Methanococcus jannaschii* and *Methanobacterium thermoautotrophicum* detected 30% putative coding regions with no similarity to any sequence in other organisms (Gaasterland 1999) and most regions similar to other genomes code for proteins of unknown function. Notably, members of the order *Methanosarcinales* have a much larger genome than other methanogens. The sequencing of a first member of the *Methanosarcinales*, *Methanosarcina mazei*, is presently in progress.

Present Knowledge on Microbes Involved in AOM

Early experiments by Zehnder and Brock (1979) indicated that methanogenic archaea are capable of oxidizing $^{14}CH_4$ to $^{14}CO_2$ under anaerobic culture conditions, although only at a fraction <0.5% of the concurrent methane production. Additionally, a small amount of ^{14}C was incorporated into archaeal biomass. The most efficient CH_4 oxidizer was *Methanosarcina barkeri*, which also produced acetate and methanol during CH_4 oxidation. Early experiments using ^{14}C-labelled methane were likely biased by impurities of ^{14}CO, which is oxidized by several anaerobic microorganisms to $^{14}CO_2$ (Harder 1997). Using pure $^{14}CH_4$, Harder (1997) showed that *Methanosarcina acetivorans*, *Methanospirillum hungatei*, and *Methanolobus tindarius* oxidized methane at low rates. None of the sulfate reducing bacteria tested with pure $^{14}CH_4$ produced significant amounts of $^{14}CO_2$ from methane. So far, all pure culture experiments (including *Methanosarcina*, *Methanosaeta* and *Methanobacterium spp.* as test organisms) employing maintenance of low H_2 pressure via gas sparging, or co-culturing of SRB (*Desulfotomaculum*) failed to reverse methanogenesis to methane consumption (Valentine and Reeburgh 2000). However, at very high methane pressure (100 atm), Shilov et al. (1999) observed a reversal of acetoclastic methanogenesis in sludge granules that consisted of mixed cultures of microorganisms dominated by *Methanosarcina* and *Methanosaeta spp.* In any case, the direct association of archaea with SRB would allow for a highly efficient transfer of intermediates by molecular diffusion compared to free-living cells. This could be a prerequisite to render the process of AOM thermodynamically favorable (Sørensen et al. in press).

Hinrichs et al. (1999) combined evidence from biomarker studies and phylogenetic analyses and thereby detected a new group of archaea possibly involved in AOM in sediments from methane seeps of the Eel River Basin. The archaea-specific lipid biomarkers isolated from the seep sediments were strongly depleted in ^{13}C ($\delta < -100$‰ *vs*. PDB), indicating that methane was the carbon source for the organisms that synthesized them. These biomarkers are known to occur in methanogenic archaea and have been isolated from members of the order *Methanosarcinales*. The ribosomal-RNA gene library yielded only two major archaeal phylogenetic groups from 176 clones analyzed. Most rDNA sequences, accounting for 148 of the 176 archaeal clones recovered, form a cluster of unique but highly related sequences (group ANME-1) (Fig. 1), each distinguishable by RFLP analysis and primary structure. Other clones were comprised of a sequence type closely related to cultivated members of the *Methanosarcinales* (ANME-2 group; Orphan et al. in press). The group ANME-1 is distinct from, but related to, methanogenic archaea of the orders *Methanomicrobiales* and *Methanosarcinales*. ANME-1 related sequences have been recovered from a variety of methane-rich locations but not from any other aquatic environments. In sediments of the Eel River Basin and Hydrate Ridge, no rRNA sequences of previously cultured methanogens like *Methanosaeta concilii* or *Methanosarcina spp.* - known producers of *sn*-2-hydroxyarchaeol - were detected, implying that the ^{13}C-depleted archaeal biomarkers are likely produced by organisms represented by sequences either within the

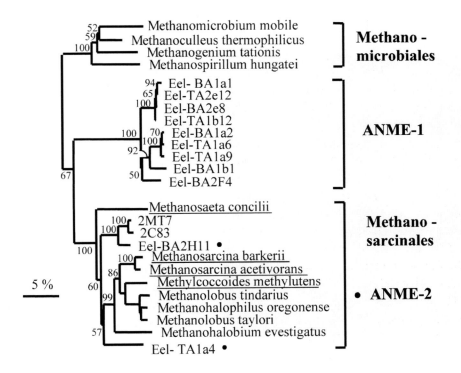

Fig. 1. Partial illustration of phylogenetic analysis of archaeal ribosomal rRNA sequences recovered from seep sediments in the Eel River Basin (modified after Hinrichs et al. 1999), showing the methanogenic archaeal orders *Methanomicrobiales*, *Methanosarcinales*, and the newly recovered groups ANME-1 and ANME-2 (anaerobic oxidation of methane). *Escherichia coli, Thermotoga maritima, Aquifex pyrophilus* and *Synechococcus* PCC6301 were used as outgroups. Scale bar represents 5 fixed mutations per 100 nucleotide sequence positions. Sequences obtained in this study have been submitted to GenBank under the accession numbers AF134380-AF134393. Sequences beginning with Eel- were obtained from samples that contain extremely ^{13}C-depleted archaeol and hydroxyarchaeol. Methanogenic archaea of the order *Methanosarcinales* that are known major producers of hydroxyarchaeols are indicated by underlined letters.

ANME-1 or ANME-2 cluster. Recent observations of conspicuous aggregates of archaea and sulfate reducing bacteria complement and strengthen earlier findings based on lipids and molecular phylogeny (Boetius et al. 2000). Fluorescence in situ hybridization (FISH) revealed that both archaea and SRB grow together in aggregates of approximately 3 μm diameter with an inner core consisting of archaeal cells and an outer shell of sulfate reducers (Fig. 2). The rRNA probes targeted specifically the ANME-2 group among the archaea and a new cluster of delta-proteobacterial SRB that is closely related to the *Desulfosarcina-Desulfococcus* group. In some sampling intervals, these aggregates comprised over 90% of the total microbial biomass and are most likely responsible for the extremely high rates of sulfate reduction (>5 μmol cm^{-3} d^{-1}) in the methane-saturated sediments.

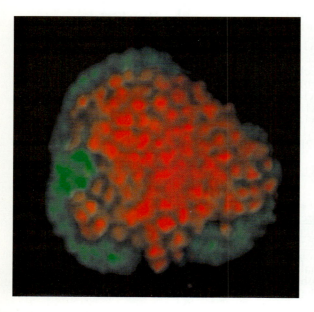

Fig. 2. An archaea/SRB consortium apparently mediating AOM. Aggregated archaea and sulfate reducing bacteria were found in sediments above gas hydrates (Hydrate Ridge, continental slope off Oregon, USA). The aggregated cells were exposed to a green-fluorescent RNA-probe targeting sulfate-reducing bacteria (*Desulfococcus-Desulfosarcina* group) and a red-fluorescent RNA-probe targeting archaea (ANME-2 group). The aggregate has a size of approxi-mately 10 μm. The image was taken by confocal laser scanning microscopy. Reprinted with permission from Nature (Boetius et al., Nature 407:623-626). Copyright (2000) Macmillan Magazines Limited.

Recent work showed that both ANME groups and the members of the *Desulfosarcina-Desulfococcus* group co-occur in other methane-rich sediments. They were found in surface sediments at deep-sea hydrothermal vents in the Guaymas Basin (Teske et al. 2001) and the Eel River Basin (Orphan et al. in press). In their study on microbial diversity in Cascadia margin deep subsurface sediments with abundant methane hydrate, Bidle et al. (1999) detected 16S rDNA sequences of SRB belonging to the *Desulfosarcina-Desulfococcus* group as well as sequences of the methyl coenzyme M reductase gene originating from members of the order *Methanosarcinales*. In sediments of the Aarhus Bay, Denmark, rDNA sequences of the ANME-1 group were detected in the AOM zone at 165 cm below the seafloor. Phylogenetic analysis of the gene of dissimilatory sulfite reductase (DSR) revealed that all retrieved sequences belonged to a novel deeply branching lineage unrelated to all previously described DSR genes (Thomsen et al. in press). Currently, it is not known whether the ANME group or any other methanogenic archaeon possesses the DSR gene and could account for the deep branching DSR.

Evidence derived from parallel biomarker chemotaxonomy and 16S rRNA or FISH probing in several environments supports the hypothesis that several phylotypes have to be considered as producers of ^{13}C-depleted archaeol and hydroxyarchaeol (for details, see chapter on lipid biomarkers). A series of studies report the predominant occurrence of phylotypes from the ANME-2 group (more closely related to *Methanosarcinales* than ANME-1) in sediments hosting active AOM communities (Boetius et al. 2000; Orphan et al. in press; Teske et al. 2001). On the other hand, we observed cases in which the ANME-1 group was predominant (Hinrichs et al. 1999) or even the exclusive archaeal group (Teske et al. 2001), but ^{13}C-depleted archaeol and hydroxyarchaeol were present as well. This indicates a considerable archaeal diversity in AOM communities. In this regard, interesting observations were made at surface sediments in the Guaymas Basin next to hydrothermal vents, where samples taken in close proximity with either ANME-1 or ANME-2 were analyzed for AOM process markers. The presence of archaeol and hydroxyarchaeol was confirmed for both sample types, but their isotopic compositions displayed significant differences. Specifically, the archaeal lipids associated with the ANME-1 assemblage were enriched in ^{13}C compared to the sample with the ANME-2 assemblage. ANME-1 was associated with archaeol and sn-2-hydroxyarchaeol being –58 and –70‰, respectively, whereas $\delta^{13}C$ was –82‰ for both compounds in association with the ANME-2 assemblage. Interestingly, the relative contribution of ^{13}C-depleted ether lipids assigned to bacterial members of the AOM consortium (Hinrichs et al. 2000) was significantly lower in the ANME-2 dominated sample.

One important question for understanding the process of AOM is whether this process obligatorily requires the syntrophy of SRB with methanogenic archaea in the form of symbiotic associations as observed in the sediments of the Hydrate Ridge. Orphan et al. (in press) found a similar consortium in sediments of the Eel River Basin. A different archaea/bacteria consortium has been detected in surface sediments above a subsurface gas hydrate layer in the Congo basin (Ravenschlag et al. unpubl. data). It is likely that different forms of syntrophic associations are responsible for AOM in such methane-rich environments. No microscopic investigations have been carried out in low energy zones like in the subsurface sulfate/methane transition zone. Some of the data of Thomsen et al. (in press) point to the existence of a single microorganism capable of AOM in subsurface sediments from Aarhus Bay, Denmark. In subsurface sediments several 100 m below the sea-floor, active sulfate-reducing bacteria are found, and concentrations of methane vary at levels suggestive of dynamic microbial control. It is possible that populations of microorganisms mediating AOM are an integral part of the deep bacterial biosphere.

Lipid Biomarkers Associated with AOM

Lipid biomarkers in the study of microbial communities allow distinction on the level of kingdoms and sometimes orders. A biomarkers' stable carbon isotopic composition bears diagnostic information on the carbon source and/or metabolic carbon fixation pathway utilized by its producer. This property so far has remained inaccessible for gene fragments recovered in environmental studies of microbial ecology. On the other hand, modern, culture-independent techniques such as the determination of molecular phylogeny based on 16S rRNA are more specific than lipid biomarkers.

The study of microbial biomarkers in sedimentary environments with active AOM led 1994 to the first evidence that anaerobic microorganisms are capable of assimilating methane-derived carbon (Bian 1994; Bian et al. 2001). Lipid biomarker of methanotrophic archaea often display $\delta^{13}C$ values of –100‰ and lower. This is indicative of more or less exclusive utilization of methane-C as carbon source for biosynthesis, because the $\delta^{13}C$ of biogenic methane is commonly lower than -60‰. Consequently, nowadays in environmental studies of the microbial ecology associated with AOM, molecular bio-markers are a crucial test for the presence of anaerobic methanotrophs and allow a circumstantial connection of other microbiological evidence to AOM (Hinrichs et al. 1999; Boetius et al. 2000; Orphan et al. in press).

Archaeal biomarker

The first observation of a biosynthetic product of anaerobic microorganisms using methane as a carbon source was restricted to crocetane (Bian 1994; Bian et al. 2001), a C_{20} isoprenoid hydrocarbon with a tail-to-tail linkage of presumed archaeal origin (Fig. 3). It was found in anoxic sediments in the Kattegat in the sulfate reduction/methane production-transition zone that exhibited maximum rates of AOM. It took several years until similar findings were made, at that time more or less simultaneously by several groups who studied lipids in ancient and modern sediments deposited in direct vicinity to active methane seeps (Elvert et al. 1999; Hinrichs et al. 1999; Pancost et al. 2000; Thiel et al. 1999). Unlike other archaeal compounds discussed in the following, crocetane has been exclusively associated with AOM, and its biological producers have never been isolated. In between, even the occurrence of sev-

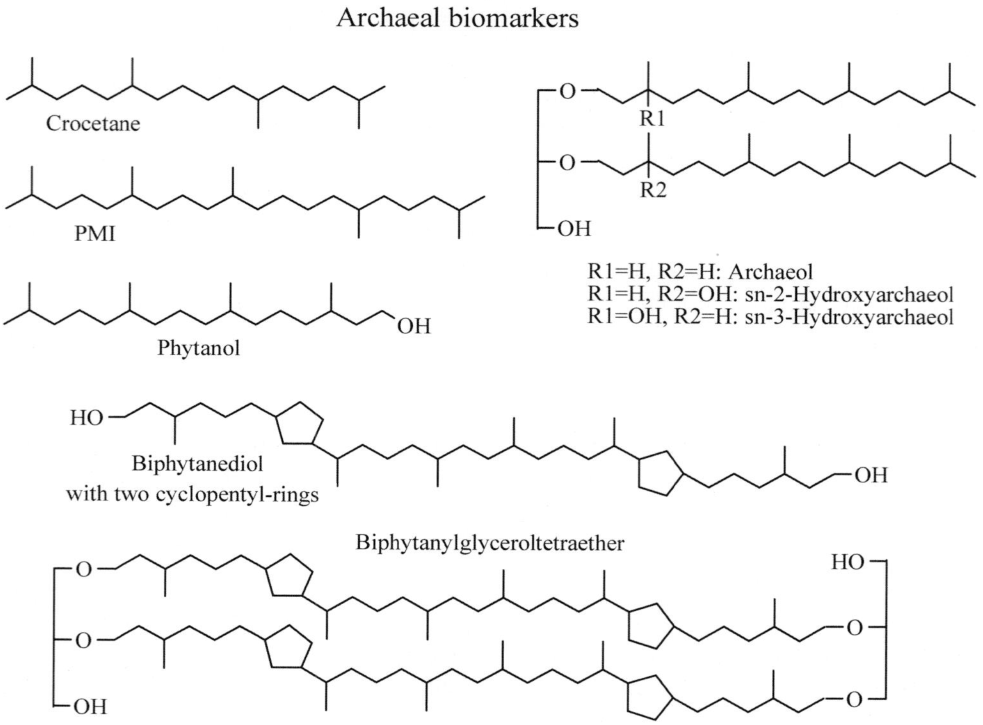

Archaeal biomarkers

Fig. 3. The most prominent archaeal biomolecules found in AOM environments. Additional compounds are unsaturated analogs of crocetane and PMI.

eral ^{13}C-depleted, unsaturated analogs of crocetane was documented (Elvert et al. 2000; Pancost et al. 2000).

The first unequivocal evidence for an involvement of archaea in AOM came from the detection of ^{13}C-depleted archaea-specific molecules in methane seep environments (Hinrichs et al. 1999). In that study, the compounds archaeol and *sn*-2-hydroxyarchaeol (Fig. 3) with $\delta^{13}C$ values of −100‰ and lower clearly testified to an involvement of archaea in AOM. Based on the chemotaxonomy of cultured organisms, the latter compound pointed to an involvement of members of the *Methanosarcinales*, with certain strains synthesizing hydroxyarchaeol in particular high abundance (Koga et al. 1998 and literature therein). However, we note that *sn*-2- and *sn*-3-hydroxyarchaeol were also observed in low amounts in selected members of the order *Methanococcales*. Archaeol and hydroxyarchaeol were subsequently found at many other methane-rich sampling locations, including Mediterranean mud

volcanoes, deep-sea sediments from the Aleutian subduction zone, and in surface sediments at Hydrate Ridge and in the Black Sea, respectively (Table 1). A representative reconstructed-ion-chromatogram of an alcohol fraction from a sediment extract from a seep in the Eel River Basin is illustrated in Figure 4 (after Orphan et al. in press) and exemplifies the imprint of the AOM activity on the inventory of sedimentary lipids.

Additional compounds of archaeal origin that are frequently found in AOM environments are 2,6,10,15,19-pentamethylicosane (PMI) and unsaturated analogs (PMIΔ), phytanol, and biphytanediols and biphytanyltetraethers with both cyclic and acyclic isoprenoid moieties (Fig. 3). Table 1 lists the isotopic compositions of the molecules most commonly associated with AOM for different environments. In this compilation, the presence or absence of data for certain archaeal lipids – in particular archaeol and hydroxyarchaeol - does not necessarily imply that these were not present, because many surveys were limited to hydro-

Environment	Crocetane	PMI, PMIΔ	Archaeol	Hydroxyarchaeol	Phytanol	CH₄[1]
Hydrate Ridge [1,2]	-118 to -108 (n=3)	-124 to -102 (PMI, n=3),	-114	-133		-72 to -62
Eel River Basin [3-5]	-92	-92, -76 (PMIΔ, n=2)	-103 to -100 (n=6)	-106 to -101 (n=6)	-88 to -81 (n=2)	-50
Black Sea [6]	-107	-101	-112	Na	Na	-58
Kattegat, transition zone [7,8]	-100 to -67 (n=5)	-47 to -32 (n=5)	Nd	Nd	Nd	-72
Guaymas Basin [9]	Nd	Nd	-81 to -58 (n=3)	-85 to -70 (n=3)	Nd	-51 to -43
Mediterranean mud volcanoes [10]	~-64 (crocetene)	-91 to -52 (n=6)	-96 to -41 (n=6)	~ -77, -58 (n=2)	Nr	Na
Aleutian SDZ [11]	-130 to -125 (n=3)	-107 to -71, (PMI, n=3), -117 to -98 (PMIΔ, n=3)	-124 (n=2)	Nd	~-120 to -82	Na
Santa Barbara Basin [4]	-119	-129 (PMI: 2)	-119	-128	-120	Na

[1]: δ¹³C of methane are data cited in referenced papers. Cited references: [1] Elvert et al. 1999; [2] Boetius et al. 2000; [3, 4] Hinrichs et al. 1999, 2000, respectively; [5] Orphan et al. in press; [6] Thiel et al. 2001; [7] Bian, 1994; [8] Bian et al. 2001; [9] Teske et al. in prep.; [10] Pancost et al. 2000; [11] Elvert et al. 2000. Na = not analyzed, Nd = not detected, Nr = not reported.

Table 1. Overview on isotopic compositions of selected archaeal lipids in different sedimentary environment hosting active AOM communities.

carbons. The distribution and isotopic signatures of archaeal lipids suggest that the archaeal AOM communities are diverse and may differ from site to site. Several studies indicate that crocetane and archaeol/hydroxyarchaeol have different archaeal producers. For example, cro-cetane was present in the transition zone in Kattegat, but the analysis of ether-bound lipids did not reveal any indication for the presence of archaeol and hydroxy-archaeol (Bian et al. 2001). PMI was detected and displayed very different isotopic compositions than crocetane, suggesting that in Kattegat sediments these two compounds originate from different archaea. At some seeps in the Eel River Basin, only archaeol and hydroxyarchaeol were abundant while crocetane and PMI were absent (Hinrichs et al. 1999, 2000). Other seeps contained all four compounds, but δ¹³C values differed significantly from each other (Orphan et al. in press). Moreover, at some sites crocetane and/or PMI are more depleted in ¹³C than archaeol and hydroxyarchaeol, while the opposite was observed at other sites. This lack of systematic isotopic relationships between

compounds is consistent with at least a partially different origin rather than biosynthesis-related differences in fractionation in a single organism. Similarly, relative concentrations of these compounds at different sites do not follow any systematic rules. An isotopic difference between archaeol and hydroxyarchaeol is observed frequently and may also indicate an at least partially different origin. In most samples from several different environments with both compounds present, hydroxyarchaeol is more depleted in ¹³C relative to archaeol by 4 to 19‰.

Additional chemotaxonomic evidence for the diversity of archaea involved in AOM is available when comparing biomarker data from different environments. For example, Pancost et al. (2000) and Thiel et al. (2001) report the presence of strongly ¹³C-depleted biphytanediols and biphytanyltetraethers (Fig. 3) in methane-laden environments in the Mediterranean and the Black Sea (Table 1). In contrast, the biphytanediols found at methane seeps from the Californian continental margin display δ¹³C values of around -20‰

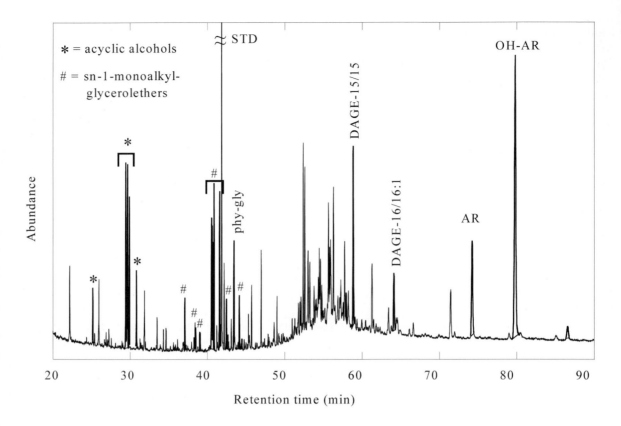

Fig. 4. Reconstructed-ion-chromatogram of alcohol fraction from a sediment sample taken at 6-9 cm depth at an active methane seep in the Eel River Basin (adapted from Orphan et al. in press). Labeled peaks designate compounds with isotopic compositions indicating a partial or exclusive derivation from anaerobic methanotrophic microbes. * = C_{14} to C_{17} acyclic alcohols from bacteria, # = sn-1-monoalkylglycerolethers with ether linked C_{14} to C_{18} acyclic alcohol moieties from bacteria. Text symbols: DAGE designates dialkylglycerolether with carbon numbers of ether-linked alkyl moieties expressed as numbers, e.g. DAGE-15/15 = diether with two C_{15}-alkyl moieties, phy-gly = sn-1-monophytanylglycerolether (archaea), AR = archaeol (archaea), OH-AR = sn-2-hydroxyarchaeol (archaea).

(K. Hinrichs, unpubl. data), which is consistent with their derivation from archaeal plankton (e.g. Schouten et al. 1998). This suggests that certain members of the archaeal methanotrophic community present in the Mediterranean and Black Sea do not participate in AOM communities at the Californian continental margin. An additional indication for the archaeal diversity is the detection of sn-3-hydroxyarchaol as a product of archaeal AOM members in Mediterranean Sea sediments (Pancost et al. 2000). This compound, originally found in *Methanosaeta concilii,* was not ob-served in other environments, where only the *Methanosarcina*-type sn-2-hydroxyarchaeol was found (Hinrichs et al. 1999; 2000; Orphan et al. in press; Teske et al. 2001).

Bacterial biomarker

In most of the previously discussed surveys of lipid inventories associated with AOM communities, ^{13}C-depleted archaeal biomarker co-occur with ^{13}C-depleted compounds of presumed bacterial origin. The bacterial compounds are usually slightly enriched in ^{13}C relative to their archaeal counterparts with typical $\delta^{13}C$ values between -100 and –50‰, depending on compound and environment. Similarly to archaeal products, this range of values indicates a partial to exclusive incorporation of methane-derived carbon. The structural distribution of these lipids is not specific for, but consistent with, their derivation from sulfate-reducing bacteria.

These lipids encompass fatty acids with 14 to 18 carbon atoms, glycerolether derivatives with one or two non-isoprenoidal alkyl chains that show strong structural resemblance to the concurrently occurring fatty acids (e.g. identical number of carbon atoms, position of double bonds, and methyl substitution), and alcohols with structural features very similar to those of the two aforementioned classes of compounds (Fig. 5). Figure 4 illustrates a high relative contribution of ^{13}C-depleted alcohols and ether lipids to an extract from a sediment at an active methane seep. The fatty acids associated with AOM communities display typical features of fatty acids from sulfate reducing bacteria, i.e., high relative amounts of methyl-branched and cyclopropyl isomers and double-bond positions typical for fatty acids from sulfate reducers

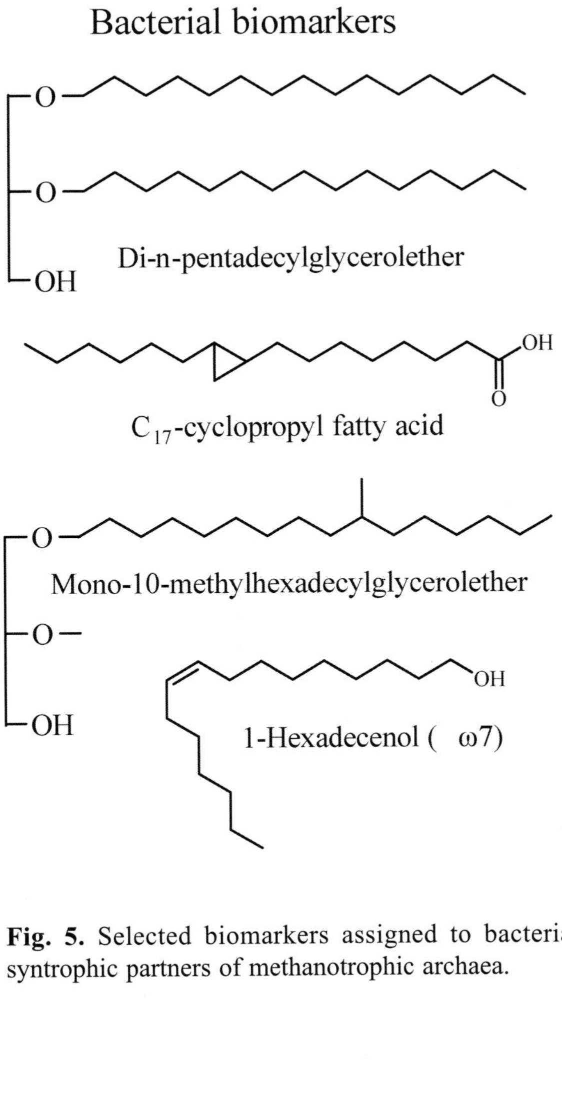

(see Hinrichs et al. 2000, for a review of structural and isotopic features of these compounds in seep sediments).

Little is known about the biological precursors of mono- and dialkyglycerolethers (MAGE and DAGE, respectively) and alcohols with similar alkyl structures (Fig. 5). Their structures are distinct from those of the ether lipids produced by archaea. The latter synthesize exclusively isoprenoidal ether lipids whereas these compounds utilize n-alkyl and methyl-alkyl substituents. Among bacteria, the only known producers of non-isoprenoidal mono- and dialkyl-ether lipids are the phylogenetically deeply branching autotrophic phyla *Aquifex spp.* and *Thermodesulfotobacterium commune* (Huber et al. 1992 and literature therein). However, these ether lipids have not been isolated from any mesophilic or psychrophilic sulfate reducer. Both are (hyper)thermophiles restricted to hydrothermally active environments. In seep sediments with abundant bacterial ether lipids, phylotypes associated with these known producers were not detected in bacterial 16S rRNA libraries generated from seep sediments (Orphan et al. in press; Teske et al. 2001). Nevertheless, the chemotaxonomic link suggests that psychrophilic or mesophilic bacterial members of the methanotrophic community have their closest relatives among thermophiles that occupy functionally well-defined niches in complex hydrothermal microbial mat systems. For instance, the only prior environmental occurrence of di-n-pentadecyl-glycerolether, a major compound in Eel River Basin seep sediments (Hinrichs et al. 2000) (see also Fig. 3) (DAGE 15/15), was noted in Yellowstone hot spring bacterial mats (Zeng et al. 1992). These compounds may be a key to further information on bacterial AOM members with unknown biochemical capabilities, e.g. as hypothesized by Thomsen et al. (in press).

The carbon isotopic compositions of individual lipids related to different AOM community members are consistent with existing hypotheses concerning the trophic structure. In general, maximum ^{13}C-depletion is observed in biolipids derived from methanotrophic archaea (Fig. 6), followed by slightly smaller depletions in the non-isoprenoidal ether lipids (Hinrichs et al. 2000). This isotopic

Bacterial biomarkers

Di-n-pentadecylglycerolether

C_{17}-cyclopropyl fatty acid

Mono-10-methylhexadecylglycerolether

1-Hexadecenol (ω7)

Fig. 5. Selected biomarkers assigned to bacterial syntrophic partners of methanotrophic archaea.

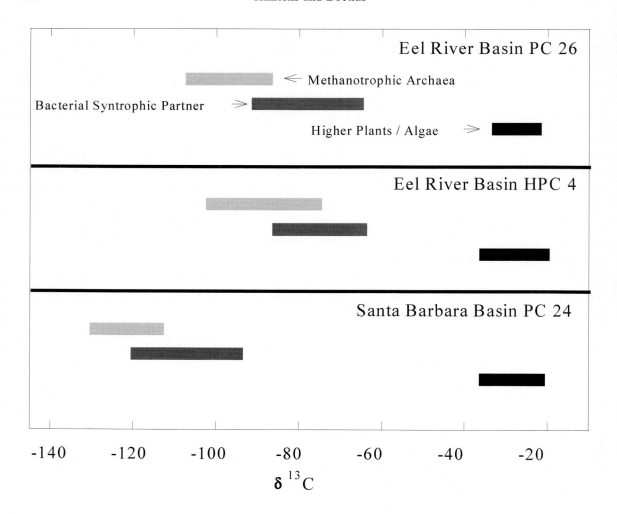

Fig. 6. Ranges of carbon isotopic compositions of sedimentary lipids assigned to different sources from three different methane seeps. Non-isoprenoidal ether- lipids were used for ranges for the bacterial syntrophic partner (adapted from Hinrichs et al. 2000).

stratification is consistent with, but not proof for, an initial oxidation of methane by methanotrophic archaea, which transfer their oxidation products (inorganic carbon plus H_2, or organic C_1 or C_2 compounds) to an intimately associated bacterial partner (e.g. Fig. 2). This partner utilizes pre-dominantly the carbon supplied by the archaeon, as indicated by the strong ^{13}C-depletion of the ether lipids. Under the assumption that the ether-lipid producers are autotrophs, this mechanism requires formation of microenvironments with strong physicochemical gradients so that diffusion of porewater CO_2 to the reaction center would be much slower than microbial utilization of CO_2 from the oxidation of methane.

Relationship between environmental factors and the δ ^{13}C of biomarker lipids

In general, the carbon isotopic composition of the lipid products of methanotrophic microorganisms appears to be controlled by the following factors:

• The carbon isotopic composition of methane.
• The composition of the microbial community, i.e., different organisms fractionate differently during lipid biosynthesis.
• Environmental conditions such as temperature, substrate limitations and ^{13}C-enrichment in residual methane in predominantly diffusion-controlled sediment columns.

For data from methane seep environments, the general influence of $\delta^{13}C$ of methane on that of lipid biomarkers assigned to methanotrophic archaea is evident and is most consistently reflected in the isotopic compositions of archaeol, hydroxyarchaeol, and crocetane. In most sediments with available data on $\delta^{13}C$ of methane, archaeol is depleted in ^{13}C relative to methane by about −50‰. The relative depletions in other lipids associated with AOM appear to be more variable. For example, the difference in $\delta^{13}C$ of methane between the Eel River Basin and Hydrate Ridge is reflected in ^{13}C-depletions in archaeol of a similar magnitude. The large fractionation between substrate and biomass is consistent with findings from laboratory experiments of the methanogen *Methanosarcina barkeri* growing on methylamine (Summons et al. 1998). Notably, the difference between the isotopic composition of methanotrophic lipid products and the substrate is significantly lower at hydrothermal vent sites in the Guaymas Basin, where methane, derived from thermal and microbial decomposition of organic matter with $\delta^{13}C$ of −51 to −43‰ occurs in high concentrations in the upper sediment column. Here, an influence of the about 15°C higher temperatures in the uppermost centimeters of the sediment may reduce the carbon isotopic fractionation during lipid biosynthesis, resulting in archaeal products with $\delta^{13}C$ between −81 and −58‰.

The microbial oxidation of methane leads to an enrichment in ^{13}C in the residual methane due to a preferential utilization of ^{12}C-methane. This process can cause a "distillation" of ^{13}C-methane in upward sediment profiles and may affect the carbon isotopic composition of the microbial biomass formed upon methane oxidation, especially in sediment zones with limited, diffusion-controlled supply of methane. However, the relatively constant fractionation between methane and methanotrophy-based, biosynthetic products at most seep sites suggest that this effect is less important when the flux of methane is large compared to the fraction consumed. For example, down-core variations of $\delta^{13}C$ of archaeol and hydroxyarchaeol at an active seep are negligible and do not indicate any effect of ^{13}C-enrichment in methane on the lipid isotopic composition (Fig. 7).

These depth profiles are consistent with a constant biological fractionation under conditions of large excess of the methane substrate. In contrast, the $\delta^{13}C$ of an ether lipid assigned to a bacterial member of syntrophic AOM community decreases with depth (Fig. 7). This trend is probably related to a parallel decrease of $\delta^{13}C$ in pore water CO_2, which may be partially utilized by the bacterial syntrophic partner in addition to intermediates and products of methane oxidation that are transferred directly to the bacterial syntrophic partner. This increasing ^{13}C-depletion with depth is supported by the isotopic compositions of authigenic carbonates in the same core, with $\delta^{13}C$ values decreasing from −18 to ~ -30‰ over the same depth interval (Orphan et al. in press).

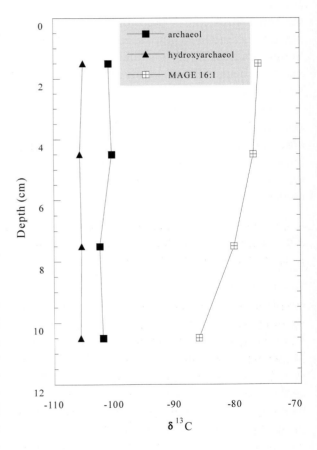

Fig. 7. Depth profile of carbon isotopic compositions of individual compounds assigned to members of the AOM consortium in a sediment core at an active seep in the Eel River Basin (data from Orphan et al. in press). MAGE 16:1 designates a monoalkylglycerolether with the alcohol moiety being $C_{16:1(\omega7)}$.

A very different situation was observed at the *Napoli mud breccia* in the Mediterranean (Pancost et al. 2000). At this site, the flux of methane was significantly lower than at other sites studied by the team. The "distillation" of ^{13}C-methane in the upward sediment profile by continuous preferential consumption of ^{12}C-methane is evident in $\delta^{13}C$ of PMI and archaeol, which increase by about 20‰ in the depth interval from 18 to 3 cm below the seafloor to values that are indistinguishable from those of products of primary production (e.g. archaeol = -20‰). The generally elevated $\delta^{13}C$ values here might be related to (a) a higher ^{13}C-content of the methane source compared to other environments, (b) a limitation of the methane substrate with possibly nearly quantitative consumption of the methane, thereby preventing the full expression of the biological fractionation during lipid biosynthesis, and (c) a continuous ^{13}C-enrichment in the residual methane, which may have been initiated already at sediment depths below those sampled by Pancost et al. (2000). A similar observation was made in Kattegat sediments, where the depth profiles of $\delta^{13}C$ of crocetane and PMI did not display any resemblance, although the isotopic composition indicated that both compounds contained at least partially methane-derived carbon (Bian et al. 2001).

The rates

First estimates of methane oxidation rates in anoxic marine sediments were obtained by numerical modeling of changes in pore water concentrations of methane and sulfate in shallow water as well as deep-sea sediments (Barnes and Goldberg 1976, Reeburgh 1976). Reeburgh (1980) and Iversen and Blackburn (1981) used ^{14}C-labeled methane for sediment incubations and subsequent measurement of [14]CO$_2$ production in coastal sediments. Both methods are still used to estimate rates of AOM in marine sediments and in the water column. For the measurement of methane concentrations in sediments, a sample is mixed into a sodium hydroxide solution. After equilibration with the headspace of the sample vial, gas is sampled with a gas tight syringe and injected into a gas chromatograph with a flame ionization

detector. The modeling approach assumes a steady state of pore water concentrations and a transport of methane by molecular diffusion only. Then, the consumption of methane (R) is balanced by the supply of methane due to molecular diffusion. Usually, the sediment zone of interest is divided into several depth intervals (z) and the simplest consumption profile is chosen which fits the concentration profile best, according to the formula

$$R = D_s \, d^2C/dz^2 \qquad (13)$$

At temperatures between 2-10°C, the diffusion coefficient (D_s) for methane is between 0.7-1 x 10^{-5} cm^2 s^{-1} (Li and Gregory 1974). One of the largest problems of this method is the accurate determination of methane concentrations (C) in zones of supersaturation of methane. In seawater, methane is soluble to a concentration of up to 1.5 mM at low temperatures and atmospheric pressure. At higher concentrations, gas bubbles form and methane is lost due to ebullition. The solubility of methane increases linearly with increasing pressure. Thus, methane can escape detection due to depressurization of deep-sea sediments during core recovery. Unfortunately, no methane sensors are yet available for *in situ* profiling in deep-sea sediments.

The radioisotope measurements are more sensitive and can detect very low methane oxidation rates. ^{14}C-methane is injected into undisturbed sediment samples and incubated for hours to days. Subsequent mixing of the samples into a solution of sodium hydroxide terminates the microbial activity. After degassing of methane from the sample, the [14]CO$_2$ produced by oxidation of radiolabeled methane is stripped from the sample by acidification and captured in a CO$_2$-adsorbing agent for measurement in a scintillation counter. The AOM rate is calculated from the amount of [14]CO$_2$ formed and the concentration and radioactivity of methane in the sample (Iversen and Blackburn 1981). Problems with this method may arise from the use of impure ^{14}C-methane (containing [14]CO or [14]CO$_2$) or due to aeration, warming, and decompression of the samples during recovery and handling. The resulting ebullition of methane causes alteration of the sediment structure and of the pore water gradients, which can lead to substantial analytical bias. No instrument for the *in situ* measurement of methane oxidation is available yet.

Other approaches to estimate AOM include numerical modeling based on pore water carbon isotope data (Burns 1998) and measurements of sulfate reduction rates (SRR; Aharon and Fu 2000; Boetius et al. 2000). However, the percentage of sulfate consumed by sulfate reduction coupled to AOM is rather variable and thus impedes the accuracy of estimates based on consumption of sulfate. In surface sediments of non-seep environments, AOM generally accounts for around 10% of sulfate reduction, with increasing proportions towards the subsurface methane-sulfate transition zone. Here, several studies detected contributions of AOM to total SRR of 40-100% (Borowski et al. 2000; Fossing et al. 2000; Niewöhner et al. 1998). To date, only two studies determined the contribution of AOM rates to SRR at methane seeps. The proportion of sulfate reduction coupled to AOM was 8% at methane seeps in Eckernförde Bay (Bussmann et al. 1999) and around 30% at the gas hydrate bearing caldera of the Arctic Haakon-Mosby mud vulcano (Lein et al. 2000). However, the comparison of SRR at methane seeps with those at close-by control stations indicates that AOM is largely responsible for sulfate reduction at seeps. In the Gulf of Mexico, SRR was 600 times higher at methane seeps than at reference stations (Aharon and Fu 2000). At Hydrate Ridge, SRR were extremely high above gas hydrate (ca. 5 μmol cm^{-3} d^{-1}), but below detection limit at a nearby reference station.

Table 2 shows an overview on published rates of AOM measured in different areas of continental margins. In each depth zone, modelled AOM rates are lower than AOM rates determined with ^{14}C-methane. In a comparative study using both methods, the modelled rates were 4-fold lower than measured rates (Iversen and Jørgensen 1985). Despite all the uncertainties related to the comparability of methods and to their intrinsic errors, the range of rates (less than two orders of magnitude) is not different to that found in SRR or oxygen consumption rates. In contrast to O$_2$ consumption, there is only a slight decrease in AOM

rates with water depth. The decrease in O$_2$ consumption with water depth in marine sediments is explained by the decrease of the input of POC to the seafloor via sedimentation. In the case of methane turnover, one would also assume a link to POC sedimentation rates, since methane is mostly derived from the microbial degradation of POC. However, this link is not obvious for the process of AOM. The calculation of the relative amounts of methane consumed via AOM in four different depth zones of continental margins indicates about equal contributions of each zone to the global budget of methane consumption in marine sediments (Table 3). In contrast, the distribution of aerobic POC degradation (equivalent to oxygen uptake) is an order of magnitude higher on the shelf than on the slope (Jørgensen 1983). According to our estimate, AOM in marine sediments sums up to 300 Tg a^{-1}, which is nearly four times higher than the previous estimate by Reeburgh (1996). The revised budget compares to 10% of the total O$_2$ consumption in marine sediments, i.e., to 12% of the aerobic POC oxidation (applying the Redfield ratio).

These calculations do not include AOM rates at methane seeps, which are at least an order of magnitude higher than AOM rates in non-seep sediments. Unfortunately, reliable estimates of the total area of sediments affected by methane seepage are not yet available. In addition, conditions similar to those at cold seeps are likely to prevail at hydrothermally active, marine sedimentary environments as indicated by observations of diagnostic biomarkers and phylotypes at hydrothermal vent sites in the Guaymas Basin (Teske et al. in 2001). A variety of different seep types are known from active and passive margins. Methane seeps may be caused by hydrocarbon deposits, brine fluids, sediment compaction, or land slides (Sibuet and Olu 1998). Also, methane see-page is often related to subsurface gas hydrate deposits which can continuously feed methane seeps at active and passive continental margins. Furthermore, the dissociation of surficial methane hydrates can support methane seepage in those zones of continental margins where the hydrates are at their stability limit (500-1000 m water depth). Areas with gas hydrates exposed at the

Area	Location	Water depth (m)	Sediment depth (cm)	AOM (nmol cm⁻³ d⁻¹)	Average (nmol cm⁻³ d⁻¹)	Integrated (mmol m⁻² d⁻¹)	Integrated (mmol m⁻² a⁻¹)	Method	Source
Inner shelf 0-50 m	Kysing Fjord	1	12	0.01-0.27	0.07	0.01	3	tracer exp.	Iversen and Blackburn 1985
	Norsminde Fjord	1	40	0-17	7	2.80	1022	tracer exp.	Hansen et al. 1998
	Cape Lookout	10	35	0-19	5	1.75	639	tracer exp.	Hoehler et al. 1994
	Aarhus Bay	16	300	0-1		0.05	16	tracer exp.	Thomsen et al. in press
	Eckernförde Bay	25	50	0-34	0.6	0.29	106	tracer exp.	Bussmann et al. 1999
Outer shelf 50-200m	Kattegat	65	170	0-6	0.5	0.83	303	tracer exp.	Iversen and Jørgensen 1985
	Skan Bay	65	35	0-9	3	1.14	415	tracer exp.	Reeburgh 1980
	Skan Bay	65	40	0-10	2	0.88	321	tracer exp.	Alperin and Reeburgh 1985
	Black Sea	130	350			0.11	41	modelling	Jørgensen et al. in press
		181	350			0.10	35		
Upper margin 200-1000m	Skagerak	200	110	0-12	1.0	1.16	423	tracer exp.	Iversen and Jørgensen 1985
	Saanich Inlet	225	27	0-0.75	5	1.26	460	tracer exp.	Devol 1983
	Black Sea	396	350			0.08	30	modelling	Jørgensen et al. in press
		1176	350			0.05	18		
Lower margin 1000-4000m	Cariaco trench	1300	120			0.44	159	modelling	Reeburgh 1976
		1400	120			0.15	55	modelling	
	Namibia slope	1312	70			0.08	30	modelling	Niewöhner et al. 1998
		2060	70			0.14	52		
	Namibia slope	1373	400			0.22	80	modelling	Fossing et al. 2000
		2065	600			0.15	55		
	Amazon Fan	2700	5000	0-0.3	0.004			isotopes	Burns 1998
	Blake Ridge	3000	2300	0-0.015		0.01	3	modelling	Borowski et al. 2000
	Zaire fan	3950	1550			0.10	36	modelling	Zabel and Schulz in press
Methane seeps	Eckernförde Bay	25	50	0-50	10	5	1880	tracer exp.	Bussmann et al. 1999
	Eel River Basin	520	10		32		11,500	isotopes	Hinrichs subm.
	Gulf of Mexico	590	30			4	1338	SRR	Aharon and Fu 2000
	Hydrate Ridge	750	15	0-3000	300	45	16,425	SRR	Boetius et al. 2000
	HMMV Caldera	1250	20	1-70	26	5	1898	tracer exp.	Lein et al. 2000

[a] For the calculation of AOM from SRR we used the average AOM/SRR ratio of all investigations (30%). Most likely, this leads to an underestimate of AOM for reasons discussed in the text.

Table 2. Rates of anaerobic oxidation of methane in sediments at continental margins and at methane seeps.

seafloor (e.g. Hydrate Ridge; Suess et al. 1999) are characterized by similar chemosynthetic communities and carbonate structures as other methane seeps. The data compiled in Table 3 indicate that, even if the area affected by methane seepage at continental margins is below 1%, this might have a significant impact on the total methane budget. Thus, a more accurate mapping of seepage areas is extremely important for realistic calculations of methane consumption in the sea. Recent investigations by Suess et al. (1999) show that the same might be true for oxygen consumption at seeps, which might significantly contribute to the total oxygen demand of continental margin sediments.

The Potential Role of AOM in the Evolution of Biogeochemical Cycles on Earth

Studies of carbon isotopic compositions of kerogens and organic carbon suggest that around 2.7 Gyr ago methanotrophy was an important pathway of carbon assimilation in earth's biogeochemical cycle (Hayes 1994). Until recently, it was accepted as common sense that this isotopic excursion indicates a period in Earth history when increases in the concentration of free oxygen in the ocean and atmosphere allowed aerobic methanotrophic bacteria to radiate. Due to the abundant methane in ocean and atmosphere, methanotrophs were able to produce significant amounts

of microbial biomass that were subsequently buried in the rock record and had decreased the isotopic composition of the buried organic carbon. An anaerobic mechanism was not seriously considered because the anaerobic, sulfate-dependent oxidation of methane was believed to be very inefficient in biomass production, at least compared to the pathway involving methanotrophic bacteria. AOM was therefore thought to be inadequate to explain large contributions by methane assimilating microbes to buried organic carbon on the order of 30 to 40% (Hayes 1994).

New findings, however, of unexpectedly large quantities of methane-derived carbon from anaerobic, methanotrophic microbial consortia in recent marine sediments have motivated a reassessment of the paleoenvironmental conditions during the era of methanotrophy 2.7 Gyr ago (Hinrichs 2002). Comparisons of the accumulation rates of methane-derived carbon in anoxic modern sedimentary environments to those in the Late Archaean Tumbiana Formation served as the basis for this reassessment. The Tumbiana Formation is a member of the Fortescue Group with consistently ^{13}C depleted organic carbon. The comparison indicated that the rates of accumulation of methane-derived carbon in rocks from the Tumbiana Formation were significantly lower than rates at modern methane seeps. Instead, they compared well to rates observed in the methane/sulfate transition zone of modern anoxic sediments at Kattegat, where AOM occurs at moderate concen-

	AOM (mmol m^{-2} d^{-1})	O_2 uptake (mmol m^{-2} d^{-1})	Area (10^{12} m^2)	AOM (10^{12} mol a^{-1})	O_2 uptake (10^{12} mol a^{-1})
Inner shelf	1.0	20	13	4.6	95
Outer shelf	0.6	10	18	4.0	66
Upper margin	0.6	3	15	3.5	16
Lower margin	0.2	0.3	106	6.9	12
			Sum: 152	19 (304 Tg CH_4)	
Seepage areas	18	[a]471	0.5%= 0.75 10^{12} m^2	4.9	129

[a] Average of oxygen uptake rates measured at different seeps as reported by Suess et al. (1999).

Table 3. Anaerobic oxidation of methane in different depth zones of continental margins. The areas and O_2 uptake rates are adapted from Jørgensen (1983). The values represent averages of the data shown in Table 2. The calculation of the contribution of areas affected by methane seepage is purely speculative (indicated in italics). 10^{12} mol CH_4 is equivalent to 16 Tg CH_4.

trations of both methane and sulfate (Iversen and Jørgensen 1985; Bian et al. 2001). This finding indicates that, in principle, AOM could provide a mechanism for the formation of isotopically light organic carbon 2.7 Gyr ago.

This alone does not explain the temporary nature (probably ~ 0.1 Gyr duration) of a hypothetical era of methanotrophy based on AOM. An aerobic mechanism as proposed (Hayes 1994) would be consistent with the evidence that oxygenic photosynthesis was already established at that time (see Summons et al. 1999 and literature therein). Accordingly, the termination of the "isotopic event" was related to the continuous increase of atmospheric oxygen concentrations to levels that drove methanogens into more restricted environments with poorer supplies of fermentable organic matter, thereby progressively decreasing the amounts of methane produced.

On the other hand, two points argue against the aerobic mechanism: (1) In the present marine environment under well aerated conditions, methane oxidation and in particular the accumulation of methane-derived carbon are clearly dominated by AOM, even at methane seeps with a large interface between methane-rich fluids and oxygenated waters. (2) Although oxygen was obviously produced, there is little consensus on the actual degree of oxygenation of the Earth's surface around that time. However, it appears well established that a significant accumulation of oxygen commenced not later than around 2.1 Gyr. Several recent studies suggest that concentrations of sulfate were significantly lower than those at present (Canfield et al. 2000; Farquhar et al. 2000) and that significant changes in the sulfur cycle occurred with the onset of the Proterozoic. In that light, an anaerobic pathway of methane-derived carbon fixation may well mark an important transition in the biogeochemical cycles of carbon and sulfur.

Accordingly, the sequence of possible events can be described in three stages (Fig. 8). (1) Prior to ~2.7 Gyr, accumulations of free oxygen and sulfate were probably minimal and the remineralization of organic carbon occurred in anoxic habitats, presumably by fermentative pathways with methane and CO_2 being the terminal products. Due to low

levels or absence of biologically utilizable electron acceptors, methane was probably stable at saturation levels in pore waters of sediments and in large parts of water bodies. (2) With increasing oxygenation of the Earth's surface as a result of oxygenic photosynthesis, sulfate concentrations increased locally and/or globally above certain thresholds that led to more favorable conditions for sulfate reducers and consequently for AOM. An excellent analog for this temporal transition in the sulfur-methane cycle can be found in a vertical sequence of a modern anoxic sediment (Fig. 8). Moving upwards, at the intersection of the zones of methanogenesis and sulfate-reduction, the an-aerobic consumption of methane becomes the dominant sulfate-consuming microbial process and leads to stoichiometric consumption of sulfate. Similarly, in certain modern environments, almost all sulfate flows to the anaerobic oxidation of methane (e.g. Boetius et al. 2000; Aharon and Fu 2000). In analogy to these modern examples, the 2.7 Gyr excursion possibly marks an geological era during which sulfate reduction is predominantly coupled to AOM. (3) The termination of the era of methanotrophy was caused by the continuous increase of oxygenation, leading to larger inventories of sulfate. For the microbial breakdown of organic carbon the consequences of such a transition are an increasing importance of anaerobic respiratory processes and a decreasing importance of fermentative pathways. Simultaneously with this transition, the fraction of methane formed from each mole of organic carbon decreases. Again, the vertical sedimentary sequence displays model character for the characterization of this transition in the modes in carbon remineralization in Earth's history. For example, in sediments containing sulfate-rich pore waters, methanogens are not successful in the competition with sulfate reducers for H_2 or acetate, preventing significant formation of methane in this zone.

In conclusion, an anaerobic mechanism for the deposition of unusually ^{13}C-depleted organic carbon at 2.7 Gyr ago appears feasible. Accordingly, the isotopic anomaly marks a period of increasing importance of microbial sulfate reduction for Earth's carbon cycle.

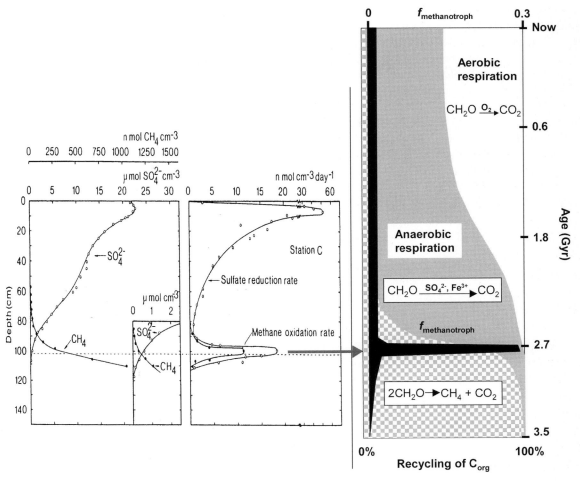

Fig. 8. Speculative comparison of a typical vertical distribution of biogeochemical zones in coastal marine sediments with a hypothetical, temporal evolution of biogeochemical cycles in Earth's history (Hinrichs 2002). Modified after Iversen and Jørgensen (1985), with permission by American Society of Limnology and Oceanography.

Acknowledgements

This manuscript profited considerably from discussions with Marcus Elvert, Lisa Levin, Vicki Orphan, Andreas Teske, Tina Treude, and David Valentine. We thank Marcus Elvert, Bo Barker Jørgensen, and Rudi Amann for their helpful reviews. KH gratefully acknowledges the support provided by a fellowship at the Hanse Institute of Advanced Study in Delmenhorst. This is Woods Hole Oceanographic Institution contribution 10488.

References

Aharon P, Fu B (2000) Microbial sulfate reduction rates and sulfur and oxygen isotope fractionations at oil and gas seeps in deepwater Gulf of Mexico. Geochim Cosmochim Acta 64:233-246

Alperin MJ, Reeburgh, WS (1985) Inhibition experiments on anaerobic methane oxidation. Appl Environ Microbiol 50:940-945.

Barnes RO, Goldberg ED (1976) Methane production and consumption in anoxic marine sediments. Geol 4:297-300

Bian L (1994) Isotopic biogeochemistry of individual compounds in a modern coastal marine sediment (Kattegat, Denmark and Sweden). MSc thesis, Indiana University

Bian L, Hinrichs K-U, Xie T, Brassell SC, Iversen N, Fossing H, Jørgensen BB, Hayes JM (2001) Algal and archaeal polyisoprenoids in a recent marine sediment: Molecular-isotopic evidence for anaerobic oxidation of methane. Geochem Geophys Geosyst 2:#2000GC000112

Bidle K, Kastner M, Bartlett DH (1999) A phylogenetic analysis of microbial communities associated with methane hydrate containing marine fluids and sediments in the Cascadia margin (ODP site 892B). FEMS Microbiol Lett 177:101-108

Boetius A, Ravenschlag K, Schubert C, Rickert D, Widdel F, Gieseke A, Amann R, Jørgensen BB, Witte U, Pfannkuche O (2000) A marine microbial consortium apparently mediating anaerobic oxidation of methane. Nature 407:623-626

Borowski WS, Paull CK, Ussler III W (1996) Marine pore-water sulfate profiles indicate in-situ methane flux from underlying gas hydrate. Geol 24:655-658

Borowski WS, Hoehler TM, Alperin MJ, Rodrigez NM, Paull CK (2000) Significance of anaerobic methane oxidation in methane-rich sediments overlying the Blake Ridge sediments. Proc ODP Sci Res 164:87-99

Burns SJ (1998) Carbon isotopic evidence for coupled sulfate reduction-methane oxidation in Amazon Fan sediments. Geochim Cosmochim Acta 62:797-804

Bussmann I, Dando PR, Niven SJ, Suess E (1999) Groundwater seepage in the marine environment: Role for mass flux and bacterial activity. Mar Ecol Prog Ser 178:169-177

Canfield DE, Habicht KS, Thamdrup B (2000) The Archaean sulfur cycle and the early history of atmospheric oxygen. Science 288:658-661

Devol AH (1983) Methane oxidation rates in the anaerobic sediments of Saanich Inlet. Limnol Oceanogr 28:738-742

Elvert M, Suess E, Whiticar MJ (1999) Anaerobic methane oxidation associated with marine gas hydrates: Superlight C-isotopes from saturated and unsaturated C_{20} and C_{25} irregular isoprenoids. Naturwiss 86:295-300

Elvert M, Suess E, Greinert J, Whiticar MJ (2000) Archaea mediating anaerobic methane oxidation in deep-sea Sediments at cold seeps of the eastern Aleutian subduction zone. Org Geochem 31:1175-1187

Farquhar J, Bao H, Thiemens M (2000) Atmospheric influence of Earth's earliest sulfur cycle. Science 289:756-758

Fossing H, Ferdelman TG, Berg P (2000) Sulfate reduction and methane oxidation in continental margin sediments influenced by irrigation (South-East Atlantic off Namibia). Geochim Cosmochim Acta 64:897-910

Gaasterland T (1999) Archaeal genomics. Curr Opin Microbiol 2:542-547

Hansen LB, Finster K, Fossing H, Iversen N (1998) Anaerobic methane oxidation in sulfate depleted sediments: Effects of sulfate and molybdate additions. Aquat Microb Ecol 14:195-204

Harder J (1997) Anaerobic methane oxidation by bacteria employing ^{14}C-methane uncontaminated with ^{14}C-carbon monoxide. Mar Geol 137:13-23

Hayes JM (1994) Global methanotrophy at the Archaean-Proterozoic transition. In: Bengtson S (ed) Early Life on Earth. Nobel Symposium No. 84, Columbia University Press, pp 220-236

Hinrichs K-U, Hayes JM, Sylva, SP, Brewer PG, DeLong EF (1999) Methane-consuming archaebacteria in marine sediments. Nature 398:802-805

Hinrichs K-U, Summons RE, Orphan V, Sylva SP, Hayes JM (2000) Molecular and isotopic analyses of anaerobic methane-oxidizing communities in marine sediments. Org Geochem 31:1685-1701

Hinrichs (2002) Microbial fixation of methane carbon at 2.7Ga: Was an anaerobic mechanism possible? Geochem Geophys Geosys Vol. 3, 10.1029/2001GC000286, http://www.g-cubed.org

Hoehler TM, Alperin MJ, Albert DB, Martens CS (1994) Field and laboratory studies of methane oxidation in an anoxic marine sediments: Evidence for methanogen-sulfate. Glob Biogeochem Cycl 8:451-463

Hoehler TM, Alperin MJ (1996) Anaerobic methane oxidation by a methanogen-sulfate reducer consortium: geochemical evidence and biochemical considerations. In: Lidstrom ME, Tabita FR (eds) Microbial Growth on C_1 Compounds. Kluwer Academic Publishers, Dordrecht, pp 326-333

Huber R, Wilharm T, Huber D, Trincone A, Burggraf S, Rachel R, Rockinger I, Fricke H, Stetter KO (1992) Aquifex pyrophilus gen. nov. sp. nov., represents a novel group of marine hyperthermophilic hydrogen-oxidizing bacteria. Syst Appl Microbiol 15:340-351

Iversen N, Jørgensen BB (1985) Anaerobic methane oxidation rates at the sulfate-methane transition in marine sediments from Kattegat and Skagerrak (Denmark). Limnol Oceanogr 30:944-955

Iversen N, Blackburn TH (1981) Seasonal rates of methane oxidation in anoxic marine sediments. Appl Environ Microbiol 41:1295-1300

Jørgensen BB (1983). Processes at the Sediment-Water Interface. In: Bolin B, Cook RC (eds) The Major Biogeochemical Cycles and their Interactions. SCOPE, pp 477-509

Jørgensen BB, Weber A, Zopfi J (in press) Sulfate reduction and anaerobic methane oxidation in Black Sea sediments. Deep-Sea Res

Koga Y, Morii H, Akagawa-Matsushita M, Ohga M (1998) Correlation of polar lipid composition with 16S rRNA phylogeny in methanogens. Further Analysis of lipid component parts. Biosci Biotech Biochem 62:230-236

Lein A, Pimenov NV, Savvichev AS, Pavlova GA, Vogt PR, Bogdanov YA, Sagalevich, AM, Ivanov, MV (2000) Methane as a source of organic matter and carbon dioxide of carbonates at a cold seep in the Norway Sea. Geochem Int 38:232-245

Li YH, Greogory, S (1974) Diffusion of ions in seawater and in deep-sea sediments. Geochim Cosmochim Acta 38:703-714

Niewöhner C, Hensen C, Kasten S, Zabel M, Schulz HD (1998) Deep sulfate reduction completely mediated by anaerobic methane oxidation in sediments of the upwelling area off Namibia. Geochim Cosmochim Acta 62:455-464

Orphan VJ, Hinrichs K-U, Paull CK, Taylor LT, Sylva SP, Delong EF (in press) Comparative analysis of methane-oxidizing archaea and sulfate-reducing bacteria in anoxic marine sediments. Appl Environ Microbiol

Pancost RD, Sinninghe Damsté JS, Lint SD, van der Maarel MJEC, Gottschal JC, Shipboard Scientific Party (2000) Biomarker evidence for widespread anaerobic methane oxidation in Mediterranean sediments by a consortium of methanogenic archaea and bacteria. Appl Environ Microbiol 66:1126-1132

Reeburgh WS (1976) Methane consumption in Cariaco Trench waters and sediments. Earth Planet Sci Lett 28:337-344

Reeburgh WS (1980) Anaerobic methane oxidation: Rate depth distribution in Skan Bay sediments. Earth Planet Sci Lett 47:345-352

Reeburgh WS (1996) "Soft spots" in the global methane budget. In: Lidstrom ME, Tabita FR, Microbial Growth on C_1 Compounds. Kluwer Academic Publishers, Dordrecht, pp 334-342

Schouten S, Hoefs MJL, Koopmans MP, Bosch H-J, Sinninghe Damsté JS (1998) Structural characterization, occurrence and fate of archaeal ether-bound acyclic and cyclic biphytanes and corresponding diols in sediments. Org Geochem 29:1305-1319

Shilov AE, Koldasheva, EM, Kovalenko SV, Akenteva NP, Varfolomeev SV, KalynzhnYSV, Sklyar VI (1999) Methanogenesis is a reversible process: The formation of acetate under the carboxylation of methane by bacteria of methane biocenosis. Rep Acad Sci 367:557-559 (in Russian)

Sibuet M, Olu K (1998) Biogeography, biodiversity and fluid dependence of deep-sea cold-seep communities at active and passive margins. Deep-Sea Res II 45:517-567

Sørensen KB, Finster K, Ramsing NB (in press) Thermodynamic and kinetic requirements in anaerobic methane oxidizing consortia exclude hydrogen, acetate and methanol as possible electron shuttles. Microb Ecol

Suess E, Torres ME, Bohrmann G, Collier RW, Greinert J, Linke P, Rehder G, Trehu A, Wallmann K, Winckler G, Zuleger E (1999) Gas hydrate destabilization: Enhanced dewatering, benthic material turnover and large methane plumes at the Cascadia convergent margin. Earth Planet Sci Lett 170:1-15

Summons RE, Franzmann PD, Nichols PD (1998) Carbon isotopic fractionation associated with methylo-trophic methanogenesis. Org Geochem 28:465-475

Summons RE, Jahnke LL, Hope JM, Logan GA (1999) 2-Methylhopanoids as biomarkers for cyanobacterial oxygenic photosynthesis. Nature 400:554-557

Teske A, Hinrichs K-U, Edgcomb V, Kysela D, Sogin ML (2001) Novel Archea in Guaymas Basin hydrothermal vent sediments: Evidence for anaerobic methanotrophy. In: Boss et al. (eds) Abstracts, General Meeting of the NASA Astrobiology Institute: Washington DC, NASA Astrobiology Institute, p 138-139

Thiel V, Peckmann J, Seifert R, Wehrung P, Reitner J, Michaelis W (1999) Highly isotopically depleted isoprenoids: Molecular markers for ancient methane venting. Geochim Cosmochim Acta 63:3959-3966

Thiel V, Peckmann J, Richnow HH, Luth U, Reitner J, Michaelis W (2001) Molecular signals for anaerobic methane oxidation in Black Sea seep carbonates and a microbial mat. Mar Chem 73:97-112

Thomsen T, Finster K, Ramsing NB (in press) Biogeochemical and molecular signatures of anaerobic methane oxidation in a marine sediment. Appl Environ Microbiol

Valentine DL, Reeburgh WS (2000) New perspectives on anaerobic methane oxidation. Environ Microbiol 2:477-484

Zabel M, Schulz HD (in press) Effects of a submarine land-slide on a pore water system at the lower Zaire (Congo) deep-sea fan. Mar Geol

Zehnder AJ, Brock TD (1979) Methane formation and methane oxidation by methanogenic bacteria. J Bacteriol 137:420-32

Zeng YB, Ward DM, Brassell SC, Eglinton G (1992) Biogeochemistry of hot spring environments 3. Apolar and polar lipids in the biologically active layers of a cyanobacterial mat. Chem Geol 95:347-360

Microbial Systems in Sedimentary Environments
of Continental Margins

A. Boetius[1,2*], B.B. Jørgensen[2], R. Amann[2], J.P. Henriet[3], K.U. Hinrichs[4,5],
K. Lochte[6], B.J. MacGregor[2], G. Voordouw[7,5]

[1] Alfred Wegener Institut für Polar- und Meeresforschung, Am Handelshafen 12,
27515 Bremerhaven, Germany
[2] Max Planck Institut für Marine Mikrobiologie, Celsiusstr. 1, 28359 Bremen, Germany
[3] RCMG University of Gent, 9000 Gent, Belgium
[4] Woods Hole Oceanographic Institution, Woods Hole, MA 02543, USA
[5] Hanse - Wissenschaftskolleg, Lehmkuhlenbusch 4, 27753 Delmenhorst, Germany
[6] Institut für Meereskunde an der Universität Kiel, 24105 Kiel, Germany
[7] University of Calgary, Calgary, Alberta T2N1N4, Canada
* corresponding author(e-mail): aboetius@awi-bremerhaven.de

Abstract: The zone of continental margins is most important for the ocean's productivity and nutrient budget and connects the flow of material from terrestrial environments to the deep-sea. Microbial processes are an important "filter" in this exchange between sediments and ocean interior. As a consequence of the variety of habitats and special environmental conditions at continental margins an enormous diversity of microbial processes and microbial life forms is found. The only definite limit to microbial life in sedimentary systems of continental margins appears to be high temperatures in the interior earth or in fluids rising from the interior. Many of the catalytic capabilities which microorganisms possess are still only incompletely explored and appear to continuously expand as new organisms are discovered. Recent discoveries at continental margins such as the microbial life in the deep sub-seafloor, microbial utilization of hydrate deposits, highly specialized microbial symbioses and the involvement of microbial processes in the formation of carbonate mounds have extended our understanding of the Earth's bio- and geosphere dramatically. The aim of this paper is to identify important scientific issues for future research on microbial life in sedimentary environments of continental margins.

Introduction

Microbial processes at continental margins: Importance to science and society

Continental margins, which are defined here as the area from the outer regions of the shelf extending to the deep-sea floor, represent a large habitat on earth which connects the flow of material from terrestrial environments to the deep-sea. This juncture between continents and ocean is of great social and economic significance to mankind. Margins are a repository for anthropogenic wastes which are transported by river discharges, currents, winds and rain. Human exploitation of resources at continental margins is steadily increasing as both fisheries and hydrocarbon prospection are progressing from the outer shelves down the slopes. The zone of continental margins is also most important for the ocean's productivity and nutrient budget. It is often associated with high productivity due to upwelling or shelf break fronts providing unusually large inputs of organic materials which are deposited in this area. Tectonic activity at continental margins, such as subduction, drives the rise of reduced fluids from the deep sediments and

From WEFER G, BILLETT D, HEBBELN D, JØRGENSEN BB, SCHLÜTER M, VAN WEERING T (eds), 2002,
Ocean Margin Systems. Springer-Verlag Berlin Heidelberg, pp 479-495

provides thus additional energy for microbial communities. These processes are partly the cause for the abundant occurrence of gas hydrates at continental slopes. Microbial processes are an important "filter" in this exchange between sediments and ocean interior. As a consequence of these diverse and special conditions at continental margins an unusually wide diversity of habitats and microbial processes is found. The aim of this paper is to identify important scientific issues for future research on microbial life in sedimentary environments of continental margins.

Microorganisms play crucial roles in all biogeochemical processes that sustain the biosphere, especially in the remineralization of organic material and recycling of elements (Jørgensen 2000). Recent estimates based on molecular biological methods suggest that less than 5% of marine microbial species have been identified yet (Amann et al. 1995). Furthermore, most of the dominant microorganisms of the different functional groups in sedimentary systems have not yet been obtained in culture. Thus, a variety of biochemical pathways and key enzymes remain to be discovered. Throughout earth's history, prokaryotic physiology has evolved towards a versatile use of chemical energy from potential redox reactions, leading to a high biodiversity of microorganisms in the sea. This enormous diversity of catalytic capabilities which microorganisms possess is still only incompletely explored and appears to continuously expand as new organisms are discovered. New incentives for this exploration have come from recent discoveries at continental margins such as the involvement of microbial processes in the formation of carbonate mounds, microbial utilization of hydrate deposits and microbial life in the deep sub-seafloor (Henriet et al. this volume; Hinrichs and Boetius this volume; Parkes et al. 2000). New biochemical pathways mediated by not-yet-cultured microorganisms and symbiotic associations with benthic invertebrates found in specific habitats at continental margins add further aspects to future scientific interest (Faulkner 2000). In each case, the important discoveries were made using new methodologies for sampling and detecting microbial organisms, especially with the help of molecular methods.

The key topics discussed here are based on the expertise of participants at the Hanse Workshop in Delmenhorst, Germany, November 2000. The topics fall into four main areas of microbiological research and represent major challenges for future studies of microbiology and biogeochemistry at continental margins. For each of the topics a background paper has been submitted to provide specific examples of our current state of knowledge and methodology (Henriet et al.; Hinrichs and Boetius; Lochte and Pfannkuche; MacGregor et al. this volume). We are well aware that the selection of interesting and urgent issues in the study of microbial systems at continental margins discussed here is by far not complete. For example, the important fields of microbial processes in the pelagic realm or in the benthic nepheloid layer have not been included, despite their potential significance to life at continental margins.

The realm of continental margins is typically governed by moderate to high hydrostatic pressure, low temperature, and a sparse energy input in the form of sedimenting particulate organic matter (POM). Whereas the water column at continental margins is generally oxic, most of the sediment column is anoxic, except from a thin upper layer extending a few mm to cm. In particular, margins underlying upwelling areas are characterised by very low oxygen content in the sediments. The lack of oxygen excludes most metazoan animals, yet a large variety of microorganisms thrive in anoxic environments. The finding of microbial life in deep sub-seafloor sediments extended the known biosphere on Earth dramatically (Parkes et al. 2000). The only definite limit to microbial life in sedimentary systems of continental margins appears to be high temperatures in the interior earth or in fluids rising from the interior. A variety of habitats at continental margins are governed by special environmental conditions like gas and oil seepage, high temperatures at hydrothermal vents, high deposition rates in "depo-centres" or erosion on ridges. Such "extreme" habitats are generally populated by highly adapted organisms such as the methane and sulfur oxidizing communities associated with deposits of methane hydrates and the coral reefs at carbonate mounds (Hinrichs and Boetius this volume; Freiwald this volume).

The seafloor of continental margins functions like a great bioreactor which harbors a vast diversity of microorganisms involved in the processing of organic and inorganic matter. Most organic material sinking to the seafloor is remineralized by benthic animals and bacteria into nutrients which re-enter the marine food web in upwelling areas. A minor part of the deposited organic matter is buried and may be ultimately transformed into oil and natural gas which are the main energy sources for human society today. Some microbial processes accelerate the precipitation of minerals (Douglas and Beveridge 1998) which may form large structures and shape the sea floor at continental margins.

Depending on the type of sedimentary environment or geological formation, microbes can extract energy from various sources, with the help of a highly diverse inventory of enzymes catalyzing the biochemical reactions. Several novel biogeochemical pathways mediated by microorganisms have recently been discovered. Exploration of the phylogenetic and functional diversity of microorganisms and of their enzymatic capacities in continental margin sediments may open a new source of biomolecules for technological and pharmaceutical applications. The diverse habitats of continental margins ranging from pelagic sediments to gas hydrates and carbonate mounds select for a wide variety of different enzymatic capacities and physiologies of microbes in the environment. The many interactions between invertebrates and microbes, which range from obligate symbiosis in chemosynthetic communities to loose associations through microbial gardening in guts and slimes of invertebrates, provide diverse ecological niches. Such prokaryote-invertebrate associations are particularly in the focus of biotechnology, because of the likelyhood of finding adaptations such as the biosynthesis of unique metabolic products (Bull et al. 2000). Many basic ecological questions remain open, e.g. regarding chemical communication,

selection of host and symbionts, coevolution of species and biogeography of symbiotic bacteria.

Earlier studies of benthic microbial systems at continental margins succeeded in detecting microbial communties in low and high energy environments of the deep-sea, in the isolation and cultivation of several marine strains of bacteria, in the quantification of microbial carbon, sulfur and other element cycles in sediments, as well as in the recognition of fast benthic responses to surface water processes. However, these studies were often hampered by inadequate sampling methods and the difficulties to isolate and cultivate microorganisms from continental margins. The combination of classical methods in biological oceanography and biogeochemistry with new *in situ* rate measurements, stable isotope analyses of biomarkers, 16S rRNA probing and genome sequencing will rapidly advance our understanding of the evolution and impact of microbial life at continental margins (MacGregor et al. this volume).

Microbial systems connecting geosphere and biosphere

Until a century ago, life was considered to be restricted to a thin surface layer of our planet, where temperatures and chemical conditions were agreeable to the known living organisms and where the biosphere was nourished by photosynthetic biomass production. The discovery of rich communities of animals and microorganisms around hydrothermal vents along oceanic spreading centers has changed our perception of the limits to life on earth. It was a completely new concept, that also the geosphere could provide the driving force for life and that thermocatalytic reactions in the oceanic crust could produce substrates useful for prokaryotic energy production. The more recent finding of bacteria at high abundance down to the deepest explored parts of the sea floor and even in basement rock, both considered abiotic, has again shifted our perspective. It is now estimated that microscopic organisms in deep sediment deposits up to several million years old constitute about 10% of the total living biomass on earth (Parkes et al.

2000). It will be an important challenge for future research to explore how this deep biosphere may influence major geosphere processes in the ocean floor.

Deep subsurface sediments and volcanic rock

The microbial biosphere detected deep inside the ocean floor has opened up a range of important questions. These deeply buried organisms may be related to those living at the surface, or they may have undergone independent evolution during several million years of isolation. There could be microorganisms adapted to completely different conditions than those prevailing on the surface. The energy supporting them could be provided by unknown geochemical processes, the understanding of which expand our knowledge of the origin and frontiers of life on the earth as well as on other planets. New methods and new technologies must be developed to approach these questions. The task will require multidisciplinary research and international collaboration in order to establish facilities for drilling cruises especially equipped for microbiological work and for the proper handling of samples without microbiological or chemical contamination.

Key question: How does the deep biosphere function?

Specific topics:
• What is the phylogenetic and functional diversity of subsurface microorganisms?
• What are the main energy-providing reactions?
• How can the, probably very low, rates of growth and metabolism be quantified?
• How can sampling methods be improved? (sterility, non-invasiveness, imaging of 3-D structures, in situ measurements)
• What is the biotechnological use of deep biosphere organisms?

Gas hydrates and cold seeps

Most of the methane formed thermogenically or biologically by slow degradation of organic carbon in the sea floor is oxidized back to carbon dioxide as it rises towards the sediment surface.

Only a minor fraction is trapped in the form of solid gas hydrate. Yet, the total inventory of hydrates exceeds the global reserve of all other fossil fuels. Prokaryotic consortia have recently been discovered which can oxidize methane by the use of sulfate from sea water and which are particularly abundant around gas hydrates (Boetius et al. 2000). They play a key role as a barrier against methane escape from dissolving hydrates, a process which could otherwise feed this effective greenhouse gas into the atmosphere. These and other bacteria, living in symbioses, are the basis for rich animal communities fuelled by methane and hydrogen sulfide (Fig. 1).

Key question: How are microbes shaping the special ecosystem at hydrate deposits and cold seeps?

Specific topics:
• What are the mechanisms of methane production and loss at hydrate deposits and cold seeps?
• How much methane is retained and oxidized by the microbial communities at hydrate deposits and seeps?
• How much of this energy is transferred via bacteria to higher trophic levels?
• Are there new microbial processes linked to the turnover of reduced fluids from seeps and hydrates?

Carbonate mounds

A newly discovered feature of continental margins are large clusters of giant carbonate mounds (Henriet et al. this volume). Their origin, morphology and effect on the benthic environment will be studied within the next several years in large European programs (ACES, ECOMOUND, GEOMOUND). Carbonate mounds represent sinks for carbon, which is locked in the seabed at discrete stages of climatic cycles. Microbial processes such as sulfate reduction can regulate the dissolution and precipitation of minerals, and may induce lithification (Warthmann et al. 2000). There is evidence that certain types of carbonate mounds and other large mineral structures form in methane-rich environments, probably in connection with microbial processes such as the oxidation of methane. Other microbial mounds,

Fig. 1. Chemosynthetic communities above marine gas hydrate. The left image shows a mat of giant sulfur-oxidizing bacteria (courtesy of P. Linke, GEOMAR), the right image shows an accumulation of the symbiotic clam *Calyptogena* which harbours sulfide-oxidizing bacteria in its gills (courtesy of U. Witte, MPI). These communities occur in high densities above surficial methane hydrate deposits at the Hydrate Ridge (Cascadia margin) off Oregon (http://www.geomar.de/projekte/tecflux_e/index.html).

e.g. stromatolithic mud-mounds, include photo-autotrophic microorganisms like cyanobacteria, which can have a wide depth range (Reid et al. 2000). A typical feature of carbonate mounds are microbial biofilms covering the surface of the mineral particles and building a barrier towards the surrounding seawater. Such biofilms likely change the diffusion of fluids through the minerals and thus affect biogeochemical processes at the interfaces of the mounds. Furthermore, it has been proposed that the formation of biofilms may be crucial for the stabilisation of slope sediments and, thus, may be the basic process preceding the formation of carbonate mounds. Microbial biofilms also precede the colonization of mounds by higher benthic organisms and are the basis for the rich macrobiotic communities which grow on mounds (Freiwald this volume). To study the impact of microbial processes on the formation of carbonate mounds, techniques for non-invasive sampling and measurements of microbial biofilms growing on submerged carbonate mounds still need to be developed. However, some of the complex processes leading to calcification in microbial mats and biofilms could be studied on parallel systems, e.g microbial mats from shallow water environments and saline lakes.

Key question: How do microorganisms contribute to the formation of carbonate mounds?

Specific topics:
• What initiates the growth of carbonate mounds and which (microbial) processes maintain it?
• What are the main microbial processes leading to calcification and what is the role of biofilms in these processes?
• Which are the dominant groups of microorganisms occurring on carbonate mounds and what is their function?
• What is the chemical structure and function of microbial biofilms on carbonate mounds?

Role of microorganisms in sedimentary biogeochemical cycles

Particulate matter, which is not recycled in the water column, is ultimately deposited on the seafloor. Hence, processing or storage of organic material at the seafloor in general has a global significance for the element cycles (Hedges et al. 2000). Due to the high productivity and deposition at continental margins they gain an over-proportional importance in oceanic element cycles. There is an enormous degradative capacity in the diverse communities of benthic microorganisms (Lochte and Pfannkuche this volume). Chemical gradients in the pore water of marine sediments can show where certain chemical species are formed and where they react with each other. This always raises questions as to the potential involvement of microorganisms and their ability to catalyze the specific reaction. However, in studies of the cycle of carbon, nitrogen, sulfur and other elements in the sea, microbial populations and their functions have often been treated as a black box (Fig. 2a), and only few marine bacteria are available in pure culture (Fig. 2b).

Progress in geochemistry has often led to progress in microbiology and vice versa. New methods are available to identify bacteria and their contribution to organic matter degradation and the "cleaning up" of the sea. Yet, we probably do not know most of the microorganisms which are responsible for the cycling of C, N, S, Mn, Fe and Si at continental margins. It is possible that a few key microbes dominate element cycling, just as a few algal species are responsible for a large fraction of primary production and POM export in the pelagic realm. Such key microbial groups and their enzymatic capacities are easier to study in shallow water than in remote, bathyal areas of continental margins. However, since the composition and supply of marine POM as well as anthropogenic waste changes substantially from the shelf to the deep slope of continental margins, the key players as well as the regulation of the relevant microbial processes may also change from shallow to deep waters.

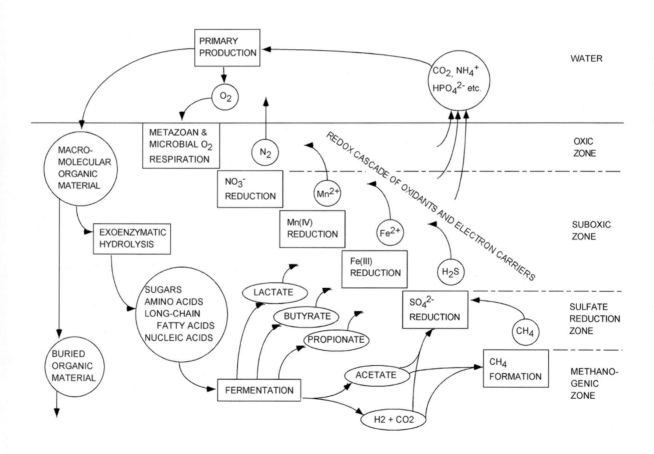

Fig. 2. Microbial turnover of elements in the ocean. **a)** A schematic illustration of the transformation of organic carbon during aerobic and anaerobic decomposition in marine sediment, and some aspects of the nitrogen and sulfur cycles (courtesy of BB Jørgensen, MPI).

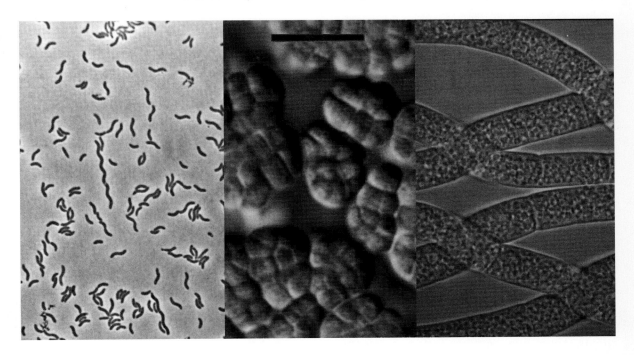

Fig. 2. cont. b) Sulfate reducing bacteria isolated from different aquatic environments. From left to right: *Desulfovibrio desulfuricans*, *Desulfosarcina variabilis*, *Desulfonema magnus* (courtesy of F. Widdel, MPI). The scale bar is 10 μm.

POM degradation

Microbes interact with the high particle load from terrestrial sources and the biogenic particles from primary production and are very efficient in the degradation of organic matter and the recycling of elements (Fig. 3). Nevertheless, many coastal environments are affected by eutrophication and the input of large amounts of anthropogenic waste and toxic substances. A substantial fraction of particulate organic matter is buried in the sediments of continental margins, but the factors regulating OM burial are still not well understood (Hedges et al. 2000). A variety of xenobiotic compounds entering the sea are resistant to microbial degradation and accumulate at continental margins.

The degradative capacities of microbes and their regulation in sedimentary systems of continental margins must be understood in order to predict the capacity of microbial systems to transform waste and xenobiotic compounds. A large fraction of energy-rich, labile molecules have been removed from sedimenting POM before it reaches the sea floor. Deposited polymers are rather refractory; their fate in the benthic food chain depends on the catalytic capacities of sedimentary microbes and probably also on physical factors such as adsorption of organic matter to minerals (Hedges and Keil 1995).

There are many open questions regarding the key microbial species responsible for the degradation of polymers in the marine environment and the regulation and degradative capacities of their hydrolytic enzymes, in both aerobic and anaerobic sedimentary systems (Boetius and Lochte 1996). The seafloor below upwelling areas is generally depleted in oxygen, and high amounts of organic matter are buried in the sediments (Fig. 4). The further degradation of such organic-rich deposits then relies on the activity of fermentative microorganisms. Most sulfate reducing bacteria do not possess enzymes for the degradation of polymers, and can only take up fermentation products (e.g. fatty acids). Hence, fermenting microorganisms are the link between the carbon and sulfur cycles in marine sediments. The identification of key microorganisms and their enzymes in

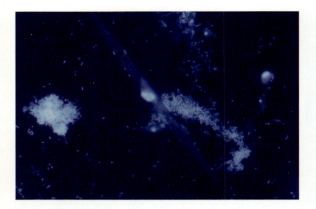

Fig. 3. Marine detritus particles densly colonized by bacteria. The bacteria are stained by the fluorescent dye DAPI which is specific for nucleic acids (courtesy of K. Lochte, IfM).

the process of fermentation is necessary to understand the degradation of POM in anaerobic sediments. There may be other factors determining the ratio between the degradation and preservation of organic matter in both oxic and anoxic environments. We do not know how long hydrolytic enzymes are intact and active in sedimentary systems. We need a better idea of the chemical and structure of polymers buried in sediments.

Little is known about the interaction of microbes with minerals, e.g whether there are microbial mechanisms for the desorption of OM from mineral particles. Experiments with particles from the pelagic environment have shown that silica dissolution can be enhanced by bacterial enzymes during POM degradation (Bidle and Azam 1999). The impact of microbes on the recycling of silica in sediments might have been completely overlooked.

Key question: How is POM degradation in sedimentary systems regulated?

Special topics:
• Which microorganisms are responsible for the hydrolysis of organic polymers in oxic and anoxic environments?
• What are the dominant groups of fermenters in marine sediments?
• Which factors determine the preservation of POM at continental margins?
• Which microbial processes and which organisms are capable of degrading recalcitrant substances in the sediments?
• Which anthropogenic chemicals are persistant to microbial degradation in sediments of continental margins?

Methane cycle

There have only been a few measurements of the production and consumption of methane in the sea. Methane is an important greenhouse gas, and recent geological data show that there have been significant variations in the atmospheric methane concentration during the history of the earth. Methanogenesis is thought to be restricted to anoxic zones in the sea. It is still debated whether microbial biofilms and aggregates could provide anoxic microzones in oxic environments of the ocean where methanogenesis could occur (Reeburgh 1996). Organic matter which becomes buried in sediments of continental margins continues to be degraded over long time scales, and will ultimately be converted into methane. The methane rising through the sediments could escape to the water column, but geochemical profiles show that nearly all methane is effectively removed before it reaches the oxic zone. The anaerobic oxidation of methane associated with sulfate reduction is a good example of the versatility of microorganisms even under thermodynamically "unfavorable" conditions (Hinrichs and Boetius this volume). A new

Fig. 4. Sediment core from the continental shelf off Peru showing interbedding of organic rich oliv-green silt with organic poor grey sand (water depth 184 m, 12°03.002'S 77°39.862'W; courtesy of BGR).

microbial consortium mediating the anaerobic oxidation of methane has recently been discovered in a gas hydrate bearing environment (Boetius et al. 2000). Oil is another end-product of refractory OM which turns out to be degradable under certain (unknown) conditions. Immense microbial activity has been detected at petroleum seeps in the Gulf of Mexico. Deep sub-seafloor oil fields may be infected by archaeal microorganisms similar to those found at hydrothermal vents (Huber et al. 2000). The fossil fuel industry has a great interest in understanding the processing of oil and other hydrocarbons in the sea, and some of the microorganisms involved in these processes may be very useful in bioremediation.

Key question: What are the main pathways of the microbial methane cycle at continental margins?

Specific topics:

• What regulates the microbial conversion of organic matter to methane?

• Where are the hot spots of methanogenesis at continental margins?

• How much methane is consumed in anoxic and oxic zones of continental margins?

• How is anaerobic methane oxidation catalysed?

• How important is acetogenesis in marine sediments and how much of the acetate produced is converted into methane?

Nitrogen cycle

Shelf areas receive substantial amounts of nitrogenous substances from anthropogenic sources (Herbert 1999). A further source of new nitrogen compounds in the sea is nitrogen fixation. The anoxic sediments of margins and shelves are the major sites of denitrification in the ocean and, hence, essential for maintaining a balance between input and losses of bound nitrogen. However, the rate of denitrification in sediments is one of the major unknowns in global nitrogen cycling (Balzer et al. 1998). There are indications that permeable sandy sediments, which are flushed by waves and advective pore water flow, sustain much higher denitrification rates than muddy sediments. It was estimated that denitrification in some continental margins is fuelled by the high nitrogen concentrations of deep ocean water

flowing onto the shelf. In order to understand the role of shelf sediments in removal of bound nitrogen, this microbial process must be quantified in different types of sediments and related to their geographical distribution.

The export of organic nitrogen from the shelf to deeper parts of continental margins is driven by the lateral transport of particulate material. However, several aspects of the degradation and remineralization of particulate organic nitrogen as well as of dissolved nitrogen species are still unclear. The increase in extracellular protease activity with water depth in surface sediments points to a role for these enzymes in the sedimentary organic nitrogen cycle. It may indicate a shortage of reduced nitrogen (in the form of amino acids or ammonium) in deep-sea sediments, but this is still highly speculative (Lochte and Pfannkuche this volume).

Some continental margin sediments show a diffusional interface between dissolved ferrous iron and nitrate. This suggests that iron may be readily oxidized by nitrate, a process presumably catalyzed by bacteria (Benz et al. 1998). It took nearly two decades of effort before such bacteria were isolated and could be studied in laboratory cultures (Fig. 5). We now know that they are widespread in marine sediments and probably play an significant role in iron and nitrogen cycling. Most of the ammonium excreted during catabolic metabolism is rapidly reoxidized at the oxic-anoxic interface in marine sediments. Recent laboratory experiments show that ammonium may also be reoxidized anaerobically (Jetten et al. 1998). However, the significance of this process in the field is still unknown.

Key question: What are the main microbial processes in the cycling of nitrogen in sedimentary systems?

Specific topics:

• What is the quantitative role of denitrification on margins and shelves in the oceanic nitrogen budget of sediments?

• How important is the uptake of amino acids compared to inorganic nitrogen sources to sedimentary microbes?

• Which are the key microorganisms mediating ammonium oxidation in sediments?

488   Boetius et al.

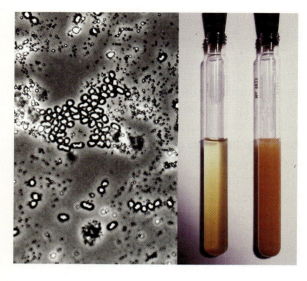

Fig. 5. The iron oxidizing bacterium *Thiodictyon siderooxidans*. The left image shows the cells containing gas vehicles, and Fe(III) minerals. The right image shows the change in colour of the culture medium due to the conversion of ferrous iron (left) to ferric iron (right) by the bacteria (courtesy of F. Widdel, MPI).

Sulfur cycle

Significant advances in linking bacterial diversity to their biogeochemical function have been made with regard to the sulfur cycle. Giant sulfur-oxidizing bacteria such as *Thioploca* and *Thiomargarita* are found at redox interfaces of continental margin sediments (Fig. 6). These bacteria use nitrate to oxidize sulfur, thus coupling the nitrogen and sulfur cycles in the sediments (Schulz et al. 1999). A new phylogenetic cluster of sulfate reducing bacteria, closely related to the *Desulfosarcina/Desulfococcus* group, has been detected in Arctic sediments and in tidal flats (Ravenschlag et al. 2000). They obviously dominate sulfate reduction in both habitats. These bacteria cannot yet be maintained in culture, and are thus not yet available for the physiological studies which should help elucidate their environmental role.

Progress in microbiology has often led to a new understanding of marine geochemistry. One such example was the discovery of a widespread ability among laboratory cultures of sulfate reducing bacteria to disproportionate inorganic sulfur compounds (Finster et al. 1998). By such dispropor-

tionation, which can be considered an inorganic fermentation, elemental sulfur or thiosulfate may be simultaneously oxidized to sulfate and reduced to sulfide. Within the last few years it has been demonstrated that this special type of redox reaction plays a key role in the marine sulfur cycle.

Key question: What are the key players of the microbial sulfur cycle in sediments of continental margins?

Specific topics:
• Which are the dominant sulfate reducing bacteria in different zones of continental margins?
• Which microbes link the sulfur cycle to the cycle of other elements such as carbon, nitrogen, Fe and Mn?
• Which factors determine the distribution and population size of the different groups of giant sulfur-oxidizing bacteria?

Microbial diversity and biotechnology

The study of the phylogenetic and functional diversity of microorganisms is not driven only by our curiosity about the evolution and functioning of life. An important applied aspect is the search

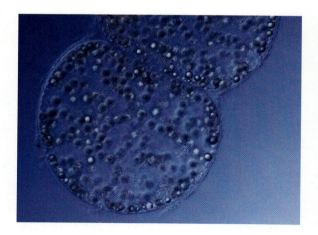

Fig. 6. A giant sulfur oxidizing bacterium. Terminal cell in a chain of *Thiomagarita namibiensis*, a large sulfur bacterium from Namibian sediments, as it occurs in the light microscope. Sinlge cells can grow up to nearly 1 mm in diameter. The cytoplasm is restricted to a thin outer layer while most of the cell appears hollow due to a large central vacuole used for storing nitrate. The granular structures are sulfur deposits in the cell. Photo by Heide Schulz.

for biomolecules with potential biotechnological ap-plications (Devries and Beart 1995). Marine microbes adapted to extremes of temperature or to growth on recalcitrant carbon sources may harbor many useful synthetic and degradative enzymes. The best-known enzymes derived from hot-spring and thermal-vent microorganisms to date are the heat-stable DNA polymerases used in the polymerase chain reaction (PCR). Thermostable enzymes are also needed for industrial processes, such as paper bleaching, which run at high temperature. The enzymes of psychrophilic microorganisms can have high catalytic rates at low temperature. These enzymes are also attractive for industrial applications, because manufacturing processes can be run at lower temperatures, thus saving energy costs. They can be also used in food processing, where low temperature helps maintain food quality and retards the growth of contaminant bacteria. Prokaryotes capable of degrading toxic anthropogenic organic compounds, such as polycyclic aromatic hydrocarbons (PAHs), can be found in marine sediments (MacRae and Hall 1998). Their distribution and activity control the persistence of these chemically recalcitrant compounds in the oceans. Attempts to increase microbial degradation rates along polluted shorelines by nutrient addition or bioaugmentation have thus far had limited success, but given the dearth of alternative methods, continued investigations seem well-warranted. The deep biosphere represents a completely new source of bacteria which may be useful in oil processing and recovery; in waste processing; or as a source of unique biomolecules (Parkes et al. 2000). Important reasons for studying this new environment and the biodiversity of its inhabitants are discussed above. In the following, we will focus on the ecology of microbial associations with invertebrates, a research field which is moving rapidly into the focus of modern biotechnology.

Associations of microbes and invertebrates

Marine microorganisms contribute to the nutrition, behavior (e.g. via bioluminscence) and self-defense of many marine invertebrates. These associations are interesting from both evolutionary and biotechnological viewpoints. Two of the principal concepts of symbiosis research are that the associated partners have evolved together, and that bacterial associations with a host provide ample opportunities for special adaptations, including the biosynthesis of unique metabolic products (Bull et al. 2000). Striking examples of symbiotic relationships between marine invertebrates and microbes are found at hydrothermal vents, e.g. between sulfide-oxidizing bacteria and the tube worm *Riftia pachyptila* (Fisher 1990). The evolution of these associations and their dispersal between widely separated vent sites are still not completely understood, in part because of our inability to grow the symbionts in culture. New molecular biological methods, such 16S rRNA-based phylogenetic identification, have helped in advancing this field of research (MacGregor et al. this volume). Similar chemosynthetic symbioses have recently been detected in a variety of special habitats at continental margins, such as cold seeps, or in the vicinity of surficial gas hydrates, mud vulcanoes and pockmarks. However, symbiotic associations of microorganisms and invertebrates in the benthic realm of continental margins might be much more widespread than previously anticipated. Most higher organisms (including humans) depend on the uptake of preformed organic molecules such as amino acids, unsaturated fatty acids, and phospholipids, which are often produced by microorganisms in more or less specific association with the host species. This interaction has not yet been studied in much detail for marine invertebrates, but modern molecular methods should make it possible to at least identify specific associations (McInerney et al. 1995). It seems intuitive that deposit feeders such as holothurians (Fig. 7) might rely on internal production of essential nutrients by harboring microbes, rather than external supply of compounds likely to be in short supply in oligotrophic sediments. The use of polymers, such as chitin for invertebrate food depends on processing by microbes. Furthermore, the types of interaction between the small sediment infauna such as foraminifera, nematodes, harpacticoids (crustaceans) and bacteria are still unclear and may cover the whole range from graz-

ing on bacteria to the specialized harboring of bacteria in animal slimes, guts or even epidermis.
Marine sponges, a typical inhabitant of continental margins, are frequently associated with bacteria,
making them interesting in the search for new
useful biomolecules (Friedrich et al. 1999). Some
sponges are already an important source of secondary metabolites with antifungal, antibacterial,
or antifouling activities, which in at least some
cases are produced by sponge-associated bacteria.
However, almost nothing is known on the function
of such associations. Sponges are difficult to grow
in culture, particularly under conditions favoring
production of the useful compounds. An alternative
approach would be to identify the associated bacteria and relevant enzymatic pathways, so that the
symbionts might be grown directly or the enzymes
introduced into more easily-grown strains.

Key question: What are the specific associations of microbes and invertebrates at continental
margins?

Specific topics:
• Do benthic invertebrates depend on microbially-
processed food?
• What are the specific traits of invertebrate-as-
sociated microbes and how are they selected by
their host?
• How species-specific are associations of microorganisms and invertebrate hosts?
• Which associations of microbes and invertebrates
are maintained by repeated reinfection and which
by direct transfer to the offspring?

*Emerging concepts and technologies in the
study of microbial communities in continental margin sediments*

It is insufficient to treat the microbial biosphere
of marine sediments as a "black box" which re-
mineralizes deposited POM to simple nutrients and
gases. Discrete populations of Bacteria and
Archaea often catalyze very specific steps in the
degradation of natural and anthropogenic compounds (Fig 2). The rates at which these metabolic
activities proceed in the different environments
determine to what extent essential nutrients are
recycled, how much carbon dioxide is removed
from the atmosphere, and whether a new indus-

Fig. 7. Dissection of a holothurian gut stuffed with
sediment. Holothurians (sea cucumber) are the most
abundant deposit feeders in the deep-sea (courtesy of
M. Nitsche, IOW).

trial product is persistent and may potentially bio-
accumulate in the marine food chain. We still
lack basic knowledge about key microbial populations and biochemical pathways, and their
regulation. This is due to three main problems:
(i) microorganisms in marine sediments are small
and difficult to identify, (ii) most of them are extremely difficult to cultivate in the laboratory, and
(iii) sampling destroys the fine structure of chemical
gradients in sediments which facilitate the processes and regulate them.

Ocean margin sediments contain sharp interfaces and steep chemical gradients which are
both caused by and inhabited by microorganisms.
There are now instruments, microsensors, to monitor these zones at the sea-floor (Kühl and Revsbech 2000). At the same time, with the advent of
nucleic acid probes, it has become feasible to
accurately determine the population size and localization of defined microorganisms (Ramsing et al.
1993), independent of our ability to isolate and
cultivate them (Amann et al. 1995). Furthermore,
microbial identification can be combined with determination of growth substrates, e.g. by using
radiotracers and microautoradiography (Lee et al.
1999) or by analyzing the stable isotope signature

of biomarker compounds (Orphan et al. in press). For the study of microbial growth on hard surfaces, in biofilms and in association with gas bubbles, we will need non-invasive techniques as well as 2- and 3-D sensoring (Fig. 8). Such new methods set the stage for understanding the habitat and function of discrete microbial populations, and should be used concurrently with molecular biological tools like FISH to identify individual microbial cells in these microenvironments (Fig. 9).

Based on the results of these cultivation-independent studies, more directed efforts to obtain pure cultures of important microorganisms can be initiated. This will remain a difficult and time-consuming undertaking. Therefore, it is very encouraging that, with the cloning and subsequent sequencing of large DNA fragments directly from marine sediments, even microorganisms which are as yet uncultured can be studied for their genetic potential. We foresee that molecular biological techniques, most notably the massive sequencing capacities originating from the human genome project will profoundly influence the way the complex microbial communities in ocean margin

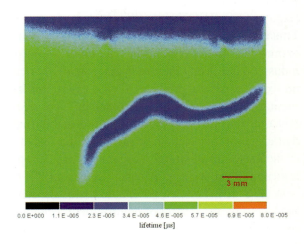

Fig. 8. 2-D sensoring of oxygen distribution in wadden sea sediment. Increased oxygen concentrations (blue) where recorded in a worm burrow. A planar optode measures the lifetime of luminescent dye, responding to oxygen concentrations, with a CCD-camera system (courtesy of U. Franke and G. Holst, MPI).

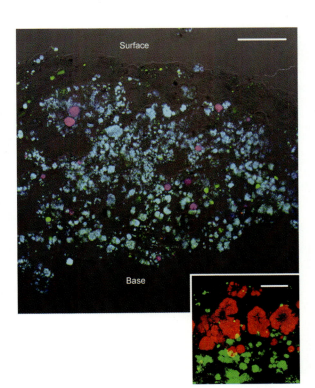

Fig. 9. Confocal image of a nitrifying biofilm after multiple probe hybridization (courtesy of A. Gieseke, MPI). The multiple hybridization in combination with confocal microscopy for exact colocalization of signals allows the unequivocal investigation of the microbial community structure. Upper micrograph: this vertical thin section shows various coexisting ammonia-oxidizing populations (scale bar is 50 μm). The three probes used are specific to most of the ammonia-oxidizing b-proteobacteria (Nso1225, displayed in green), the *Nitrosomonas europaea/eutropha/halophila* subgroup (NEU, shown in blue), and the species *Nitroscoccus mobilis* (NmV, displayed in red). The bright blue color indicates the affiliation to the *N. europaea/eutropha/halophila* group (binding of both probes Nso1225 and NEU), magenta is caused by double hybridization with probes NEU and NmV, indicative for *Nitrosococcus mobilis*. Insert: Spatial reconstruction based on an image stack (44 images; 19 μm thickness) showing the close arrangement of members of the *N. europaea/eutropha/halophila* group (green), and spherical clusters of *Nitrosococcus mobilis* (red), which appear hollow in the center (scale bar is 20 μm).

systems are studied. The genome sequencing of important marine bacteria has recently been started (e.g. project REGX; http://www.regx.de. vu/). This will be the basis for a detailed study of gene expression and its regulation. The biotechnology industry has a strong interest in this potentially rich pool of new natural products. Given a swift implementation of the emerging molecular biological concepts and technologies it seems feasible that by the end of this decade the presence and activity of the most important microorganisms can be rapidly monitored by e.g. a marine sediment DNA chip (Fig. 10). This would enable us for the first time to study the impact of environmental change on ocean margin sediments in which microorganisms control processes of crucial importance. A project of this size requires a well-coordinated, joint international effort in which defined tasks like the analysis of selected functional groups of microorganisms are allocated to different countries.

International cooperation is also beneficial when it comes to the exploration of the more remote areas of continental margins such as the adjacent deep-sea and deep sub-seafloor in general or ice covered margin regions. Exploration of special geological structures like canyons, sea mounts, mud vulcanoes, pockmarks, carbonate mounds, seeps and hydrate deposits require also specific efforts. Projects of this type will rely strongly on advanced sampling technologies, some of which are not easily available or maintained. Some of the major requirements for future microbiology studies in sedimentary systems are
• visually-guided sampling
• small scale sampling of solid structures, sediment interfaces and pore waters
• maintenance of *in situ* conditions such as pressure, temperature, and anoxia during sampling and experimental manipulation
• *in situ* experimental work in benthic chambers
• combination of activity measurements and phylogenetic analyses
• avoidance of contamination during drilling and handling of samples

Recently, a substantial improvement in this regard has been made by installing and equipping a microbiology laboratory on the research vessel *JOIDES Resolution* of the international Ocean Drilling Program (http://www-odp.tamu.edu/sciops/labs/microbiol/index.html#equip). Table 1 gives an overview of important concepts and technologies for future microbiological studies which go hand in hand with the main research challenges described in the previous sections.

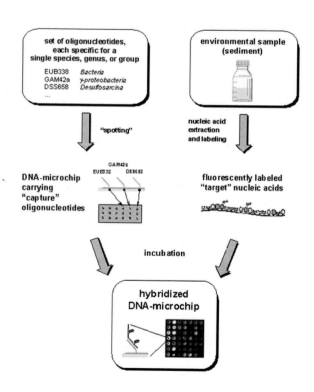

Fig. 10. Schematic illustration of the probing of environmental sediment samples with a DNA microarray (courtesy of J. Peplies, MPI).

Topic	Task	Approach
sampling design and communication	long term investigations	monitoring studies in key areas exposed to disturbances as well as in Marine Protection Areas
	databases and libraries	libraries of cultured marine microorganisms, gene sequences, microbial biomarkers, collection of environmental microbiology data (e.g. ADPED, JGOFS, OMEX),
	cultivation and screening	cultivation of key organisms, search of key enzymes and biomolecules for application in biotechnology, genomics, proteomics
	education and information	international training courses for students and teachers, educative web sites and programs
in situ technology	microsensors	new biosensors for key components such as hydrogen, methane, sulfate, acetate, DOC
	function and phylogeny	combined FISH and microsensor/activity measurements, MAR-FISH, DNA chips
	2-D and 3-D monitoring systems	imaging systems, optodes, fluorometers
	intelligent chamber landers	simulation of environmental conditions e.g. current flow, oxygen concentrations, capability of manipulation and experimentation
	long-term observation systems	imaging, sampling and fixation, controlled movement, on-line data transmission and other communication systems
sampling technology	modular instruments for submersibles, ROVs and landers	fine scale sampling, maintenance of pressure and temperature, fixation and manipulation of samples
	microbial sampling technology for deep ocean drilling	sampling without contamination, pressure retaining sampling from deep drilling

Table 1. Concepts and technologies for future studies of microbial systems on continental margins.

References

Amann RI, Ludwig W, Schleifer KH (1995) Phylogenetic identification and in situ detection of individual microbial cells without cultivation. Microbiol Rev 59:143-169

Balzer W, Helder W, Epping E, Lohse L, Otto S (1998) Benthic denitrification and nitrogen cycling at the slope and rise of the NW European Continental Margin (Goban Spur). Prog Oceanogr 42:111-126

Benz M, Brune A, Schink B (1998) Anaerobic and aerobic oxidation of ferrous iron at neutral pH by chemoheterotrophic mitrate-reducing bacteria. Arch Microbiol 169:159-165

Bidle KD, Azam F (1999) Accelerated dissolution of diatom silica by marine bacterial assemblages. Nature 397:508-512

Boetius A, Lochte K (1996) The effect of organic matter

composition on hydrolytic potentials and growth of benthic bacteria in deep-sea sediments. Mar Ecol Prog Ser 140:235-250

Boetius A, Ravenschlag K, Schubert C, Rickert D, Widdel F, Gieseke A, Amann R, Jørgensen BB, Witte U, Pfannkuche O (2000) A marine microbial consortium apparently mediating anaerobic oxidation of methane. Nature 407:623-626

Bull AT, Ward AC, Goodfellow M (2000) Search and discovery strategies for biotechnology: The paradigm shift. Microb Molec Biol Rev 64:573-606

Devries DJ, Beart PM (1995) Fishing for drugs from the sea. – Status and strategies. Trends Pharmacol Sci 16:275-279

Douglas S, Beveridge TJ (1998) Mineral formation by bacteria in natural microbial communities. FEMS Microb Ecol 26:79-88

Faulkner DJ (2000) Marine natural products. Nat Prod Rep 17:7-55

Finster K, Liesack W, Thamdrup B (1998) Elemental sulfur and thiosulfate disproportionation by *Desulfocapsa sulfoexigens sp nov*, a new anaerobic bacterium isolated from marine surface sediment. Appl Environ Microb 64:119-125

Fisher CR (1990) Chemoautotrophic and methanotrophic symbiosis in marine invertebrates. Rev Aquat Sci 2:399-436

Freiwald A (2002) Reef-forming cold-water corals. In: Wefer et al. (eds) Ocean Margin Systems. Springer, Berlin pp 365-385

Friedrich AB, Merkert H, Fendert T, Hacker J, Proksch P, Hentschel U (1999) Microbial diversity in the marine sponge *Aplysina cavernicola* (formerly *Verongia cavernicola*) analyzed by fluorescence *in situ* hybridization (FISH). Mar Biol 134:461-470

Hedges JI, Eglinton G, Hatcher PG, Kirchman DL, Arnosti C, Derenne S, Evershed RP, Kögel-Knabner, de Leeuw JW, Littke R, Michaelis W, Rullkötter J (2000) The molecularly-uncharacterized component of nonliving organic matter in natural environments. Org Geochem 31:945-958

Hedges JI, Keil RG (1995) Sedimentary organic matter preservation: an assessment and speculative synthesis. Mar Chem 49:81-115

Henriet JP, Guidard S & the ODP "Proposal 573" Team (2002) Carbonate mounds as a possible example for microbial activity in geological processes. In: Wefer et al. (eds) Ocean Margin Systems. Springer, Berlin pp 439-455

Herbert RA (1999) Nitrogen cycling in coastal marine ecosystems. FEMS Microbiol Rev 23:563-590

Hinrichs KU, Boetius A (2002) The anaerobic oxidation of methane: new insights in microbial ecology and biogeochemistry. In: Wefer et al. (eds) Ocean Margin Systems. Springer, Berlin pp 457-477

Huber R, Huber H, Stetter KO (2000) Towards the ecology of hyperthermophiles: biotopes, new isolation strategies and novel metabolic properties. FEMS Microbiol Rev 24:615-623

Jetten MSM, Strous M, van de Pas-Schoonen KT, Schalk J, van Dongen UGJM, van de Graaf AA, Logemann S, Muyzer G, van Loosdrecht MCM, Kuenen JG (1998) The anaerobic oxidation of ammonium. FEMS Microbiol Rev 22:421-437

Jørgensen BB (2000) Bacteria and marine biogeochemistry. In: Schulz HD, Zabel M (eds) Marine Geochemistry. Springer, Berlin pp 173-207

Kühl M, Revsbech NP (2000) Biogeochemical microsensors for boundary layer studies. In: Boudreau BB, Jorgensen BB (eds) The Benthic Boundary Layer. Oxford University Press, New York pp 180-210

Lee N, Nielsen PH, Andreasen KH, Juretschko S, Nielsen JL, Schleifer KH, Wagner M. (1999) Combination of fluorescent in situ hybridization and microautoradiography - a new tool for structure-function analyses in microbial ecology. Appl Environ Microbiol 65:1289-1297

Lochte K, Pfannkuche O (2002) Processes driven by the small-sized organisms at the water-sediment interface. In: Wefer et al. (eds) Ocean Margin Systems. Springer, Berlin pp 405-418

MacGregor BJ, Ravenschlag K, Amann R (2002) Nucleic acid based techniques for analyzing the diversity, structure, and function of microbial communities in marine waters and sediments. In: Wefer et al. (eds) Ocean Margin Systems. Springer, Berlin pp 419-438

MacRae JD, Hall KJ (1998) Biodegradation of Polycyclic Aromatic Hydrocarbons (PAH) in marine sediment under denitrifying conditions. Water Sci Techn 38:177-185

McInerney JO, Wilkinson M, Patching JW, Embley TM, Powell R (1995) Recovery and phylogenetic analysis of novel archaeal rRNA sequences from a deep-sea deposit feeder. Appl Environ Microbiol 61:1646-1648

Orphan V, House CH, Hinrichs KU, McKeegan KD, DeLong EF (in press) Coupled isotopic and phylogenetic analyses of single cells: direct evidence for methane-consuming archaeal/bacterial consortia. Science

Parkes RJ, Cragg BA, Wellsbury P (2000) Recent studies on bacterial populations and processes in subseafloor sediments: A review. Hydrogeol J 8:11-28

Ramsing NB, Kühl M, Jørgensen BB (1993) Distribution of sulfate-reducing bacteria and O_2-H_2S in biofilm determined by oligonucleotide probes and microelectrodes. Appl Environ Microbiol 59:3840-3849

Ravenschlag K, Sahm K, Knoblauch C, Jorgensen BB, Amann R (2000) Community structure, cellular rRNA content, and activity of sulfate-reducing bacteria in marine Arctic sediments. Appl Environ Microb 66:3592-3602

Reeburgh WS (1996) "Soft spots" in the global methane budget. In: Lidstrom ME, Tabita FR (eds) Microbial Growth on C1 Compounds. Kluwer Academic Publishers, Netherlands pp 334-342

Reid RP, Visscher PT, Decho AW, Stolz JF, Bebout BM, Dupraz C, MacIntyre IG, Paerl HW, Pinckney JL, Prufert-Bebout L, Steppe TF, DesMarais DJ (2000) The role of microbes in accretion, lamination and early lithification of modern marine stromatolites. Nature 406:989-992

Schulz HN, Brinkhoff T, Ferdelman TG, Marine MH, Teske A, Jorgensen BB (1999) Dense populations of a giant sulfur bacterium in Namibian shelf sediments. Science 284:493-495

Warthmann R, van Lith Y, Vasconcelos C, McKenzie JA, Karpoff AM (2000) Bacterially induced dolomite precipitation in anoxic culture experiments. Geol 28:1091-1094

Printing and Binding: Stürtz AG, Würzburg